Land Modifications and Impacts on Coastal Areas

Land Modifications and Impacts on Coastal Areas

Editors

Pietro Aucelli
Angela Rizzo
Rodolfo Silva Casarín
Giorgio Anfuso

MDPI • Basel • Beijing • Wuhan • Barcelona • Belgrade • Manchester • Tokyo • Cluj • Tianjin

Editors

Pietro Aucelli
University of Naples
Parthenope
Italy

Angela Rizzo
University of Bari Aldo Moro
Italy

Rodolfo Silva Casarín
Universidad Nacional
Autónoma de México
Mexico

Giorgio Anfuso
University of Cádiz
Spain

Editorial Office
MDPI
St. Alban-Anlage 66
4052 Basel, Switzerland

This is a reprint of articles from the Special Issue published online in the open access journal *Land* (ISSN 2073-445X) (available at: https://www.mdpi.com/journal/land/special_issues/land_modifications_impacts_coastal_areas).

For citation purposes, cite each article independently as indicated on the article page online and as indicated below:

LastName, A.A.; LastName, B.B.; LastName, C.C. Article Title. *Journal Name* **Year**, *Volume Number*, Page Range.

ISBN 978-3-0365-6135-6 (Hbk)
ISBN 978-3-0365-6136-3 (PDF)

Cover image courtesy of Angela Rizzo
The picture was taken in Spain at Puerto Real (Cádiz)

© 2023 by the authors. Articles in this book are Open Access and distributed under the Creative Commons Attribution (CC BY) license, which allows users to download, copy and build upon published articles, as long as the author and publisher are properly credited, which ensures maximum dissemination and a wider impact of our publications.

The book as a whole is distributed by MDPI under the terms and conditions of the Creative Commons license CC BY-NC-ND.

Contents

About the Editors vii

Preface to "Land Modifications and Impacts on Coastal Areas" ix

Juan Carlos Alcérreca-Huerta, Cesia J. Cruz-Ramírez, Laura R. de Almeida, Valeria Chávez and Rodolfo Silva
Interconnections between Coastal Sediments, Hydrodynamics, and Ecosystem Profiles on the Mexican Caribbean Coast
Reprinted from: *Land* **2022**, *11*, 524, doi:10.3390/land11040524 1

Mauro Bonasera, Ciro Cerrone, Fabiola Caso, Stefania Lanza, Giandomenico Fubelli and Giovanni Randazzo
Geomorphological and Structural Assessment of the Coastal Area of Capo Faro Promontory, NE Salina (Aeolian Islands, Italy)
Reprinted from: *Land* **2022**, *11*, 1106, doi:10.3390/land11071106 23

Laura Borzì, Giorgio Anfuso, Giorgio Manno, Salvatore Distefano, Salvatore Urso, Domenico Chiarella and Agata Di Stefano
Shoreline Evolution and Environmental Changes at the NW Area of the Gulf of Gela (Sicily, Italy)
Reprinted from: *Land* **2021**, *10*, 1034, doi:10.3390/land10101034 49

Maurizio D'Orefice, Piero Bellotti, Tiberio Bellotti, Lina Davoli and Letizia Di Bella
Natural and Cultural Lost Landscape during the Holocene along the Central Tyrrhenian Coast (Italy)
Reprinted from: *Land* **2022**, *11*, 344, doi:10.3390/land11030344 71

Philippes Mbevo Fendoung, Mesmin Tchindjang and Aurélia Hubert-Ferrari
Weakening of Coastlines and Coastal Erosion in the Gulf of Guinea: The Case of the Kribi Coast in Cameroon
Reprinted from: *Land* **2022**, *11*, 1557, doi:10.3390/land11091557 107

Juan J. Santos-Vendoiro, Juan J. Muñoz-Perez, Patricia Lopez-García, Jose Manuel Jodar, Javier Mera, Antonio Contreras, Francisco Contreras, et al.
Evolution of Sediment Parameters after a Beach Nourishment
Reprinted from: *Land* **2021**, *10*, 914, doi:10.3390/land10090914 143

Pedro Aguilar, Edgar Mendoza and Rodolfo Silva
Interaction between Tourism Carrying Capacity and Coastal Squeeze in Mazatlan, Mexico
Reprinted from: *Land* **2021**, *10*, 900, doi:10.3390/land10090900 159

Carlos Mestanza-Ramón, Selene Paz-Mena, Carlos López-Paredes, Mirian Jimenez-Gutierrez, Greys Herrera-Morales, Giovanni D'Orio and Salvatore Straface
History, Current Situation and Challenges of Gold Mining in Ecuador's Litoral Region
Reprinted from: *Land* **2021**, *10*, 1220, doi:10.3390/land10111220 183

Alexis Mooser, Giorgio Anfuso, Hristo Stanchev, Margarita Stancheva, Allan T. Williams and Pietro P. C. Aucelli
Most Attractive Scenic Sites of the Bulgarian Black Sea Coast: Characterization and Sensitivity to Natural and Human Factors
Reprinted from: *Land* **2022**, *11*, 70, doi:10.3390/land11010070 201

Isabel Bello-Ontiveros, Gabriela Mendoza-González, Lizbeth Márquez-Pérez and Rodolfo Silva
Using Spatial Planning Tools to Identify Potential Areas for the Harnessing of Ocean Currents in the Mexican Caribbean
Reprinted from: *Land* **2022**, *11*, 665, doi:10.3390/land11050665 237

Stephen Kankam, Adams Osman, Justice Nana Inkoom and Christine Fürst
Implications of Spatio-Temporal Land Use/Cover Changes for Ecosystem Services Supply in the Coastal Landscapes of Southwestern Ghana, West Africa
Reprinted from: *Land* **2022**, *11*, 1408, doi:10.3390/land11091408 261

Angela Rizzo, Francesco De Giosa, Antonella Di Leo, Stefania Lisco, Massimo Moretti, Giovanni Scardino, Giovanni Scicchitano, et al.
Geo-Environmental Characterisation of High Contaminated Coastal Sites: The Analysis of Past Experiences in Taranto (Southern Italy) as a Key for Defining Operational Guidelines
Reprinted from: *Land* **2022**, *11*, 878, doi:10.3390/land11060878 285

Lindsey S. Smart, Jelena Vukomanovic, Paul J. Taillie, Kunwar K. Singh and Jordan W. Smith
Quantifying Drivers of Coastal Forest Carbon Decline Highlights Opportunities for Targeted Human Interventions
Reprinted from: *Land* **2021**, *10*, 752, doi:10.3390/land10070752 315

About the Editors

Pietro Aucelli

Pietro P.C. Aucelli is a Professor of Physical Geography and Geomorphology at University of Naples "Parthenope" (Italy). Pietro Aucelli's research interests concern thematic in the fields of geomorphology and physical geography, with a specialisation in applied geomorphology, GIS, and Quaternary geology. He is a member of the Teaching Committee for the PhD in "Environmental Phenomena and Risks" at University of Naples "Parthenope". During his scientific activity, he has participated in numerous national research projects and activities of scientific coordination within the bounds of several national projects and agreements between different institutions and universities. He is a member of the Governing Council of the Italian Association of Physical Geography and Geomorphology (AIGeo). Since 2019, he has been the Coordinator of the AIGeo Working Group "Coastal Morphodynamic". Pietro P.C. Aucelli has authored numerous papers published on international and national journals, as well as several contributions in books. He is author of over 125 contributions published on the Scopus database.

Angela Rizzo

Angela Rizzo is a post-doc researcher at University of Bari Aldo Moro (Italy) at the Department of Earth and Environmental Sciences. She achieved her PhD in Coastal Geomorphology in 2017 at University of Naples "Parthenope" (Italy). Her research activity focuses on the analysis of physical and environmental processes affecting coastal areas, with special attention on mobile coastal systems. She is currently working on the evaluation of potential coastal morphological evolution under different sea-level scenarios, as well as on coastal risk assessments. Furthermore, she has technical competence in methodological and operative approaches for the analysis of beach litter distribution and impacts by carrying out direct in situ surveys and applying innovative analysis approaches based on remote sensing data (e.g., UAV images). During her career, she has worked at the CMCC foundation, the Italian research centre on climate change. She collaborates with international colleagues from Spain, Malta, and Colombia. She has co-authored 29 peer reviewed publications in international journals and is a Guest Editor for other Special Issues and Volume (published by MDPI, Elsevier, and Springer). Since 2014, she has been a member of the Italian Association of Physical Geography and Geomorphology (AIGeo).

Rodolfo Silva Casarín

Rodolfo Silva Casarín is a Senior Researcher and Professor of Civil Engineering at the Institute of Engineering of the National University of Mexico (UNAM). Rodolfo has a PhD in Coastal and Port Engineering, achieved at the University of Cantabria (Spain). Since 1995, Rodolfo has been the Head of the Coastal and Oceanographic Group at UNAM and the Mexican Centre for Ocean Renewable Energies. He has been the Mexican delegate (since 2011) and the Latin American Regional Coordinator (2017–2019) in the Excellence Center for Development Cooperation–Sustainable Water Management (EXCEED), and the Mexican Delegate in the Ocean Energy System (OES-IEA). He has authored or co-authored over 650 publications and more than 100 technical reports produced for local and national government authorities in Mexico, as well as companies and organizations in several countries.

Giorgio Anfuso

Giorgio Anfuso, PhD in Marine Science, has been a coastal researcher and Full Professor at University of Cadiz (Spain) in the Faculty of Marine and Environmental Sciences since 1999. For approximately 25 years, he has studied coastal processes at different time scales, from hours (by sediment transport and disturbance depth campaigns), days, months (by field surveys and monitoring programs), to decadal coastal evolution by means of aerial photos and satellite image analysis. He has also worked on management issues, environmental sensitivity maps for oil spilling, coastal scenery, and litter characteristics and distribution, carrying out investigations in Spain, Italy, Morocco, Ireland, Colombia, Cuba, Mexico, and Ecuador, among other countries. He has participated in different national and international projects focused on beach morphology, evolution, and responses to storm impacts. He has co-authored more than 100 peer-reviewed publications in international journals.

Preface to "Land Modifications and Impacts on Coastal Areas"

Due to their favourable geomorphological characteristics and intrinsic environmental and scenic value, coastal areas have attracted humans and related settlements and activities. The contemporary population density is significantly higher in coastal areas than in continental areas, and most of the world's megacities are located in the coastal zones. According to the most recent European data, approximately 40% of the population of the European Union lives within 50 km from the sea, and less than 1% of the Mediterranean coast remains relatively unaffected by human activities. Coastal areas also play a remarkable role in supporting local economies, providing resources, assets, and opportunities for commercial, industrial, and cultural activities, as well as for marine transport and trade. Worldwide coastal zones are home to important industrial plants, including chemical, petrochemical, metallurgical, steel, mechanical, and cement facilities, as well as shipyards, military arsenals, and harbour areas with high maritime traffic. Due to intense urbanization, the morphodynamic evolution of coastal areas is directly and indirectly affected by human-related activities, which can act at both the local and watershed scales. In addition, coastal morphoevolutive dynamics are a result of the interactions among oceanic, terrestrial, and weather-driven factors, which act at different spatial (local, regional, and global scale) and temporal scales (short, medium, and long time scales). Interdisciplinary studies carried out during the last few decades have highlighted that low-lying coasts worldwide are currently subject to erosion and flooding processes, and these phenomena are expected to increase in intensity and frequency because of ongoing climate change. The long-term maintenance of all ecosystem functions offered by coastal areas and associated habitats can provide relevant support in the conservation of those ecosystem services which are able to enforce coastal resilience against extreme marine events and sea level rise; therefore, the protection of natural and anthropogenic assets of economic interests, as well as coastal infrastructure, is necessary.

Based on the assumption that accurate knowledge of the natural and anthropogenic factors acting on the coastal environments is of great significance for their effective management and for the sustainable exploitation of coastal resources, the main aim of this Special Issue, entitled "Land Modifications and Impacts on Coastal Areas", was to collect studies showing examples of the interconnection among marine/coastal and anthropogenic processes and evolution of the local landscape, shoreline modification, and coastal geo-environmental changes.

The volume includes case studies from Bulgaria, Cameroon, Ecuador, Ghana, Italy, Mexico, North Carolina (USA), and Spain, demonstrating methodological approaches applied in different regions across the world.

The first group of papers (1–6) is focused on the assessment of geomorphodynamic and hydrodynamic changes and associated effects in terms of shoreline variation and erosion, cliff retreat, Holocene landscape modification, and ecosystem distribution. The second group of papers (7–9) analyses the interactions between coastal zones and anthropogenic activities (e.g., tourism and mining) from a management perspective. Finally, the last group of papers (10–13) provides examples of how to address environmental issues (e.g., land use, forest carbon decline, and sediment pollution).

Hopefully, the scientific collection proposed here will be of interest for different categories of professionals involved in coastal studies and management to favour integrated research aimed at the sustainable use of resources offered by coastal environments. Lastly, as Guest Editors, we would like to kindly thank all the authors for their participation and contribution to the volume as well as all the reviewers that have strongly contributed to the high-quality papers published. We also strongly appreciated the support provided by the *Land* journal editorial staff.

Pietro Aucelli, Angela Rizzo, Rodolfo Silva Casarín, and Giorgio Anfuso
Editors

Article

Interconnections between Coastal Sediments, Hydrodynamics, and Ecosystem Profiles on the Mexican Caribbean Coast

Juan Carlos Alcérreca-Huerta [1], Cesia J. Cruz-Ramírez [2], Laura R. de Almeida [2], Valeria Chávez [2,*] and Rodolfo Silva [2]

[1] Departamento de Observación y Estudio de la Tierra, la Atmósfera y el Océano, Consejo Nacional de Ciencia y Tecnología (CONACYT-ECOSUR), Chetumal 77014, Mexico; jcalcerreca@conacyt.mx
[2] Instituto de Ingeniería, Universidad Nacional Autónoma de México, Mexico City 04510, Mexico; ccruzr3@gmail.com (C.J.C.-R.); lauraribas.a@gmail.com (L.R.d.A.); rsilvac@iingen.unam.mx (R.S.)
* Correspondence: vchavezc@iingen.unam.mx

Abstract: The interconnections between hydrodynamics, coastal sediments, and ecosystem distribution were analysed for a ~250 km strip on the northern Mexican Caribbean coast. Ecosystems were related to the prevailing and extreme hydrodynamic conditions of two contrasting coastal environments in the study area: Cancun and Puerto Morelos. The results show that the northern Mexican Caribbean coast has fine and medium sands, with grain sizes decreasing generally, from north of Cancun towards the south of the region. Artificial beach nourishments in Cancun have affected the grain size distribution there. On beaches with no reef protection, larger grain sizes ($D_{50} > 0.46$ mm) are noted. These beaches are subject to a wide range of wave-induced currents (0.01–0.20 m/s) and have steeper coastal profiles, where sediments, macroalgae and dune-mangrove systems predominate. The coastline with the greatest amount of built infrastructure coincides with beaches unprotected by seagrass beds and coral reefs. Where islands or coral reefs offer protection through less intense hydrodynamic conditions, the beaches have flatter profiles, the dry beach is narrow, current velocities are low (~0.01–0.05 m/s) and sediments are finer ($D_{50} < 0.36$ mm). The results offer a science-based description of the interactions between physical processes and the role played by land uses for other tropical coastal ecosystems.

Keywords: sediment transport; sedimentary environments; ecosystem distribution; coastal dynamics; coastal profiling; physical processes; anthropogenic pressure

Citation: Alcérreca-Huerta, J.C.; Cruz-Ramírez, C.J.; de Almeida, L.R.; Chávez, V.; Silva, R. Interconnections between Coastal Sediments, Hydrodynamics, and Ecosystem Profiles on the Mexican Caribbean Coast. *Land* **2022**, *11*, 524. https://doi.org/10.3390/land11040524

Academic Editor: Richard C. Smardon

Received: 9 March 2022
Accepted: 2 April 2022
Published: 4 April 2022

Publisher's Note: MDPI stays neutral with regard to jurisdictional claims in published maps and institutional affiliations.

Copyright: © 2022 by the authors. Licensee MDPI, Basel, Switzerland. This article is an open access article distributed under the terms and conditions of the Creative Commons Attribution (CC BY) license (https://creativecommons.org/licenses/by/4.0/).

1. Introduction

The enormous biodiversity of coastal ecosystems in the tropics has long been appreciated and is now recognised scientifically. However, mangroves, foredunes, beaches, seagrass meadows and coral reefs are all currently subject to environmental degradation aggravated by anthropogenic drivers, extreme hydrometeorological events, and climate warming [1–3]. Coastal dynamics are influenced by the ecosystems that exist in any locality. They affect the morphology, sediment transport, and sediment features of the environment where they are found.

Coral reefs enhance wave dissipation [4], as do seagrass meadows. This, in turn, contributes to sedimentation and the entrapment of sediment particles [5,6]. Dunes and beaches are interconnected by the wave energy that reaches the coastline [7,8] and the sediment budget along the beach profile [9]. Both beaches and dunes are also dependent on the characteristics of adjacent ecosystems [10–12].

The spatial arrangement of coastal ecosystems is associated with topo-bathymetric profile changes, waves, and currents, that are also linked with sediment transport patterns and sediment distribution [13]. Low hydrodynamic energy conditions promote the accumulation of fine sediments, whereas higher energy levels lead to coarse sediments [14]. Grain size and sediment type contribute to shaping the coverage of submerged aquatic vegetation and its distribution. This contributes to reducing sediment transport capacity [15] and alters the flow dynamics [16,17] and the coastal morphology [18]. Thus, the complex interaction of hydrodynamics and sedimentary processes in coastal dynamics, under different sets of forcing conditions (waves, winds, tide, storm, etc.), affects the evolution of the environment [19].

In this regard, Gillies et al. [20] state that factors such as hydrodynamic energy and sedimentation, together with light availability, nutrient loads, and herbivore numbers are critical for the growth and establishment of tropical ecosystems (e.g., mangrove forests, seagrass beds and coral reefs).

Natural and anthropogenic changes impact directly on the two-way ecological –geomorphological linkages [21], and the stability of any ecosystem depends on the combination of the threats they face [22]. The resilience of coastal ecosystems is also dependent on physical maritime processes, geomorphology, and environmental characteristics, with coral reef-seagrass-mangrove systems being more resilient than open beaches with dune-mangrove systems [23]. For example, a study in Belize [12] showed that the presence of coral reefs, seagrass meadows and mangroves together, substantially moderates incoming wave energy, inundation levels and loss of mud sediment. They also found that, although mangroves alone can offer the coastal protection services mentioned, corals and seagrasses also influence the nearshore wave climate.

In the Caribbean, the coastal biodiversity is modulated by calm wave and wind conditions for most of the year, with extreme events of short duration from time to time (i.e., hurricanes and tropical storms) [23,24]. During these storms, currents can cause significant sediment transport, resulting in coastal erosion and accumulation [25]. Temporal and spatial distribution patterns of sediment, textural variations and biophysical characteristics of ecosystems, alongshore and cross-shore, provide information on coastal hydrodynamics [26].

On the northern Mexican Caribbean coast, tourist destinations, such as Cancun and the Riviera Maya, have been built along the coastal strip of sandy beaches, interspersed with a few low-lying rocky coastal terraces. However, the attractiveness of the area has been negatively affected by the impacts of increasing anthropisation, including both tourism and urban infrastructure. This has caused fragmentation and degradation of ecosystems, as well as disassociation from traditional and sustainable practices [27]. A range of studies on the distribution of marine ecosystems and sediment analysis (e.g., [28,29]), or hydrodynamics and sediment transport features in this area, are available. Tools such as numerical modelling [18] and statistical trend analysis [30] have also been used to improve the understanding of the hydro-sedimentary processes, including temporal and seasonal variations. More recent work in the northern Mexican Caribbean by [23,31,32] has related hydrodynamics to ecosystems, describing their fragmentation, distribution, systemic interactions, connectivity and resilience. However, analyses using detailed data are still required to improve the understanding of coastal behaviour.

This paper aims to analyse the connection between hydrodynamics (average and extreme conditions), sediments and ecosystems along the northern Mexican Caribbean coast. Wave climate, sediments and the distribution of coastal ecosystems were examined with a focus on two morphodynamically different beaches, one with, and the other without, coral reefs and seagrass meadows. The physical properties of 228 coastal sediment samples from the area, collected between 2005 and 2018, and kept at the Institute of Engineering, UNAM, were analysed. In addition, two contrasting coastal environments in the area, Cancun and Puerto Morelos, were analysed for a more comprehensive description of their sediment distribution and ecosystem profiles in relation to coastal hydrodynamic behaviour, for which numerical modelling was performed.

2. Materials and Methods

2.1. Study Area

The study area is in the Mexican state of Quintana Roo, along ~250 km of the northern Mexican Caribbean coast, between Cabo Catoche in the north (21.60583° N, 87.10334° W) and Punta Allen in the south (19.7976° N, 87.47430° W) (Figure 1). Mangroves, foredunes, beaches, seagrass meadows and coral reefs interact along the coast, forming a complex ecological system that influences the flow of matter and marine energy on the very narrow continental shelf [33,34]. Between Cancun and Tulum, the main touristic corridor in Mexico covers ~220 km² of the coastal zone. Most of the population and infrastructure in Quintana Roo are concentrated here [31]. It has 913,179 inhabitants and had an annual population growth rate of ~2.54% by 2015 [35]. In this area, scores of tourist developments of varying size alternate with urbanised areas such as Cancun, Puerto Morelos, Playa del Carmen, Akumal and Tulum (Figure 1a).

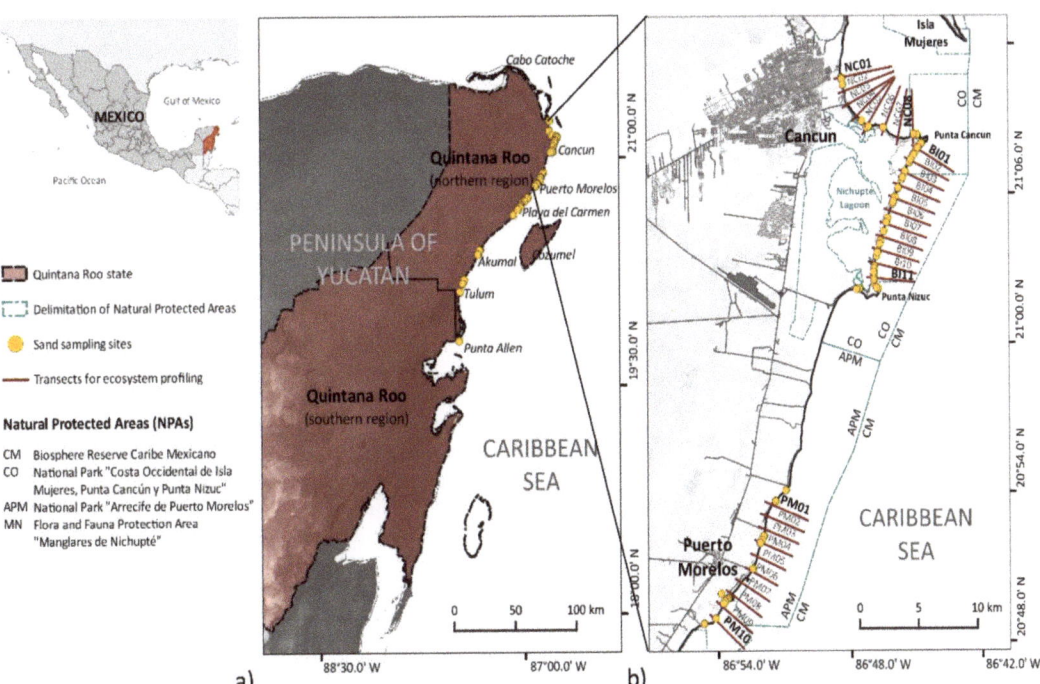

Figure 1. (a) Study area in the state of Quintana Roo, Mexico, and location of sand sampling sites. (b) Transects for coastal ecosystem profiling, sand sampling sites and surrounding Natural Protected Areas at the case study locations of Cancun and Puerto Morelos.

The area is affected by hurricanes and tropical storms that induce strong waves, winds, and coastal flooding in the summer months [36,37]. Trade winds also affect the area, as well as intense winds from the north. The microtidal conditions have a tidal range of 0.16–0.25 m [38]. Parallel to the coast, ocean currents flow with a southwest-northeast direction parallel to the deviation of the Yucatan landmass [39], but the nearshore currents, generated by the trade winds, produce southerly, littoral currents with a longshore sediment drift to the south [40].

In front of the Quintana Roo shore, a series of discontinuous coral reefs run parallel to the coast, particularly between Tulum and Puerto Morelos. These are part of the Mesoamerican Reef System, and form shallow coastal lagoons separating the reef from the shoreline [41]. The reef lagoons have sandy bottoms usually covered by seagrass meadows. The dominant foundation species is *Thalassia testudinum*, along with *Syringodium filiforme* or *Halodule wrightii* and algae [42]. The combination of coral reefs, shallow reef lagoons and seagrass meadows provides most of the coast with natural protection by dissipating wave energy, producing a less hydrodynamic, sheltered, stable shoreline [11,12,43]. However, the international tourist resort of Cancun differs from other areas as there are no coral reefs and reef lagoons offshore, so the beach receives the direct impact of waves, making it more dynamic and variable than other beach areas [34,44].

Many protected areas and natural parks have been established along the Mexican Caribbean but, again, the beach front at Cancun is the exception (Figure 1b). The barrier island, between Punta Cancun and Punta Nizuc, separates the sea from the Bojorquez-Nichupte lagoon system, an important leftover section of wetland (Figure 1b). Over the last 50 years there has been massive tourism development over the former dunes, and canalisation of the lagoon inlets.

2.2. Sampling Techniques, Numerical Modelling and Data Analysis

2.2.1. Regional Analysis

Sand samples were collected in the study area from 2005 to 2018. The sampling sites were ~1 km apart, with special focus on beaches that have different types of coastal ecosystems and degrees of anthropisation, such as Cancun, Puerto Morelos, Playa del Carmen, Akumal and Tulum (Figure 1a). At each site, three sand samples were taken; at the backshore, foreshore and at a water depth of 0.5 m. A total of 228 sand samples were collected, of which 54.3%, 20.2%, and 13.6% were from Cancun, Puerto Morelos and Playa del Carmen, respectively. Most of the sand samples were from 2006–2007 (56.6%), with 15.8% taken in 2010, following the beach nourishment projects in Cancun subsequent to the impacts of Hurricane Wilma in 2005 [34].

Grain size distribution, sphericity (SPHT) and shape factor (SF) were measured with a CAMSIZER P4 particle analyser based on a dynamic image analysis method for particle size identification and the Krumbein–Sloss chart for particle shape parameters [45]. A pycnometer method was used to determine the specific gravity and density of sand samples. A descriptive statistical analysis of grain size (D_{10}, D_{50} and D_{90}), SPHT, SF and density is presented for all the sand samples in the study area.

The variation in the grain size, D_{50}, as a function of the longshore distance (D_{LS}) was analysed, taking the most northerly location in the study area, Costa Mujeres (21.24112° N, 86.80213° W), as a reference point, to Punta Allen in the south (19.79760° N, 87.47430° W). This information allows changes in the mechanical properties of sediments to be identified and the determination of any possible directions of sediment transport along the coastline [46,47].

Wave climate conditions were examined using the output dataset from hourly estimates of wave parameters from ERA5 reanalysis at the location 21.0° N, 86.5° W from 1979 to 2021 [48]. The ERA5 dataset provides wave climate conditions over a regular latitude-longitude global grid, with spatial spacing of 0.5°, and with the selected location being the most suitable for the analysis of wave climate conditions in the northern region. Wave rose diagrams and density histograms of wave height and wave period were produced to define the wave height (H_S), wave period (T) and wave direction (θ) under prevailing and extreme wave conditions for further numerical modelling. Extreme wave-conditions were defined considering an h_{crit} threshold, exceeded for a minimum period of 12 h, with h_{crit} being 1.5 times the annual average significant wave height based on the existing hourly records of wave climate conditions (h_{crit} = 1.5 $H_{S,\,annual}$) [49].

The spatial distribution of coastal environments in northern Quintana Roo was assessed for a coastal strip, 5 km from the shoreline, using an existing ecosystem classification (i.e., mangrove, coastal dunes, coral reefs, seagrass and macroalgae) obtained from [31,50,51]. The areas covered by human settlements, including their number of inhabitants, were also included, based on [52,53]. Considering the length of each beach, the ecosystem and human settlement densities were calculated for Cancun, Puerto Morelos, Playa del Carmen, Akumal and Tulum.

2.2.2. Local-Scale Analysis

Cancun and Puerto Morelos were taken as specific case studies due to their contrasting coastal morphology, hydrodynamics and coastal ecosystem types (Figure 1b). The spatial grain size distribution for 150 m wide strips on the shore for both cases was assessed through the interpolation of D_{50}-values, to obtain the cross-shore distributions of grain sizes. The raster of interpolated D_{50}-values for each location was overlaid with the numerically modelled wave-induced currents, for prevailing and extreme wave events.

The WAPO model [54] was used to obtain the surface wave propagation and the wave-induced currents in the two areas. This model couples the REF/DIF model [55] to resolve the parabolic form of the mild slope equation including wave refraction-diffraction effects in intermediate waters [55,56]. Using the REF/DIF results, the WAPO model calculates the wave-induced currents by resolving the long wave equations. The numerical domains were defined as 20.79–20.91° N and 86.94–86.82° W (14.5 km × 18.4 km) for Cancun, and 21.01–21.27° N, 86.84–86.70° W (12.6 km × 13.0 km) for Puerto Morelos. The bathymetric information for the numerical domains was based on data from nautical charts of the Mexican Navy (SM 900; SM922; SM 922.1; SM 922.3) [57], bathymetric data from WorldView-2 satellite images [50], and local bathymetries obtained in 2008 and 2010. Both numerical domains were discretised in regular grids of 20-m spatial resolution.

Using the available topo-bathymetry data and including information on their coastal ecosystems, 19 beach profiles were assessed for Cancun, perpendicular to the coastline, and 10 for Puerto Morelos, each with a spacing of 1 km. For Cancun there were two areas of profiles: those north of Cancun and those on the barrier island. The classification detailed anthropised areas (urbanisation, hotel development, population, and roads), natural areas (subaerial vegetation other than mangroves), bodies of water (coastal lagoons), mangroves, coastal dunes, type of coast (beach, cliffs or rocky shoreline and artificial protection structures), seagrass and macroalgae, sediment accumulation and coral reefs. The profiles extend 500 m landward from the shore. On the seaside, the profiles extend up to a water depth of 20 m, depending on the information available on the coastal ecosystems. The length of each ecosystem in each profile was obtained through the intersection between the geographical position of the ecosystems, mangrove, reef, seagrass/macroalgae [50,51] and dunes [31], as well as the location of urban infrastructure, the type of coastline and water bodies [52,53]. The identification of the ecosystems in each section was also corroborated with Google Earth satellite images. Subsequently, the extension of each ecosystem in the profile was obtained and its sedimentary and hydrodynamic characteristics were associated with it.

Characteristic profiles were developed to provide a summary of the distinctive features that are shared between all profiles within a specific area of the case studies and to visualise the typical extent and distribution of coastal environments easily. Each profile was normalised by dividing the cross-shore distances (D_{CS}) by the total length of each profile (L) to obtain unit length profiles with the normalised cross-shore distance (D_{CS}*) given by Equation (1).

$$D_{CS}* = \frac{D_{CS}}{L} \tag{1}$$

In this way, the topo-bathymetric information could be averaged to produce two characteristic unit profiles for Cancun and one for Puerto Morelos. The ecosystem type within the characteristic unit profile was found from the statistical mode of the ecosystems at each corresponding location (D_{CS}*) in the unit profiles. The normalised length of the characteristic profile was then scaled, using the average length of the n-profiles considered (i.e., $L_{AV} = \Sigma L_i/n$, with L_{AV} as the scale factor), to give the three characteristic profiles for the ecosystems in each case study.

3. Results

3.1. Sediments, Wave Climate, and Coastal Environment

3.1.1. Sediment Characterisation

Table 1 shows the sediment features from all the sand samples examined and individually for the case studies. The coefficients of variation (CoV) based on the standard deviation and the mean describe low regional variation (CoV < 13.3%) for the density, SF, and SPHT, but not for the grain size distribution for which variation is higher (CoV > 48.4%).

Table 1. Grain size (D_{10}, D_{50} and D_{90}), density, shape factor and sphericity for the sand samples from the study area.

	Regional Results				[a] Case Studies	
Variable	Min.	Max.	Mean	Std. Dev.	Cancun	Pto. Morelos
Grain size, D_{10} [mm]	0.050	0.727	0.219	0.106	0.254	0.214
Grain size, D_{50} [mm]	0.093	1.433	0.391	0.200	0.451	0.368
Grain size, D_{90} [mm]	0.196	4.754	0.875	0.702	0.992	0.754
Density [kg/m^3]	1632	3232	2443	324	2565	2393
Shape factor, SF = b/L [-]	0.653	0.744	0.692	0.018	0.693	0.693
Sphericity, SPHT [-]	0.679	0.897	0.831	0.029	0.836	0.832

[a] The values provided correspond to the mean values of each variable in the case study areas.

Most of the samples were calcareous deposits of biogenic origin. However, higher density values (3232 kg/m^3) were found at backshore sites close to the rocky points of Punta Cancun and Punta Nizuc. In contrast, lower density sediment samples were found at Tulum, 1632 kg/m^3, possibly related to the porous calcite content in the samples. The SF and SPHT show values associated with fine and medium beach sands. The highest SPHT-values were for the northern beaches of Cancun, with SPHT > 0.87.

The grain size distribution in the study area was 0.219–0.875 mm for the mean values of D_{10} and D_{90}. However, less than 25% of the sand samples had a D_{50} < 0.219 mm or a D_{50} > 0.875 mm. Sampling sites in the northern part of Cancun and at Punta Brava had a wide sediment distribution, with D_{90}-values of 4.562 and 4.754 mm, respectively, which correspond to granular sediment and pebbles. On the other hand, sediment with >10% silt was found in Punta Allen, where D_{10} < 0.0627 mm. Less than 5% of the sand samples had such extreme values of grain size distribution.

Table 1 shows the differences between the case studies of Cancun and Puerto Morelos. The grain size distribution (D_{50} and D_{90}), density and SPHT in Puerto Morelos were lower than that of the region, the opposite of Cancun, where the sediment features were higher than in the region as a whole.

Findings worth noting are two sand samples from Cancun (Figure 2): a 2005 sample from Cancun, before the beach nourishment, and a sand sample from the same location, but after the beach nourishment. The first sample showed a remarkable degree of well-sorted sediment, as seen in Figure 2a, and a D_{90} = 0.436 mm and a D_{10} = 0.252 mm, both close to the D_{50} = 0.317 mm, giving one of the most homogeneous grain size distributions of all the sand samples collected. In contrast, the second samples showed similar sand parameters but a significantly higher grain size distribution and a low degree of sorting (Figure 2b). With this great difference in the grain size distributions, it is to be expected that the morphodynamic behaviour of the beach under present conditions is different than before the beach nourishment.

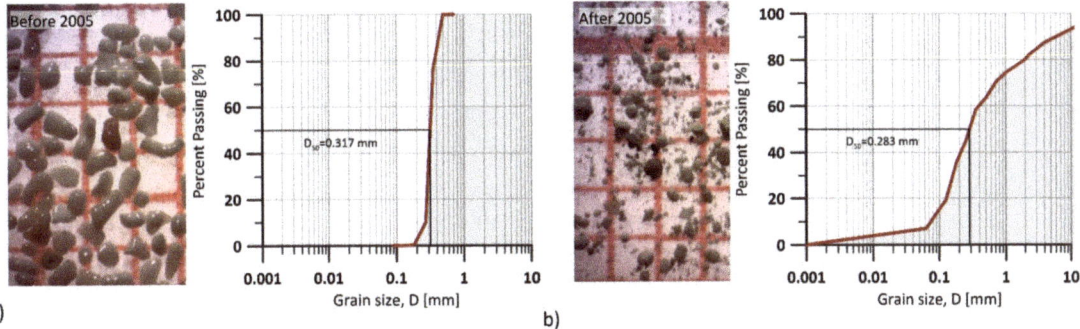

Figure 2. Cumulative grain size distribution and microscope images (1 mm grid) of sand samples at Cancun (21.172913° N, 86.805336° W) (**a**) before and (**b**) after the 2005 beach nourishments in that area following hurricane Wilma.

Changes in sediment grain size for each of the three parts on the beach profiles are given in Figure 3. The coarsest backshore sands were from the southern half of the Cancun barrier island, with a grain size $0.430 < D_{50} < 0.747$ mm for ~69.0% of the sand samples. In contrast, 26.6% of the sand samples in the north of Cancun had a grain size of over 0.430 mm. Although some peaks in the grain size were seen at specific sites, such as Punta Brava ($D_{50} = 0.616$ at $D_{LS} = 61.3$ km), north of Playa del Carmen ($D_{50} = 0.483$ at $D_{LS} = 79.1$ km), and at Akumal ($D_{50} = 0.631$ at $D_{LS} = 139.2$ km), there was generally a decrease in grain size towards the south, for nearshore, foreshore and backshore samples (Figure 3a–c). The finest sands were from the backshore at Tulum ($D_{LS} = 169.4$ km) and Punta Allen ($D_{LS} = 221.1$ km) with $D_{50} < 0.101$ mm.

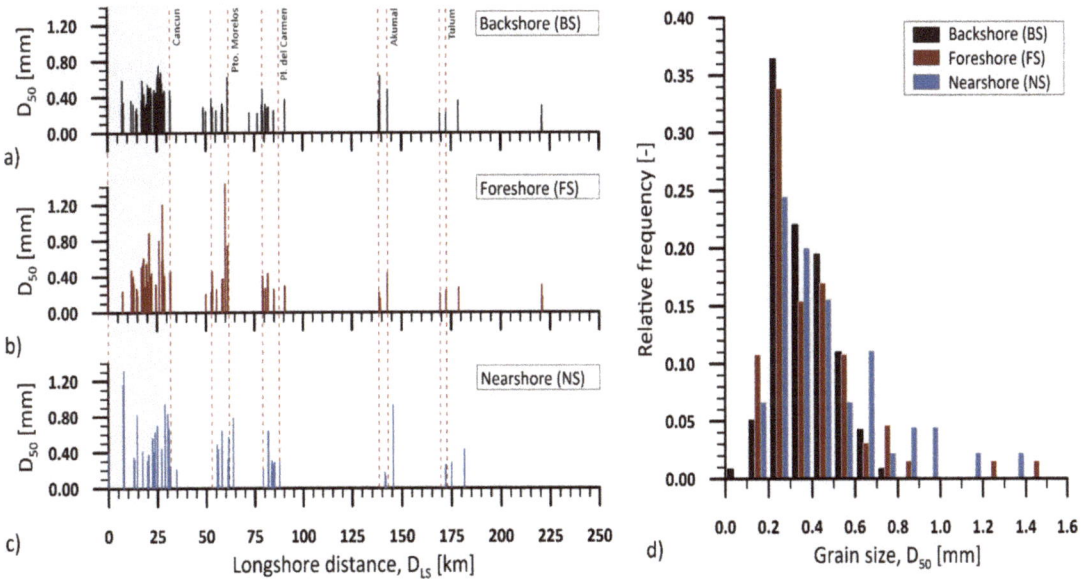

Figure 3. Grain size variations for the different beach profile sections: (**a**) BS, (**b**) FS and (**c**) NS. The zero-reference point D_{LS} is at 21.24112° N, 86.80213° W, with DLS values increasing parallel to the coast and in a southerly direction. (**d**) Histogram for the grain size distribution for the different parts of the profiles for all the sample sites.

The grain size distribution in the foreshore and nearshore samples showed medium to fine sands, with a grain size that had higher D_{50}-values than those of the backshore (Figure 3d). The nearshore had the largest grain size ($D_{50,mean}$ = 0.464 ± 0.26 mm) with a wider variation along the coast. This was followed by the foreshore ($D_{50,mean}$ = 0.401 ± 0.23 mm) and backshore ($D_{50,mean}$ = 0.358 ± 0.13 mm). About 13.45% of the nearshore sand samples exceeded D_{50} = 0.800 mm; however, this percentage decreased to 4.6% for the foreshore and to zero for the backshore.

The coarsest sands on the northern Mexican Caribbean were from specific sites in Cancun and Punta Brava (D_LS = 7.89 km and D_LS = 61.3 km), regardless of which part of the beach profile they were taken from. In summary, fine to very coarse sands were found on the northern Mexican Caribbean coast, according to the Udden–Wenworth grain size scale, with a predominance of fine and medium sand, with 75% of sands having D_{50} < 0.479 mm.

3.1.2. Wave Climate Conditions

The most frequent wave climate in the area between 1979 and 2020 had waves with a predominant ESE-WNW direction (θ = 112.5°) (Figure 4a). Nearly 95.5% of the waves had heights of HS < 2.0 m, of which 41.7% were of HS < 1.0 m, with peak wave periods of TP = 4.0–8.0 s. The period of prevailing wave TP = 6.0 s had an individual probability of occurrence > 10%. About 4.5% of the waves were over HS = 2 m and only 0.3% of the wave height exceeded HS = 3 m in the prevailing wave conditions.

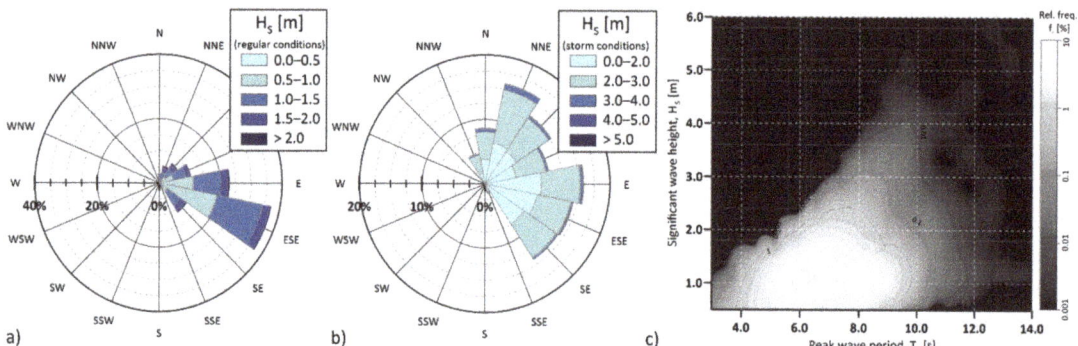

Figure 4. Wave rose diagrams for incident waves in (**a**) prevailing and (**b**) extreme conditions. (**c**) Bivariate histogram of H_S and T_P for wave climate conditions from ERA5 at 21.0° N, 86.5° W.

For extreme wave conditions, the critical wave height threshold was h_{crit} = 1.71 m. Wave heights were above this threshold, but below 2.0 m for 48.8% of the storm conditions, whereas 46.5% were 2.0 < HS < 3.0 m (Figure 4b). Severe conditions, with HS > 3.0 m, occurred 4.7% of the time, with waves coming mainly from a NNE direction (θ = 22.5°) and up to HS = 11.28 m, associated with wave periods of TP = 7–8 s. Therefore, extreme wave conditions for the numerical modelling of wave propagation in the case studies were defined for a critical condition of HS = 3.0 m, θ = 22.5° (Figure 4b) and a TP = 8.0 s (Figure 4c).

The average storm data for every five-year period is shown in Table 2, including the mean storm duration, total number of storms, and the yearly averaged minima and maxima within the period. The maximum significant wave height reached within the quinquennial period is also provided.

The average wave height was $H_{S-A}V$ = 2.04 ± 0.02 m with wave periods of TP = 7–8 s. Maximum wave heights of H_{S-Max} = 10.67 m and H_{S-Max} = 11.28 m were recorded with Hurricane Gilbert (1988, H5) and Hurricane Wilma (2005, H4), respectively, which made landfall in the north of the peninsula. Other very energetic waves were associated with hurricanes that approached without making landfall on the peninsula (e.g., hurricanes

Allen 1980, H5; Ivan 2004, H5; and Michael, H1). In addition, the maximum wave heights in 1991 and 1995 occurred in stormy, not hurricane conditions.

Table 2. Mean and maximum wave height during extreme wave conditions, storm duration and number of storms for every five-year period.

Period	Wave Height		Storm Duration [h]			Number of Storms		
	$H_{S\text{-}AV}$ [m]	[a] $H_{S\text{-}Max}$ [m]	Mean	[a] Min	[a] Max	Mean	[a] Min	[a] Max
1980–1985	2.03	7.43 (1980)	35.2	31.4 (1984)	39.3 (1981)	19.8	14 (1982)	25 (1980)
1985–1990	2.04	11.28 (1988)	35.5	29.0 (1985)	41.3 (1989)	19.8	14 (1986)	26 (1988)
1990–1995	2.04	4.02 (1991)	39.0	31.4 (1992)	44.0 (1990)	17.0	14 (1992)	22 (1990)
1995–2000	2.06	5.69 (1995)	46.3	31.3 (1995)	55.2 (1999)	22.0	14 (1997)	28 (1996/1998)
2000–2005	2.03	7.13 (2004)	36.6	30.5 (2001)	45.5 (2004)	21.6	19 (2000)	28 (2001)
2005–2010	2.08	10.67 (2005)	38.7	31.9 (2007)	44.1 (2008)	23.6	21 (2006)	27 (2007)
2010–2015	2.02	4.28 (2014)	34.2	31.9 (2013)	37.4 (2012)	22.2	17 (2011/2012)	33 (2010)
2015–2020	2.03	3.71 (2018)	37.7	26.7 (2016)	53.4 (2015)	21.8	17 (2019)	29 (2016)
Average			37.9	30.5	45.0	21.0	16.3	27.3

[a] The year of occurrence is given in parentheses.

The duration of storms was variable, with no clear pattern, being generally of between 30.5 and 45.0 h, with maximums in 1999 and 2015. For 1995–2000 the duration of storms was longer than for other periods, up to 46.3 h. The historical minimum for the number of storms per year was reached in 1982, 1986, 1992, and 1997 (14), with a variable average storm duration, while in 2010 and 2016, there were more storms than in other years, but with low persistence.

In general, there has been a trend to more storminess, with the maximum number of storms, 22–26 in 1980–1995, increasing to 27–28 for 1995–2010 and up to 33 after 2010. There has also been a rise in the minimum number of storms in a year, from 14 before 2000 to more than 17 after 2010.

3.1.3. Land Use

The main characteristics of land use in the ~250 km coastal trip from Cabo Catoche to Punta Allen are shown in Table 3. Of these, about 31.8, 9.0, 8.4, 4.4 and 3.1 km of the anthropised waterfront of Cancun, Puerto Morelos, Playa del Carmen, Akumal and Tulum, respectively, were examined in more detail (see Figure 1a). The coastal tourist corridor, from Cancun to Tulum, has 79.4% of the total population living in the coastal area of the Mexican Caribbean, of which 96.4% live in Cancun and Playa del Carmen (i.e., 780,882 inhabitants).

An area of 22,544 ha of the study area is covered by mangroves, with the most extensive being between Cabo Catoche and northern Cancun (in the north), and between Tulum and Punta Allen (in the south). Cancun has a large area of mangrove, with a density similar to that of Puerto Morelos, but 4–10 times higher than that of Playa del Carmen, Tulum or Akumal. Coastal dunes are virtually absent in the anthropised areas analysed. Seagrass and macroalgae are also present in the northern region, with a similar density to that found off Cancun and Puerto Morelos, but far lower than off Playa del Carmen, Akumal and Tulum. The coral reefs are a discontinuous ecosystem, most dense between the south of Punta Nizuc (Cancun) and Puerto Morelos, and with some sections, such as the urban/tourist zone of Cancun, where there are no reef remnants.

Table 3. Characteristics of land use in the northern Mexican Caribbean.

[a] Variable		Northern Region	Anthropized Areas in the Northern Region				
			Cancun	Pto. Morelos	Pl. del Carmen	Akumal	Tulum
Inhabitants	total per km	818,876 (3276)	630,959 (19,842)	9188 (1021)	149,923 (17,848)	1310 (293)	18,233 (5881)
Human settlements	ha ha/km	17,313 (69)	11,803 (371)	186 (21)	3543 (422)	36 (8)	386 (125)
Mangrove	ha ha/km	22,544 (90)	3153 (99)	1055 (117)	212 (25)	41 (9)	29 (9)
Coastal dune	ha ha/km	1719 (7)	27 (1)	15 (2)	8 (1)	— —	0.1 (0.0)
Seagrass and macroalgae	ha ha/km	33,327 (133)	4381 (138)	1222 (136)	642 (76)	203 (46)	254 (82)
Coral reefs	ha ha/km	4589 (18)	123 (4)	237 (26)	47 (6)	51 (12)	50 (16)

[a] Hectares (1 ha = 10,000 m^2) and the density in hectares per kilometre of coastline (ha/km).

3.2. Case Studies

The detailed results for Cancun and Puerto Morelos are used to show the links between beach sediments features, wave climate and land use, in a local-scale analysis. The chosen locations mainly differ in the presence, and lack of, a fringing reef, as well as the extent of urban infrastructure nearby.

The wave-induced currents and grain size variation in the sediments are described, along with the land use spatial distribution and their associated beach profiles. For each location, two scenarios were modelled using the wave climate results for the northern Mexican Caribbean for: a) prevailing wave conditions with H = 1.0 m, T = 6.0 s and ESE direction (θ = 112.5°), and b) extreme conditions with H = 3.0 m, T = 8.0 s and NNE direction (θ = 22.5°).

3.2.1. Case Study 1: Cancun

For Cancun, the land use, distribution of sediment grain size and wave induced currents, with their velocity, magnitude and direction, are shown in Figure 5. In Figure 5a land use and bands showing grain size variations are shown, while Figure 5b shows currents induced by prevailing waves, and Figure 5c shows these currents in extreme waves.

The southern section of the beach in Cancun (~21.04–21.06° N) has the coarsest sediment found in the case study (Figure 5a). Another area with coarse sediment is the central part of the barrier island (~21.10° N). A proxy of sediment transport paths can be inferred, as they normally run from areas of coarse material towards zones with finer sediments, resulting in a south-north direction in the north of the beach (~21.10° N to ~21.13° N) and a north-south direction in the south of the beach (~21.04° N to ~21.06° N). This observed sediment transport path is closely related to the wave-induced currents under prevailing conditions (Figure 5b), where the magnitude along the barrier island (~0.05 m/s) increases close to the shoreline and at the nearshore, allowing the settlement of coarse particles (>0.600 mm), according to the Hjulström curve. Under extreme conditions, there is a dominant north-south direction along the barrier island with velocities >0.20 m/s (Figure 5c) that, based on the Hjulström curve, might lead to erosional processes which might modify the prevailing sediment distribution in the area. It is worth noting that, although the wave-induced currents are small, even under extreme conditions (<0.3 m/s), they can still induce transport and erosional processes for the grain sizes found in the study area.

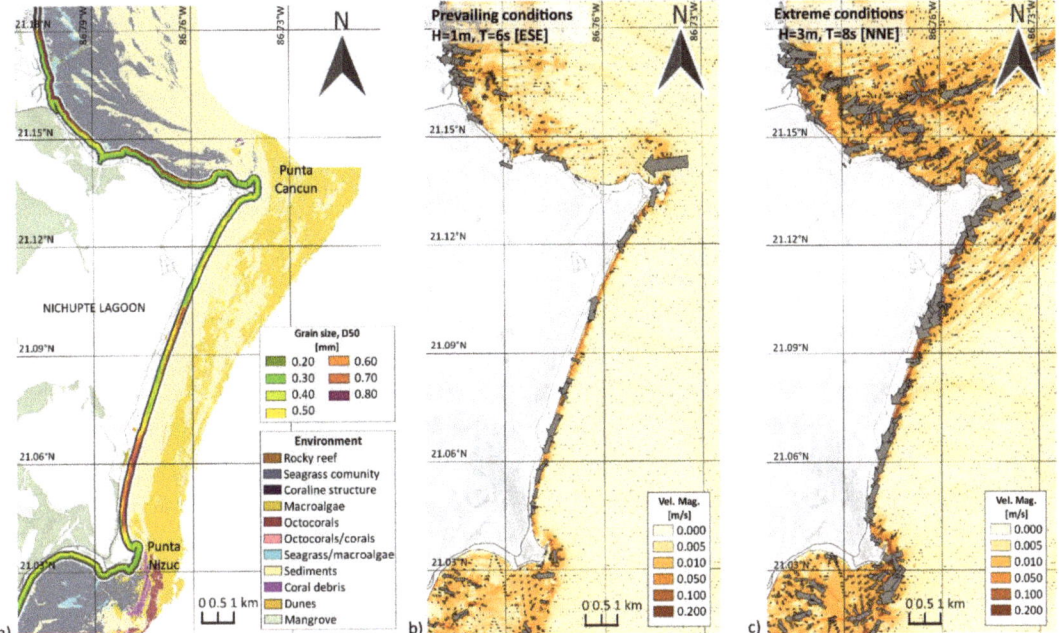

Figure 5. Results for Cancun showing (**a**) land use and variation of the sediment grain size (D_{50}). Sediment grain size variation is shown for a 150 m width strip along the shoreline, with sand samples from the backshore, nearshore and foreshore zones. Speed and direction of wave-induced currents are also shown for (**b**) prevailing [H_S = 1 m, T_P = 6 s, θ = 112.5° (ESE)] and (**c**) extreme conditions [H_S = 3 m, T_P = 8 s, θ = 22.5° (NNE)].

The medium and coarse sands observed in the south of the barrier island (D_{50} > 0.40 mm) are among the coarsest sand samples in the northern Mexican Caribbean. This could be linked to the wave-induced currents, but also to the possible degradation of a lithified Pleistocene dune located at ~21.06° N, that extends over the Nichupte Lagoon, and that could be submerged seaward (Figure 5a). Between the two areas of coarse sediment on the barrier island, there is an area of finer material near to a coastal dune (Playa Delfines), This material is of the finest grain size found in the study area with D_{50} = 0.154 mm and D_{50} = 0.169 mm at the dune crest and the berm, respectively. This sediment is possibly becoming a reservoir that can be transported to other beach segments under extreme wave conditions.

Sediment and macroalgae are the main bottom cover on the submerged beach of the barrier island of Cancun. The coarse sands found here are linked to intense currents very close to the shoreline, where most of the wave-energy dissipation and continuous high turbulent velocities due to wave breaking are expected to occur. At ~21.11–21.13° N, under extreme conditions, the velocity magnitude increases close to the shore, with velocity magnitudes of >0.05 m/s spreading seaward, over patches of sediment interspersed with areas of macroalgae. There are low velocity magnitudes (<0.010 m/s) over the sediment and macroalgae in front of the barrier island (Figure 5) where the greater water depth lessens the impact of the waves.

The wave-induced currents around Punta Nizuc flow southwards due to the diffraction of the waves by the rocky point. The higher velocities develop close to the corals, possibly related to turbulence and wave breaking. In the reef lagoon south of Punta Nizuc, velocity magnitudes of ~0.010 m/s develop over a wide area of backreef, induced by the shallower water in the reef lagoon. Finer sediments in the shoreline were found as the velocity magnitudes decrease over the shoreline. Mangroves coincide where there is finer sediment and where seagrass communities are also found.

In the northern part of the beach of Cancun, the dominant bottom covers are seagrass and sedimentary deposits, which give rise to submarine dunes. Over these, in prevailing conditions, velocity magnitudes (<0.05 m/s) increase, especially at 21.15–21.18° N, induced by the shallower waters. Sediment size is greater at the nearshore ($D_5 0 > 0.800$ mm), and finer sediment is found in the backshore. This area coincides with the development of seagrass meadows and macroalgae patches. The wave-induced currents from Punta Cancun, with a W-E direction, follow the longitudinal direction of the submarine dunes over the seagrass areas. This effect is also seen in extreme conditions, with the wave-induced currents being more intense at the transition between areas of seagrass and sediments. This is because the sedimentary deposits lie at depths about 1 m less than where seagrass meadows are found. It is important to mention that the wave propagation model does not consider wave energy attenuation from seagrass meadows, nor from coral reefs, so that the currents reaching the coast could be less strong than those indicated by the model.

Isla Mujeres is located near the northern beaches of Cancun (Figure 1). This island provides shelter from wave effects, thereby decreasing the intensity with which sediment is transported. In the long term, this protection may have created the conditions that explain why the northern part of Cancun has shallower waters, as it produces a less dynamic area, where sediment can be deposited. Nowadays, wave diffraction-refraction processes could lead to the wave-induced currents observed in Cancun's northern beaches. Similarly, the island could induce the conditions needed for the establishment of seagrass meadows—waters with low turbidity, sheltered from waves. Once they have become established, seagrasses decrease wave power and current energy, increase sedimentation and fix bottom sediments, due to the structure of their leaves, roots and rhizomes [6,58,59]. The protected conditions of the northern beaches could also explain the reduction in grain size compared to that of the barrier island.

The beach profiles are shown in Figures 6 and 7, where the distribution of coastal ecosystems, their segmentation, and land use in the north of Cancun (NC), and on the barrier island (BI), are detailed.

The anthropised areas in the subaerial profiles are about 210–235 m in width. The maximum anthropised width is ~424 m, at NC02, limited by terrestrial or marine water bodies and occupying the entire subaerial portion of the profile. On the barrier island, the maximum extent of anthropised areas is ~344 m, also limited by water bodies.

The profiles for northern Cancun (NC01 to NC08) show mangroves and seagrass communities. The latter are present in all the submarine profiles, covering 8.4–52.1% of the profile length. There are no coral structures in the northern part of Cancun, nor in front of the barrier island, although south of Punta Nizuc, there is a coral reef community (Figure 5a). Off the barrier island, macroalgae covers 10.7–38.3% of the profile length, in combination with marine sediments (6.1–54.3%), and there are no seagrass meadows. Here, an average 35% of the profile length is marine sediments, but these could cover up to 2064 m of the sea bottom, as at BI03. It is worth noting that the presence of macroalgae on the seabed in front of the barrier island occurs at depths of over 10 m, except at profile BI11 (Punta Nizuc). Therefore, the drag and roughness effects of macroalgae on wave propagation are limited.

The data show that the subaerial landscape of Cancun barrier island is dominated by anthropised areas, mangrove ecosystems and water bodies. However, this barrier island was entirely composed of dunes prior to the large-scale tourist development of the 1970s, which practically eliminated dunes [34], leaving traces of the dune field [10]. Prior to the urbanisation, eroded beach sediment used to be recovered from dunes following extreme meteo-oceanographic conditions, a natural process that cannot occur nowadays [34].

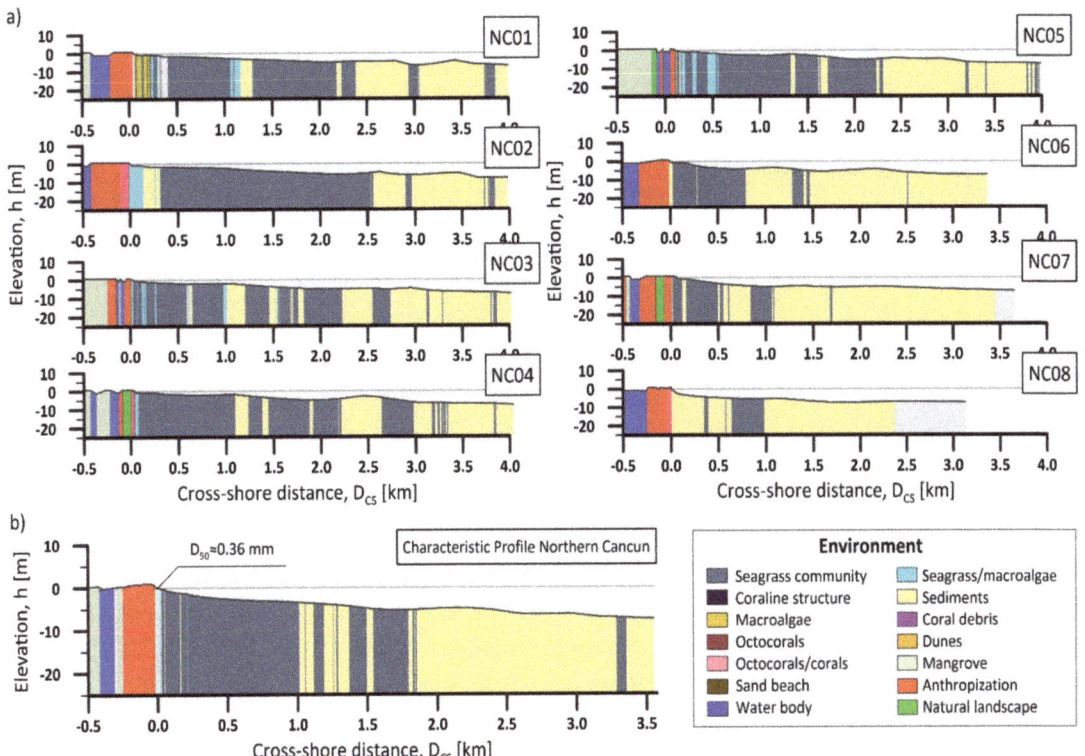

Figure 6. Beach profiles in northern Cancun, extending from 500 m inland, and up to 4 km seaward. The shoreline is located at a cross-shore distance $D_CS = 0$ m. The profiles (**a**) are numbered as in Figure 1. The land use for northern Cancun (**b**) corresponds to profiles NC01-NC08.

Almost all the coastal profiles at Cancun show sandy beaches, except NC03 and NC05. The beaches are 17.4 m wide, on average, in the north, and 45 m wide on the barrier island, with a maximum, BI09, of 65.9 m, where the only existing coastal dune is found and the greatest grain sizes (Playa Delfines). South of BI09, towards Punta Nizuc, the profiles also show mangrove areas, similar to the north of Cancun, with an average extent landward of ~170.5 m (i.e., ~34.1% of the subaerial landscape analysed). These mangrove areas are part of the Bojorquez–Nichupte lagoon system (Figure 1b). The connection between the lagoon and the sea is now limited to two rigidised inlets, meaning that the connectivity between the mangrove and the coastal zone, as well as the sediment balance once offered by intermittent inlets, has been drastically modified [34].

The characteristic beach profiles obtained for northern Cancun (Figures 6b and 8a) and the barrier island (Figures 7b and 8b) differ in their morphology and the ecosystems found. For the north of Cancun, the profile is mainly wide areas of shallow waters (h < 10 m at 0 < D_CS < 4 km) where seagrass meadows predominate at cross-shore distances of 0–2 km, with marine sediments beyond. The effect of the shallow water contributes to the slight increase in wave-induced velocities here (~0.05 m/s). On the other hand, the subaerial profile for the barrier island is anthropised, and limited by the Nichupte Lagoon. The submarine part of the profile shows rapid deepening, with water depths of >10 m at a cross-shore distance of 500 m from the beach. Sandy beaches are found close to the shore, then marine sediments up to a cross-shore distance of ~950 m, beyond which there are macroalgae ecosystems.

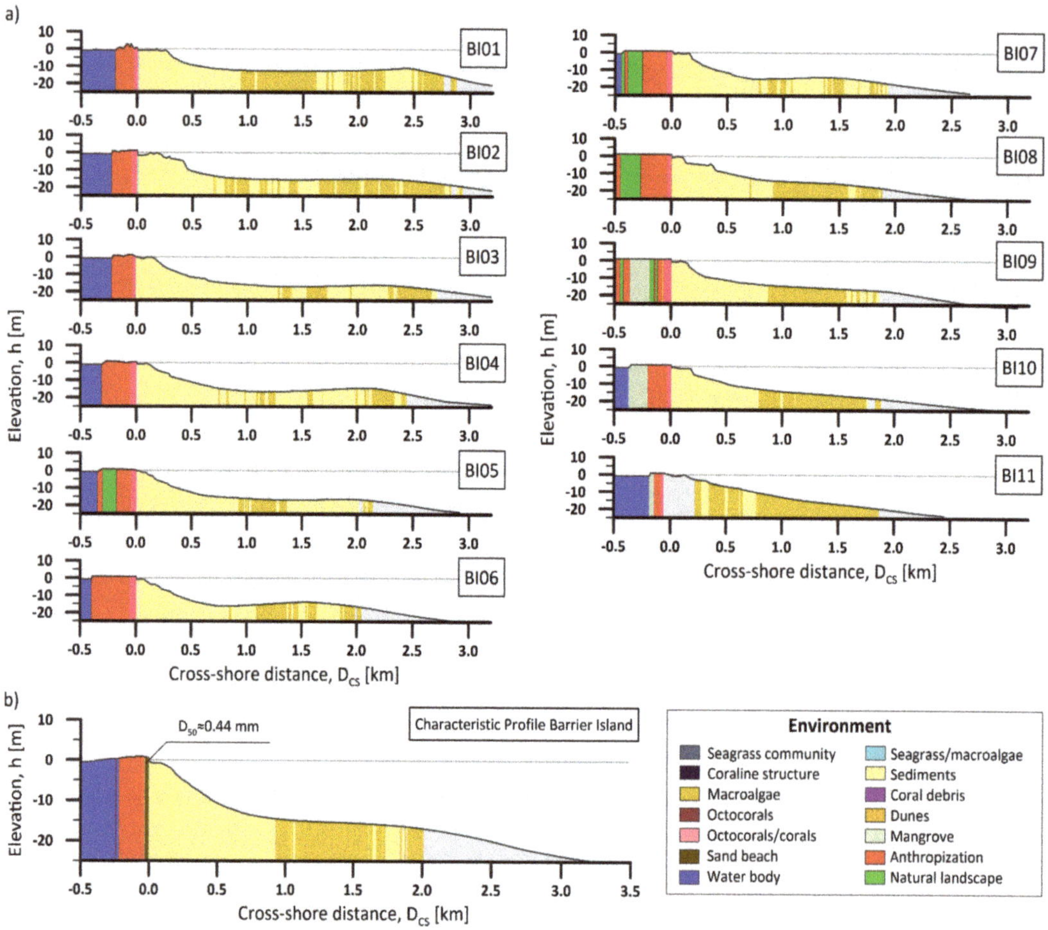

Figure 7. Beach profiles at the Cancun barrier island, extending from 500 m inland, towards a water depth of 20 m. The shoreline is located at a cross-shore distance $D_{C}S = 0$ m. The profiles (**a**) are numbered from north to south as in Figure 1. The land use on the Cancun barrier island (**b**) corresponds to profiles BI01-BI11.

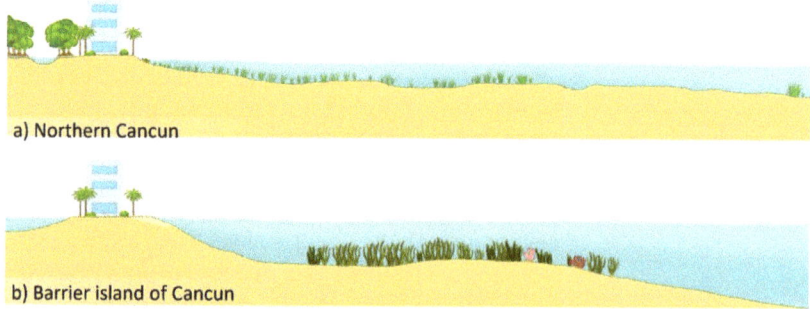

Figure 8. Schematisation of the distribution of coastal environments and characteristic coastal infrastructure at (**a**) northern Cancun and (**b**) the barrier island of Cancun.

3.2.2. Case Study 2: Puerto Morelos

Figure 9 shows the sediment grain sizes, land use and wave-induced currents for prevailing and extreme wave conditions along the Puerto Morelos coast. The effects of the nearby fringing reef are clear, with substantial changes in grain size, wave-induced currents and ecosystem distribution.

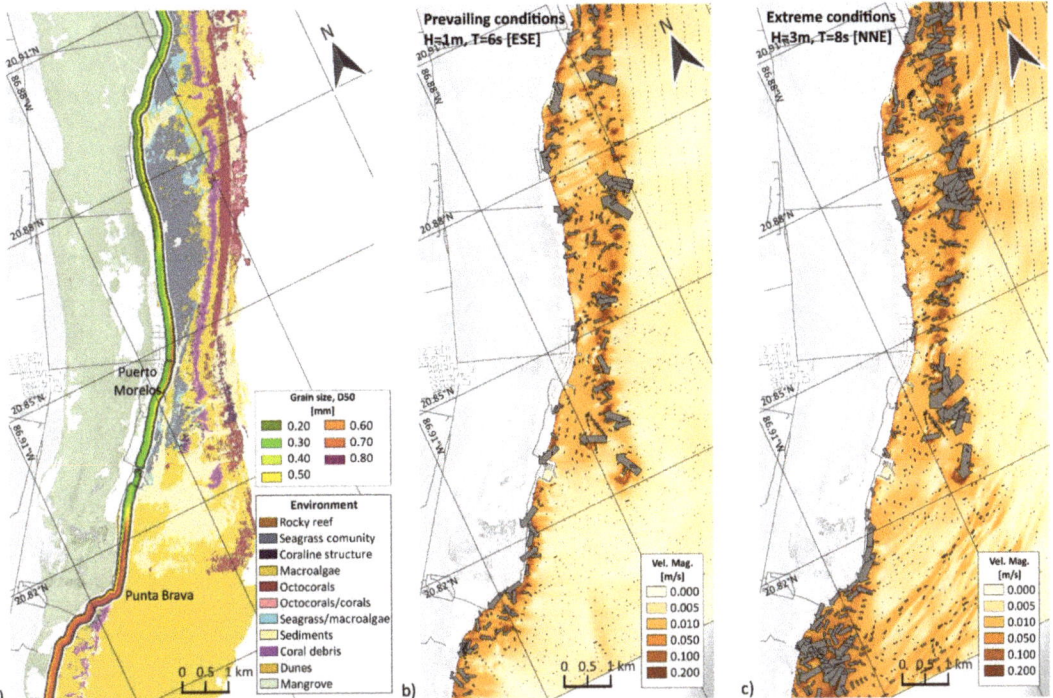

Figure 9. Results for Puerto Morelos showing (**a**) land use and variation of the sediment grain size (D_{50}). Sediment grain size variation is depicted over a strip of 150 m width along the shoreline considering sand samples at the backshore, nearshore and foreshore zones. The speed and direction of wave-induced currents is also shown for (**b**) prevailing [HS = 1 m, TP = 6 s, θ = 112.5° (ESE)] and (**c**) extreme conditions [HS = 3 m, TP = 8 s, θ = 22.5° (NNE)].

The highest velocities, >0.200 m/s, are where coral reef formations and coral debris are found, independent of wave conditions. Other areas with high velocities are seen where there is no reef, such as Punta Brava, where there is also an increase in the sediment grain size, D_{50} > 0.750 mm. Near to the town of Puerto Morelos, the backreef lagoon is narrower and the sediment grain size is greater at the nearshore. In the lee of the reef, seagrass communities and macroalgae are found, along with unvegetated areas of finer sediments, D_{50} < 0.400 mm. To the south, there are areas of submerged sediment where the velocity magnitude is low, less than ~0.005 m/s. In general, the unvegetated areas are not sheltered by reefs, but velocity magnitudes are low, which favours sediment deposition.

North of the town there is a wider area between the reef and the shore, where the velocity magnitudes are > 0.010 m/s. The sea bottom is mainly covered by seagrass and sediments (Figure 9a). Like northern Cancun, and in the reef lagoon south of Punta Nizuc, these velocities are lower than those impacting the shoreline where there is no wave protection from coral reefs (Figure 9b,c).

At open beaches, without reef protection (Figure 9a), there are large areas covered by macroalgae, similar to the beaches along the Cancun barrier island (Figure 5a). Macroalgae settle in areas of low velocity magnitudes, <0.005 m/s in prevailing conditions (Figure 9b) and <0.050 m/s in extreme conditions.

In prevailing conditions, the wave-induced currents are perpendicular to the coral reef barrier. However, a longshore current is seen, close to the shoreline, with a north-south direction. The longshore current continues south to Punta Brava in both prevailing and extreme wave conditions. South of Punta Brava, in extreme wave conditions, a small area covered by coral debris and octocorals induces a very complex pattern of wave-induced currents, although the direction is predominantly southward.

In Puerto Morelos the extensive mangrove forests have no direct connection with the sea [60], which means that they do not directly influence the size of the sediment found on the beach. In the reef lagoon the small sediment grain size (D_{50}< 0.300 mm) is probably due to the seabed being covered by seagrass communities that are protected by the reefs (Figure 9).

The coastal profiling results for Puerto Morelos are given in Figure 10. In every profile all the ecosystems are represented, except for P10 in the southernmost part of Puerto Morelos, close to Punta Brava, where macroalgae is predominant. Mangrove forests cover ~5–12% of the length of each profile, about 275 m of the subaerial landscape. Inland, the mangrove area is almost continuous in all the profiles, only disrupted by anthropised areas, in general, with a cross-shore distance $0 < D_CS < 300$ m. In profiles P01-P08, the anthropisation lies next to the beach, with no buffer area (Figure 9a). Profiles P03 and P05-P07 have most anthropisation, ~200–310 m of the profile length, representing ~40–60% of the subaerial profile.

The dune ecosystem is minimal, 9.8 m on average, mainly in the south (P09 and P10), where it extends for up to 50 m. The foredunes at Puerto Morelos are ~4 m high and degraded in many areas due to the construction of infrastructure [33]. In contrast, all the profiles have sandy beaches, with an average width of ~17 m, increasing to 30–50 m northward, except for P08, where there are structures protecting the port infrastructure.

In the marine section of the profiles, macroalgae bottom cover predominates, with an average coverage of ~930 m, although in the south, macroalgae covers up to ~3050 m (P10). Marine sediments are common in the south, as shown in P07 to P09, providing an average coverage of ~465 m and reaching cross-shore distances of up to ~1110 m.

The seagrass communities are mostly between P02 and P07, with ~420 m on average, reaching a maximum in P02 and P03 of ~1 km. In contrast to the macroalgae ecosystems, seagrass communities are almost continuous, as can be observed in most of the profiles (e.g., P02 and P03). Coral reef structures, coral debris and octocorals cover distances of ~390.4 m on average, with octocorals being the most common, especially in P01 and P03, where they cover a cross-shore distance of up to 540 m. The fragmentation of the ecosystems is more pronounced in profiles P01, and P07 to P09 (close to the port), being usually interspersed with seagrass, corals and marine sediments. These profiles have the greatest land use segmentation with ~71–75 interwoven patches of different ecosystems, marine sediments and macroalgae being the most common.

The characteristic profile for Puerto Morelos is shown in Figures 10b and 11. The sub-aerial landscape is mainly composed of mangroves. Anthropised areas limited by the beach are the second most common land use (Figure 11). There is a narrow band of sediments, ~30 m, seaward from the shoreline, with a grain size of $D_{50} \approx 0.30$ mm, interspersed with areas of macroalgae and occasional seagrass patches. Seaward of $D_CS = 175$ m, a continuous area of seagrass communities predominates, but with segments of marine sediments and macroalgae intercepted. Beyond $D_CS = 880$ m, the predominant environment is macroalgae, interrupted by octocoral segments at $1650 < D_CS < 1850$ m, over the forereef, at water depths of 10 m. It is worth noting that coral debris and coraline structures form a mound in almost all profiles landward of the forereef (Figure 10a).

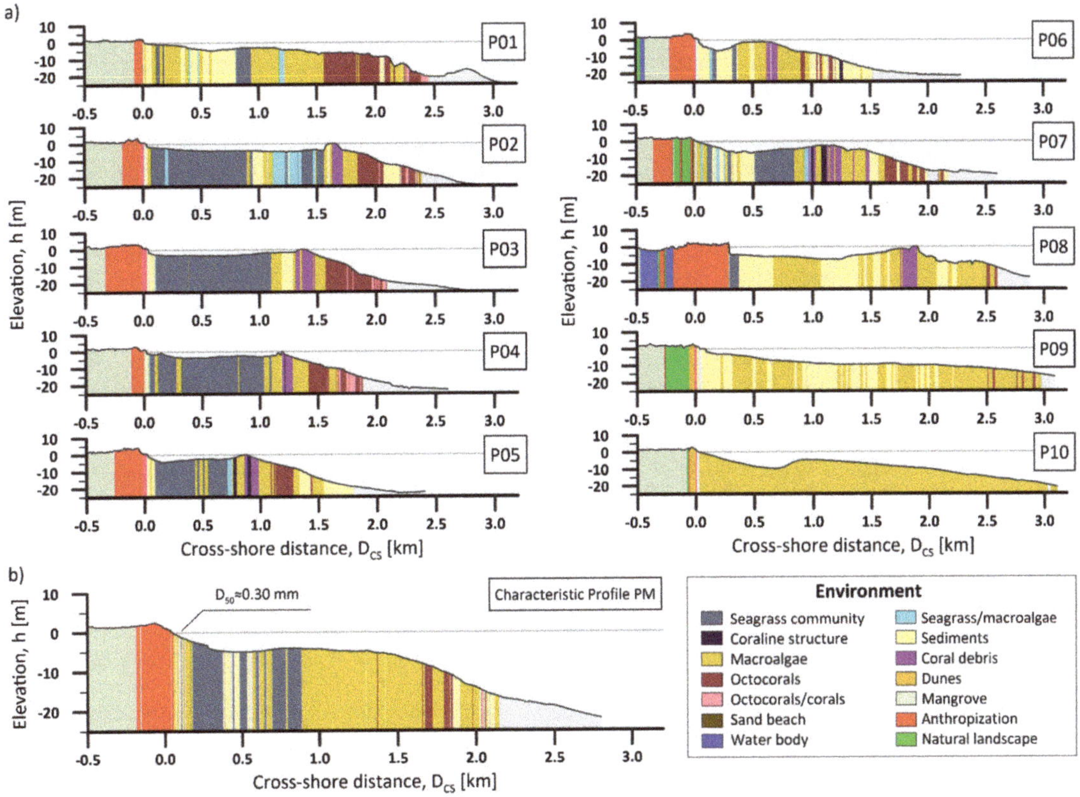

Figure 10. Beach profiles at Puerto Morelos (PM), extending from 500 m inland, towards a water depth of 25 m. The shoreline is located at a cross-shore distance D_{CS} = 0 m. The profiles (**a**) are numbered from north to south as in Figure 1. The land use at Puerto Morelos (**b**) corresponds to profiles P01–P10.

Figure 11. Schematisation of the distribution of coastal environments and characteristic coastal infrastructure at Puerto Morelos.

4. Discussion and Concluding Remarks

The dynamics of any coast involve a complex exchange of matter and energy that comes from continually changing ecological-geomorphological linkages. When evaluating the sedimentary characteristics of the northern Mexican Caribbean coast, it has been possible to verify that, on a large spatial scale, the sedimentary flow is from north to south. However, at local scale, the evaluation of detailed coastal dynamics, characteristics of coastal ecosystems (types and extent) and land uses is essential to understand the complex and interconnected relationships between these factors and local geomorphology.

The evaluation of case studies at a local level has shown that coastlines that are hydrodynamically protected by islands or coral reefs have shallow, flatter profiles created by sedimentation. This protection also contributes to the development of seagrass meadows. The interconnectivity between coral reefs and seagrasses is recognised: coral reefs dissipate the force of currents and wave energy through breaking and bottom friction [61], creating a suitable environment for seagrass. In turn, the seagrasses control turbidity and nutrients, providing water conditions that encourage the growth of coral reefs [62]. Since seagrasses also attenuate currents and waves, sediment particles are deposited in the reef lagoon, so that, over time, a shallow area develops that further contributes to wave attenuation [63]. This condition has been verified in Puerto Morelos. As a consequence, backshore/foreshore sediments are finer (~0.2–0.3 mm), dry beaches are narrower (~17 m) and more stable [44], with smaller foredunes [10] than nearby areas without this protection. In the study area north of Cancun, despite not presenting a barrier reef but with the protection of Isla Mujeres, similar geomorphological characteristics have been found, mainly due to shallow and flat nearshore, dominated by seagrasses. In these study cases, the characteristics of the profile are related to coastal ecosystem development. However, despite the extensive mangrove areas in Puerto Morelos, they seem not to be linked to hydro-sedimentary dynamics. This is because they are basin forest mangroves, a very specific type of mangrove that are not directly connected to the coastal zone [23,60]. For a better understanding of these relationships and their influence on the morphodynamics of the beaches affected by the ecosystems, the different values of bottom friction should be included in hydrodynamic models, depending on the various bottom covers (reef, sediment, seagrass, algae), so advanced models capable of making this type of evaluation are recommended.

On the other hand, beaches unprotected by seagrass and coral reefs, as found on the barrier island of Cancun, have large grain sizes (D_{50}~0.4 mm), wider beaches (~45 m) and have high dunes. These conditions are associated with steep, deep coastal profiles dominated by interspersed segments of sediment and macroalgae. Since macroalgae induce little wave friction and turbulence, their effect on reducing sediment transport is limited, as well as their capacity for sediment retention [43]. In these cases, the characteristics of the beach profile basically depend on the characteristics of the available sediment and the energy of the waves [64]. Foredunes are important ecosystems, especially on the exposed beaches, as they act as a buffer to flooding and coastal erosion, as well as serving as a sediment reservoir. However, the degradation of these ecosystems, seen most clearly on the barrier island of Cancun, means that this environmental service has been lost. Building on the dune area of exposed beaches interrupts the dynamic interaction between the dunes, beach and foreshore. As a result, the coastal profile steepens, the sediment coarsens, the habitat for seabed vegetation deteriorates, the wave attenuating effect of the vegetation vanishes, the coast becomes more exposed, and vulnerability to extreme wave climate events increases, developing into a vicious cycle of environmental deterioration that directly impacts human activities and livelihoods.

The northern Mexican Caribbean coast is exposed to both natural and human disturbances that affect the spatial distribution of sediments, hydrodynamics and ecosystem health, which shape the coastal landscape. This area is predominantly urban, with anthropogenic pressures on the environment, often fractioning subaerial ecosystems (e.g., coastal dunes), submerged aquatic vegetation and coral reefs. Water pollution, infrastructure development or water-based tourism activities are among the pressure drivers in the region [10,41,65]. These cause the coastal ecosystems to be unable to perform their primary functions, reducing their resilience in episodic or chronic events, and thereby decreasing their capacity to protect the coast, and increasing their vulnerability to extreme events and climate change [66,67]. For example, the artificial beach nourishments in Cancun have affected grain size distribution there. According to the analysis of the large sediment samples collected, no changes to sedimentary or environmental features over time were seen. However, the information presented might serve as a baseline for future comparisons.

Global and local changes in frequency, persistence and intensity of atmospheric and marine conditions seem to be outpacing the adaptive capacity of ecosystems. A growing storm number was identified for the northern Mexican Caribbean; therefore, further study is recommended to strengthen the analysis of wave conditions at locations near to the region and indices that reflect anomalies in climate patterns. In tropical coastal zones, long periods of calm allow ecosystems to establish, recover and mature, while extreme-event-induced pulses allow species turnover and simultaneous hydrodynamic interconnection of neighbouring ecosystem. Through the morphodynamic study of two beaches on the Mexican Caribbean that are close geographically, but have very different environments, it is possible to highlight the strong interrelationships between hydrodynamics, land use and sedimentation processes.

Through the use of information associated with sediment characteristics and land use (e.g., coastal ecosystems, human occupation data), spatial and cross-sectional representations of the study area of these three-dimensional connections and anthropogenic pressures have allowed us to identify trends in the coastal dynamics that may improve the coastal management of the beaches studied. Continuous monitoring might provide hard evidence of spatial and temporal changes in coastal dynamics, hydro-sedimentary processes, and the effects of natural and human disturbances [68]. Long-term monitoring of wave climate conditions, sediment and environmental changes is important, as seen in Cancun, where changes in the grain size distribution testify to the human intervention from beach nourishments following Hurricane Wilma in 2005.

Finally, tourism in the Caribbean is based on the sun, sea and sand concept. Its sustainability is dependent on the health of the ecological environment and the balance between anthropised areas and open beaches. Therefore, the use of green infrastructure for adaptive coastal management must be based on an understanding and diagnosis of the state of the interconnections of these ecosystems.

Author Contributions: Conceptualization, J.C.A.-H., C.J.C.-R., L.R.d.A., V.C. and R.S.; methodology, J.C.A.-H., C.J.C.-R., L.R.d.A. and V.C.; software, V.C. and R.S.; validation, J.C.A.-H., C.J.C.-R., L.R.d.A., V.C. and R.S.; formal analysis, J.C.A.-H., C.J.C.-R. and L.R.d.A.; investigation, J.C.A.-H., C.J.C.-R., L.R.d.A. and R.S.; resources, R.S.; data curation, J.C.A.-H. and C.J.C.-R.; writing—original draft preparation, J.C.A.-H., C.J.C.-R. and L.R.d.A.; writing—review and editing, V.C. and R.S.; visualisation, J.C.A.-H. and C.J.C.-R.; supervision, J.C.A.-H. and R.S.; project administration, R.S.; funding acquisition, R.S. All authors have read and agreed to the published version of the manuscript.

Funding: The APC was funded by CEMIE-Océano (CONACYT -SENER-Fondo de Sustentabilidad Energética project: FSE-2014-06-249795 "Centro Mexicano de Innovación en Energía del Océano CEMIE-Océano").

Data Availability Statement: Restrictions apply to the availability of the sedimentological data. This data was obtained from Instituto de Ingeniería, UNAM and is available from the authors with the permission of Instituto de Ingeniería, UNAM. Further datasets are available at [50] and [53].

Acknowledgments: The authors would like to acknowledge the access to the sand sample collection and database of the Instituto de Ingeniería, UNAM. Acknowledgements are given to the CONACYT (Consejo Nacional de Ciencia y Tecnología) program 'Investigadoras e Investigadores por México' project 761. The program of postdoctoral grants from DGAPA-UNAM for the third author is also recognised.

Conflicts of Interest: The authors declare no conflict of interest. The funders had no role in the design of the study; in the collection, analyses, or interpretation of data; in the writing of the manuscript, or in the decision to publish the results.

References

1. Hughes, T.P.; Barnes, M.L.; Bellwood, D.R.; Cinner, J.E.; Cumming, G.S.; Jackson, J.B.C.; Kleypas, J.; van de Leemput, I.A.; Lough, J.M.; Morrison, T.H.; et al. Coral reefs in the Anthropocene. *Nature* **2017**, *546*, 82–90. [CrossRef] [PubMed]
2. Larcombe, P.; Costen, A.; Woolfe, K.J. The hydrodynamic and sedimentary setting of nearshore coral reefs, central Great Barrier Reef shelf, Australia: Paluma Shoals, a case study. *Sedimentology* **2001**, *48*, 811–835. [CrossRef]
3. van den Hoek, L.S.; Bayoumi, E.K. Importance, destruction and recovery of coral reefs. *IOSR J. Pharm. Biol. Sci.* **2017**, *12*, 59–63. [CrossRef]
4. Sheppard, C.; Dixon, D.J.; Gourlay, M.; Sheppard, A.; Payet, R. Coral mortality increases wave energy reaching shores protected by reef flats: Examples from the Seychelles. *Estuar. Coast. Shelf Sci.* **2005**, *64*, 223–234. [CrossRef]

5. Christianen, M.J.A.; van Belzen, J.; Herman, P.M.J.; van Katwijk, M.M.; Lamers, L.P.M.; van Leent, P.J.M.; Bouma, T.J. Low-canopy seagrass beds still provide important coastal protection services. *PLoS ONE* **2013**, *8*, e62413. [CrossRef]
6. Maxwell, P.S.; Eklöf, J.S.; van Katwijk, M.M.; O'Brien, K.R.; de la Torre-Castro, M.; Boström, C.; Bouma, T.J.; Krause-Jensen, D.; Unsworth, R.K.F.; van Tussenbroek, B.I.; et al. The fundamental role of ecological feedback mechanisms for the adaptive management of seagrass ecosystems—A review. *Biol. Rev.* **2017**, *92*, 1521–1538. [CrossRef]
7. Cohn, N.; Hoonhout, B.; Goldstein, E.; De Vries, S.; Moore, L.; Durán Vinent, O.; Ruggiero, P. Exploring marine and aeolian controls on coastal foredune growth using a coupled numerical model. *J. Mar. Sci. Eng.* **2019**, *7*, 13. [CrossRef]
8. Pellón, E.; de Almeida, L.R.; González, M.; Medina, R. Relationship between foredune profile morphology and aeolian and marine dynamics: A conceptual model. *Geomorphology* **2020**, *351*, 106984. [CrossRef]
9. Delgado-Fernandez, I. Meso-scale modelling of aeolian sediment input to coastal dunes. *Geomorphology* **2011**, *130*, 230–243. [CrossRef]
10. de Almeida, L.R.; Silva, R.; Martínez, M.L. The relationships between environmental conditions and parallel ecosystems on the coastal dunes of the Mexican Caribbean. *Geomorphology* **2022**, *397*, 108006. [CrossRef]
11. Franklin, G.; Mariño-Tapia, I.; Torres-Freyermuth, A. Effects of reef roughness on wave setup and surf zone currents. *J. Coast. Res.* **2013**, *165*, 2005–2010. [CrossRef]
12. Guannel, G.; Arkema, K.; Ruggiero, P.; Verutes, G. The power of three: Coral reefs, seagrasses and mangroves protect coastal regions and increase their resilience. *PLoS ONE* **2016**, *11*, e0158094. [CrossRef] [PubMed]
13. Golshani, A.; Baldock, T.E.; Mumby, P.J.; Callaghan, D.; Nielsen, P.; Phinn, S. Climate impacts on hydrodynamics and sediment dynamics at reef islands. In Proceedings of the 12th International Coral Reef Symposium, James Cook University, Townsville, Australia, 9–13 July 2012.
14. Elfrink, B.; Baldock, T. Hydrodynamics and sediment transport in the swash zone: A review and perspectives. *Coast. Eng.* **2002**, *45*, 149–167. [CrossRef]
15. Liu, C.; Shen, Y. Flow structure and sediment transport with impacts of aquatic vegetation. *J. Hydrodyn.* **2008**, *20*, 461–468. [CrossRef]
16. Moki, H.; Taguchi, K.; Nakagawa, Y.; Montani, S.; Kuwae, T. Spatial and seasonal impacts of submerged aquatic vegetation (SAV) drag force on hydrodynamics in shallow waters. *J. Mar. Syst.* **2020**, *209*, 103373. [CrossRef]
17. Nakayama, K.; Shintani, T.; Komai, K.; Nakagawa, Y.; Tsai, J.W.; Sasaki, D.; Tada, K.; Moki, H.; Kuwae, T.; Watanabe, K.; et al. Integration of submerged aquatic vegetation motion within hydrodynamic models. *Water Resour. Res.* **2020**, *56*, 8. [CrossRef]
18. Amoudry, L.O.; Souza, A.J. Deterministic coastal morphological and sediment transport modeling: A review and discussion. *Rev. Geophys.* **2011**, *49*, RG2002. [CrossRef]
19. Storlazzi, C.D.; Elias, E.; Field, M.E.; Presto, M.K. Numerical modeling of the impact of sea-level rise on fringing coral reef hydrodynamics and sediment transport. *Coral Reefs* **2011**, *30*, 83–96. [CrossRef]
20. Gillis, L.; Bouma, T.; Jones, C.; van Katwijk, M.; Nagelkerken, I.; Jeuken, C.; Herman, P.; Ziegler, A. Potential for landscape-scale positive interactions among tropical marine ecosystems. *Mar. Ecol. Prog. Ser.* **2014**, *503*, 289–303. [CrossRef]
21. Stallins, J.A. Geomorphology and ecology: Unifying themes for complex systems in biogeomorphology. *Geomorphology* **2006**, *77*, 207–216. [CrossRef]
22. Perry, C.T.; Kench, P.S.; Smithers, S.G.; Riegl, B.; Yamano, H.; O'Leary, M.J. Implications of reef ecosystem change for the stability and maintenance of coral reef islands. *Glob. Chang. Biol.* **2011**, *17*, 3679–3696. [CrossRef]
23. Odériz, I.; Gómez, I.; Ventura, Y.; Díaz, V.; Escalante, A.; Gómez, D.T.; Bouma, T.J.; Silva, R. Understanding drivers of connectivity and resilience under tropical cyclones in coastal ecosystems at Puerto Morelos, Mexico. *J. Coast. Res.* **2020**, *95*, 128–132. [CrossRef]
24. Salazar-Vallejo, S.I. Huracanes y biodiversidad costera tropical. *Rev. Biol. Trop.* **2002**, *50*, 2.
25. Vincent, C.E.; Young, R.A.; Swift, D.J. Bed-load transport under waves and currents. *Mar. Geol.* **1981**, *39*, M71–M80. [CrossRef]
26. Klein, A.H.d.F.; Miot da Silva, G.; Ferreira, O.; Alveirinho-Dias, J. Beach sediment distribution for a headland bay coast. *J. Coast. Res.* **2005**, *42*, 285–293.
27. Murray, G. Constructing paradise: The impacts of big tourism in the Mexican Coastal Zone. *Coast. Manag.* **2007**, *35*, 339–355. [CrossRef]
28. Calva-Benítez, L.G.; Torres-Alvarado, R. Organic carbon and textural characteristics of sediments in areas with turtlegrass Thalassia testudinum in coastal ecosystems of the southeastern gulf of Mexico. *Univ. Ciencia. Tróp. Húmed.* **2011**, *27*, 133–144.
29. Ouillon, S. Why and how do we study sediment transport? Focus on coastal zones and ongoing methods. *Water* **2018**, *10*, 390. [CrossRef]
30. Poizot, E.; Anfuso, G.; Méar, Y.; Bellido, C. Confirmation of beach accretion by grain-size trend analysis: Camposoto beach, Cádiz, SW Spain. *Geo-Mar. Lett.* **2013**, *33*, 263–272. [CrossRef]
31. Guimarais, M.; Zúñiga-Ríos, A.; Cruz-Ramírez, C.J.; Chávez, V.; Odériz, I.; van Tussenbroek, B.I.; Silva, R. The conservational state of coastal ecosystems on the Mexican Caribbean coast: Environmental guidelines for their management. *Sustainability* **2021**, *13*, 2738. [CrossRef]
32. Silva, R.; Martínez, M.L.; van Tussenbroek, B.I.; Guzmán-Rodríguez, L.O.; Mendoza, E.; López-Portillo, J. A framework to manage coastal squeeze. *Sustainability* **2020**, *12*, 610. [CrossRef]
33. Franklin, G.L.; Torres-Freyermuth, A.; Medellin, G.; Allende-Arandia, M.E.; Appendini, C.M. The role of the reef–dune system in coastal protection in Puerto Morelos (Mexico). *Nat. Hazards Earth Syst. Sci.* **2018**, *18*, 1247–1260. [CrossRef]

34. Martell-Dubois, R.; Mendoza, E.; Mariño-Tapia, I.; Odériz, I.; Silva, R. How effective were the beach nourishments at Cancun? *J. Mar. Sci. Eng.* **2020**, *8*, 388. [CrossRef]
35. SEGOB-COESPO. *Atlas Sociodemográfico Quintana Roo*. Available online: https://qroo.gob.mx/index2.php/segob/coespo/atlas-sociodemografico (accessed on 1 December 2021).
36. Jáuregui, E. Climatology of landfalling hurricanes and tropical storms in Mexico. *Atmósfera* **2003**, *16*, 193–204.
37. Ojeda, E.; Appendini, C.M.; Mendoza, E.T. Storm-wave trends in Mexican waters of the Gulf of Mexico and Caribbean Sea. *Nat. Hazards Earth Syst. Sci.* **2017**, *17*, 1305–1317. [CrossRef]
38. Kjerfve, B. Tides of the Caribbean Sea. *J. Geophys. Res.* **1981**, *86*, 4243. [CrossRef]
39. Cetina, P.; Candela, J.; Sheinbaum, J.; Ochoa, J.; Badan, A. Circulation along the Mexican Caribbean coast. *J. Geophys. Res. C Ocean.* **2006**, *111*, C8. [CrossRef]
40. Kjerfve, B. *Coastal Oceanographic Characteristics: Cancun-Tulum Corridor, Quintana Roo*; ECOMAR-EPOMEX: Campeche, Mexico, 1994.
41. Rioja-Nieto, R.; Álvarez-Filip, L. Coral reef systems of the Mexican Caribbean: Status, recent trends and conservation. *Mar. Pollut. Bull.* **2019**, *140*, 616–625. [CrossRef]
42. Jordán-Dahlgren, E.; Rodríguez-Martínez, R.E. The Atlantic coral reefs of Mexico. In *Latin American Coral Reefs*; Cortés, J., Ed.; Elsevier: Amsterdam, The Netherlands, 2003; pp. 131–158, ISBN 9780444513885.
43. James, R.K.; Silva, R.; van Tussenbroek, B.I.; Escudero-Castillo, M.; Mariño-Tapia, I.; Dijkstra, H.A.; van Westen, R.M.; Pietrzak, J.D.; Candy, A.S.; Katsman, C.A.; et al. Maintaining tropical beaches with seagrass and algae: A promising alternative to engineering solutions. *Bioscience* **2019**, *69*, 136–142. [CrossRef]
44. Mariño-Tapia, I.; Enríquez-Ortiz, C.; Silva, R.; Mendoza, E.; Escalante-Mancera, E.; Ruiz-Renteria, F. Comparative morphodynamics between exposed and reef protected beaches under hurricane conditions. *Coast. Eng. Proc.* **2014**, *1*, 55. [CrossRef]
45. Retsch Technology. *Operating Instructions/Manual. Particle Size Analysis System CAMSIZER*; Doc.Nr. CA.; Jenoptik Germany: Haan, Germany, 2007.
46. Anfuso, G.; Pranzini, E.; Vitale, G. An integrated approach to coastal erosion problems in northern Tuscany (Italy): Littoral morphological evolution and cell distribution. *Geomorphology* **2011**, *129*, 204–214. [CrossRef]
47. Frihy, O.E.; Dewidar, K.M. Patterns of erosion/sedimentation, heavy mineral concentration and grain size to interpret boundaries of littoral sub-cells of the Nile Delta, Egypt. *Mar. Geol.* **2003**, *199*, 27–43. [CrossRef]
48. Hersbach, H.; Bell, B.; Berrisford, P.; Biavati, G.; Horányi, A.; Muñoz-Sabater, J.; Nicolas, J.; Peubey, C.; Radu, R.; Rozum, I.; et al. ERA5 Hourly Data on Single Levels from 1979 to Present. Available online: https://cds.climate.copernicus.eu/cdsapp#!/dataset/10.24381/cds.bd0915c6?tab=overview (accessed on 3 January 2022).
49. Boccotti, P. *Wave Mechanics for Ocean Engineering*; Elsevier Science: Amsterdam, The Netherlands, 2000; ISBN 9780080543727.
50. Cerdeira-Estrada, S.; Martell-Dubois, R.; Heege, T.; Rosique-De La Cruz, L.O.; Blanchon, P.; Ohlendorf, S.; Müller, A.; Silva, R.; Mariño-Tapia, I.J.; Martínez-Clorio, M.I.; et al. *Cobertura Bentónica de los Ecosistemas Marinos del Caribe Mexicano: Cabo Catoche—Xcalak 2018*; Scale 1:4; Comisi´on Nacional para el Conocimiento y Uso de la Biodiversidad (CONABIO): Mexico City, Mexico, 2018.
51. CONABIO. *Distribución de los Manglares en México en 2020*; 1:50000; Comisión Nacional para el Conocimiento y Uso de la Biodiversidad (CONABIO): Mexico City, Mexico, 2020.
52. Cruz, C.J.; Mendoza, E.; Silva, R.; Chávez, V. Assessing degrees of anthropization on the coast of Mexico from ecosystem conservation and population growth data. *J. Coast. Res.* **2019**, *92*, 136. [CrossRef]
53. INEGI. *Mapa de Uso de Suelo y Vegetación*; 1:50000; Instituto Nacional de Estadística y Geografía: Mexico City, Mexico, 2017.
54. Silva, R.; Baquerizo, A.; Losada, M.Á.; Mendoza, E. Hydrodynamics of a headland-bay beach—Nearshore current circulation. *Coast. Eng.* **2010**, *57*, 160–175. [CrossRef]
55. Kirby, J.T.; Dalrymple, R.A. A parabolic equation for the combined refraction–diffraction of Stokes waves by mildly varying topography. *J. Fluid Mech.* **1983**, *136*, 453. [CrossRef]
56. Battalio, B.; Chandrasekera, C.; Divoky, D.; Hatheway, D.; Hull, T.; O'Reilly, B.; Seymour, D.; Srinivas, R. *FEMA Coastal Flood Hazard Analysis and Mapping Guidelines—Wave Transformation*; Focused Study Report; Federal Emergency Management Agency (FEMA): Washington, DC, USA, 2005.
57. Secretaría de Marina. *Catálogo de Cartas y Publicaciones Náuticas 2022*; Secretaría de Marina: Mexico City, Mexico, 2022.
58. Koch, E.W.; Ailstock, S.; Stevenson, J.C. Beyond light: Physical, geological and chemical habitat requirements. In *Chesapeake Bay Submerged Aquatic Vegetation Water Quality and Habitat-Based Requirements and Restoration Targets: A Second Technical Synthesis*; Environmental Protection Agency: Washington, DC, USA, 2000; pp. 71–94.
59. Madsen, J.D.; Chambers, P.A.; James, W.F.; Koch, E.W.; Westlake, D.F. The interaction between water movement, sediment dynamics and submersed macrophytes. *Hydrobiologia* **2001**, *444*, 71–84. [CrossRef]
60. Adame, M.F.; Zaldívar-Jimenez, A.; Teutli, C.; Caamal, J.P.; Andueza, M.T.; López-Adame, H.; Cano, R.; Hernández-Arana, H.A.; Torres-Lara, R.; Herrera-Silveira, J.A. Drivers of mangrove litterfall within a karstic region affected by frequent hurricanes. *Biotropica* **2013**, *45*, 147–154. [CrossRef]
61. Lowe, R.J. Spectral wave dissipation over a barrier reef. *J. Geophys. Res.* **2005**, *110*, C04001. [CrossRef]
62. Barbier, E.B.; Hacker, S.D.; Kennedy, C.; Koch, E.W.; Stier, A.C.; Silliman, B.R. The value of estuarine and coastal ecosystem services. *Ecol. Monogr.* **2011**, *81*, 169–193. [CrossRef]

63. Koch, E.W.; Barbier, E.B.; Silliman, B.R.; Reed, D.J.; Perillo, G.M.; Hacker, S.D.; Granek, E.F.; Primavera, J.H.; Muthiga, N.; Polasky, S.; et al. Non-linearity in ecosystem services: Temporal and spatial variability in coastal protection. *Front. Ecol. Environ.* **2009**, *7*, 29–37. [CrossRef]
64. Dean, R.G. *Equilibrium Beach Profiles: US Atlantic and Gulf Coasts*; Department of Civil Engineering and College of Marine Studies, University of Delaware: Newark, NJ, USA, 1977.
65. van Tussenbroek, B.I.; Cortés, J.; Collin, R.; Fonseca, A.C.; Gayle, P.M.H.; Guzmán, H.M.; Jácome, G.E.; Juman, R.; Koltes, K.H.; Oxenford, H.A.; et al. Caribbean-wide, long-term study of seagrass beds reveals local variations, shifts in community structure and occasional collapse. *PLoS ONE* **2014**, *9*, e90600. [CrossRef]
66. Osorio-Cano, J.D.; Osorio, A.F.; Peláez-Zapata, D.S. Ecosystem management tools to study natural habitats as wave damping structures and coastal protection mechanisms. *Ecol. Eng.* **2019**, *130*, 282–295. [CrossRef]
67. Alcérreca-Huerta, J.C.; Montiel-Hernández, J.R.; Callejas-Jiménez, M.E.; Hernández-Avilés, D.A.; Anfuso, G.; Silva, R. Vulnerability of Subaerial and Submarine Landscapes: The Sand Falls in Cabo San Lucas, Mexico. *Land* **2020**, *10*, 27. [CrossRef]
68. Li, W.; Gong, P. Continuous monitoring of coastline dynamics in western Florida with a 30-year time series of Landsat imagery. *Remote Sens. Environ.* **2016**, *179*, 196–209. [CrossRef]

Article

Geomorphological and Structural Assessment of the Coastal Area of Capo Faro Promontory, NE Salina (Aeolian Islands, Italy)

Mauro Bonasera [1], Ciro Cerrone [2,*], Fabiola Caso [3], Stefania Lanza [4], Giandomenico Fubelli [1] and Giovanni Randazzo [4]

1. Department of Earth Sciences, University of Turin, Via Valperga Caluso 35, 10125 Turin, Italy; mauro.bonasera@unito.it (M.B.); giandomenico.fubelli@unito.it (G.F.)
2. Department of Earth, Environmental and Resources Science—DiSTAR, University of Naples Federico II, Via Vicinale Cupa Cintia 21, 80126 Naples, Italy
3. Department of Earth Sciences "A. Desio", University of Milan, Via Mangiagalli 34, 20133 Milan, Italy; fabiola.caso@unimi.it
4. Department of Mathematical and Computer Science, Physical Sciences and Earth Sciences, University of Messina, Via F. Stagno d'Alcontres 31, 98166 Messina, Italy; stefania.lanza@unime.it (S.L.); giovanni.randazzo@unime.it (G.R.)
* Correspondence: ciro.cerrone@unina.it

Citation: Bonasera, M.; Cerrone, C.; Caso, F.; Lanza, S.; Fubelli, G.; Randazzo, G. Geomorphological and Structural Assessment of the Coastal Area of Capo Faro Promontory, NE Salina (Aeolian Islands, Italy). *Land* 2022, 11, 1106. https://doi.org/10.3390/land11071106

Academic Editors: Pietro Aucelli, Angela Rizzo, Rodolfo Silva Casarín and Giorgio Anfuso

Received: 27 June 2022
Accepted: 15 July 2022
Published: 19 July 2022

Publisher's Note: MDPI stays neutral with regard to jurisdictional claims in published maps and institutional affiliations.

Copyright: © 2022 by the authors. Licensee MDPI, Basel, Switzerland. This article is an open access article distributed under the terms and conditions of the Creative Commons Attribution (CC BY) license (https:// creativecommons.org/licenses/by/ 4.0/).

Abstract: Capo Faro Promontory, located in Salina (Aeolian Islands, southern Italy), is a popular summer destination due to its volcanic morphologies, seaside, and enogastronomy. A flat area, right behind the scarp edge of a coastal cliff, hosts the Capo Faro Estate, one of the most renowned vineyards and residences on Salina Island. The promontory has been characterised in terms of geomorphological features. Remote sensing analysis, after nadir and off–nadir UAV flights, supports the field activities to explore the hazard to which the area is subjected. In particular, the coastal cliff turns out to be affected by a rapid retreat inducing landslides. Therefore, the cliff area has been investigated through a detailed stratigraphic and structural field survey. Using the generated high–resolution Digital Elevation Model, bathymetric–topographic profiles were extracted along the coastline facing the cliff. The thickness of volcanic deposits was evaluated to obtain a geological model of it. The main rock mass discontinuities have been characterised to define the structural features affecting the stability of the rock wall. The obtained results prove the contribution of such research fundamental in planning risk mitigation measures.

Keywords: hazard evaluation; coastal evolution; slope structural analysis; structure for motion; Salina Island

1. Introduction

With approximately 2300 residents, Salina is the second–most populated island of the Aeolian Archipelago (Sicily, southern Italy). It is overrun by mass tourism during the summer, reaching up to 15,000 visitors a day (based on media information and personal evaluation around the middle of August 2021). Tourists are interested in its worth for volcanic morphology, its three quaint coastal villages, and its enogastronomy, the reasons why Salina and the other six islands of the Archipelago are included in the World Heritage List. Being the most fertile of the Aeolian Islands, high–quality grapes are grown, from which the renowned wine "Malvasia" is obtained and exported all over the world. Furthermore, local people and visitors flock to the Salina beaches, often of limited width to manage all the bathers. As with the other six "sister" islands, Salina presents a considerable number of embayed pocket beaches [1–3], which are highly attractive due to the difficulty of access that often offers more discretion. The pocket beaches are backed by vertical cliffs a dozen metres high, as a consequence of the erosive process due to waves and weathering [4]. In this context, the geomorphological evolution induces a rapid cliff retreat, which triggers rockfalls and topples exploiting the discontinuities network [5–7] in the volcanic rock mass (e.g., joints and faults). Techniques in all fields of geology have been already explored in similar cases in the southern Italian peninsula [8–13], to define the hazardous conditions that could even provoke injuries or fatalities. Assessing these conditions, it is possible to find the best solution to address the risk mitigation measures or to fix them where they become ineffective, in a broad scenario of coastal management [14–16].

In this paper, the approach to this issue integrates geomorphology, morpho–stratigraphy, and structural geology. It deals with the hazard assessment of a coastal cliff in Capo Faro Promontory, in the north–eastern part of Salina Island, in the vicinity of Capo Faro Estate residence and vineyard. The aim was to provide valid support to future risk mitigation interventions, suggesting prevention tools to slow down the natural retreat process and to simultaneously secure the beach from sudden gravitational phenomena.

2. Geological Setting
2.1. The Aeolian Geodynamics

Salina belongs to the Aeolian volcanic province, consisting of seven islands and several seamounts. The volcanic arc extends for about two hundred kilometres, around the seamount Marsili and the homonymous basin, with the concave part pointing towards the centre of the Tyrrhenian Sea (Figure 1). This constitutes a key area to unravel the geodynamic framework of this Mediterranean sector, the regional tectonic trends, and Africa–Europa subduction–related processes. The subaerial zones of the eruptive complex (the islands) have formed in the last 250,000 years, while the submerged parts are older: the oldest, about 1.3 million years, is the submarine volcano Sisifo, to the northwest of the island of Alicudi. Volcanism is still active at Stromboli and Vulcano and dormant at Panarea and Lipari; in the other islands the volcanic activity ceased between 10,000 and 30,000 years ago [17,18].

The geodynamic interpretation of Aeolian volcanism is controversial and is still debated today. According to some authors [19–22], the Aeolian islands represent a volcanic arc in an advanced evolutionary stage due to the geochemical affinity and the presence of very deep seismicity (>500 km) in the southern Tyrrhenian Sea. The arc is linked to the active subduction of the Ionian plate under the Calabrian arc. Conversely, other authors [23–26] state that the subduction stopped about 1 My ago, when both the Calabrian arc and the Apennine chain were affected by extensional tectonics, which lead to the opening of the Tyrrhenian back–arc basin, and by a general uplift. From this perspective, the Aeolian Islands would represent volcanism linked to post–collisional extension processes in a compressive margin. Three different sectors can be identified by different volcanic age and structural domains [27]: (i) the Western sector (1.3 My–0.05 My) located near the Sisifo–Alicudi Fault System (SA in Figure 1; WNW–ESE direction and transpressional regime) [28]; (ii) the Central sector (0.4 My–current) around the Tindari–Letojanni Fault System (TL in Figure 1; NNW–SSE direction and mixed regime, transpressional in its north–western sector and transtensional in the south–eastern one; (iii) the Eastern sector (0.8 My–current) characterised by a NE–SW fault system (extensional regime). The TL crosscut the SA and the extensional NE–SW fault system. The Moho discontinuity is shallower moving eastward, at depths greater than 25 km [17]. The distribution of the earthquake hypocenters and, above all their depth, suggests the presence of a Benioff plane inclined ~50–60° and dipping toward NW, located along the Ionic edge of Calabria. In addition, tomographies performed in the southern Tyrrhenian Sea show the presence of a cold lithosphere (anomalous positive velocities) diving towards NW [29]. Two domains can be distinguished from the seismological point of view: one characterised by shallow and deep seismicity (to the east) and one with exclusively shallow seismicity (to the west) [30].

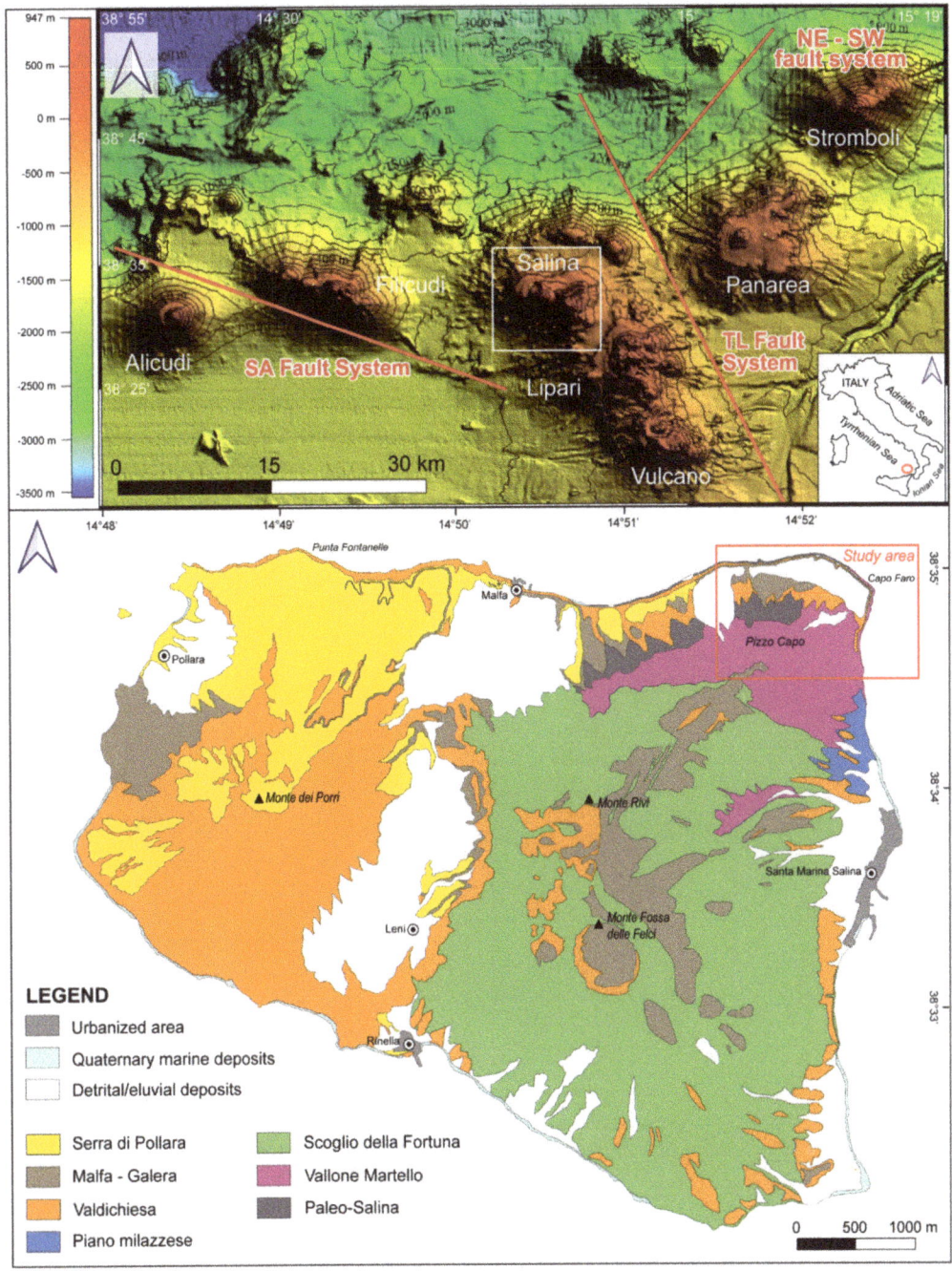

Figure 1. The geographical framework of the Aeolian Islands in the southern Italian peninsula in the upper panel. Salina is in the Central sector of the archipelago and is the subaerial expression of a mostly submerged volcanic system at the intersection between SA and TL fault systems. In the lower panel, simplified geological map with the main synthems of Salina Island (modified from [31]).

2.2. Salina Volcanological Evolution

The island, with a total surface area of 26.4 km^2 and an altitude of 968 m a.s.l. (Monte Fossa delle Felci), is the subaerial expression of a volcanic complex (80–85% of the entire volume), situated in the Central sector (see Section 2.1), at the intersection between the arc–shaped structure of the archipelago and the NNW–SSE elongated Salina–Lipari–Vulcano volcanic belt, in correspondence with TL (Figure 1). The volcanic activity of Salina Island is controlled by both the local and regional tectonic framework [17,27,32–36]. The dominant geomorphological feature is represented by two twin stratocones, Monte dei Porri (859 m a.s.l.), located in the western sector, and Monte Fossa delle Felci, located in the south–eastern one, separated by a low–level area oriented N–S, with a rather complex structure. The two stratocones preserve a regular conical shape, similar in size and topography, giving the island a peculiar morphology.

The subaerial activity has evolved through six eruptive epochs during the Middle–Late Pleistocene [32], individuated by the identification and dating of marine terraces, erosional surfaces, and chronostratigraphic guide levels, by means of which geological maps of the island have been produced [31,37]. Five of the six periods belong to a central stratovolcano: (I) Pizzo Corvo (n.d.), (I–II) Pizzo Capo (ca. 244–226 ky), in Capo Faro Promontory, and (III) Monte Rivi (ca. 160–131 ky) which are barely recognisable from a morphological point of view, while (IV) Monte Fossa delle Felci (ca. 147–121 ky) and (V) Monte dei Porri (ca. 70–57 ky) are both almost perfectly preserved [38,39]. The most recent sixth eruption occurred on the north–western corner of the island between 30 and 15.6 ky ago and formed a semi–circular crater near the small village of Pollara. It represents an explosive large crater of about 1.5 km in diameter [40,41], whose activity produced widespread pumice deposits. Half of the tuff ring lies just above sea level. The only remains of the endogenic activities are post–volcanic phenomena such as gurgling and thermal springs, caused by the emission of underwater hydrogen sulphide and vapours. An uplift of the seafloor may occur at the peak of their activity.

2.3. Capo Faro Promontory Structural Features and Stratigraphy

The study area is in the NE coastal sector of Salina, at the Capo Faro Promontory, whose area is mostly covered by the lithological products of the Pizzo Capo and Monte dei Porri eruptive periods. Among the earlier volcanic events affecting Salina Island, the Pizzo Capo one originated from a NE–SW fissure belonging to the NE–SW extensional system, that affected the entire eastern sector of the Aeolian archipelago (see Section 2.1) [17,27,32]; alternatively, other authors [42] considered the Pizzo Capo activity as an expression of a radial dyke propagating from a central conduit. The Monte dei Porri activity developed in a tectonic context compatible with the TL fault system [33,43]. Moreover, the position of the later Pollara crater (NW of Salina) may suggest an NNW–SSE alignment with the Monte dei Porri cone, thus enforcing the hypothesis of the tectonic influence of the TL in the volcanic evolution (i.e., the westward shifting of the eruptive events; [32]). Furthermore, the fault systems present in this area might have been the cause of several collapse events which affected the Salina Island: (i) the collapse of the Salina Island from Monte Rivi to Pizzo Capo along a NE–SW structural discontinuity causing an asymmetric morphology of the edifice and (ii) the NW–dipping sector collapse of the NW flank of Monte dei Porri [42,43]. Specifically, in Capo Faro Promontory eleven units crop out, as described in [31,32] (Table 1).

Table 1. Stratigraphy of Capo Faro Promontory (modified from [31]).

Eruptive Epoch	Synthem	Volcanic Vent	Formation	Description	Age
VI	–	(Vulcano Island)	Piano Grotte dei Rossi	Ash tuffs in a massive brownish shape up to 2 m of thickness	n.d.
VI	Serra di Pollara	Pollara	Punta di Fontanelle	Pumiceous pyroclastic deposits (30–40 m thick)	27.5 ky
V	Valdichiesa	Monte dei Porri	Serra di Sciarato	Two members: (i) massive CA basaltic andesite lava flow, 5–10 m thick; (ii) scoriaceous deposits, up to several metres thick	n.d.
V	Valdichiesa	Monte dei Porri	Rocce di Barcone	Pyroclastic deposits 50–70 m–thick, made of two facies: (i) massive proximal with lots of lithics; (ii) distal with stratified lapilli/tuffs in thin layers and planar to cross–stratified tuffs	72.7–67.9 ky
V	–	(Vulcano Island)	Pianoconte	Distal fallout deposits made of massive ash tuff, 5–7 m of thickness	n.d.
Q.P. [1]	Fontanelle	–	Punta Brigantino	Poorly sorted, a coarse conglomerate with rounded pebbles and boulders up to 1.5 m in size (3–4 m–thick). The erosional basal contact is referable to the marine transgression of MIS 5c and 5a	100–81 ky
Q.P.	Piano Milazzese	–	Serro dell'Acqua	Volcanic re–worked debris deposits up to 20 m–thick	110–105 ky
III	Scoglio della Fortuna	Monte Rivi	Vallone del Castagno	Pyroclastic succession made up of alternated thin layers of incoherent, massive pyroclastic breccias of reddish scoria, and planar to cross–stratified lapilli–tuff beds up to 50 m–thick. In Capo Faro Promontory crops out as a massive lava flow	ca. 168 ky
II	Vallone Martello	Pizzo Capo	Portella	Up to 120 m–thick successions of scoria with the alternation of metre–thick layers with planar–stratified fallout deposits	ca. 240 ky
II	Vallone Martello	Pizzo Capo	Piano del Serro del Capo	Scoriaceous pyroclastic succession up to 15 m–thick, made by an alternation of massive lithic–rich beds with tuff–breccias and lava flows	n.d.
I	Paleo–Salina	Pizzo Capo	Torricella	Poorly bedded fallout and volcanic debris deposits with discontinuous interbedded massive lava flow	n.d.

[1] Q.P.: quiescence period.

3. Materials and Methods

An area of about 1.3 km^2 in the north–eastern sector of Salina Island has been investigated through a geomorphological survey flanked by UAV flights in a 0.31 km^2 target area in proximity of the coastal cliff. A morpho–stratigraphic characterization and a structural analysis of discontinuities were performed to model the cliff. The collected data were processed to understand the triggering conditions of geomorphological hazards.

3.1. Field Survey and UAV Flights

The field activities took place during the end of summer 2021 along the Capo Faro Promontory based on the identification of the outcropping units according to [31]. The geomorphological field survey has been conducted using the 1:10,000 topographic base of CTR (Carta Tecnica Regionale, "Sezione n. 581020bis Isola di Salina, Regione Siciliana 1994") coupled with a 10 m resolution DEM (Digital Elevation Model, TINITALY; [44] (and references therein), from which the slope acclivity has been extracted in QGIS environment. Gravitational phenomena affecting or that could potentially occur in the area have been described by referring to the [45] classification.

After drawing a geomorphological map, the collection of information about morphology and geology has been necessary for the proximity of the cliff to evaluate the hazard affecting the rocky coastline. Hence, the stratigraphy of the outcropping units in terms of lithologies and their resistance properties has been evaluated. Due to the difficulties to investigate on the field a vertical cliff 40–60 metres high, nadir and off–nadir UAV flights have been conducted with good light exposure to be sure to have the same parts of the cliff in the shadow. The results of such stratigraphical investigations allow us to refine the known–in–literature thickness of deposits to obtain a geological model of the cliff as accurate as possible.

Structure–from–Motion (SfM) and Multi–View–Stereo (MVS) approaches have been used to obtain a 3D reconstruction of the tip of the Capo Faro Promontory. The two methods are based on combining photogrammetric notions and vision algorithms to compute 3D images as better explained in [46]. A dataset of overlapping pictures of the study cliff and its surface is necessary to run them. About seven hundred pictures have been shot with a Drone, a Phantom 4 Pro equipped with a 20 M pixels camera, 1 CMOS sensor with an 84° 8.8 mm/24 mm Field of View (FoV), from an altitude of 60 m and with a Ground Sampling Distance of 1.60 cm/pixel (Figure 2a). The collected data has been post–processed by Agisoft Metashape software. By using the SfM algorithm [47] 552 pictures have been aligned producing (i) a dense point cloud of the investigated cliff (Figure 2b); (ii) the internal calibration parameters [48] and (iii) the spatial distribution of the pictures. This procedure allowed the building of the 50 cm/pixel resolution ortho–mosaic, the 12 cm cell resolution Digital Elevation Model (DEM) and the textured 3D model.

From the high–resolution DEM, six profiles have been extracted through the QGIS plug–in Profile Tool. They have been traced longitudinally to the coastline and are partly topographic, partly bathymetric. For two of them, two geological cross–sections have been drawn using the collected stratigraphic field data.

3.2. Structural Analysis

The structural analysis has been conducted starting from a structural field survey by measuring the discontinuities orientations (i.e., dip direction and dip) on vertical coastal–cut exposures. The dataset consists of 148 faults with various displacements. The structural elements have been plotted on a lower hemisphere equal angle stereograph using the software Dips® 6 (RocScience; University of Toronto, Canada) dividing the planes into different families, each one named with the suffix K–. Moreover, with the above–mentioned software, the Markland Test [49] on two different orientated sub–vertical cliff–exposures has been performed to assess the influence of the measured discontinuities on gravitational phenomena triggering. The Markland Test was aimed at the evaluation of rock–fast events such as planar and wedge sliding, and toppling (both direct and oblique). These gravitational phenomena occur when an equilibrium condition is exceeded. Planar sliding is the failure of a rock mass along a plane, while wedge sliding occurs along the intersection line between two discontinuities. All the discontinuities and intersection lines between planes exceeding the frictional angle (φ') are potential causes of planar and wedge sliding. The toppling to occur needs: (i) discontinuity dip direction parallel to the slope (20° of interval); (ii) discontinuity dipping toward the slope; (iii) the poles of the discontinuities must have a dip minor than the dip slope minus the frictional angle of the toppling planes [50,51]. This could be better explained by the overcoming of the frictional angle (see the Appendix A for calculation details).

 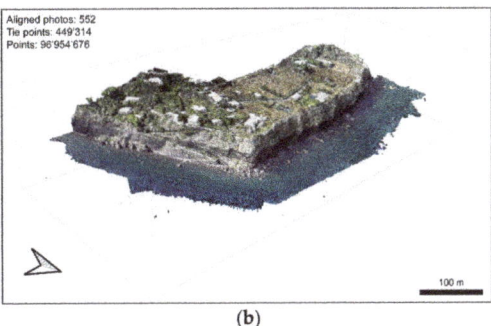

(a) (b)

Figure 2. (a) The Phantom 4 Pro drone during its take–off in the flat courtyard of Capo Faro Estate; (b) Dense point cloud of the flight area in RGB colour, processed with the Agisoft Metashape software.

4. Results

4.1. Geomorphological Features

The Capo Faro Promontory is characterised by a strip of flat surface where buildings, a lighthouse and the Capo Faro Estate with its vineyard are located (Figure 3). This small plateau is interrupted to the north–east and south–east by a 40–60 m–high coastal cliff (Figure 4a), and to the south–west by the north–eastern slope of the Pizzo Capo stratovolcano.

The intensity of morphogenetic processes relates to the recent dynamic context of the active volcanic area. The volcanic products are characterised by high erosion rates. The water and the gravity are the prevalent morphogenetic agents, as testified by a high–energy hydrographic network and by debris deposits widespread over the slope of Pizzo Capo. Both the watercourses and the detrital accumulation zones feature a radial pattern around the Pizzo Capo cone and reveal an accelerated erosion. Then, in the study area, the river pattern has an extremely low drainage density, influenced by the high permeability of the outcropping rocks. The watercourses are mostly characterised by narrow and elongated first–order talwegs which converge to form short second–order creeks. The main basin areas are limited in size (Figure 3). The incised deep gullies become less accentuated only in the proximity of the plateau facing the cliff.

The gravitational phenomena, dominant on slope acclivity >35°, are: (i) rockfalls in the sub–vertical head of the narrow channels and along their sides; (ii) debris flows along the channels with the associated debris cones in sub–aerial environment, often truncated by the active coastal cliff; (iii) rockfalls, topplings (Figure 4b) and rock avalanches (Figure 4c) affecting the frontal portion of the cliff (Figure 4b). The presence of detachment surfaces and cavities on the cliff, in addition to the disseminated boulders of remarkable size standing on the beach (Figure 4d), suggests the system is in retreat. On the other hand, the coastal area of the Capo Faro Promontory is susceptible to storm surges from N–NW which generate a very etched swash line 4–5 m above the foot of the escarpments (Figure 5). The speed of the phenomenon of dismantling is evident in the rapid retreat of the drainage system (Figure 6a). In particular, the morphological jumps of the riverbeds (hanging valleys) near the cliff edge (Figure 6b,c) show how this process is sometimes faster than the incision of the ditches (Figure 6d). They thus remain suspended.

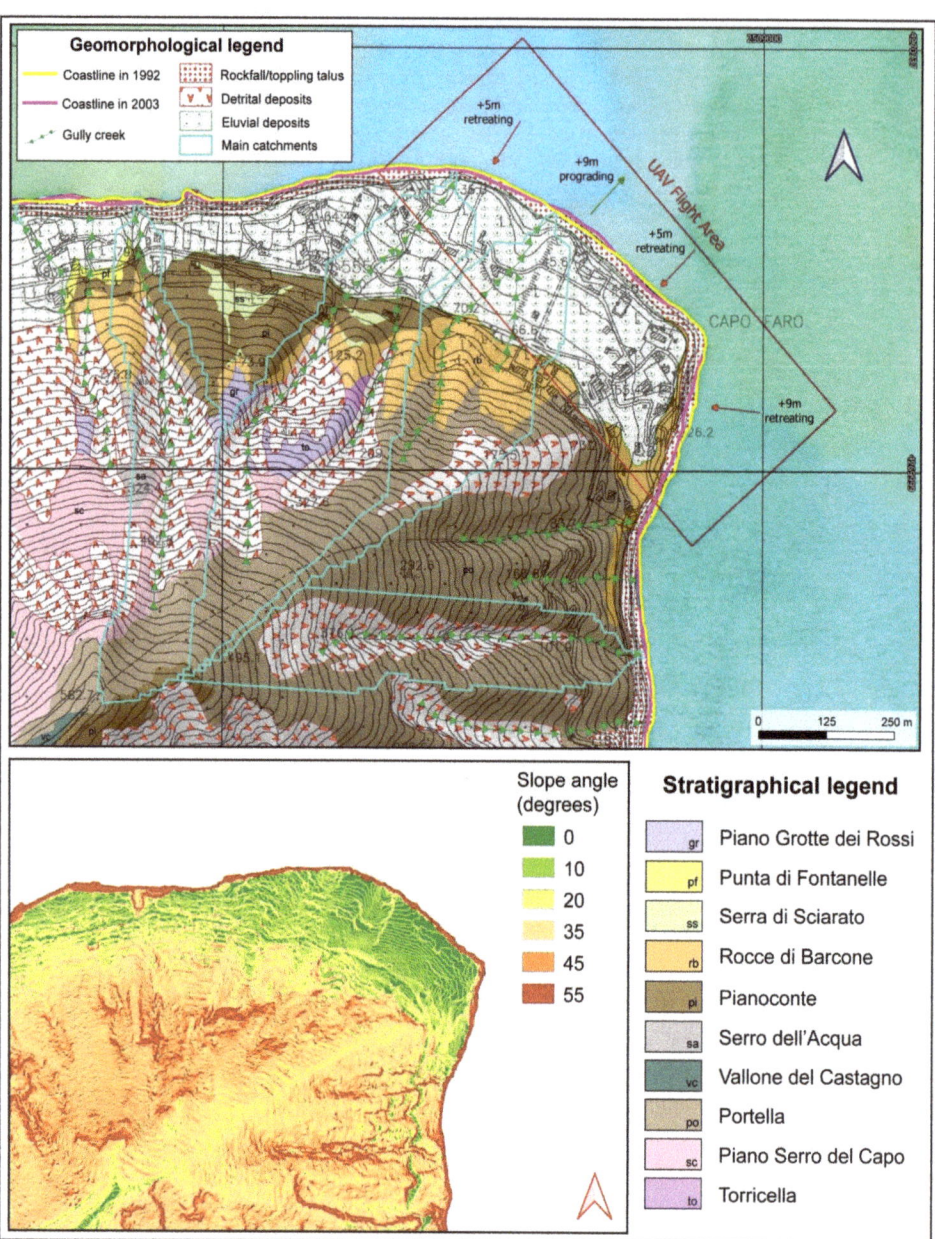

Figure 3. The geomorphological map focused on coastal rockfall/toppling talus, on the detrital cover characterising the slope which shows the hydrography of Capo Faro Promontory. The boundaries of the river basins that encompass the main creeks have been traced. The evolution of the coastline between 1992 and 2003 has been shown with the estimate of its variation (taken from Piano Assetto Idrogeologico (PAI), https://www.sitr.regione.sicilia.it/pai/, accessed on 26 June 2022). The red arrows show the retreating next to Capo Faro cliff, the green ones show the prograding coastline in correspondence with the mouth of a little impluvium. The stratigraphy has been taken from [31]. The lower–left of the figure shows the slope gradient map.

Figure 4. (**a**) The 3D model, generated after dense point cloud extraction by UAV photogrammetric flight through Agisoft Metashape, shows the flat area on which vineyards and buildings are located. It is bounded by an unstable coastal cliff with gravitational phenomena threatening the pocket beaches. (**b**,**c**) shooting areas are indicated by red squares; the yellow circle shows the zoomed 3D of Figure 6a; (**b**) Two large detachment surfaces and cavities of recent rockfalls characterising the cliff in question. The boulders in the foreground have an estimated size (through the dense point cloud) of about 1000 m^3; (**c**) The eastern part of the sea cliff from the boat shows the impressive rock avalanche with the subsequent 30 m retreating at the tip of Capo Faro Promontory occurred in 2011. (**d**) shooting area is indicated by a red square; (**d**) A projecting big boulder (dashed yellow line) on the top of the scarp where a recent rockfall creates a detachment surface (dashed green line) parallel to the scarp. A red line indicates an open fracture in pyroclastic deposits.

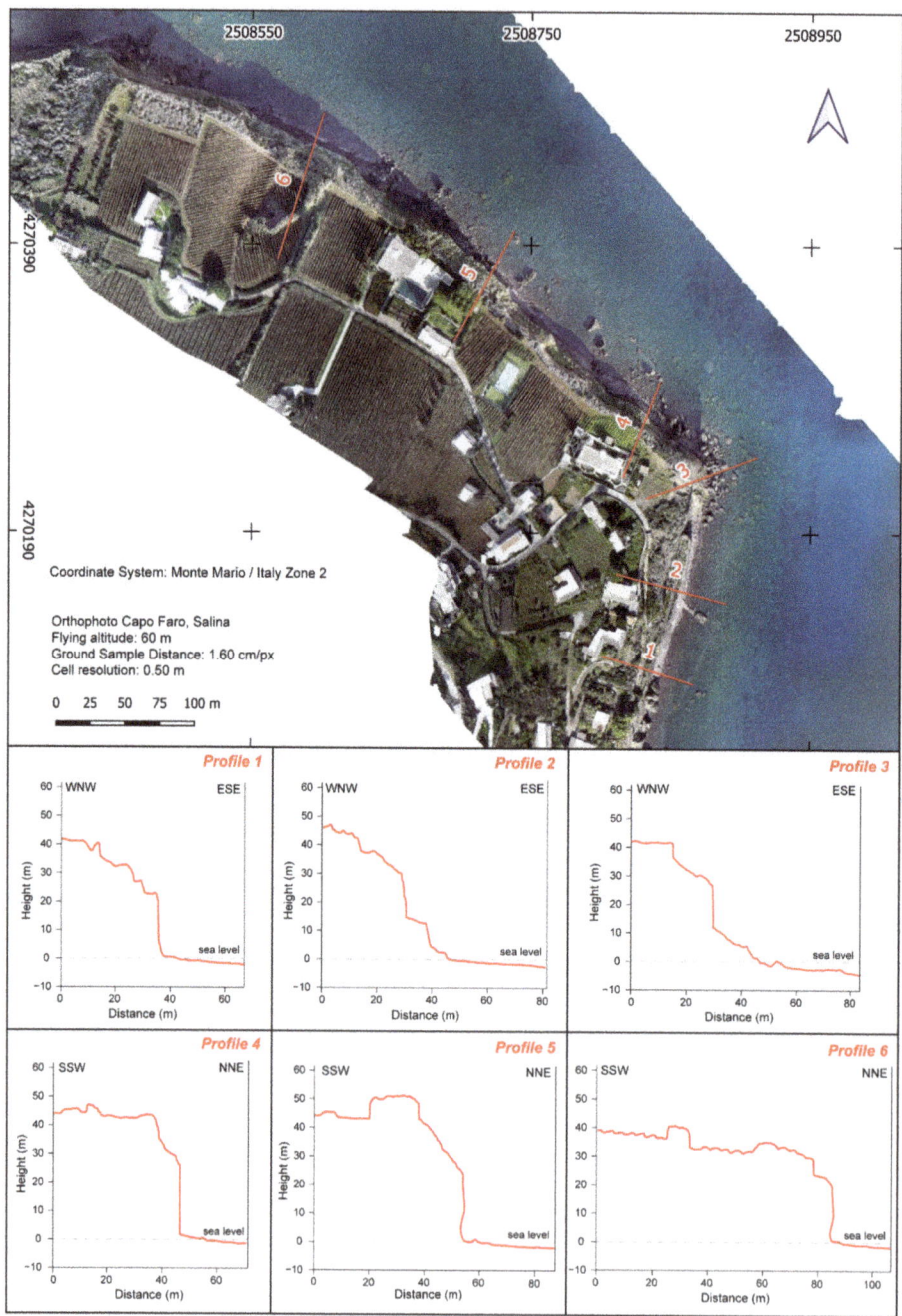

Figure 5. The 2D orthophoto, generated by a 60 m altitude UAV flight, on which six topographic profiles have been traced. As highlighted by profiles 4 and 5, the relief energy is higher in correspondence with the Capo Faro vineyard and the Capo Faro Estate, respectively. Profiles 3, 4 and 5 show a clear vertical cliff. Profiles 5 and 6 evidence the etched swash line at about 5 m a.s.l.

Figure 6. (**a**) 3D textured model focused on the intense gully erosion affecting the flat land right behind the edge of the coastal cliff. (**b**–**d**) shooting areas are indicated by red squares; (**b**) Hanging valley at the outlet of a wide channel. The erosive action of water caused the alteration of the pyroclastic deposits of the cliff; (**c**) The vineyard of Capo Faro Estate gently slopes into the hanging valley of (**b**); (**d**) The deep groove of a gully caused by runoff water.

4.2. Cliff Stratigraphy and Structural Assessment

The multidisciplinary approach allows to precisely model the cliff, investigating its topographic, stratigraphical, and structural features. The local thickness of the outcropping formations has been evaluated for the construction of two geological cross–sections (Figure 7a,b).

Overlapped pyroclastic layers referable to Portella Formation (*po*), emplaced during the II eruptive epoch (c.a. 240 ky, Pizzo Capo volcanic vent; Table 1), crop out on the lower part of the analysed cliff. These deposits, whose thickness in the study area ranges from 35.5 to 40 m show different degrees of cementation and are surmounted by the products of the III and the V eruptive epoch (Monte Rivi and Monte dei Porri volcanic vents; Table 1). Regarding the III epoch, a lenticular massive lava flow body, of basaltic–andesitic to andesitic–dacitic composition, crops out only at the tip of Capo Faro Promontory. It is associated with the Vallone del Castagno formation (*vc*) and has a thickness ranging from 2 to 7 m (Figure 7a). The marine conglomerate of Punta Brigantino Formation (*pb*) crops out overlying the *vc* or the *po* Formations through an erosional depositional surface. Its thickness ranges from 1.5 to 2 m (Figure 7a,b). As for the V epoch, the distal fallout deposits of the Pianoconte Formation (*pi*), made of massive ash tuff, are 2 to 6.5 m–thick. The Rocce di Barcone Formation (*rb*) are pyroclastic deposits whose thickness ranges from 9.5 to 5 m (Figure 7a,b). At the top, the eluvial deposits of the plateau (1.5 m–thick) close the stratigraphic sequence.

Permeability, porosity, and rheological properties of the different deposits influence the development of faults and fractures. Indeed, the permeability is different among eluvial, pyroclastic deposits, and massive lavas. The eluvial and pyroclastic deposits are characterised by high and medium–to–low primary permeability, respectively. On the contrary, due to the presence of fractures, high secondary permeability affected massive lavas. Eluvial deposits, made of clastic gravelly–sandy sediments and placed at the top of the cliff stratigraphy, behave as an incoherent material. For this reason, only the friction angle influences their rheological behaviour.

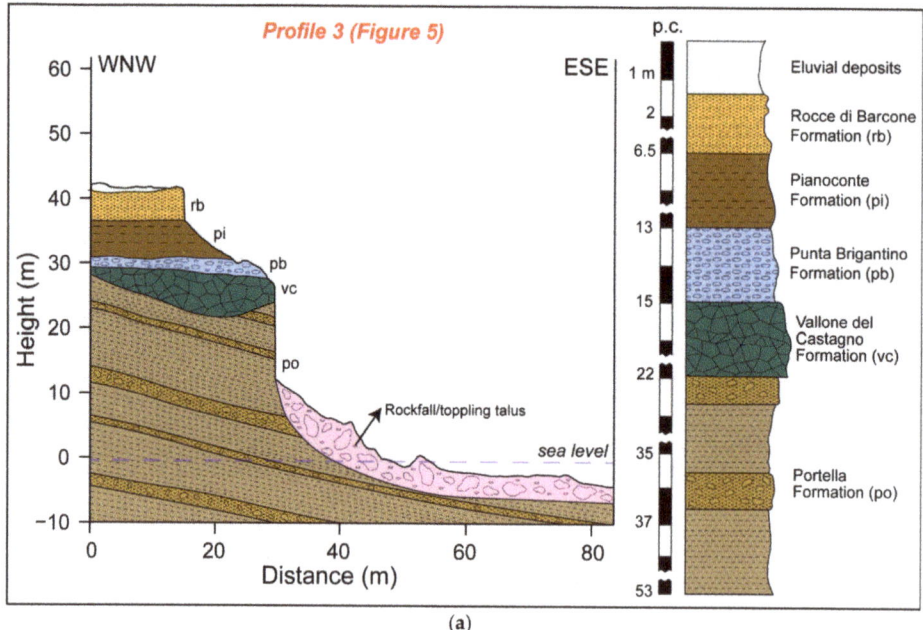

(a)

Figure 7. Cont.

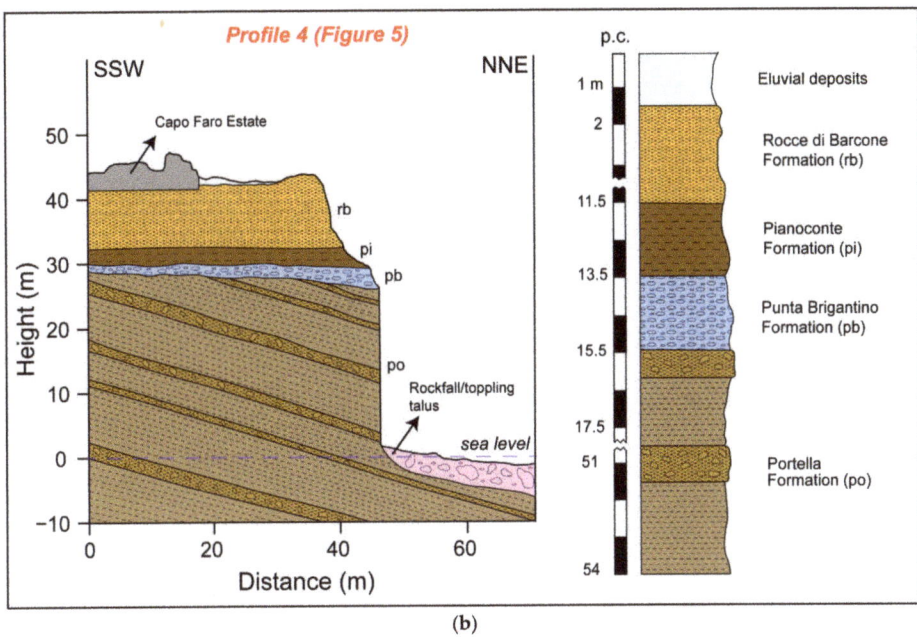

Figure 7. (a) Geological cross–section showing the main stratigraphic units (topographic profile 3 in Figure 5); (b) Geological cross–section showing the main stratigraphic units (topographic profile 4 in Figure 5).

Three main discontinuity systems have been measured on the field and divided into five subsystems, considering the conjugated subfamilies. The systems have been projected on equal area stereographs and all the intersections among the planes have been calculated (Figure 8a,b). The K1 system has been subdivided into K1a and K1b, which are conjugate faults dipping toward NW and SE, with dip angles ranging between 60 and 85°. The K2 discontinuity system has been also divided into K2a and K2b conjugated subfamilies, respectively dipping toward N and S with dip angles ranging between 60 and 80°. The K3 system dips toward E–NE with dip angles ranging between 70–90° (Figure 8a). Based on field observation (Figure 9a), K1 and K2 systems correspond to faults with a dip–slip normal movement, which forms structures with a displacement between 1–10 cm, well recognizable in the pyroclastic deposits of the *po* Formation. Its lithological elements are characterised by a low degree of cementation and strong compositional and chromatic variations between different layers (i.e., the scoria and the planar stratified fallout deposits), which allow a reliable estimation of the cm–displacement produced by normal faulting (Figure 9b,c). The K3 system corresponds to joints, without any displacement or filling. These joints produce fractures with an opening of about 1 cm, which crosscut the faults of the K1 and K2 systems (Figure 9e). The conjugated discontinuities of K1 and K2 systems often intersect among them forming pluri–centimetric to decametric wedges, well visible in the deposits of the *po* Formation (Figure 9b–d). The *vc* lava flow is a more competent and massive formation and it is only affected by sub–vertical open joints (K3 system) fracturing the rock mass in isolated metric blocks.

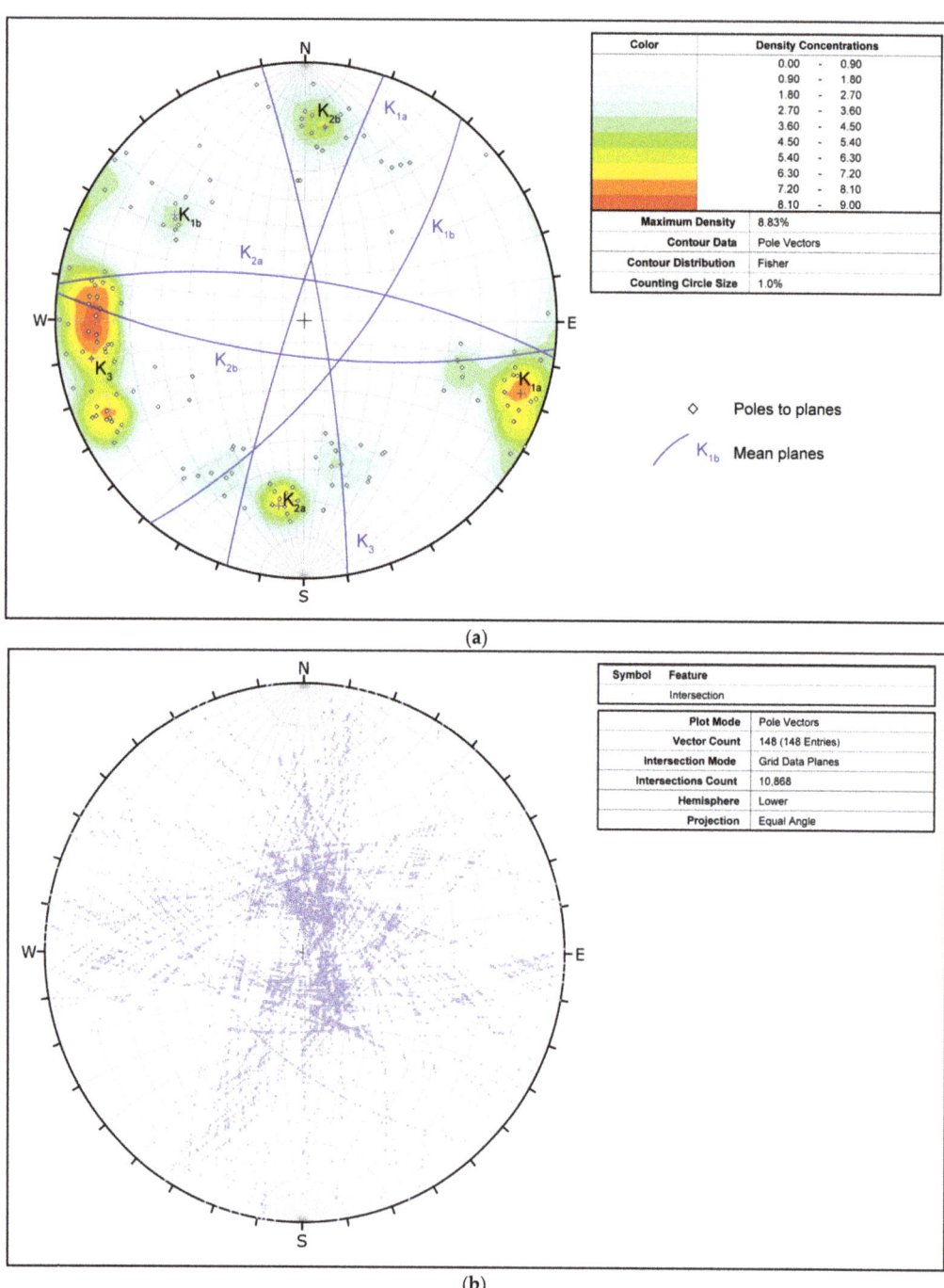

Figure 8. (**a**) Equal angle projection (lower hemisphere) of the measured discontinuities (poles) and contours. The blue lines are the mean planes of each recognized system; (**b**) Equal angle projection of all the intersections among planes (blue dots).

Figure 9. (**a**) 3D textured model focused on the vertical coastal cliff (red squares indicate the described outcrops). (**b–e**) shooting areas are indicated by red squares; (**b**) Pyroclastic deposits of the *po* affected by normal faulting of the K1 conjugated system (yellow arrow indicated the sense of displacement); (**c**) Boat–view of the cliff affected by K2a and K2b conjugated faults, intersecting forming a wedge; (**d**) K2a and K2b conjugated faults with a normal sense of displacement (indicated in yellow); (**e**) K3 discontinuity cutting the alternance of pyroclastic layers with different cementation degree of the *po*.

4.3. Slope Stability Analysis

The slope stability assessment (i.e., Markland test) has been conducted near Capo Faro. The two main exposures (Slope 1 and Slope 2) of the coastal cliff with different orientations intersecting in correspondence with Capo Faro were analysed. Slope 1 orientation has been approximated to 38/85° (dip direction, dip) and Slope 2 is oriented 104/85°. The analysis has been made with the Dips®6 software for planar, wedge sliding and toppling (both direct and oblique). The thick black line is the slope orientation, and the red circle of 34° represents the friction angle (φ') (see Appendix A) measured for pyroclastic deposits and used as an input by the software for the test. The blue lines are the mean planes of the different systems. Planar sliding for both slopes, as provided by the software, gave thirteen critical poles (8.78% of the total) for Slope 1 and 19 (12.84% of the total) for Slope 2 (critical poles are in red in Figure 10a,b). For wedge sliding and toppling, Dips® projected all the intersections between the discontinuities (grey squares), marking with red and yellow areas those critical for triggering gravitational phenomena. Regarding wedge sliding, 5550 over 10,868 intersections (50.61%) have been evaluated as critical (Slope 1), thus falling in the reddish area, corresponding to the intersections between K1a and K3, K1b and K2a, K2a and K3, K1a and K2a systems (Figure 10c). In correspondence to Slope 2, 3988 intersections over 10,868 (36.69%) are critical for wedge sliding and correspond to K1b and K2a, K2b and K3, K1b and K2b, K1b and K3 (Figure 10d). Concerning the toppling mechanism, a number of 1159 intersections in Slope 1 (10.66% of the total) could be responsible for direct toppling (20° of lateral limits with respect to the slope dip direction; red area in Figure 10e), and 1778 (16.36%) for oblique toppling (yellow area; Figure 10e). The intersections critical for toppling in Slope 1 are between K1a and K2b, K1b and K2b, K3 and K1b, K3 and K2b. In Slope 2, 543 intersections (5.00%) are critical for direct toppling (red area with 20° of lateral limits; Figure 10f) and 2666 (24.53%) for oblique toppling (yellow area; Figure 10e). They correspond to the intersection between K1a and K2a, K1a and K2b, K1a and K3, K2a and K3.

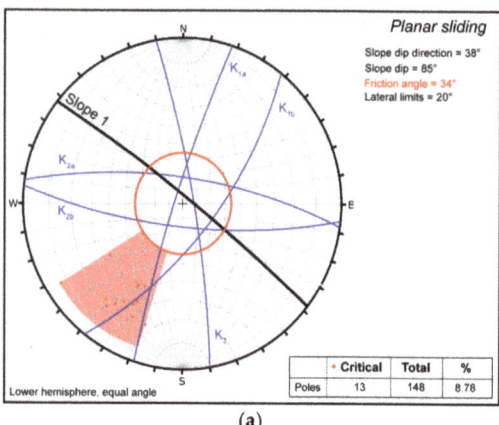

(a)

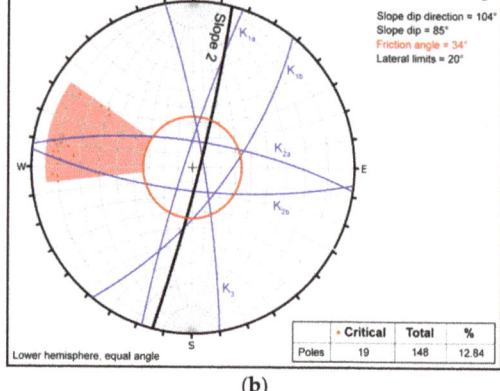

(b)

Figure 10. *Cont.*

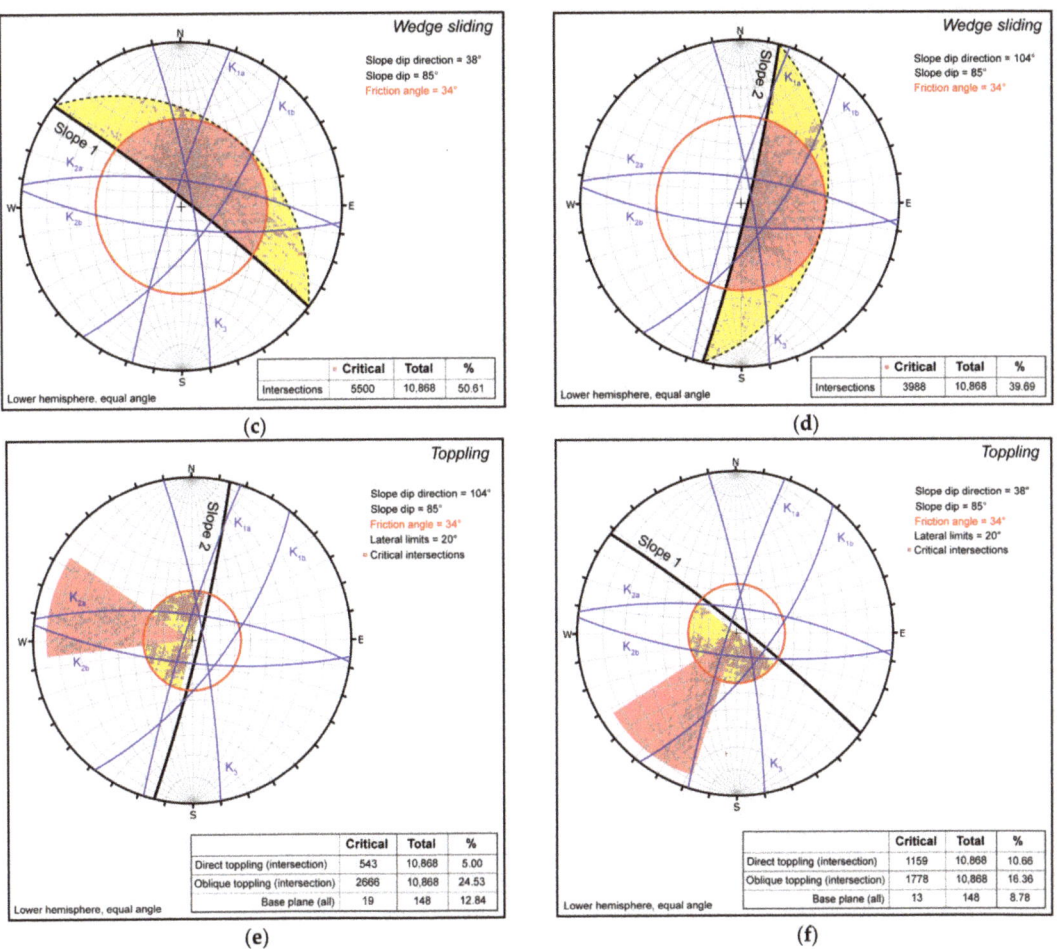

Figure 10. Stereographic equal angle plots (lower hemisphere) with the critical zones (red and yellow areas) for failure mode. (**a**) Planar sliding for Slope 1; (**b**) Planar sliding for Slope 2; (**c**) Wedge sliding for Slope 1; (**d**) Wedge sliding for Slope 2; (**e**) Direct and oblique toppling for Slope 1; (**f**) Direct and oblique toppling for Slope 2.

5. Discussions

The evolution of the coast in the study area represents a constant hazard factor for the buildings and the economy, linked to the high–quality vineyards upstream of the Capo Faro cliff. The retreating of the coastline, around 50–80 cm/y (PAI between 1992 and 2003), is directly caused by the succession of landslide events. It is still in progress nowadays, as testified by the relevant rock avalanche event in 2011 (Figure 4c).

The clearest predisposing factor is the severe undercutting by water mass pressure and debris of the sea waves, especially during the storms in winter, directly responsible for the swash line at the basis of the coastal scarp (Figure 5). Erosion is also favoured by the heterogeneous composition of the outcropping rocks (Figure 7a,b) and by the chemical action of the saltwater infiltrating the fractures. Whenever the erosion deepens the swash line, the overlying wall weakens and collapses. Debris produced by topplings (Figure 11a) form a sort of reef that temporarily protects the cliff till it is dismantled by the constant action of the wave motion. The nearby beaches are fed by the remodelled products deriving from the dismantling of the pyroclastic rocks of the *po* Formation (Figure 11b). This lithotype constitutes about 60–70% of the sheer walls. This phenomenon could be limited by reinforcing structures such as wire meshes or anti–erosion blankets, whose adhesion to the walls is kept by soil nailing and steel rods. Furthermore, coastal works (e.g., brushes, artificial reefs) must be positioned aiming at the dissipation of the energy of the waves.

Another mechanism that facilitates collapses triggering is the establishment of high–tension forces due to the superposition of lithologies with different degrees of cementation, deformability, and rigidity. That is the case of the lava flow (*vc*) superimposed on pyroclastic rocks (*po*) in correspondence with the tip of Capo Faro Promontory (Figure 11c). The consequence is the formation of vertical fractures (often quite deep tension cracks), parallel to the slope (Figure 11d). The run–off water tends to infiltrate within these cracks, triggering hydraulic thrusts and detensioning the rocks of the cliff causing rockfalls. Furthermore, the run–off water flowing down the slopes leads to the formation of small hanging valleys. The result is the ongoing erosion of the facing scarp which contributes to the retreating of the cliff. To face this problem, a system for collecting, conveying, and draining the run–off water should be realised upstream of the edge of the cliff and flanked by sub–horizontal drains to prevent water infiltration in the tension cracks.

The series of discontinuities characterising the cliff, besides being a preferential path for run–off water, could trigger planar and wedge slides and topple events. In particular, the K3 system, as already observed in Slope 2 (and in minor amounts in Slope 1; Figure 9b,e) and resulting from the Markland Test (Figure 10) is almost parallel to the analysed slopes and therefore could potentially trigger planar slidings. The test highlighted that some discontinuities of K2a and K2b systems in Slope 1 and K1a, K2b and K3 in Slope 2 could trigger these types of gravitational events (Figure 9a,b and *vc* Formation block in Figure 11d). Discontinuities in pyroclastic deposits of the *po* Formation both in slopes 1 and 2 intersect each other (K1a with K1b and K2a with K2b), making wedges of different dimensions (from metric– to decametric–scale). The Markland test (Figure 10c,d) confirms that these intersections among discontinuities previously observed on the field (Figure 9b–d) may trigger wedge sliding. Subvertical and horizontal discontinuities with a strike parallel to the cliff intersect originating several plurimetric parallelepipedal blocks which could topple (Figure 11d). This, again, has been confirmed by the Markland test for direct and oblique toppling (K2a and K2b for Slope 1 and K1a, K1b and K3 for Slope 2; Figure 10e,f).

(a)

(b)

Figure 11. *Cont.*

(c)   (d)

Figure 11. (a) The red arrow indicates the detachment cavity on the vertical cliff; (b) The deposits of the rockfalls and topplings coming from the cliff feed the pocket beach in front; (c) The lava flow referable to Vallone del Castagno Formation (*vc*) overlying the pyroclastic deposits of the Portella Formation (*po*). The left red arrow indicates the detachment area of the rock avalanche, the right one points to a fracture in the lava layer, warning of an incipient failure; (d) The cliff is retreating: the dashed green line indicates the crown of the rock avalanche in the vertical cliff; the yellow one the incipient detaching block, probably falling onto rock avalanche deposits. Detrital blocks maximum dimensions are about 70 m^3, assessed with the dense point cloud.

6. Conclusions

The Capo Faro Promontory in north–eastern Salina Island has been assessed from a geomorphological, stratigraphical, and structural point of view. This volcanic area is characterised by a steep slope and a flat plateau on the coastal strip around 40–60 m a.s.l. The coast is rocky and includes a narrow beach backed by a sheer cliff, on which the volcanic deposits succession outcrops. Economic activities of noteworthy importance for tourism (the presence of the Capo Faro Estate, dedicated to summer housing and for Malvasia wine production) are placed on the plateau, not so distant from the edge. Considering the area in question as one of the riskiest in Salina, a detailed field survey has been performed to comprehend the hazard factors. To model the cliff precisely, a drone flight has been carried out. The fracturing conditions of the rock mass have been unravelled by a structural survey. The first results were the redaction of a geomorphological map and the generation of a DEM, with 12 cm spatial resolution, of a 50 cm/pixel resolution orthophoto and a textured 3D model. The area turned out to be characterised by recent gravitational phenomena such as rock avalanches, rockfalls, and topplings. They represent the most dangerous types of failure, both for triggering speed and for unforeseeable nature. The abrupt landslides can stress the pocket beaches used by bathers and they directly cause the dismantling of the cliff, with frequent, great magnitude events provoking up to 30 m retreating. All the field observations on the rock walls have been validated by the Markland test, which confirms the strong control exerted by different intersecting discontinuity patterns affecting the coastal cliff. In particular, a total number of 148 discontinuities has been measured (i.e., their dip direction and dip) on two different exposures of the Capo Faro cliff. The measured discontinuities have been grouped into three main systems and conjugated subsystems (i.e., K1a and K1b, K2a and K2b and K3) to ease the slope stability analysis. The Markland test analysed discontinuity poles orientation for planar sliding, and intersections among planes for wedge sliding and toppling mechanisms. The test highlighted that: (i) the 8.78% (Slope 1) and the 12.84% (Slope 2) of the total discontinuities could trigger planar sliding; (ii) the 50.61% (Slope 1) and the 39.69% (Slope 2) of the total intersections cause wedge sliding; (iii) the 10.66% (Slope 1) and 5.00% (Slope 2) might cause direct toppling; (iv) the 16.36% (Slope 1) and 24.53% (Slope 2) might cause oblique toppling instead.

In essence, the buildings and economical facilities of the area undergo very high–risk conditions with possible involvement of people during summer. Taking into account the interaction between morpho–stratigraphical and structural features of the cliff, the realisation of mitigation measures would be unavoidable to safeguard people's safety, civil buildings, Capo Faro Estate vineyards and residences.

Author Contributions: Conceptualization and investigation, M.B. and C.C.; methodology, validation, data curation and writing—original draft preparation, M.B., C.C. and F.C.; software, M.B. and F.C.; resources, S.L. and G.R.; writing—review and editing, M.B., C.C. and F.C.; supervision, G.F. and G.R.; project administration, M.B. and G.R.; funding acquisition, G.R. All authors have read and agreed to the published version of the manuscript.

Funding: This research received no external funding.

Institutional Review Board Statement: Not applicable.

Data Availability Statement: The data that supports the findings of this study are freely available within the article or from the first author, [M.B.], upon reasonable request. Furthermore, the coastline evolution data and the topographic base are provided by the Geoportale Regione Siciliana Infrastruttura dati territoriali (https://www.sitr.regione.sicilia.it/geoportale (accessed on 26 June 2022)).

Acknowledgments: Many thanks must be expressed to Alessandro Petroccia for his precious suggestions during the survey, data acquisition and manuscript preparation. Antonio Crupi and Mario Vitti are also thanked for UAV flight planning and surveying. Furthermore, we also thank the Capo Faro Estate for encouraging this research and for the helpfulness in accessing the study area. Orazio Barbagallo is warmly acknowledged for dealing with laboratory data available in the Appendix A.

Conflicts of Interest: The authors declare no conflict of interest.

Appendix A

To obtain information on the rheological behaviour of the rock mass of Capo Faro Promontory, geotechnical analysis has been carried out on 10 samples of pyroclastic materials outcropping along the studied cliff.

The 10 samples have been sent to the MTR laboratory of Troina (Enna, Sicily, Italy), to determine the Point Load resistivity, while on four of them the apparent density has been calculated (Table A1).

Table A1. Determination of the apparent density.

	Sample 1	Sample 2	Sample 3	Sample 4	Mean Value
Apparent density (Mg/m^3)	1.31	1.32	1.30	1.29	1.31
Apparent density (kN/m^3)	12.85	12.94	12.75	12.65	1.80

Point Load Measurements

The Point Load Test is widely used to determine the resistivity index of rocks due to the easy and simple use of the instrument and the relatively low cost. The test measures the uniaxial compressive resistivity of rock samples. The test measures the resistivity of rock samples *Is* (50) to break them by adding a load concentrated in a point. *Is* varies depending on the diameter of the sample and/or on the equivalent diameter if the samples have an irregular shape. Hence, a correction is required to obtain a unique value for each rock type. In fact, such value may be used to classify the rocks. The correct value of resistivity *Is* (50) of a sample is defined as the *Is* values respect to a standard sample with a diameter

D = 50 mm. If there are only irregular shape samples, as in our case, the shape correction is given by ISRM (1972) formula:

$$Is(50) = F \times Is \quad (A1)$$

The correction factor may be obtained also through graphical method (Figure A1) or from the following equation:

$$F = De \div 50^{0.45} \quad (A2)$$

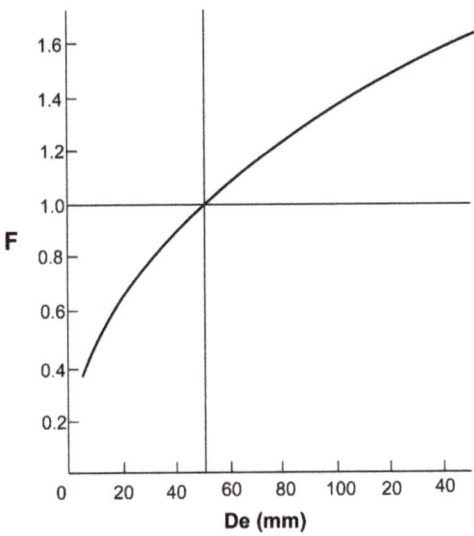

Figure A1. Graphical method to derive k correction factor.

When the Is (50) has been calculated, the Uniaxial Compressive Strength (*UCS*) may be expressed by the relationship:

$$UCS = k \times Is(50) \quad (A3)$$

where k is the transformation factor from the literature [52]. The results of the Point Load Test are reported in Table A2.

Table A2. Point Load Test Results.

	Sample	Is (kPa)	F	Is (50)	C0 (kPa)	C0 (MPa)	C0 (Kg/cm²)
	1	9.35	1.39	12.97	269.78	0.27	2.75
	2	12.06	1.31	15.8	328.64	0.33	3.35
	3	13.33	1.4	18.7	388.96	0.39	3.97
	4	19.8	1.23	23.48	488.38	0.49	4.98
	5	9.06	1.33	12.03	250.22	0.25	2.55
	6	9.44	1.38	130.6	271.65	0.27	2.77
	7	11.66	1.26	14.63	304.3	0.3	3.1
	8	12.38	1.38	17.09	355.47	0.36	3.62
	9	9.39	1.28	12	249.6	0.25	2.55
	10	8.27	1.22	10.09	209.87	0.21	2.14
Simple mean				15	311.7	0.3	3.2
Standard deviation				4	28.2	0.1	1.1
Corrected mean				14.5	302.3	0.3	3.1
Corrected standard deviation				2.5	51.4	0.1	0.5

The *UCS* values have been used to calculate the shear strength. To do that, the non-linear method proposed by [53] has been applied. In this case, the shear strength of a rock mass, on the plane σ_3–σ_1, may be calculated from the following Equation (A4):

$$\sigma'_1 = \sigma'_3 + \sqrt{m\sigma_c\sigma_3 + m\sigma_c^2} \qquad (A4)$$

where:

σ'_1 = principal stress, maximum to break;
σ'_3 = principal stress, minimum to break;
σ_{ci} = uniaxial compressive strength of the rock;
m_b = rock mass parameter, in the case of intact rocks: $m_b = m_i$;
α = rock mass parameter, in the case of intact rocks: $\alpha = 0.5$;
s = rock mass parameter, in the case of intact rocks $s = 1$;

After [54], such a method has been modified by the application of a regression procedure, which consists of overlapping the linear failure criterion of Mohr–Coulomb with the curve generated from the previous formula when $\sigma_3{'}$ values are between σt and σ_{3max}. Finally, c' and φ' have been calculated (Table A3) from:

$$\varphi' = \sin^{-1}\left[\frac{6\alpha m_b(s + m_b\sigma'_{3m})^{\alpha-1}}{2(1+\alpha)(2+\alpha) + 6\alpha m_b(s + m_b\sigma'_{3n})^{\alpha-1}}\right] \qquad (A5)$$

and

$$c' = \frac{\sigma_{ci}[(1+2\alpha)s + (1-\alpha)m_b\sigma'_{3n}](s + m_b\sigma'_{3n})^{\alpha-1}}{(1+\alpha)(2+\alpha)\sqrt{1 + \frac{6m_b(s + m_b\sigma'_{3n})^{\alpha-1}}{(1+\alpha)(2+\alpha)}}} \qquad (A6)$$

where: $\sigma'_{3n} = \sigma'_{3max}/\sigma_{ci}$ ([53] suggested for a general case, a σ_{3max} value equal to 0.25 σ_{ci});

$$m_b = m_i e^{\left(\frac{GSI-100}{28-14D}\right)} \qquad (A7)$$

$$S = e^{\left(\frac{GSI-100}{9-3D}\right)} \qquad (A8)$$

$$\alpha = \frac{1}{2} + \frac{1}{6}\left(e^{-\left(\frac{GSI}{15}\right)} - e^{-\left(\frac{20}{3}\right)}\right) \qquad (A9)$$

D is 0 in an undisturbed rock mass or 1 for a disturbed rock mass.

Table A3. Rock mass resistivity characterization.

	Hoek Brown Classification	
σ_{ci}	5	Mpa
GSI *	61	
m_i	10	
D	0	
E_i	1375	
	Hoek Brown criterion	
m_b	2.484	
s	0.013	
a	0.503	
	Cover break range	
σ_{3max}	1.25	MPa

Table A3. *Cont.*

Mohr–Coulomb overlap		
c'	0.3	MPa
φ'	33.74	degrees
Rock mass parameters		
σ_t	−0.026	MPa
σ_c	0.566	MPa
σ_{cm}	1.118	MPa

* The *GSI* (*Geological Strenght Index*) value, used for the non–linear method [53], derives from the Bieniawsky classification [55–57].

$\varphi' = 33.74$ is the friction angle for pyroclastic deposits obtained from the Mohr–Coulomb overlap. It has been approximated to 34 for the stability analysis calculation (i.e., the Markland test; Section 4.3).

References

1. Randazzo, G.; Cascio, M.; Fontana, M.; Gregorio, F.; Lanza, S.; Muzirafuti, A. Mapping of Sicilian Pocket Beaches Land Use/Land Cover with Sentinel–2 Imagery: A Case Study of Messina Province. *Land* **2021**, *10*, 678. [CrossRef]
2. Short, A.D.; Masselink, G. Embayed and Structurally Controlled Beaches. In *Handbook of Beach and Shoreface Morphodynamics*; Short, A.D., Ed.; Wiley: Hoboken, NJ, USA, 1999; pp. 230–250.
3. Bini, M.; Mascioli, F.; Pranzini, E. Geomorphological Hazard and Tourist Use of Rocky Coasts in Tuscany (NW Italy). In Proceedings of the 12th European Geoparks Conference, Pisa Coastal Plain, NW Tuscany, Italy, 4–6 September 2013; pp. 34–39.
4. Violante, C. Rocky coast: Geological constraints for hazard assessment. *Geol. Soc. Lond. Spec. Publ.* **2009**, *322*, 1–31. [CrossRef]
5. Bieniawski, Z.T. Classification of Rock Masses for Engineering: The RMR System and Future Trends. In *Comprehensive Rock Engineering*; Hudson, J.A., Ed.; Pergamon Press: Oxford, UK, 1993; Volume 3, pp. 553–573.
6. Carter, R.W.G. *Coastal Environments*; Academic Press: London, UK, 1988.
7. Petroccia, A.; Bonasera, M.; Caso, F.; Nerone, S.; Morelli, M.; Bormioli, D.; Moletta, G. Structural and geomorphological framework of the upper Maira Valley (Western Alps, Italy): The case study of the Gollone Landslide. *J. Maps* **2020**, *16*, 534–542. [CrossRef]
8. Redweik, P.; Matildes, R.; Marque, F.; Santos, L. Photogrammetric Methods for Monitoring Cliffs with Low Retreat Rate. *J. Coast. Res.* **2009**, *2*, 1577–1581.
9. Ruberti, D.; Marino, E.; Pignalosa, A.; Romano, P.; Vigliotti, M. Assessment of Tuff Sea Cliff Stability Integrating Geological Surveys and Remote Sensing. Case History from Ventotene Island (Southern Italy). *Remote Sens.* **2020**, *12*, 2006. [CrossRef]
10. Di Crescenzo, G.; Santangelo, N.; Santo, A.; Valente, E. Geomorphological Approach to Cliff Instability in Volcanic Slopes: A Case Study from the Gulf of Naples (Southern Italy). *Geosciences* **2021**, *11*, 289. [CrossRef]
11. Randazzo, G.; Italiano, F.; Micallef, A.; Tomasello, A.; Cassetti, F.P.; Zammit, A.; D'Amico, S.; Saliba, O.; Cascio, M.; Cavallaro, F.; et al. WebGIS Implementation for Dynamic Mapping and Visualization of Coastal Geospatial Data: A Case Study of BESS Project. *Appl. Sci.* **2021**, *11*, 8233. [CrossRef]
12. Barbano, M.; Pappalardo, G.; Pirrotta, C.; Mineo, S. Landslide triggers along volcanic rock slopes in eastern Sicily (Italy). *Nat. Hazards* **2014**, *73*, 1587–1607. [CrossRef]
13. Matano, F.; Caccavale, M.; Esposito, G.; Fortelli, A.; Scepi, G.; Spano, M.; Sacchi, M. Integrated dataset of deformation measurements in fractured volcanic tuff and meteorological data (Coroglio coastal cliff, Naples, Italy). *Earth Syst. Sci. Data* **2020**, *12*, 321–344. [CrossRef]
14. Lanza, S.; Randazzo, G. Tourist–beach protection in north–eastern Sicily (Italy). *J. Coast. Conserv.* **2013**, *17*, 49–57. [CrossRef]
15. Lanza, S.; Randazzo, G. Improvements to a Coastal Management Plan in Sicily (Italy): New approaches to borrow sediment management. *J. Coast. Res.* **2011**, *64*, 1357–1361.
16. Randazzo, G.; Lanza, S. Regional Plan against Coastal Erosion: A Conceptual Model for Sicily. *Land* **2020**, *9*, 307. [CrossRef]
17. De Astis, G.; Ventura, G.; Vilardo, G. Geodynamic significance of the Aeolian volcanism (Southern Tyrrhenian Sea, Italy) in light of structural, seismological, and geochemical data. *Tectonics* **2003**, *22*, 1040. [CrossRef]
18. Santo, A.P.; Clark, A.H. Volcanological Evolution of Aeolian Arc (Italy): Inferences from 40Ar/39Ar ages of Filicudi Rocks. In Proceedings of the IAVCEI Congress, Ankara, Turkey, 12–16 September 1994.
19. Barberi, F.; Innocenti, F.; Ferrara, G.; Keller, J.; Villari, L. Evolution of Eolian Arc volcanism (Southern Tyrrhenian Sea). *Earth Planet. Sci. Lett.* **1974**, *21*, 269–276. [CrossRef]
20. Beccaluva, L.; Rossi, P.L.; Serri, G. Neogene to Recent volcanism of the Southern Tyrrhenian–Sicilian area: Implications for the geodynamic evolution of the Calabrian Arc. *Earth Evol. Sci.* **1982**, *3*, 222–238.
21. Beccaluva, L.; Gabbianelli, G.; Lucchini, F.; Rossi, P.L.; Savelli, C. Petrology and K/Ar ages of volcanic dredged from the Eolian seamounts: Implications for geodynamic evolution of the Southern Tyhrrenian basin. *Earth Planet. Sci. Lett.* **1985**, *74*, 187–208. [CrossRef]

22. Ferrari, L.; Manetti, P. Geodynamic framework of the Tyrrhenian volcanism: A review. *Acta Vulcanol.* **1993**, *3*, 1–10.
23. Westaway, R. Quaternary uplift of southern Italy. *J. Geophys. Res.* **1993**, *98*, 21741–21772. [CrossRef]
24. Hippolyte, J.; Angelier, J.; Roure, F. A major change revealed by Quaternary stress patterns in the Southern Apennines. *Tectonophysics* **1994**, *230*, 199–210. [CrossRef]
25. Milano, G.; Vilardo, G.; Luongo, G. Continental collision and basin opening in southern Italy: A new plate subduction in the Tyrrhenian Sea? *Tectonophysics* **1994**, *230*, 249–264. [CrossRef]
26. Carminati, E.; Wortel, M.J.R.; Spakman, W.; Sabadini, R. The role of slab detachment processes in the opening of the western-central Mediterranean basins: Some geological and geophysical evidence. *Earth Planet. Sci. Lett.* **1998**, *160*, 651–665. [CrossRef]
27. Ventura, G. Kinematics of the Aeolian volcanism (Southern Tyrrhenian Sea) from geophysical and geological data. In *The Aeolian Islands Volcanoes*; Lucchi, F., Peccerillo, A., Keller, J., Tranne, C.A., Rossi, P.L., Eds.; Memoirs Geological Society: London, UK, 2013; Volume 37, pp. 3–11. [CrossRef]
28. Bortoluzzi, G.; Ligi, M.; Romagnoli, C.; Cocchi, L.; Casalbore, D.; Sgroi, T.; Cuffaro, M.; Tontini, F.C.; D'Oriano, F.; Ferrante, V.; et al. Interactions between Volcanism and Tectonics in the Western Aeolian Sector, Southern Tyrrhenian Sea. *Geophys. J. Int.* **2010**, *183*, 64–78. [CrossRef]
29. Neri, G.; Barberi, G.; Orecchio, B.; Mostaccio, A. Seismic strain and seismogenic stress regimes in the crust of the southern Tyrrhenian region. *Earth Planet. Sci. Lett.* **2003**, *213*, 97–112. [CrossRef]
30. D'Agostino, N.; Selvaggi, G. Crustal motion along the Eurasia–Nubia plate boundary in the Calabrian Arc and Sicily and active extension in the Messina Straits from GPS measurements. *J. Geophys. Res.* **2004**, *109*, B11402. [CrossRef]
31. Lucchi, F.; Tranne, C.A.; Keller, J.; Gertisser, R.; Forni, F.; De Astis, G. Geological Map of the Island of Salina, Scale 1:10,000 (Aeolian Archipelago). In *The Aeolian Islands Volcanoes*; Lucchi, F., Peccerillo, A., Keller, J., Tranne, C.A., Rossi, P.L., Eds.; Memoirs Geological Society: London, UK, 2013; Volume 37, pp. 155–211.
32. Lucchi, F.; Gertisser, R.; Keller, J.; Forni, F.; De Astis, G.; Tranne, C.A. Eruptive History and Magmatic Evolution of the Island of Salina (Central Aeolian Archipelago). In *The Aeolian Islands Volcanoes*; Lucchi, F., Peccerillo, A., Keller, J., Tranne, C.A., Rossi, P.L., Eds.; Memoirs Geological Society: London, UK, 2013; Volume 37, pp. 155–211.
33. Mazzuoli, R.; Tortorici, L.; Ventura, G. Oblique rifting in Salina, Lipari and Vulcano Islands (Aeolian Islands, Southern Tyrrhenian Sea, Italy). *Terra Nova* **1995**, *7*, 444–452. [CrossRef]
34. Continisio, R.; Ferrucci, F.; Gaudiosi, G.; Lo Bascio, D.; Ventura, G. Malta escarpment and Mt. Etna: Early stages of an asymmetric rifting process? Evidences from geophysical and geological data. *Acta Vulcanol.* **1997**, *9*, 39–47.
35. Lanzafame, G.; Bousquet, J.C. The Maltese escarpment and its extension from Monte Etna to Aeolian Islands (Sicily): Importance and evolution of a lithospheric discontinuity. *Acta Vulcanol.* **1997**, *9*, 121–135.
36. Romagnoli, C.; Casalbore, D.; Bortoluzzi, G.; Bosman, A.; Chiocci, F.L.; D'Oriano, F.; Gamberi, F.; Ligi, M.; Marani, M. Bathymortphological setting of the Aeolian islands. In *The Aeolian Islands Volcanoes*; Lucchi, F., Peccerillo, A., Keller, J., Tranne, C.A., Rossi, P.L., Eds.; Memoirs Geological Society: London, UK, 2013; Volume 37, pp. 27–36. [CrossRef]
37. Keller, J. Die Geologie der Insel Salina. Ph.D. Thesis, Albert–Ludwigs–Universitat Freiburg, Breisgau, Germany, 1966; *Unpublished*.
38. Keller, J.; Ryan, W.B.F.; Ninkovich, D.; Altherr, R. Explosive volcanic activity in the Mediterranean over the past 200.000 yr as recorded in deep–sea sediment. *Geol. Soc. Am. Bull.* **1978**, *89*, 591–604. [CrossRef]
39. Morche, W. Tephrochronologie der Äolischen Inseln. Ph.D. Thesis, Albert–Ludwigs–Universitat Freiburg, Breisgau, Germany, 1988; *Unpublished*.
40. Lucchi, F.; Tranne, C.A.; De Astis, G.; Keller, J.; Losito, R.; Morche, W. Stratigraphy and significance of Brown Tuffs on the Aeolian Islands (southern Italy). *J. Volcanol. Geoth. Res.* **2008**, *177*, 49–70. [CrossRef]
41. Keller, J. The Island of Salina. *Rend. Della Soc. Ital. Di Mineral. E Petrol.* **1980**, *36*, 489–524.
42. Quareni, F.; Ventura, G.; Mulargia, F. Numerical modelling of the transition from fissure– to central–type activity on volcanoes: A case study from Salina Island, Italy. *Phys. Earth Planet. Inter.* **2001**, *124*, 213–221. [CrossRef]
43. Barca, D.; Ventura, G. Evoluzione vulcano–tettonica dell'isola di Salina (arcipelago delle Eolie). *Mem. Della Soc. Geol. Ital.* **1991**, *47*, 401–415.
44. Tarquini, S.; Nannipieri, L. The 10–m resolution TINITALY DEM as a trans–disciplinary basis for the analysis of the Italian territory: Current trends and new perspectives. *Geomorphology* **2017**, *281*, 108–115. [CrossRef]
45. Cruden, D.M.; Varnes, D.J. Landslide types and processes. In *Landslides Investigation and Mitigation. Transportation Research Board*; Turner, A.K., Schuster, R.L., Eds.; Special Report 247; US National Research Council: Washington, DC, USA, 1996; Chapter 3, pp. 36–75.
46. Carrivick, J.L.; Smith, M.W.; Quincey, D.J. Structure from Motion in Practice. In *Structure from Motion in the Geosciences*; John Wiley & Sons, Ltd.: New York, NY, USA, 2016; pp. 60–96. [CrossRef]
47. Ullman, S. The interpretation of Structure from Motion. *Proc. R. Soc. Lond. Ser. B Biol. Sci.* **1979**, *203*, 405–426. Available online: https://www.jstor.org/stable/77305 (accessed on 26 June 2022).
48. Casella, E.; Rovere, A.; Pedroncini, C.P.; Stark, M.; Casella, M.; Ferrari, M.; Firpo, M. Drones as tools for monitoring beach topography changes in the Ligurian Sea (NW Mediterrenean). *Geo–Mar. Lett.* **2016**, *36*, 151–163. [CrossRef]
49. Markland, J.T. A useful technique for estimating the stability of rock slopes when the rigid wedge slide type of failure is expected. *Imp. Coll. Rock Mech. Res. Repr.* **1972**, *19*, 1–10.

50. Goodman, R.E.; Bray, J.W. Toppling of Rock Slopes. In Proceedings of the Specialty Conference Rock Engineering for Foundations and Slopes, American Society of Civil Engineers, Boulder, CO, USA, 15–18 August 1976; pp. 201–234.
51. Goodman, R.E. *Introduction to Rock Mechanics*; Wiley: Hoboken, NJ, USA, 1980; p. 478.
52. Rusnak, J.; Mark, C. Using the Point Load Test to Determine the Uniaxial Compressive Strength of Coral Measure Rock. In Proceedings of the 19th International Conference on Ground Control in Mining, Morgantown, WV, USA, 8–10 August 2000; pp. 362–371.
53. Hoek, E.; Brown, E.T. Empirical strength criterion for rock masses. *J. Geotech. Eng. Div.* **1980**, *106*, 1013–1035. [CrossRef]
54. Hoek, E.; Carranza, C.; Itasca, T. Hoek–Brown Failure Criterion–2002 Edition. In Proceedings of the North American Rock Mechanics Society Meeting, Toronto, ON, Canada, 7–10 July 2002.
55. Bieniawski, Z.T. Rock mass Classification in Rock Engineering. In *Exploration for Rock Engineering*; Balkema: Cape Town, South Africa, 1976; Volume 1, pp. 97–106.
56. Bieniawski, Z.T. Determining rock mass deformability. *Int. J. Rock Mech. Min. Sci.* **1978**, *15*, 335–343. [CrossRef]
57. Bieniawski, Z.T. *Engineering Rock Mass Classifications: A Complete Manual for Engineers and Geologists in Mining, Civil, and Petroleum Engineering*; Wiley–Interscience: New York, NY, USA, 1989; pp. 40–47, ISBN 0-471-60172-1.

Article

Shoreline Evolution and Environmental Changes at the NW Area of the Gulf of Gela (Sicily, Italy)

Laura Borzì [1,*], Giorgio Anfuso [2], Giorgio Manno [3], Salvatore Distefano [1], Salvatore Urso [1], Domenico Chiarella [4] and Agata Di Stefano [1]

1. Department of Biological, Geological and Environmental Sciences, University of Catania, Corso Italia, 57, 95129 Catania, Italy; salvodist82@unict.it (S.D.); salvatore.urso@unict.it (S.U.); agata.distefano@unict.it (A.D.S.)
2. Department of Earth Sciences, Faculty of Marine and Environmental Sciences, University of Cádiz, Polígono del Río San Pedro s/n, 11510 Puerto Real, Spain; giorgio.anfuso@uca.es
3. Department of Engineering, University of Palermo, Viale delle Scienze, Bd. 8, 90128 Palermo, Italy; giorgio.manno@unipa.it
4. Clastic Sedimentology Investigation (CSI), Department of Earth Sciences, Royal Holloway, University of London, Egham TW20 0EX, UK; Domenico.Chiarella@rhul.ac.uk
* Correspondence: laura.borzi@unict.it

Abstract: Coastal areas are among the most biologically productive, dynamic and valued ecosystems on Earth. They are subject to changes that greatly vary in scale, time and duration and to additional pressures resulting from anthropogenic activities. The aim of this work was to investigate the shoreline evolution and the main environmental changes of the coastal stretch between the towns of Licata and Gela (in the Gulf of Gela, Sicily, Italy). The methodology used in this work included the analysis of: (i) shoreline changes over the long- and medium-term periods (1955–2019 and 1989–2019, respectively), (ii) dune system fragmentation and (iii) the impact of coastal structures (harbours and breakwaters) on coastal evolution. The shoreline change analysis mainly showed a negative trend both over the long- and medium-term periods, with a maximum retreat of 3.87 m/year detected over the medium-term period down-drift of the Licata harbour. However, a few kilometres eastward from the harbour, significant accretion was registered where a set of breakwaters was emplaced. The Shoreline Change Envelope (SCE) showed that the main depositional phenomena occurred during the decade between 1955 and 1966, whereas progressive and constant erosion was observed between 1966 and 1989 in response to the increasing coastal armouring.

Keywords: shoreline changes; DSAS; dune fragmentation; coastal armouring

Citation: Borzì, L.; Anfuso, G.; Manno, G.; Distefano, S.; Urso, S.; Chiarella, D.; Di Stefano, A. Shoreline Evolution and Environmental Changes at the NW Area of the Gulf of Gela (Sicily, Italy). Land 2021, 10, 1034. https://doi.org/10.3390/land10101034

Academic Editor: Le Yu

Received: 10 September 2021
Accepted: 28 September 2021
Published: 2 October 2021

Publisher's Note: MDPI stays neutral with regard to jurisdictional claims in published maps and institutional affiliations.

Copyright: © 2021 by the authors. Licensee MDPI, Basel, Switzerland. This article is an open access article distributed under the terms and conditions of the Creative Commons Attribution (CC BY) license (https://creativecommons.org/licenses/by/4.0/).

1. Introduction

Natural coastal landscape modelling is an interactive complex phenomenon ruled by several dynamic processes, all linked in a non-linear way. Shorelines are dynamic in nature, and coastal behaviour is the result of natural and anthropic processes occurring and interacting on a variety of time and spatial scales [1–3] The coastal sediment budget can be altered both by such physical processes, including waves, currents, tides, storm surges, seasonal fluctuations, aeolian transport and relative changes in sea level and by human actuations, such as the construction of inland infrastructure (e.g., dams) and coastal structures (e.g., protection structures and ports/harbours) [4]. However, decreased sediment river load and altered longshore drift appear to be pivotal factors in sediment coastal changes, and therefore manmade actuations can be the main causes of coastal erosion [5,6].

Most of the world's major cities are in coastal regions, and 40% of all people on the planet live within 100 km of a coastal zone [5]. Coastal settlements are often planned with insufficient attention to natural hazards such as coastal erosion and flooding, and often coastal structures built too close to eroding shorelines experience wave inundation and damage [7,8]. In response, sea walls and revetments are often built to protect coastal homes, hotels and infrastructure, but their emplacement can lead to the total loss of the existing beaches [8].

As a result, numerous shoreline studies aim to quantify trends in shoreline change [8–13]. Shoreline studies are thus vital to the early stages of the decision-making process for planned coastal developments to mitigate the potential loss of buildings and infrastructure as well as to preserve natural environments such as beaches and dunes [14,15].

However, shoreline evolution studies can be challenging, and the computation of the rates of change should consider the processes that have affected the coast over time [16]. Nowadays different and varied cartographic datasets (i.e., aerial photographs, orthophotos, satellite imagery) provide useful tools to assess shoreline changes over the past decades, but the chosen spatial and temporal scales must be planned in advance to enhance the reliability of the data analysis. Indeed, long-term data (>60 years) produce more predictable trends, filtering out variations ranging from days to seasons typical of short-term fluctuations (<10 years) [16,17]. Consistency of the dataset and reliability of shoreline change analysis should be the basis of forecasting future shoreline position and wisely planning construction setbacks, especially in those areas of increasing economic growth [18].

About 20,000 km of coast, corresponding to 20% of the whole European coast length, faced serious impacts in 2004, and 15,100 km were actively retreating at the time, some of them despite the emplacement of coastal protection works [19]. Sicily (Southern Italy) is one of the regions with the highest coastal urban development in Europe, and about 900,000 inhabitants live within that area; more than 27% of the Sicilian coast has been still experiencing erosional phenomena, and more than 110 km of the coast have been armoured [19].

This paper represents the first analysis of the coastal evolution of one of the most human-impacted areas of Southern Sicily, i.e., the Gulf of Gela. This coast is highly valued because of its recreational, ecological, economic and cultural aspects, which strongly depend on its good status [20]. Therefore, investigation of local and regional coastal changes is of high priority for local managers to adopt sound management strategies to counteract erosion problems essentially related to inland human actuations and the emplacement of hard protection structures on the coast. Thus, shoreline migration in response to the coastal and inland changes was studied. Rate-of-change statistics were computed on long- and medium-term periods by means of the ESRI ArcGIS© Digital Shoreline Analysis System (DSAS), and coastal environmental variations were quantified using the coastal armouring index [21] and the new dune fragmentation index proposed for the first time by Molina et al. [5].

2. Study Area

The area investigated in this paper, located on the Southern coast of Sicily (Italy), plays a key role in the economic strategies of the regional management. The area includes long and wide sandy beaches that enhance the development of leisure activities and infrastructures; greenhouse-crop systems are quite widespread thanks to mild weather conditions; on the coast of the Gela town stands one of the largest petrol-chemical poles in Europe that has been recently converted into a biorefinery, and some of the coastal regional strategic infrastructures insist on this wide coastal sector. Such strategic economic activities and human settlements are significantly affected by coastal erosional phenomena [20,22–24]. Therefore it is mandatory to understand past and actual shoreline trends to adopt sound management strategies to preserve coastal ecosystems and economic activities.

At a local scale, the Southern coast of Sicily is identified as a coastal sub-cell of I order by the Regional Plan against Coastal Erosion [20]. This cell has been split into II- and III-order sub-cells, and, in this paper, the shoreline evolution and the environmental changes of the third-order coastal sub-cell no. 4.2.2 were studied. The coastal area under study has a total length of 24.96 km, and it was subdivided into two sectors: sector no.

1, from the Licata harbour to the Falconara Castle, and sector no. 2, from the Falconara Castle to the Gela harbour (Figure 1). Sector no. 1 is ca. 10 km long, and the main coastal human work insisting on it is the Licata harbour. The harbour, essentially devoted to fishing activities, is considered a strategic infrastructure for the local economy. It has been implemented several times since the 1940s, and last modifications were carried out in 1997. Sector no. 2 stretches between the Falconara Castle and the Gela harbour and is ca. 15 km long.

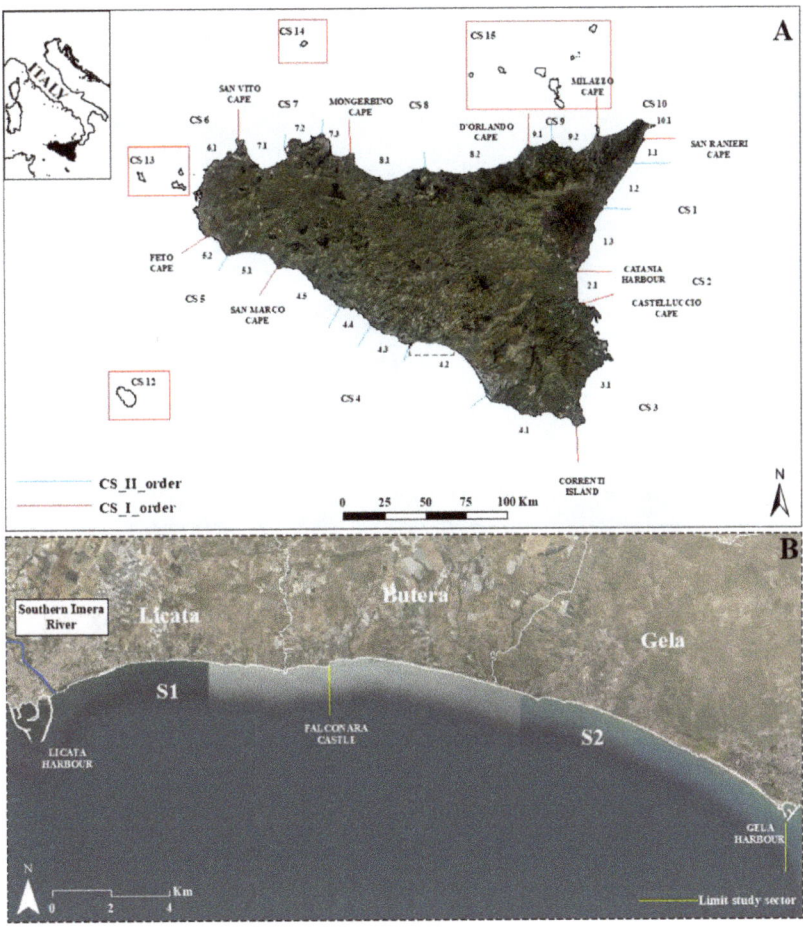

Figure 1. (**A**) The coastal sub-cell subdivision of Sicily in I- and II-order cells [20]; (**B**) The study area and its subdivision into two sectors: sector no. 1, from the Licata harbour to the Falconara Castle, and sector no. 2, from the Falconara Castle to the Gela harbour.

The Southern Imera River, set just a few metres east of the Licata harbour, is the main watercourse outflowing within the study area, and its drainage basin is the second largest one in Sicily (Figure 1) [25].

The river mouth profoundly changed during the past decades since it experienced significant lateral migration, and, consequently, the small harbour entrance has been repeatedly changed following the migration of the Southern Imera River mouth [23]. The lateral migration of the river mouth can be mainly related to the climatic modifications recorded between the 17th and 19th centuries and to the deforestation of the hinterland deriving from the expansion of croplands [22]. Over the 1940s, the harbour was significantly expanded, and the first dikes (western and eastern) were emplaced.

The wind regime of the coastal area is dominated by winds blowing from the third quadrant and partly from the fourth quadrant, both in terms of frequency and maximum velocity [25,26]. The wave motion computed is mainly made of waves coming from the west-south-west, and the most energetic waves come from the west [27,28]. The area is characterized by a microtidal regime, as well-documented by the variations in levels at the tide gauge station located at Porto Empedocle (Figure 2).

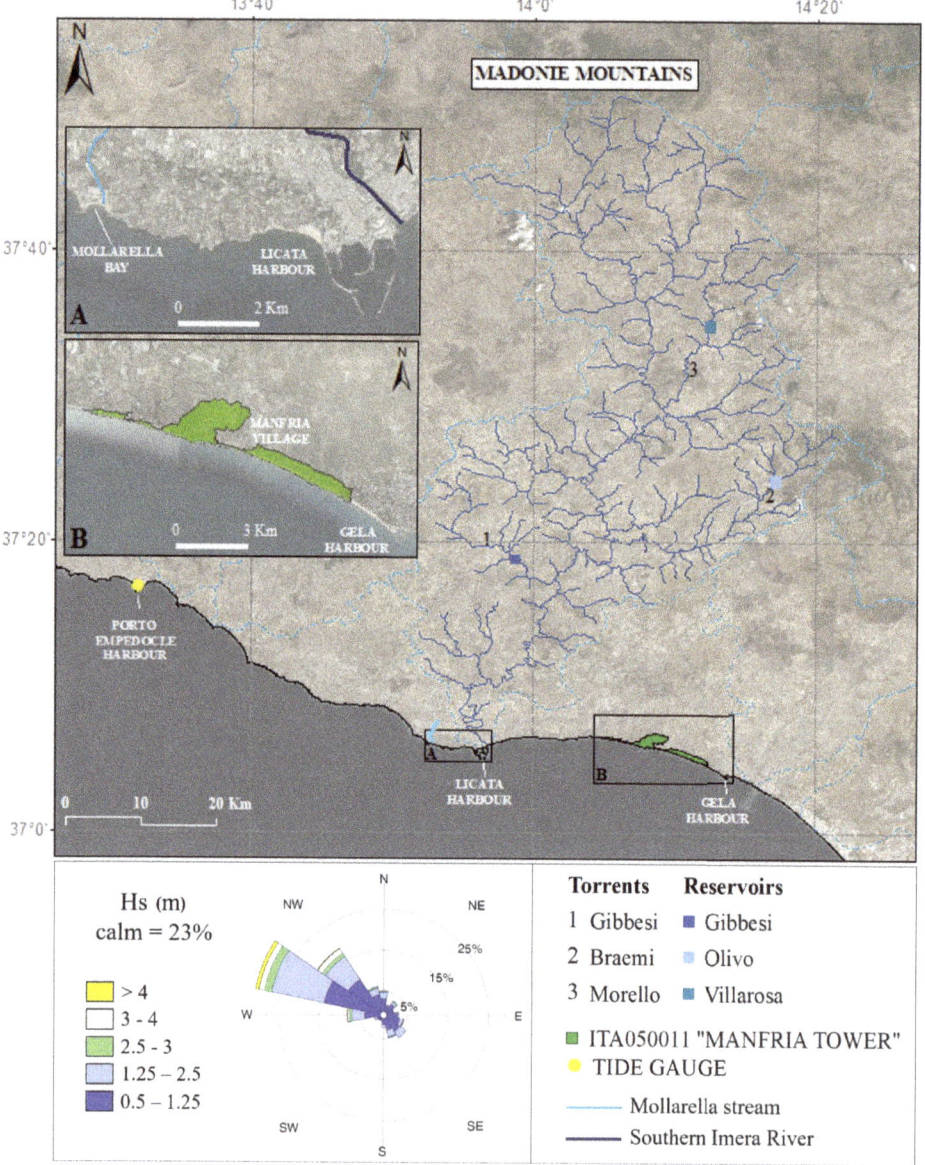

Figure 2. Southern Imera drainage basin, with tributaries and artificial reservoirs, location of the tide gauge station at Porto Empedocle (yellow dot) and the Site of Community Importance (SCI) ITA 050011 "Manfria Tower" (green polygon).

(**A**) Location of the present-day Southern Imera River mouth and the disused second river mouth, in correspondence of Mollarella Bay, ca. 6 km from the Licata harbour. (**B**) Details of the Site of Community Importance (SCI) ITA 050011 "Manfria Tower".

The coast is essentially composed of narrow sandy beaches; a few headlands of small size are located east of the Falconara Castle promontory. Beach grain-size sediment ranges between fine and medium classes, and beaches generally show intermediate to dissipative morphodynamic states.

The Southern Imera crosses in the N–S direction the island of Sicily and until the early 1900s split into two streams just 5 km from the coastline [23]: the first outflowed in correspondence of the present mouth and the second towards the Mollarella Bay, 6 km westward of the Licata town, which was abandoned over the 1950s (Figure 2) [23]. Fluvial sediment inputs to the sub-cell are minimal, even though a high number of rivers and torrents flow out within this area; most of those rivers pass through predominantly chalk catchments, resulting in high solute but low sediment loads [23]. Three main artificial reservoirs intercept the Southern Imera River course: the Villarosa reservoir that blocks the Morello torrent, the Olivo reservoir, which blocks the Braemi torrent, and the Gibbesi reservoir, which is still not in operation. The Villarosa reservoir has been built nearby the town of Villarosa between 1969 and 1973, and it has a volume of 17.16 Mm3 and an estimated silting volume of 5.00×10^6 m^3. In 1989, a bathymetric survey recorded an infilling volume of 1.37×10^6 m^3. The Olivo reservoir has a total volume of 18 Mm^3 and an estimated volume of 2.00×10^6 m^3 [25] (Figure 2). One Natura 2000 site falls within the study area. Natura 2000 is an ecological network composed of sites designated under the Birds Directive (Special Protection Areas, SPAs) and the Habitats Directive (Sites of Community Importance, SCIs, and Special Areas of Conservation, SACs). The aim is to create breeding and resting sites for rare and threatened species and to protect some rare natural habitat types [29]. At 2 km eastward of the Falconara Castle, within sector no. 2, the Site of Community Importance ITA 050011 "Manfria Tower" is set; it is considered a biotope of high naturalistic and environmental interest, where a small dune ridge east of the Manfria Tower is still well preserved [30,31].

3. Materials and Methods

The main constraining factor for diachronic analysis is the availability of data for the specific study site [32]. The cartographic dataset covered a time span of 64 years, from 1955 to 2019. The data included (i) IGMI (Istituto Geografico Militare Italiano) aerial photographs (1955, 1966), (ii) orthophotographs (1989, 2000, 2006, 2012), available online at the website http://www.pcn.minambiente.it/mattm (accessed on 16 July 2021) [33], and (iii) Google images acquired by Google Earth (2016, 2019) [34]. Cartographic dataset resolution ranged from 0.5 m (Google Earth images) to 3 m (IGMI aerial photographs). Shorelines were acquired in ArcGIS 10.3 environment using as shoreline proxy the wet/dry line [32], and common coordinate system was set (Gauss–Boaga, Monte Mario Italy 2). The study area was split into two sectors limited by natural coastal physiographic or manmade structures. For each sector, the shoreline change analysis was carried out on medium-term period, covering a time span of 30 years (1989 to 2019) and on long-term period, between 1955 and 2019 [10]. The shoreline change analysis was based on the integration of remote sensing and geographic information system techniques. Shorelines were traced from aerial photographs, ortophotos and UAV image and analysed by the Digital Shoreline Analysis System (DSAS), which is an extension to ESRI ArcGIS© that can calculate the shoreline rate-of-change statistics starting from multiple historical shoreline positions [35]. The reliability of the statistics computed by the system mainly depends on the uncertainty value related to each shoreline dataset. In order to reduce the effect of short-term variability on long-term analysis, the uncertainties were considered independent, uncorrelated and random, and, following [10,36–40], the total positional uncertainty of each dataset was calculated with the following equation:

$$\sigma_t = \sqrt{\sigma_d^2 + \sigma_p^2 + \sigma_r^2 + \sigma_{td}^2 + \sigma_{wr}^2} \quad (1)$$

The Digitizing Error (σ_d) was obtained detecting several times the same feature on the same image and calculating the error as the standard deviation of the residual value for that feature. The Pixel Error (σ_p) was assumed to be equal to the pixel size. For the Orthorectification Error (σ_r), the RMSE computed for the photogrammetric and polynomial rectification process was used as error value. The Tidal Fluctuation Error (σ_{td}) was considered as the variations in levels at the tide gauge station located at Porto Empedocle, 40 km NW from the study area. It is set in the harbour, and since 2010, the tide level is measured by a new sensor with millimetre accuracy, the SIAP+MICROS TLR. The min–max tidal range varied from 0.03 m to more than 1 m, but the average water level fluctuation was 0.04 m. The Weighted Linear Regression (WLR) rate index was classified following the methodology proposed by Molina et al. [12]. All values were normalized, and the Gaussian distribution was used to set the class limits (Table 1). More than 40% of all data ranged between ± 0.4 m, and it was assumed as the most recurrent smallest change due to seasonal oscillations and thus classified as stability state. Values between −0.4 and −0.7 m and between +0.4 and +0.7 m were considered as moderate erosion and moderate accretion, respectively, corresponding to 25% of all datasets; the remaining 34% data were grouped into 4 classes: high erosion (<−0.7 m; ≥−1.5 m), very high erosion (<−1.5 m), high accretion (>0.7 m; ≤1.5 m) and very high accretion (>1.5 m). The percentage of each beach evolution class was computed for each sector.

Table 1. Beach evolution classes defined for the present work following the methodology proposed by Molina et al. [12].

Class	m/yr
Very high erosion	<−1.5
High erosion	≥−1.5; <−0.7
Moderate erosion	≥−0.7; <−0.4
Stability	≥−0.4; ≤+0.4
Moderate accretion	>+0.4; ≤+0.7
High accretion	>+0.7; ≤+1.5
Very high accretion	+1.5

The coefficient of coastal armouring K [21] was used to assess the anthropogenic structure impact on the coastal area. It was computed dividing the total length of all emerged and visible submerged maritime structures (groins, moles, seawalls, revetments, breakwaters, etc.) by the entire length (L) of the coast under study, which was subdivided into subsectors of 500 m each. The extent of coastal armouring was "Minimal" at K = 0.0001−0.1, "Average" when K = 0.11−0.5, "Maximal" at K = 0.51−1.0 and "Extreme" if K > 1.0.

The dune fragmentation index (F) was proposed for the first time by Molina et al. [5]. It was used to assess how the dune systems have been impacted over time. The F index is expressed as the ratio between the length of all breaks (l) identified within a dune system and the whole dune toe length (L):

$$F = \frac{l}{L} \quad (2)$$

Each shorefront dune system was divided into 100 m sectors, and the F index was calculated for each. The F index can be grouped into three classes that were computed by using the Natural Breaks Function [5]. The annual level of dune fragmentation was computed as the average value for all sectors for each available photogrammetric flight. In the present paper, five classes were used to depict dune fragmentation evolution over time—limits between classes were calculated using the Natural Breaks Function applied to the total set of values obtained. Dune fragmentation level was "Null" (F = 0), "Low" (0 < F $\leq$ 0.05), "Medium" (0.05 < F $\leq$ 0.1), "High" and "Very High/Maximum" (0.1 < F $\leq$ 0.4, F > 0.4, respectively).

4. Results

4.1. Shoreline Change Analysis

The long-term shoreline change analysis was carried out over a time span covering 64 years (1955–2019). Results show that within sector no. 1, covered by an amount of 380 transects (Figure 3), the area between the Licata harbour and the set of breakwaters (first 40 transects) registered very high erosion with a maximum WLR negative value of −6.25 m/year. The Southern Imera River mouth area has significantly eroded, and some buildings have been damaged by the intense shoreline landward movement (Figure 4).

The 1000 m of the coast, corresponding to the area protected by eleven breakwaters, experienced high (7%, 25 transects) to very high accretion (6%, 23 transects). A stable trend was recorded just down-drift of the breakwaters and along with 9.5 km (58% and 222 transects). The Net Shoreline Movement index registered a huge shoreline erosion between 1955 and 2019 close to the Southern Imera River mouth and a significant seaward movement where the Licata breakwaters are today set. The Shoreline Change Envelope index revealed that high accretional phenomena occurred between 1955 and 1966, but intensive erosional phenomena took place after 1966, even though they have been partly reduced by the trap action of the breakwaters.

Over the long-term time span, sector no. 2 experienced accretion that mainly occurred between the SCI "Manfria Tower" and the Gela harbour (Figure 5). In total, 37% of the data (205 transects) showed accretion, 32% recorded stability (181 transects), 13% of the coast faced moderate erosion (71 transects) and 18% high erosion (101 transects). The Net Shoreline Movement index registered the highest seaward migration east of the Licata harbour, and the Shoreline Change Envelope confirmed that accretion occurred over the 1955–2019 period.

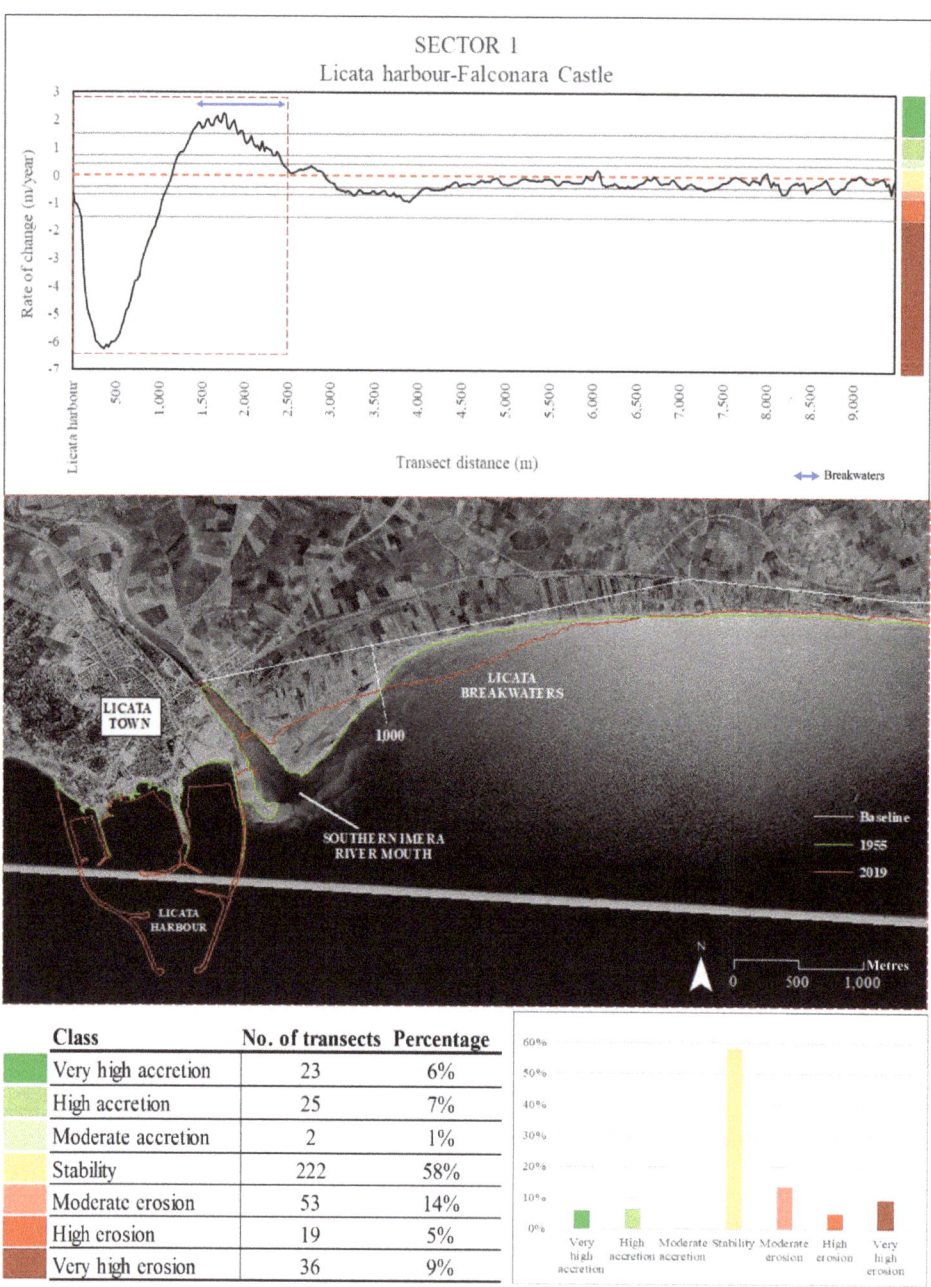

Figure 3. Rates of change within sector no. 1 over the long-term period (1955–2019). In total, 58% of the data falls within the stability state range, especially observed in the coastal area at a distance between 3 and 10 km from the harbour (highest negative WLR = −6.25 m/year), but significant sediment deposition was recorded between 1955 and 1989 where a set of breakwaters have been emplaced.

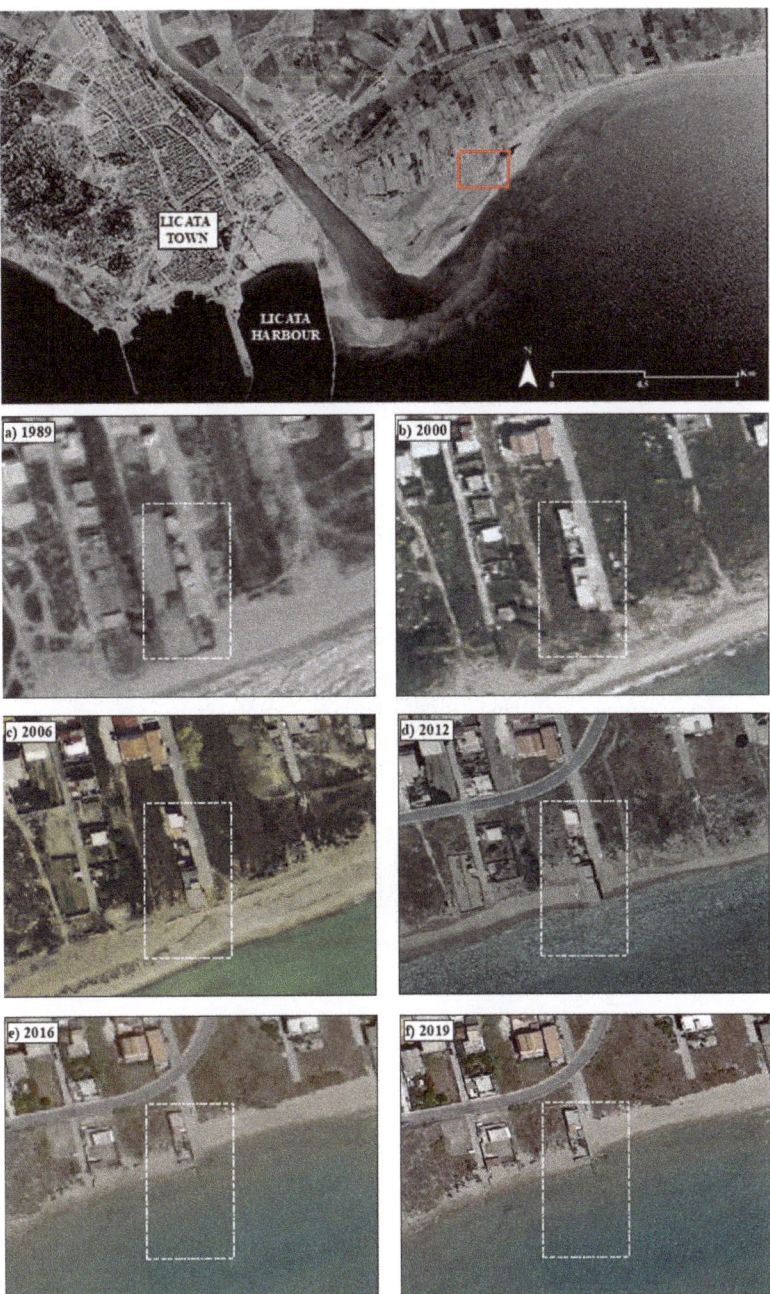

Figure 4. The area east of the Southern Imera River mouth faced severe coastal erosion. In 1955, the coastal area was undisturbed, but over the following decades, urbanization and anthropogenic disturbance have significantly increased, despite the severe erosion observed within this area, and (**a**–**f**) holidays houses have been seriously damaged and partly submerged, as shown by the white square that frames a house partly swallowed up by the sea.

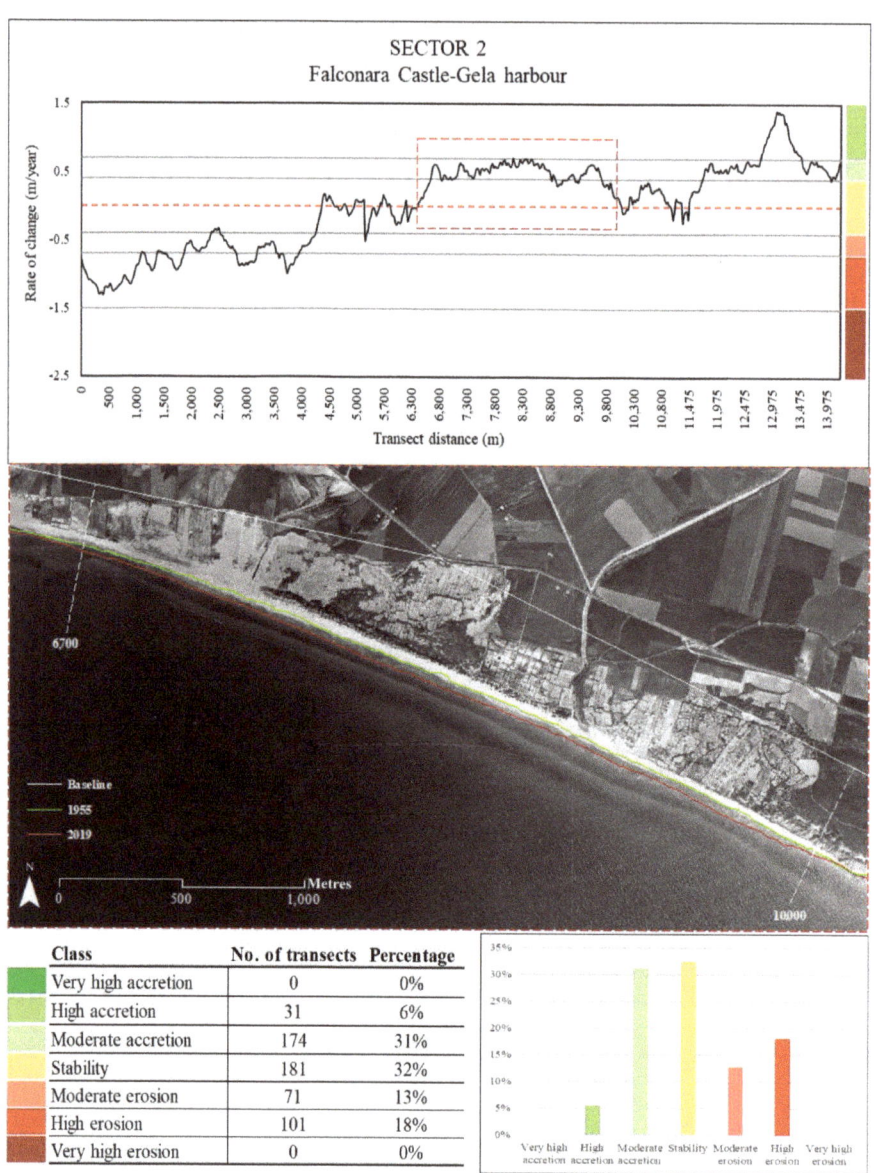

Figure 5. Shoreline evolution expressed as weighted linear regression rate within sector no. 2 over the long-term period (1955–2019). Stable trend (32%, 181 transects) was mainly recorded within sector no. 2, followed by moderate accretion (31%, 174 transects). Significant retreat occurred along 31% (172 transects) of the sector.

The medium-term shoreline change analysis was performed for the 1989–2019 time span. Within sector no. 1, 43% of the transects (165) fell within the stability state range, and erosion classes reached 48% (182 transects) with a maximum negative WLR value of 3.87 m/year registered nearby the Licata harbour. Lastly, the total accretion classes percentage was 9% (33 transects) and was only recorded at the set of breakwaters close to Licata (Figure 6).

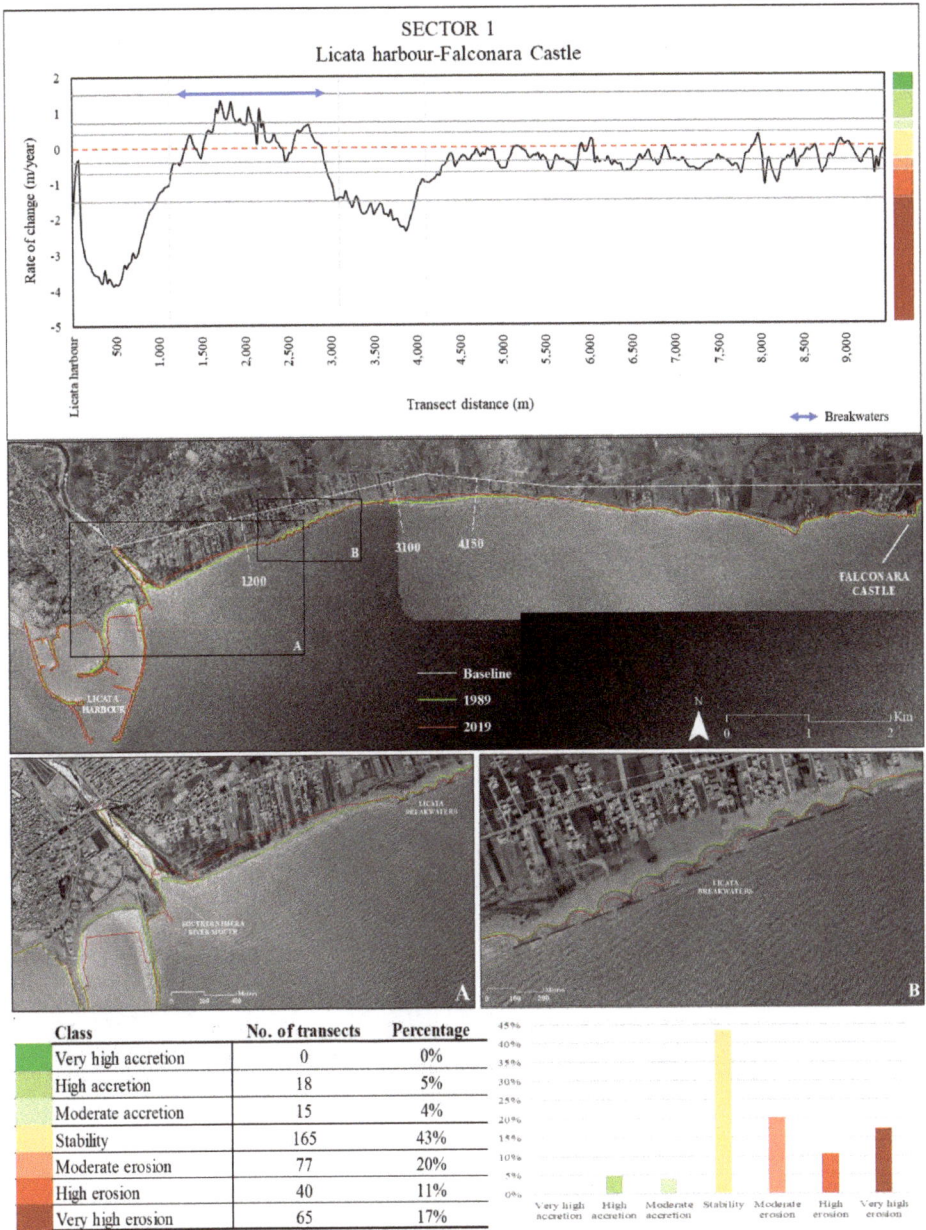

Figure 6. Shoreline changes over the mid-term period (1989–2019) within sector no. 1. Sector no. 1 is about 10 km long, from the Licata harbour to the Falconara Castle, but the rate-of-change plot indicated that the highest variability was detected within 4 km eastward of the Licata harbour. (**A**) The most severe erosion phenomena occurred right next to the Southern Imera River mouth; (**B**) accretional classes have been recorded about 1.2 km eastward the Licata harbour where 11 breakwaters have been emplaced to block the intense sediment loss. Grey dashed lines are those transects cast by the DSAS and delimit coastal areas that have recorded main changes.

Within sector no. 2, stability state class represented 38% of the data (210 transects), even though most of the stable transect values were recorded at the eastern part of the coast, and moderate accretion values (18%) were mainly found along the coast falling within the Site of Community Importance "Manfria Tower". Shoreline erosion was mostly observed at the western part of the sector, where the shoreline faced a maximum retreat of 1.73 m/year (Figure 7).

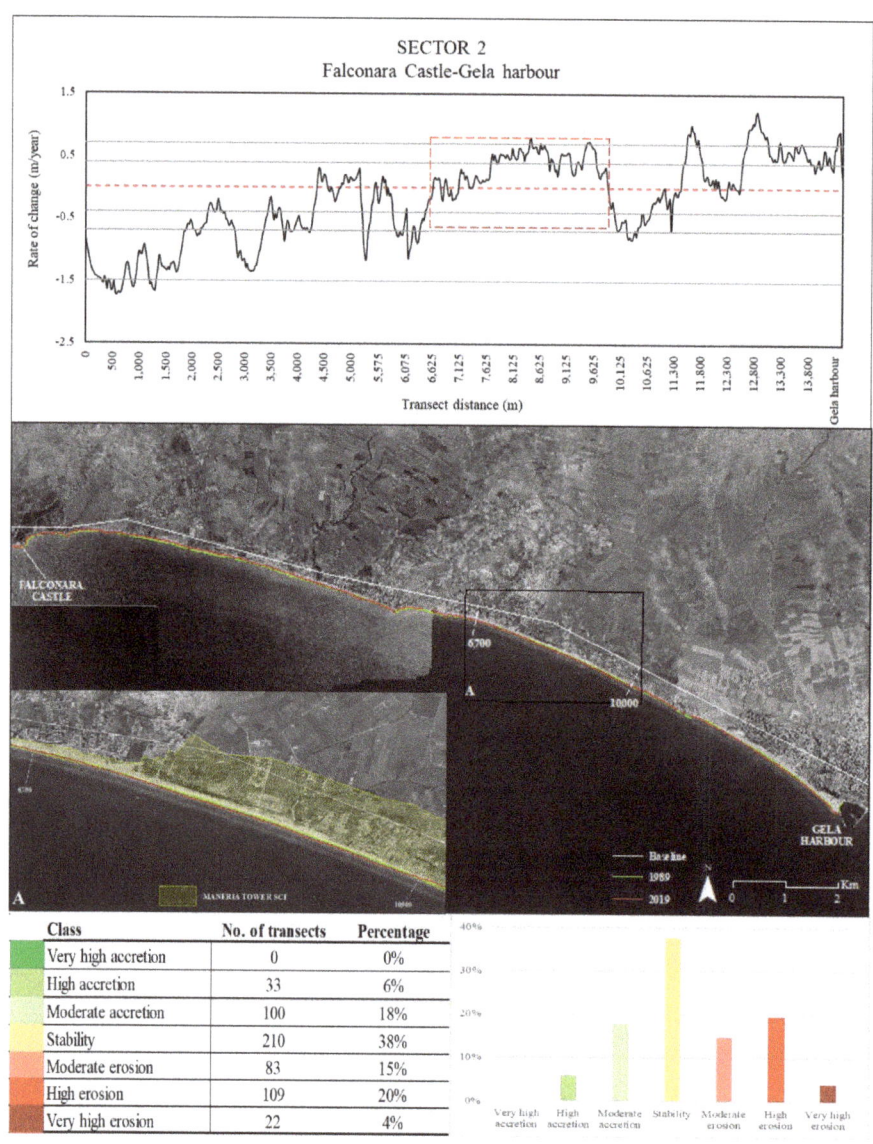

Figure 7. Shoreline evolution over the mid-term period (1989–2019) within sector no. 2. This sector mostly showed a stable trend (38%, 210 transects). In total, 39% of the coast faced moderate to very high erosion (214 transects), and 24% ranged between moderate and high accretion (133 transects). (A) Higher sediment deposition processes have been recorded in correspondence of the Site of Community Importance ITA 050011 – Manfria Tower. Grey dashed lines are the transects at 6350 m and at 10,125 m from the westernmost edge of the sector.

4.2. Dune Fragmentation Analysis

The dune system detected in the 1955 and 1966 aerial photographs presents a great alongshore continuity. Therefore, the dune fragmentation index computed for those years was null. Over time, continuity of dune system toe and associated vegetation line significantly decreased (Figure 8).

Figure 8. Dune toe detected nearby the Licata port on the 1966 aerial photo and on the 1989 orthophoto. (**a**) The dune system in 1966 is several kilometres long and only interrupted by physiographic or natural elements. (**b**) In 1989, the dune ridge significantly retreated (white square) and was partly interrupted by manmade works. Green line is the dune toe proxy.

Over the period between 1989 and 2019, nine dune systems were detected along the two sectors: two within sector no. 1 and seven within sector no. 2. The dune fragmentation index was computed for each system. Dune system no. 1 totally disappeared over the time span between 1989 and 2012 (F = 1 in 2012, "Very High/Maximum" class), but in 2016, dune system no. 1 was naturally recovered. Two of the nine systems experienced an increase in dune fragmentation, and seven of them did not change their class. As such, 56% of the dune systems were not fragmented at all in 1989, and 22% were scarcely fragmented ("Low" fragmentation class). The tendency changed after 2006 when the "Very high/Maximum" class appeared for the first time. The "Medium" class varied as well from 0% in 1989 to 22% in 2016. In 2019, the "Very high/Maximum" class was not observed anymore; most of the dune system fragmentation values fell within the range of the "Null" and "Low" classes (55%, five dune systems), but 33% of the systems were still highly fragmented (Figure 9).

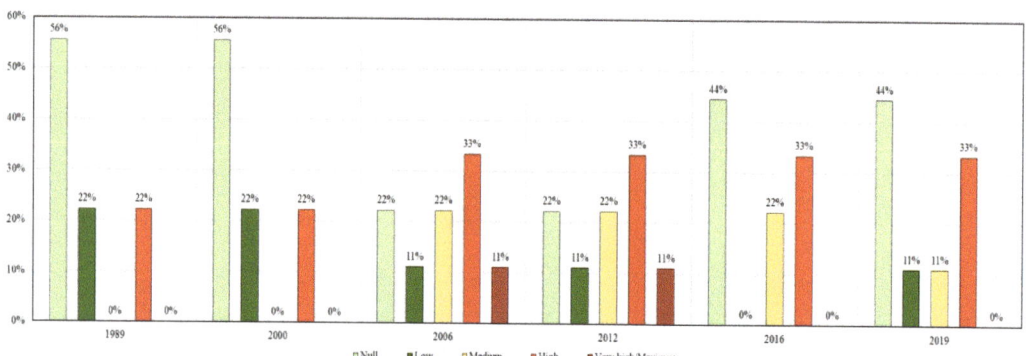

Figure 9. Percentage of F index classes per year. In 1989, the most frequent classes are the "Null" and the "Low" fragmentation ones, "Medium" fragmentation class is not represented at all, the "High" one is 21%, and the "Very High/Maximum" fragmentation class has not been found.

4.3. Coastal Armouring Analysis

The coefficient of coastal armouring (K) was computed to assess the maritime structures' impact on the shoreline evolution. This coefficient represents the ratio between the total length of all maritime hydraulic structures (groins, seawalls, dikes, jetties, etc.) and total coastal length of the study area; the study area was, however, subdivided into subsectors of 500 m each [21].

Table 2 shows the coefficient of coastal armouring (K), the total coastal length (L) per year and the total length of coastal structures (l) per year. Over the 1950s and 1960s, the Licata harbour was the only infrastructure insisting on the coast. The K value computed for sector no. 1 did not appreciably vary over those years and fell within the "Maximal" range class (0.5–0.8). A few coastal structures were detected in 1955 and 1966 within sector no. 2. In 1989, the K value of sector no. 1 resulted to be significantly higher (1.2) reaching the "Extreme" class. Sector no. 2 did not register any change. Between 1966 and 1989, the total structure length significantly increased, mainly due to the Licata harbour implementation. Eleven breakwaters were built eastward of the Licata harbour over those decades. Over the 2000–2019 time span, the coefficient K slightly increased within sector no. 1, varying from 1.3 to 1.6 (i.e., "Extreme" class). Within sector no. 2, no changes were detected.

Table 2. The coefficient of coastal armouring K computed for each sector per year. The Total Coastal Length (L) and the Total Structure Length (l) per year are shown. All measures are expressed in metres (m). Sector no. 1 experienced significant increase, with K passing from "Maximal" (K = 1.2 in 1989) to the "Extreme" class (K = 1.6 in 2019). Sector no. 2 did not face any changes over the considered time span.

Year	Sector	Total Coastal Length (L)	Total Structure Length (l)	K
1955	1	9826	7706	0.8
1966	1	10,134	8443	0.8

Table 2. Cont.

Year	Sector	Total Coastal Length (L)	Total Structure Length (l)	K
1989	1	10,216	11,969	1.2
2000	1	10,198	13,640	1.3
2006	1	9727	13,918	1.4
2012	1	9732	15,014	1.5
2019	1	9719	15,456	1.6

5. Discussion

In the following sections, the shoreline changes along the two sectors are discussed as the main erosional/accretional/stable phenomena observed. The findings on the shoreline evolution are interpreted in the light of the main environmental coastal variations, considering the main river variations, the dune fragmentation level and the coastal armouring impact.

The mid- and long-term period analyses showed that the coastal area under study mainly experienced stability (43% and 58%, respectively). However, significant shoreline changes were observed within the area of the Southern Imera River mouth, and 9% of very high erosion phenomena were recorded down-drift of the Licata harbour, where the highest WLR negative value was also detected (−6.25 m/year). Amore et al. [23] showed that significant erosional phenomena were observed at the Southern Imera River mouth earlier than the 1960s, and coastal progradation was mainly observed over the past decades. The Shoreline Change Envelope index confirmed that high accretional phenomena occurred between 1955 and 1966, and the shoreline significantly began retreating after 1966 (Figure 10). As already shown by Amore et al. and Amore and Randazzo [23,41], the higher accretion rate detected over the decade 1950–1960 can be explained by the increasing river sediment load caused by (i) the disuse of the secondary stream mouth of the Southern Imera River and (ii) the deforestation of the drainage basin area to make space for crops and cultivation [23]. Deforestation is considered as a triggered event for coastal advancement, as shown by Pranzini [42] in the area of the Arno and Ombrone River deltas in Italy and by Zengcui and Zeheng [43] in the area of the Qiantang estuary in China. After the 1960s, the sediment supplies to the two studied sectors significantly decreased due to the implementation of (i) the Villarosa, Olivo and Gibbesi artificial reservoirs, which were built over the decade 1960–1970, and (ii) the Licata harbour. The reservoirs blocked the Southern Imera River course to satisfy the huge increase in irrigation water demands, forming sediment traps and reducing the peak flood flows, thereby decreasing the sediment supply to the coast. The decrease of sediment river discharge caused by the implementation of dams is a pivotal factor in beach retreating. Rosskopf et al. [44] pointed out that the long-term shoreline retreat of the Molise coast (Italy) was primarily related to channel adjustments of the Biferno and Trigno rivers, trapping most of the rivers' solid load, affecting the sediment budget of the river mouth areas and adjacent beaches. The reduction in sediment discharge caused by the dams was also well studied by Amrouni et al. [13], who considered the implementation of artificial reservoirs as the main cause of the negative sediment budget leading to a shortage of sand sediment on the Medjerda delta (Tunisia) and the dominant erosion of the coastline. The reduced flux of sediment reaching the world's coasts because of retention within reservoirs is also well documented by Syvitski et al. [45], who estimated that over 100 billion metric tons of sediment are sequestered in reservoirs constructed largely within the past 50 years.

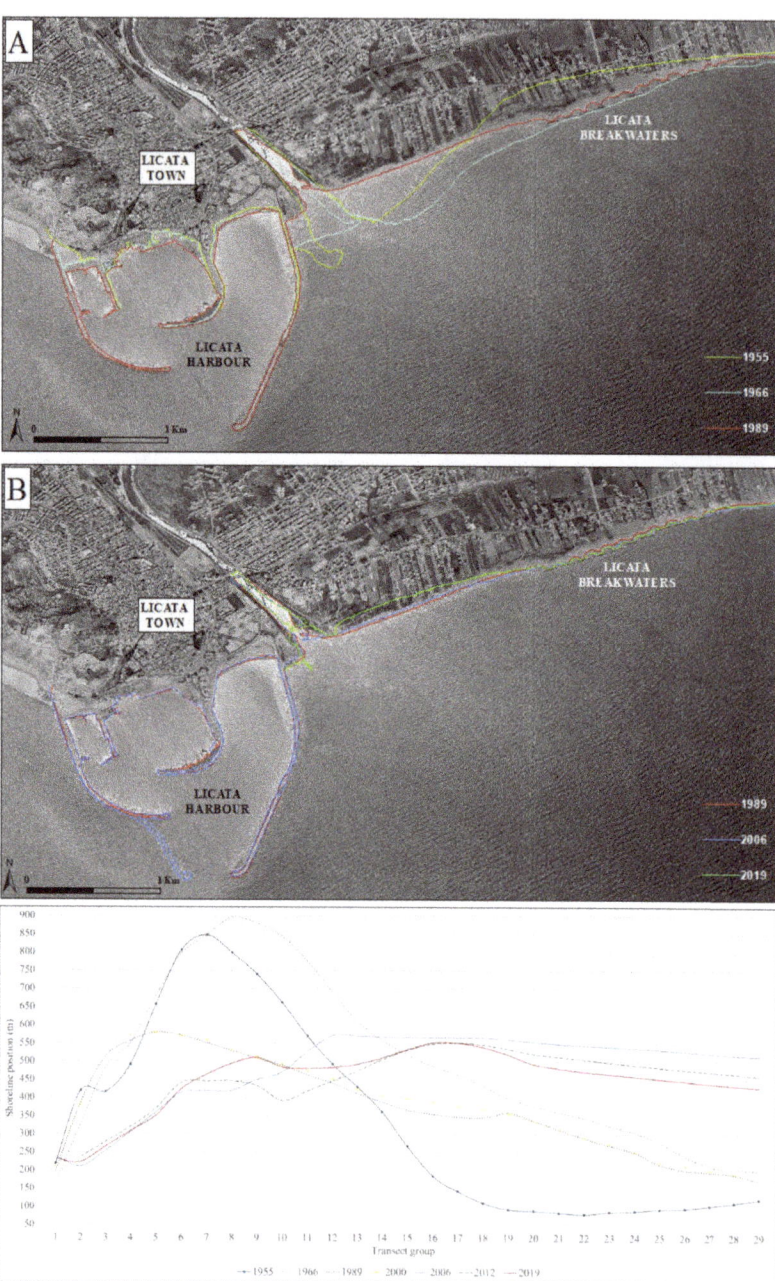

Figure 10. The shoreline evolution of the Southern Imera River mouth and the area east of the mouth. The trend seemed to be negative between 1955 and 2016, but the Shoreline Change Envelope revealed that (**A**) significant accretion has been recorded between 1955 and 1966, and very high erosion has been registered between 1966 and 1989, as shown by the interpolated plot of the shoreline changes (1 transect group = 100 m), but (**B**) shoreline moved seaward between 1989 and 2016 where a set of eleven breakwaters was emplaced.

Over the time between 1955 and 2019, the coastal area between Licata and Gela has been modified following the increasing economic growth that occurred after the Second World War. Between the 19th and the 20th centuries, the town of Licata became one of the biggest industrial centres in Europe for the refinement of sulphur which led to the implementation of a larger harbour accessible for larger ships served by a railway line [23]. As a result, (i) the harbour impounded littoral material, partly interrupting the W–E longshore sediment load [12,46,47], and (ii) the dune systems of the area nearby the Licata town have been profoundly damaged due to the increase in urbanization, as largely observed worldwide in areas of rapid economic growth [48–53].

However, the highest retreats were also recorded down-drift of coastal structures (harbours and breakwaters), and sediment deposition was mainly found in correspondence of structures (i.e., breakwaters) and along coastal areas where beaches are backed by well-preserved dune systems and breakwater barriers (Figure 11). Indeed, east of the Licata harbour, a set of eleven breakwaters was emplaced to face the huge landward migration of the Southern Imera River mouth and of the proximal coastal area. Moderate to high erosion phenomena were observed down-drift of the set of structures, whereas accretion was recorded in correspondence and up-drift of them, with a maximum value of 2.22 m/year (Figure 11). The same trend, i.e., significant accretion, was observed within sector no. 2, up-drift of two breakwaters emplaced westward of the Gela harbour. A few kilometres westward of the study area, at San Leone beach, Manno and Ciraolo [54] showed that the maritime works (port, groins and breakwaters), which were emplaced to limit the increasing shoreline retreat, changed the original coastal equilibrium, affecting the coastal sediment transport and thus causing very high variability in shoreline position up-drift and down-drift of the coastal structures. In nearby South-eastern Sicilian areas, Anfuso et al. [55] confirmed as well the common tendency of accretion/erosion at up-drift/down-drift of coastal structures, respectively. Coastline armouring is a very common engineering solution against erosion [46,56], but there are a number of adverse influences, including disturbance of cross- and long-shore sediment transport, associated with up-drift sedimentation but down-drift beach reduction, accelerated bottom erosion in front of structures, restricted public access to the beach, formation of dangerous currents for bathers and negative impact on landscape value [8,46,57–61]. Worldwide examples of anthropogenic disturbance (i.e., coastal armouring) on the coastal dynamic are well documented. Sousa et al. [61,62] showed that coastal structures are the most applied short-term solutions to coastal erosion in Massaguaçú Beach (Brazil) but not the most efficient ones, often changing the long-shore transport, affecting the beach profile and the scenic beauty of the beach, and thus increasing coastal vulnerability.

Moderate to high accretion classes were recorded in correspondence of the Site of Community Importance ITA 050011 "Manfria Tower", where the long and wide beach is backed by a well-preserved dune ridge. Such accretion processes have been probably active for several decades as it is reflected by the formation of such well-developed dune systems. Within sector no. 2, the coastal land use has not significantly changed over time, and the area did not face huge coastal changes both in terms of dune fragmentation and of implementation of coastal structures. The Site of Community Importance ITA 050011 "Manfria Tower" acted as a constraining factor to human pressure on this area. As observed in Australia, at places, recent human disturbance is the main cause of the increasing sediment supply to dune systems, often resulting in foredune formation [63]. Moreover, natural dunes often provide sediment reservoirs to the shore and act as barriers to erosion and flooding processes, as showed by Manca et al. [64] in West Sardinia (Italy), where the dune system at Maria Pia beach is still acting as a source of sand. Saye et al. [65] in the UK demonstrated that dune morphology often reflects beach morphology (width, slope, sediment grain size, etc.); indeed, high foredune ridges develop where the sediment budget of the beach and dune is in balance. However, Saye et al. [65] also found that, within an accreting coastal sector, new dune ridges can progressively form, and the maximum crest height of each ridge is limited by the rate of beach accretion and shoreline progradation.

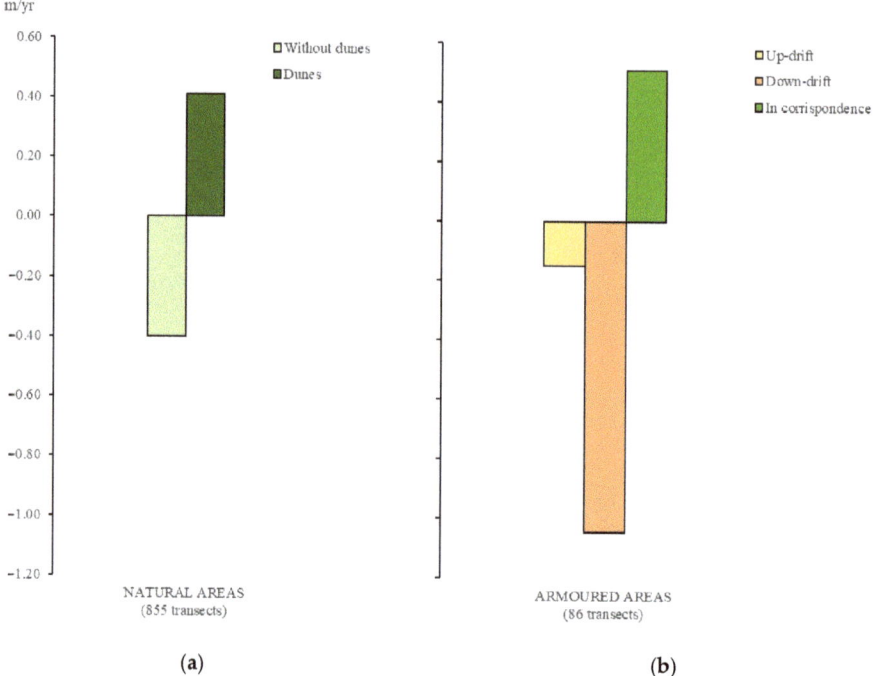

Figure 11. Shoreline evolution plot of the average WLR values found in correspondence of transects placed at (**a**) natural areas, subdivided into areas backed by dunes (93 transects) and areas without dunes (762 transects), and at (**b**) armoured areas, considering the up-drift and down-drift zones and in corrispondence of structures (48 transects).

6. Conclusions

The shoreline evolution and the environmental changes of the coastal strip between the Licata harbour and the Gela harbour (Southern Sicily, Italy) were investigated over mid- and long-term periods (1989–2019 and 1955–2019, respectively). This study showed that (i) significant shoreline retreat occurred in correspondence of the Southern Imera River mouth, where, on one hand, the construction of artificial reservoirs along the river course caused the shortage of sediment river discharge and, on the other hand, coastal environmental changes, as the progressive implementation of the Licata harbour and the significant loss of the dune ecosystem, altered the coastal sediment dynamic; (ii) coastal armouring affected longshore sediment transport, giving rise to sediment deposition in correspondence and up-drift of structures (i.e., breakwaters), while relevant retreat is generally found down-drift of coastal structures (harbours and breakwaters); and (iii) beaches backed by a well-preserved dune system experienced sedimentation or maintained a stable trend as dunes likely provide sediment reservoirs to the shore and act as barriers to erosion and flooding processes.

The lack of coastal management strategies and non-integrated decision-making processes often triggered an increase in coastal zone vulnerability. Within sector no. 1, despite the severe erosional phenomena recorded, improper land-use planning has been carried out and buildings have been seriously damaged and progressively destroyed by wave processes. As such, this study can be the starting point to forecast future shoreline trends (taking also into consideration climate change processes) and can support the local coastal management administrations to properly plan sustainable policies in a long-term scenario, i.e., assess areas at high risk of erosion and flooding, evaluate the efficiency of coastal armouring and of natural protected coastal sectors and identify best resilience solutions to stabilise and limit the impacts related to coastal erosion.

Author Contributions: Conceptualization, L.B., A.D.S. and D.C.; Methodology, L.B., G.A., G.M., A.D.S. and D.C.; Software, L.B., G.M. and G.A.; Formal analysis, L.B., A.D.S., D.C., G.A., G.M., S.D. and S.U.; Data curation, L.B., A.D.S., D.C., G.A., G.M., S.D. and S.U.; writing—original draft preparation, L.B., A.D.S., D.C., G.A., G.M., S.D. and S.U.; writing—review and editing, A.D.S., D.C., G.A., G.M., S.D. and S.U. All authors have read and agreed to the published version of the manuscript.

Funding: The study also benefited from grants assigned in the framework of the Project n. 183371—CUP E66C18001300007 "Attraction and International Mobility" (AIM) funded by the Italian MIUR (Ministry of Instruction and University and Research) (D.D. 407/27.2.2018) Action 1.2 Axis I-PON R&I 2014-2020 Blue Growth Area (Beneficiary: Salvatore Distefano; Scientific Responsible: Agata Di Stefano).

Informed Consent Statement: Not applicable.

Data Availability Statement: Not applicable.

Acknowledgments: Special thanks go to Giuseppe Ciraolo and Carlo Lo Re for the support given on the uncertainties and errors computation.

Conflicts of Interest: The authors declare no conflict of interest.

References

1. Bird, E.C.F. *Submerging Coasts: The Effects of a Rising Sea Level on Coastal Environments*; Wiley and Sons: New York, NY, USA, 1993; p. 184.
2. Pethick, J. Coastal management and sea-level rise. *Catena* **2001**, *42*, 307–322. [CrossRef]
3. Todd, P.A.; Heery, E.C.; Loke, L.H.; Thurstan, R.H.; Kotze, D.J.; Swan, C. Towards an urban marine ecology: Characterizing the drivers, patterns and processes of marine ecosystems in coastal cities. *Oikos* **2019**, *128*, 1215–1242. [CrossRef]
4. Komar, P.D. *Beach Processes and Sedimentation*, 2nd ed.; Prentice Hall: Upper Saddle River, NJ, USA, 1998; p. 544.
5. Molina, R.; Manno, G.; Re, C.L.; Anfuso, G. Dune systems' characterization and evolution in the Andalusia Mediterranean Coast (Spain). *Water* **2020**, *12*, 2094. [CrossRef]
6. Di Stefano, A.; De Pietro, R.; Monaco, C.; Zanini, A. Anthropogenic influence on coastal evolution: A case history from the Catania Gulf shoreline (eastern Sicily, Italy). *Ocean. Coast. Manag.* **2013**, *80*, 133–148. [CrossRef]
7. Wong, P.P.; Losada, I.J.; Gattuso, J.-P.; Hinkel, J.; Khattabi, A.; McInnes, K.L.; Saito, Y.; Sallenger, A.; Cheong, S.-M.; Dow, K.; et al. Coastal systems and low-lying areas. In *Climate Change 2007: Impacts, Adaptation and Vulnerability. Contribution of Working Group II to the Fourth Assessment Report of the Intergovernmental Panel on Climate Change*; Nicholls, R.J., Santos, F., Eds.; Cambridge University Press: Cambridge, UK, 2007; pp. 315–356.
8. European Environmental Agency. *The Changing Faces of Europe's Coastal Areas*; Office for Official Publications of the European Communities: Bruxelles, Belgium, 2006.
9. Romine, B.M.; Fletcher, C.H. Armoring on eroding coasts leads to beach narrowing and loss on Oahu, Hawaii. In *Pitfalls of Shoreline Stabilization*; Springer: Dordrecht, The Netherlands, 2012; pp. 141–164.
10. Crowell, M.; Leatherman, S.P.; Buckley, M.K. Historical shoreline change: Error analysis and mapping accuracy. *J. Coast. Res.* **1991**, *7*, 839–852.
11. Galgano, F.A.; Douglas, B.C.; Leatherman, S.P. Trends and variability of shoreline position. *J. Coast. Res.* **1998**, *26*, 282–291.
12. Molina, R.; Anfuso, G.; Manno, G.; Gracia Prieto, F.J. The Mediterranean coast of Andalusia (Spain): Medium-term evolution and impacts of coastal structures. *Sustainability* **2019**, *11*, 3539. [CrossRef]
13. Amrouni, O.; Sánchez, A.; Khélifi, N.; Benmoussa, T.; Chiarella, D.; Mahé, G.; Abdeljaouad, S.; & McLare, P. Sensitivity assessment of the deltaic coast of Medjerda based on fine-grained sediment dynamics, Gulf of Tunis, Western Mediterranean. *J. Coast. Conserv.* **2019**, *23*, 571–587. [CrossRef]
14. Coelho, C.; Silva, R.; Veloso-Gomes, F.; Taveira Pinto, F. A vulnerability analysis approach for the Portuguese west coast. In *Risk Analysis V: Simulation and Hazard Mitigation*; Popov, V., Brebbia, C.A., Eds.; WIT Press: Southampton, UK, 2006; pp. 251–262.
15. Anfuso, G.; Martínez Del Pozo, J.Á. Assessment of coastal vulnerability through the use of GIS tools in South Sicily (Italy). *Environ. Manag.* **2009**, *43*, 533–545. [CrossRef] [PubMed]
16. Dolan, R.; Fenster, M.S.; Holme, S.J. Temporal analysis of shoreline recession and accretion. *J. Coast. Res.* **1991**, *7*, 723–744.
17. Crowell, M.; Leatherman, S.P.; Buckley, M.K. Shoreline change rate analysis: Long term versus short term data. *Shore Beach* **1993**, *61*, 13–20.
18. Molina, R.; Manno, G.; Re, C.L.; Anfuso, G.; Ciraolo, G. A Methodological approach to determine sound response modalities to coastal erosion processes in mediterranean andalusia (Spain). *JMSE* **2020**, *8*, 154. [CrossRef]

19. Salman, A.; Lombardo, S.; Doody, P. *Living with Coastal Erosion in Europe: Sediment and Space for Sustainability*; Hydraulic Engineering Reports; Eurosion project; EUCC: Porto, Portugal, 2004; p. 21.
20. Sicilian Region—Office of the Commissioner of Government against the Hydrogeological Instability in the Sicilian Region. In *Regional Plan against Coastal Erosion 2020*; v. 1.0; Sicilian Region—Office of the Commissioner of Government: Sicily, Italy, 2020; p. 1527.
21. Aybulatov, N.A.; Artyukhin, Y.V. *Geo-Ecology of the World Ocean's Shelf and Coasts*; Hydrometeo Publishing: Saint Petersburg, Russia, 1993; p. 304.
22. Brambati, A.; Amore, C.; Giuffrida, E.; Randazzo, G. Relationship between the port structures and coastal dynamics in the Gulf of Gela (Sicily-Italy). In Proceedings of the International Coastal Congress ICC, Kiel, Germany, 6–13 September 1992; pp. 773–793.
23. Amore, C.; Geremia, F.; Randazzo, G. Historical evolution of the Salso River mouth with respect to the Licata harbour system (Southern Sicily, Italy). In *Littoral The Changing Coast*; EUROCOAST/EUCC: Porto, Portugal, 2002; pp. 253–260.
24. Martino, C.; Curcuruto, E.; Di Stefano, A.; Monaco, C.; Zanini, A. Fenomeni erosivi lungo il litorale di Marina di Butera (CL), Sicilia centro-meridionale. *Geol. Di Sicil.* **2011**, *3*, 4–15.
25. Regione Siciliana—Assessorato Territorio e Ambiente—Dipartimento Territorio e Ambiente. Servizio III Assetto del Territorio e Difesa del Suolo. Piano Stralcio di Bacino per l'Assetto Idrogeologico della Regione Siciliana; (P.A.I.). Bacino Idrografico del F. Imera Meridionale (072) Area territoriale tra il Bacino Idrografico del F. Palma e il Bacino Idrografico del F. Imera Meridionale (071). 2008. Available online: https://www.sitr.regione.sicilia.it/pai/bacini.htm (accessed on 27 September 2021).
26. Anfuso, G.; Martínez del Pozo, J.Á.; Rangel-Buitrago, N. Bad practice in erosion management: The southern Sicily case study. In *Pitfalls of Shoreline Stabilization: Selected Case Studies*; Cooper, J.A.G., Pilkey, O.H., Eds.; Springer: Berlin/Heidelberg, Germany, 2012; Volume 3, pp. 215–233.
27. Regione Siciliana—Assessorato Territorio e Ambiente, Dipartimento Territorio e Ambiente. Servizio 4, Assetto del territorio e difesa del suolo. Piano Stralcio di Bacino per l'Assetto Idrogeologico; (P.A.I.). UNITA' FISIOGRAFICA N° 8—PUNTA BRACCETTO—PORTO DI LICATA. 2007. Available online: https://www.sitr.regione.sicilia.it/pai/unitafisiografiche.htm (accessed on 30 September 2021).
28. Foti, E.; Musumeci, R.E.; Leanza, S.; Cavallaro, L. Feasibility of an offshore wind farm in the gulf of Gela: Marine and structural issues. *Wind. Eng.* **2010**, *34*, 65–84. [CrossRef]
29. European Commission. Available online: http://ec.europa.eu/environment/nature/legislation/habitatsdirective/index_en.htm (accessed on 24 March 2020).
30. Brullo, S.; Minissale, P.; Spampinato, G. Considerazioni fitogeografiche sulla flora della Sicilia. *Ecol. Mediterr.* **1995**, *21*, 99–117. [CrossRef]
31. Giusso Del Galdo, G.; Sciandrello, S. Contributo alla flora dei dintorni di Gela (Sicilia meridionale). In *98° Congresso della Società Botanica Italiana, Riassunti*; Università di Catania: Catania, Italy, 2003; p. 235.
32. Boak, E.H.; Turner, I.L. Shoreline definition and detection: A review. *J. Coast. Res.* **2005**, *21*, 688–703. [CrossRef]
33. Geoportale Nazionale. Available online: http://www.pcn.minambiente.it/mattm (accessed on 16 July 2021).
34. Google Earth. Available online: https://earth.google.com/web (accessed on 16 July 2021).
35. Thieler, E.R.; Himmelstoss, E.A.; Zichichi, J.L.; Ergul, A. *Digital Shoreline Analysis System (DSAS) version 4.0—An ArcGIS Extension for Calculating Shoreline Change*; U.S.G.S. Open-File Report; U.S. Geological Survey: Reston, VA, USA, 2008; Volume 1278.
36. Manno, G.; Lo Re, C.; Ciraolo, G. Uncertainties in shoreline position analysis: The role of run-up and tide in a gentle slope beach. *Ocean Sci.* **2017**, *13*, 661–671. [CrossRef]
37. Fletcher, C.; Rooney, J.; Barbee, M.; Lim, S.C.; Richmond, B. Mapping shoreline change using digital orthophotogrammetry on Maui, Hawaii. *J. Coast. Res.* **2003**, *38*, 106–124.
38. Genz, A.S.; Fletcher, C.H.; Dunn, R.A.; Frazer, L.N.; Rooney, J.J. The predictive accuracy of shoreline change rate methods and alongshore beach variation on Maui, Hawaii. *J. Coast. Res.* **2007**, *23*, 87–105. [CrossRef]
39. Romine, B.M.; Fletcher, C.H.; Frazer, L.N.; Genz, A.S.; Barbee, M.M.; Lim, S.C. Historical shoreline change, southeast Oahu, Hawaii; applying polynomial models to calculate shoreline change rates. *J. Coast. Res.* **2009**, *25*, 1236–1253. [CrossRef]
40. Virdis, S.G.; Oggiano, G.; Disperati, L. A geomatics approach to multitemporal shoreline analysis in Western Mediterranean: The case of Platamona-Maritza beach (northwest Sardinia, Italy). *J. Coast. Res.* **2012**, *28*, 624–640.
41. Amore, C.; Randazzo, G. First data on the coastal dynamics and the sedimentary characteristics of the area influenced by the River Irminio basin (SE Sicily). *Catena* **1997**, *30*, 357–368. [CrossRef]
42. Pranzini, E. Pandemics and coastal erosion in Tuscany (Italy). *Ocean Coast. Manag.* **2021**, *208*, 105614. [CrossRef]
43. Zengcui, H.; Zeheng, D. Reclamation and river training in the Qiantang estuary. In *Engineered Coasts*; Chen, J., Eisma, D., Hotta, H.J., Walker, H.J., Eds.; Kluwer Academic Press: Dordrecht, The Netherlands, 2002; pp. 121–138.
44. Rosskopf, C.M.; Di Paola, G.; Atkinson, D.E. Recent shoreline evolution and beach erosion along the central Adriatic coast of Italy: The case of Molise region. *J. Coast. Conserv.* **2018**, *22*, 879–895. [CrossRef]
45. Syvitski, J.P.; Vörösmarty, C.J.; Kettner, A.J.; Green, P. Impact of humans on the flux of terrestrial sediment to the global coastal ocean. *Science* **2005**, *308*, 376–380. [CrossRef]
46. Griggs, G.B. The impacts of coastal armoring. *Shore Beach* **2005**, *73*, 13–22.

47. Dugan, J.; Airoldi, L.; Chapman, M.; Walker, S.; Schlacher, T.; Wolanski, E.; McLusky, D. Estuarine and coastal structures: Environmental effects, a focus on shore and nearshore structures. In *Treatise on Estuarine and Coastal Sciencem*; Wolanski, E., McLusky, D.S., Eds.; Academic Press: Waltham, MA, USA, 2011; Volume 8, pp. 17–41.
48. Hesp, P. Foredunes and blowouts: Initiation, geomorphology and dynamics. *Geomorphology* **2002**, *48*, 245–268. [CrossRef]
49. Scarelli, F.M.; Sistilli, F.; Fabbri, S.; Cantelli, L.; Barboza, E.G.; Gabbianelli, G. Seasonal Dune and beach monitoring using photogrammetry from UAV surveys to apply in the ICZM on the Ravenna coast (Emilia-Romagna, Italy). *Remote. Sens. Appl.* **2017**, *7*, 27–39. [CrossRef]
50. Díez-Garretas, B.; Comino, O.; Pereña, J.; Asensi, A. Spatio-temporal changes (1956–2013) of coastal ecosystems in Southern Iberian Peninsula (Spain). *Mediterr. Bot.* **2019**, *40*, 111–119. [CrossRef]
51. Carrasco, A.R.; Ferreira, O.; Matias, A.; Freire, P. Natural and human-induced coastal dynamics at back-back barrier beach. *Geomorphology* **2012**, *159–160*, 30–36. [CrossRef]
52. Psuty, N.P. Sediment budget and beach/dune interaction. *J. Coast. Res.* **1988**, *3*, 1–4.
53. Sherman, D.J.; Bauer, B.O. Dynamics of beach-dune interaction Progress. *Phys. Geogr.* **1993**, *17*, 413–447. [CrossRef]
54. Manno, G.; Ciraolo, G. Diachronic analysis of the shoreline in San Leone beach (Agrigento-Sicily). In *Establishment of an Integrated Italy-Malta Cross-Border System of Civil Protection—Geological Aspects*; Aracne: Catania, Italy, 2015.
55. Anfuso, G.; del Pozo, J.Á.M. Towards management of coastal erosion problems and human structure impacts using GIS tools: Case study in Ragusa Province, Southern Sicily, Italy. *Environ. Geol.* **2005**, *48*, 646–659. [CrossRef]
56. Charlier, R.H.; Chaineux, M.C.P.; Morcos, S. Panorama of the history of coastal protection. *J. Coast. Res.* **2005**, *21*, 79–111. [CrossRef]
57. Stancheva, M.; Marinski, J. Coastal defense activities along the Bulgarian Black Sea coast-methods for protection or degradation? In *Coastal Structures, Proceedings of the 5th International Conference, Venice, Italy, 2–4 July 2007*; Franco, L., Tomasicchio, G., Lamberti, A., Eds.; World Scientific Publishing Company: Singapore, 2007; pp. 480–489.
58. Stancheva, M. Human-induced impacts along the Coastal Zone of Bulgaria. A pressure boom versus environment. *Compt. Rend. Acad. Bulg. Sci.* **2007**, *63*, 137–146.
59. Manno, G.; Anfuso, G.; Messina, E.; Williams, A.T.; Suffo, M.; Liguori, V. Decadal evolution of coastline armouring along the Mediterranean Andalusia littoral (South of Spain). *Ocean Coast. Manag.* **2016**, *124*, 84–99. [CrossRef]
60. Anfuso, G.; Dominguez, L.; Gracia, F. Short and medium-term evolution of a coastal sector in Cadiz, SW Spain. *Catena* **2007**, *70*, 229–242. [CrossRef]
61. Sousa, P.; Siegle, E.; Tessler, M. Vulnerability assessment of Massaguaçú Beach (SE Brazil). *Ocean. Coast. Manag.* **2013**, *77*, 24–30. [CrossRef]
62. Sousa, P.; Siegle, E.; Tessler, M. Environmental and anthropogenic indicators for coastal risk assessment at Massaguaçú Beach (SP) Brazil. *J. Coast. Res.* **2011**, *64*, 319–323.
63. Oliver, T.; Tamura, T.; Short, A.; Woodroffe, C. Rapid shoreline progradation followed by vertical foredune building at Pedro Beach, southeastern Australia. *Earth Surf. Process. Landf.* **2018**, *44*, 655–666. [CrossRef]
64. Manca, E.; Pascucci, V.; Deluca, M.; Cossu, A.; Andreucci, S. Shoreline evolution related to coastal development of a managed beach in Alghero, Sardinia, Italy. *Ocean. Coast. Manag.* **2013**, *85*, 65–76. [CrossRef]
65. Saye, S.E.; Van der Wal, D.; Pye, K.; Blott, S.J. Beach—Dune morphological relationships and erosion/accretion: An investigation at five sites in England and Wales using LIDAR data. *Geomorphology* **2005**, *72*, 128–155. [CrossRef]

Article

Natural and Cultural Lost Landscape during the Holocene along the Central Tyrrhenian Coast (Italy)

Maurizio D'Orefice [1], Piero Bellotti [2], Tiberio Bellotti [3], Lina Davoli [2] and Letizia Di Bella [4,*]

[1] Department for the Geological Survey of Italy, ISPRA—Italian Institute for Environmental Protection and Research, 00144 Rome, Italy; maurizio.dorefice@isprambiente.it
[2] Department of Earth Sciences, AIGeo—Italian Association of Physical Geography and Geomorphology, University of Rome, 00185 Rome, Italy; piero.bellotti@gmail.com (P.B.); lina.davoli@uniroma1.it (L.D.)
[3] Freelance Archeologist Via Capo Spartivento 13, 00122 Rome, Italy; t_b80@yahoo.it
[4] Earth Sciences Department of Sapienza, University of Rome, Piazzale Aldo Moro 5, 00185 Rome, Italy
* Correspondence: letizia.dibella@uniroma1.it

Citation: D'Orefice, M.; Bellotti, P.; Bellotti, T.; Davoli, L.; Di Bella, L. Natural and Cultural Lost Landscape during the Holocene along the Central Tyrrhenian Coast (Italy). *Land* 2022, 11, 344. https://doi.org/10.3390/land11030344

Academic Editors: Pietro Aucelli, Angela Rizzo, Rodolfo Silva Casarín and Giorgio Anfuso

Received: 9 February 2022
Accepted: 21 February 2022
Published: 25 February 2022

Publisher's Note: MDPI stays neutral with regard to jurisdictional claims in published maps and institutional affiliations.

Copyright: © 2022 by the authors. Licensee MDPI, Basel, Switzerland. This article is an open access article distributed under the terms and conditions of the Creative Commons Attribution (CC BY) license (https://creativecommons.org/licenses/by/4.0/).

Abstract: Landscape evolution over the last 8000 years in three areas located along Tuscany, Latium, and Campania coasts (central Tyrrhenian) has been deduced through a morphological, stratigraphical, and historical approach considering the physical evolution and human activity. Between 8000 and 6000 yr BP, the Sea Level Rise (SLR) dominated and, near the river mouths, inlets occurred. In the Tuscany area, Mt. Argentario was an island and to SE of the Ansedonia promontory a lagoon occurred. The areas were covered by a dense forest and the human influence was negligible. Between 6000 and 4000 yr BP, humans organized settlements and activities, and a general coastline progradation occurred. A tombolo linked Mt. Argentario to the mainland. In the Tiber and Campania areas, coastal lakes and a strand plain developed. Between 4000 and 3000 yr BP, near Mt. Argentario, two tombolos enclosed a wide lagoon. At the SE of the Ansedonia promontory, the lagoon split into smaller water bodies. In the Tiber and Campania areas, delta cusps developed. The anthropogenic presence was widespread and forests decreased. During the last 3000 years, anthropic forcing increased when the Etruscans and Romans changed the territory through towns, salt pans, and ports. After the Roman period, natural forcing returned to dominate until the birth of the Italian State and technological evolution.

Keywords: coastal evolution; cultural and land use changes; anthropic impacts; Holocene; Tyrrhenian Sea

1. Introduction

The physical coastal landscape is the most sensitive to changes in environmental parameters. It is strongly influenced, also in a relatively short time, by glacioeustasy, particularly in areas prone to load subsidence, tectonics, and/or volcanic activity [1–4]. The landscape changes can be rapid along the coastal plains or deltas. These areas are vulnerable to fast climatic change such as Bond Events or Rapid Climate Change (RCC) [5–10] that are characterized by high variability in frequency and amplitude of storms [11–13] and fluvial solid discharge [14–16]. In the historical period, anthropic forcing was added to the natural one. In fact, human activities have been often settled in coastal areas for the available food sources. Therefore, the landscape was modified over time by the construction of salt pans, fishponds, landings, and ports, and more recently by the intense urbanization often linked to tourist activity [13,17–21].

The central Tyrrhenian coast, between the Argentario promontory and the Garigliano River (Figure 1), shows different morphologies. Rocky promontories (Mt. Argentario, Ansedonia, Capo Linaro, Capo d'Anzio, Mt. Circeo, and Gaeta) and cliffs characterize the coast. The sandy coast is more widespread; it consists of coastal plains characterized by lakes and lagoons (i.e., Orbetello, Burano, Pontini Lakes), deltas (Tiber, Garigliano), and extensive marshy areas, now reclaimed, such as the Pontina Plain. Much of the coast developed during the last million years on the western edge of a volcanic area (the Vulsino, Cerite, Vicano, Sabatino, Albano, Roccamonfina systems), which today shows only late volcanic signs. In this period, a general uplift affected the area [22]. However, since the Late Pleistocene, the central northern part of this coastal stretch was tectonically stable except for some areas such as the northern sector of the Versilia coast that showed a slight uplift [1,23–26]. The southernmost sector of this coastal stretch (Pontina and Fondi Plains) shows significant subsidence [26], while its southern portion is influenced by the Phlegraean active volcanism.

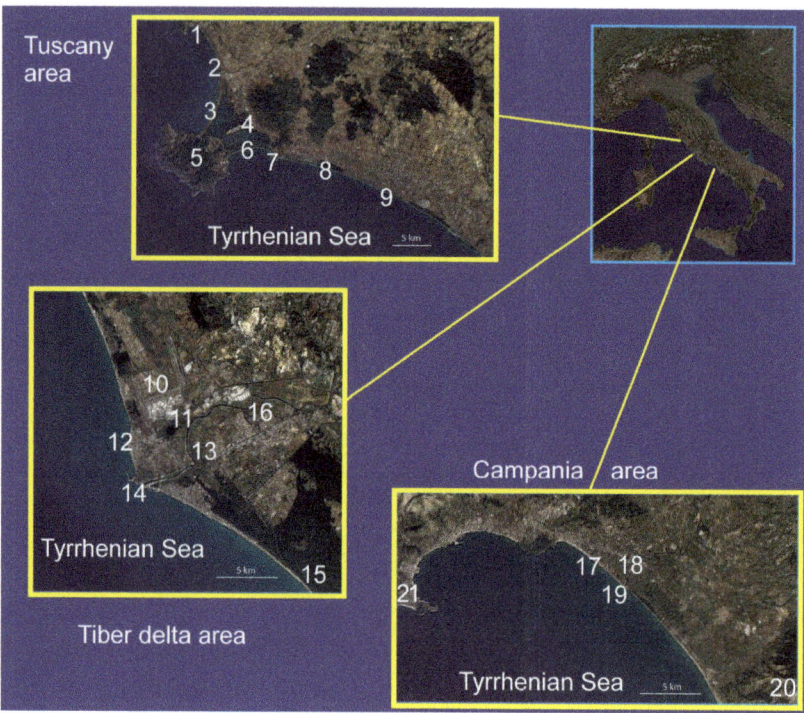

Figure 1. Location of the studied areas. *Tuscany area*: (1) Talamone; (2) Albegna River mouth; (3) Giannella Tombolo; (4) Orbetello; (5) Mt. Argentario; (6) Feniglia tombolo; (7) Ansedonia promontory; (8) Burano Lake; (9) Fosso Chiarone mouth. *Tiber delta area*: (10) Fiumicino airport; (11) *Portus* and Trajan Lake; (12) Minor Tiber River mouth; (13) *Ostia*; (14) Main Tiber River mouth; (15) *Laurentum*; (16) *Ficana*. *Campania area*: (17) Mt. d'Argento; (18) *Minturnae*; (19) Garigliano River mouth; (20) Mt. Massico; (21) Gaeta. (Image Landsat/Copernicus-Google Earth).

The current morphology is the result of severe changes since the Last Glacial Maximum (LGM). They were due to the SLR, the sub-Milankovian climatic oscillations, the amount of fluvial solid discharge and, partially, local tectonic events [1,2,9,13,14,16,21,22,27]. In addition, during the last 3000 years, anthropic impact became particularly significant. Near the LGM, the shoreline was located about 5–10 km west to its present position. Thus, promontories and some small islands were reliefs on the mainland, which were progressively isolated or submerged for the progressive SLR. At the quasi-still stand, several inlets near each river mouth characterized the coast. In the last 6000 years, the fluvial sediments, more or less abundant according to the sub-Milankovian climatic oscillations, the anthropic activity, and the coastal drift, reconnected islands (i.e., Mt. Argentario), generating lagoons (i.e., Orbetello, Maccarese, Ostia, Minturno), coastal lakes (Pontina lakes), and delta (i.e., Tiber).

During the Holocene, also the activity of humans conditioned the evolution of the landscape. In the central Tyrrhenian coast, although Neanderthal settlements were already known, i.e., Mt. Argentario [28] and Mt. Circeo [29] from the Eneolithic period [30] and mainly from the Bronze Age, the coast is dotted with more or less important settlements [31–33]. Significant coastal settlements developed with the Etruscan civilization north of the Tiber delta (i.e., *Rusel, Tarcuna, Pyrgi*) and the Latin one in the south (i.e., *Lavinium*), which exploited the lagoon areas as a harbor or salt works. The expansion of Rome progressively affected the whole coast with settlements often equipped with ports, increasing the areas devoted to salt extraction. Many coastal settlements were abandoned after the fall of Rome, and marshy areas developed, favoring malaria spreading. This situation persisted during the Renaissance when the coast became dotted with watchtowers. The coast returned to being an important economic and strategic resource only following the reclamation of the area carried out between the end of the 19th century and the beginning of the 20th.

The goal of this paper and the main novelty were to outline the evolution of the landscape over the last 8000 years in three areas (Orbetello-Burano area, Tiber Delta, Garigliano coastal plain; Figure 1) with a holistic perspective that considered, beyond the physical evolution, the presence and anthropic activity from the Bronze Age.

2. Materials and Methods

The reconstruction of the landscape is partly based on the revision of data from previous works by the authors, to which reference should be made for the lithological, palynological, faunistical, and chronological details [34–37]. Some data from the literature were also considered (recalled in the discussion of the three areas considered). A new dating (see Table 1) and a new core were added the quotas of all the cores considered were reviewed as well as the environmental interpretation of the different facies. The analysis of the historical evolution of each area described was then added. The paleoenvironmental reconstruction of the three areas was carried out by means of the following approach.

Table 1. The values of the ^{14}C ages used to plot the sea level rise curves of Figures 4, 8 and 11 are reported.

Laboratory Number	Core	Altitude	Material	Age yr BP	Age cal. yr BP
		Tuscany Area			
Lyon-14233(sacA-49741)	LB3	−5.00	Peaty clay	6335 ± 35	7329–7180
Rome-2339	BU4	−2.05	Peat	3570 ± 40	4090–3920
Rome-2342	BU7	−1.35	Peat	3550 ± 40	3900–3720
Rome-2349	BU10	−1.30	Peat	2790 ± 40	2950–2840
Rome-2360	BU10	−1.80	Peat	4320 ± 40	4970–4830
Rome-2364	BU10	−2.30	Peat	4840 ± 40	5620–5480
Rome-2361	BU10	−2.80	Peat	5310 ± 40	6270–5990
Rome-2365	BU10	−3.30	Peat	5840 ± 40	6650–6560
Rome-2350	BU10	−3.80	Peat	6115 ± 40	7160–6820
Rome-2344	BU12	−0.50	Peat	3700 ± 40	4090–3930
Rome-2345	BU12	−1.00	Peat	4560 ± 45	5320–5050

Table 1. Cont.

		Tiber delta area			
Laboratory Number	Core	Altitude	Material	Age yr BP	Age cal. yr BP
R-1198 α	150	−3.24	Peat	4710 ± 50	5575–5325
R-1198	150	−3.24	Peat	4750 ± 60	5585–5335
R-887A/α	150	−4.24	Peaty clay	4640 ± 80	5574–5090
R-888	150	−9.44	Peaty clay	7730 ± 80	8585–8420
R-889	150	−9.59	Peaty clay	7770 ± 60	8590–8455
R-890	150	−9.74	Peaty clay	7930 ± 70	8990–8605
LTL-461 a	new	−1.20	Peat	1140 ± 40	1170–975
Rome-2066	S3	−2.93	Peat	2720 ± 50	2860–2770
Rome-2067	S3	−4.03	Peat	3465 ± 55	3830–3640
Rome-2069	S5	−2.95	Peat	2555 ± 50	2760–2490
Rome-2070	S5	−4.15	Peat	3375 ± 55	3690–3480
		Campania area			
Laboratory Number	Core	Altitude	Material	Age yr BP	Age cal. yr BP
Rome-2151	P1	−2.65	Peat	2905 ± 40	3170–2990
Rome-2153	P1	−3.40	Peat	3710 ± 50	4150–3930
Rome-2154	P1	−3.95	Peat	5110 ± 55	5920–5730
Rome-2158	P2	−2.50	Peat	4355 ± 50	4980–4850
Rome-2160	P2	−4.00	Peat	5740 ± 65	6640–6450
Rome-2162	P2	−5.50	Peat	6835 ± 70	7730–7580
Rome-2164	P2	−7.05	Peat	7375 ± 70	8330–8040
Rome-2165	P3	−1.45	Peat	3250 ± 45	3560–3390
Rome-2167	P3	−4.35	Peat	6220 ± 60	7250–7020
Rome-2168	P3	−5.10	Peat	7330 ± 55	8180–8030
Rome-2257	P4	−2.00	Peat	4650 ± 50	5440–5050
Rome-2258	P4	−2.90	Peat	4710 ± 45	5580–5200
Rome-2260	P5	−1.80	Peat	4975 ± 55	5850–5610
Rome-2262	P5	−3.70	Peat	5615 ± 65	6450–6300
Rome-2261	P5	−5.20	Peat	6705 ± 65	7660–7300
Not declared	CL1	−1.50	Peat	2503 ± 40	2742–2456
Not declared	CL1	−4.30	Peat	6802 ± 45	7704–7575

Geomorphological analysis. Historical maps available for some areas since the Renaissance were considered including aerial photos starting from the Royal Air Force 1943 surveys, satellite images (Google Earth 2005–2019); LiDAR survey (only for Tiber delta), and field surveys over the last 30 years.

Stratigraphical analysis. Several drillings (by manual, rotary, and percussion mechanical system) and, locally, geophysical surveys were considered. For each sampled sediment, cores were defined as:

- Lithology. Grain-size analysis was performed for the clastic sediments (by sieving and laser diffractometry on fractions > and <62 microns, respectively). Sediments were classified according to [38]. In addition, the main sedimentary structures were considered.
- Faunistic content. Qualitative and quantitative analyses were conducted on samples where the microfaunal content (foraminifera and locally ostracoda) was recorded following the standard procedure.
- Palynology. Although the analyses were carried out by different laboratories for the different areas, they all utilized the same standard procedure for palynological processing.

- Geochronology. The ^{14}C calibrated ages were calculated using peat and wood samples and, less frequently, shell and bones. Liquid Scintillation Counting (LSC) measurement technique was normally used; but, for materials very low in organic C, Accelerator Mass Spectrometry (AMS) was also used. The conventional ages were calculated according to [39] and reported as yr BP. To take into account the reservoir effect and the past fluctuations of the tropospheric $^{14}CO_2$, the conventional ages were calibrated [40] and given as calendar yr BP time spans. The uncertainty on both conventional and calibrated ages was at the level of $\pm 1\sigma$ (68.2% of probability). Only for the Tuscany area we utilized also the Optical Stimulated Luminescence (OSL) dating of quartz extracts followed standard preparation techniques. Table 1 shows the dates of only the lagoon peaty level used for tracing the SRL curves inserted in Figures 4, 8, and 11.

The historical frame was essentially based on historical sources and on the analysis of data deriving from surveys and archaeological excavations. By comparing the formal and quantitative characteristics of the main productions and settlement traces and correlating the available data, the anthropization patterns resulting from the relationship between the communities and the ecosystem over the centuries under consideration was reconstructed. Table 2 shows the chronology of the historical ages from the Eneolithic to the Renaissance and the life span of the settlements mentioned in the text.

Table 2. Chronology of the protohistoric/historical ages and life span of the main settlements mentioned in the text. The Greek Gothic War took place between 535 and 553 AD. The Etruscan civilization started in the ninth century BC.

Age	Phases	Chronology	Tuscany Area	Tiber Delta Area		Campania Area
Eneolithic		5000–3700 BP				
Bronze	Ancient	3700–3500 BP	Grottino di Ansedonia / Poggio Terrarossa	Le Cerquete-Fianello		
	Middle	3500–3100 BP			small settlements on dune ridges	
	Recent—Late	3100–3000 BP	Punta degli Stretti	Le Vignole	small settlements in humid costal habitat	
Iron		2900–2700 BP		Ficana		
Archaic		7th–6th BC	Orbetello	Area rustica Rio Galeria	Monte d'Argento	Pre Roman Minturnae
Roman Republic		6th–1st BC				Temple of Marica
Roman Empire	Early	1st BC–1st AD	Cosa	Portus		Roman Minturnae
	Middle	2nd–3rd AD				
	Late	4th–5th AD				
Middle Ages	High	6th–10th AD		Gregoriopoli-Ostia Antica		Mons Garelianus
	Low	11th–15th AD				Turris Garelianus-Turris ad mare
Renaissance		16th–17th AD				Catrum Argenti–Traetto

3. Local Results and Discussion

3.1. Tuscany Area (Orbetello-Burano)

Two lagoon systems, (Orbetello Lagoon and Burano paleolagoon separated by Ansedonia promontory) characterize the Tuscan coast between the mouths of the Albegna River and the Fosso Chiarone (Figure 2).

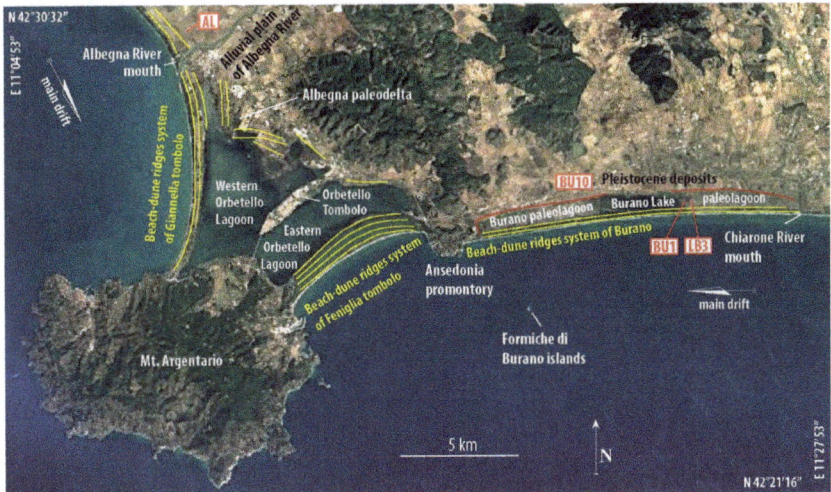

Figure 2. Morphological scheme of the Tuscany area. The yellow lines indicate the trend of the Holocene beach/dune ridges. The red line indicates the outline of the Burano paleolagoon. The labels in the white rectangles indicate the location of the boreholes, shown in Figure 3 (Image Landsat/Copernicus-Google Earth).

3.1.1. Morphological Setting

The first system (Orbetello Lagoon) is a submerged area, about 27 km^2 wide with a maximum depth of 2 meters, enclosed between two sandy tombolos that connect Mt. Argentario with the coastal plain between the Albegna river mouth and the Ansedonia promontory.

The northern tombolo (Giannella) is a *spit* 6 km long with a width ranging from about 800 m near the Albegna river mouth and about 300 m near the Argentario where it is interrupted by the only natural inlet of the lagoon (Nassa channel). It is marked by a series of locally discontinuous beach/dune ridges and the inner edge shows washover fans. The southern tombolo (Feniglia), 6 km long, is more regular, has a width close to 1 km, a more organized system of beach/dune ridges dissected by numerous blowouts, and it is rooted landwards at the Ansedonia promontory. Its inner border shows several irregularities related to washover events, which occurred in an initial phase of its formation. Two artificial canals, placed near the roots landwards of the two tombolos, connect the lagoon with the sea and the Albegna River. A third, uncomplete tombolo (Orbetello tombolo), artificially connected to Mt. Argentario, separates the lagoon into two hydraulically connected parts. Several outcrops of rocks and sediments attributable to fluvial, marshy, and paralic *latu sensu* facies characterize the coastal plain. The oldest deposits belong to the upper Pleistocene and are part of the Orbetello Syntema [41]. The Albegna River migrated during the Holocene, leaving meandering paleochannels and alluvial deposits. On the inner edge of the lagoon, a series of beach ridges between the Giannella and Orbetello tombolos mark two little paleodelta cusps connected to the aforementioned paleochannels [41]. Between the Orbetello and Feniglia tombolos only a limited development of the beach ridges occurs. Near the Orbetello tombolo and at the edge of the western lagoon there are limited marshy areas. In the western lagoon, some emerged flat areas and limited sandbanks occur.

The second system consists of a 1-km-wide depression, parallel to the coast with a WNW–ESE trend, limited landward by Pleistocene sandstone deposits and seaward by two parallel Holocene dune ridges. Before the 19th century reclamation, the depression was a wetland periodically submerged into which the Fosso Melone and the Fosso del Chiarone flowed. Currently, the depression is almost flat and at some decimeters below sea level. The depression includes the Burano Lake, the only submerged area after the wetland reclamation, 3.5 km long, 0.5 km wide, with a maximum depth of about 1 m. An inlet, located at the center of the outer side, connects the lake to the sea. Washover fans are particularly evident in the northwestern area of the lake.

3.1.2. Vegetation Frame

Between 8000 and 4000 yr BP, the coastal plain landscape was dominated by thermophilous deciduous forest composed of *Quercus cerris*, *Quercus pubescens*, *Quercus suber*, *Carpinus orientalis/Ostrya*, and *Corylus*. *Quercus ilex* is abundant in the surrounding hills followed by Ericaceae, *Olea*, *Phillyrea*, and *Pistacia*. Plants growing under greater control of local edaphic conditions include prevalently herbaceous taxa such as Amaranthaceae and Cyperaceae particularly developed in the perilacustrine areas. A strong increase in herbaceous taxa occurred after 4000 yr BP. Pollen analyses carried out in some lake basins in Tuscany and northern Latium indicated that the variations in vegetation cover in the Neolithic are related more to climatic fluctuations than to anthropic activity [33,42,43]. Only starting from the Bronze Age, the anthropogenic effect on the vegetation appears significant; in the Roman period, the arboreal cover was characterized, similarly to today, by the Mediterranean scrub mixed with oak wood [44].

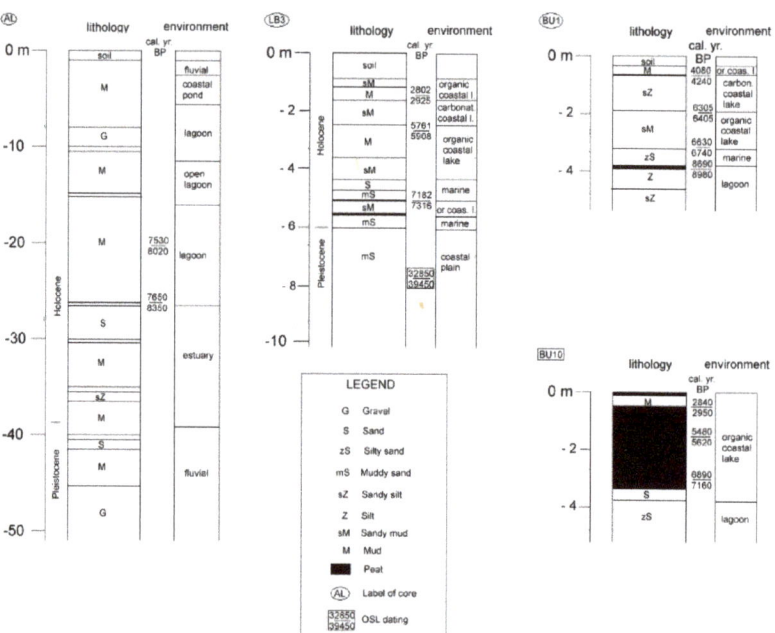

Figure 3. Lithological/environmental scheme of boreholes reported in Figure 2 regarding the Tuscany area. AL and LB3 are rotary boreholes; BU1 and BU10 are percussion boreholes. AL is modified after [44], while LB3, BU1, and BU10 are modified after [36].

3.1.3. Facies

Six different facies are recognized in this area (Figure 3).

Facies 1. In the subsoil of the Albegna coastal plain, this facies is present between −17 and −10 m. It consists of silty clay and silt with sandy, level intercalations and local thin peat lens. It contains brackish and marine mollusks (*Acanthocardia* spp. and Rissoidae) and meso-polyhaline ostracods (*Cyprideis torosa, Loxochonca elliptica, Cyprideis fischeri, Leptocythere lagunae,* and *Xestoleberis communis*). In the Burano area, the facies occurs at the bottom of the Holocene succession.

Interpretation: Towards the top, the fauna indicates a decrease in salinity (*Cerastoderma lamarcki*), and the presence of sand and fine gravel levels testifies to a greater contribution of fluvial sediment [45]. The facies is attributed to a lagoon/estuary environment.

Facies 2. This facies is well developed in the area of the Burano paleolagoon, which occurs everywhere between the Pleistocene sandstone deposits and the Holocene dune ridges. The facies consists of soft peats, sandy-silty mud, mud with abundant organic matter, and some thin, locally bioturbated silty levels with shell debris. The remains of Posidonia and undecomposed vegetal matter, occasional bivalve fragments, and thin-shell gastropods are present. Ostracods are mainly composed of both brackish (*C. torosa* and *L. elliptica*) and lagoon/coastal (*Xestoleberis dispar*) taxa with variable proportions. They are occasionally associated with a few phytal coastal (*Aurila* spp. and *Leptocythere* spp.) and/or freshwater to low brackish specimens (*Candona angulata* and *Darwinula stevensoni*). A brackish, lagoonal assemblage containing an oligospecific foraminiferal fauna is characterized by the high dominance of *Ammonia tepida* followed by *Haynesina germanica* and *Porosononion granosum* mainly. Locally, in the lower part, more marine taxa (*Nonion* spp., *Triloculina* spp., and *Quinqueloculina* spp.) are present. On the contrary, towards the top, a decrease in foraminiferal content associated with abundant gastropods such as *Hydrobia* spp. and *Planorbis* spp. is recorded. In the lower part, marine dinocysts are also commonly present (including *Spiniferites mirabilis* and *Lingulodinium machaerophorum*), which tend to decrease upward until they are completely absent at the top. Moreover, in the lower part, *Botryococcus* (Chlorophyceae, Chlorococcales/Tetrasporales), a fresh–brackish water colonial green algae, and Terrestrial Fungi are quite abundant. Fungi spores, *Pseudoschizaea, Botryococcus,* and especially *Cosmarium* algae, decrease upward but, unlike the dinocysts, they are still present at the top.

Interpretation: The facies is attributable to an organic coastal lake with a transition from brackish to freshwater conditions.

Facies 3. This facies was intercepted only in the upper part of the cores drilled in the central sector of the Burano paleolagoon. It shows a maximum thickness of 2 m. Whitish $CaCO_3$ (with subordinate gypsum)-enriched silt and scarce fine sand characterize the sediment of this facies only locally interbedded with thin, blackish, peaty levels. Freshwater gastropods (*Hydrobia* spp. and *Planorbis* spp.) and freshwater/low brackish ostracods (*C. torosa, C. angulata, D. stevensoni, Limnocythere inopinata*) are the components of the bioclastic fraction.

Interpretation: The data indicate a carbonatic coastal lake environment where the main components of the NPPs' assemblages are *Cosmarium*, indicative of prevalent freshwater conditions, and *Botryococcus*, which thrives in fresh–brackish waters. Evidence permits us to define a freshwater depositional basin with clear and oligotrophic waters in which peculiar chemical–physical conditions induced an abundant precipitation of $CaCO_3$. The development of this facies started after 6000 yr BP and ended at about 4000 yr BP.

Facies 4. The facies is present between the Albegna River and the Orbetello tombolo both in the subsoil at 3–5-m depth and in the current coastal ponds such as Lo Stagnino. It consists of clayey silt locally with organic matter and a scarce bioclastic fraction. The latter is composed of freshwater and oligohaline ostracofauna (*Cyprideis neglecta* and *Ilyocypris bradyi* with a subordinate presence of *C. torosa* and *Heterocypris salina*) and freshwater mollusks (*Valvata piscinalis, Gyraulus laevis,* and *Bithynia leachi*) [45].

Interpretation: The facies is attributable to coastal pond.

Facies 5. Medium-fine sand with marine taxa (*Cardium* spp., *Glycimeris* spp., *Venus* spp., tellinids, *Quinqueloculina* spp., *Triloculina* spp., *Elphidium* spp., and *Rosalina* spp.)

characterizes this facies. In the northern sector, it forms the Giannella and Feniglia tombolos, whose mineralogical compositions differ in the greater presence of amphiboles in the Giannella site [46].

Interpretation: The facies is attributable to a strand plain and constitutes the beach ridges of some small delta cusps, no longer active, and of the incomplete Orbetello tombolo [41]. Between Ansedonia and the Fosso Chiarone, the facies represents the beach/dune ridges that isolate the Burano depression from the sea.

Facies 6. It is essentially present behind the strand plain between the Albegna River and Orbetello. It shows thicknesses of about 3 m near the river that taper towards the tombolo. It consists of silty clay with an interbedded sandy silt level and lens. Locally calcareous concretions and oxidation tracks are present. Freshwater mollusks (as *Bithynia leachi* and *Lymnaea truncatula*) are rarely recorded.

Interpretation: The described characters permit us to define a fluvial facies.

3.1.4. Diachronic Physical Landscape Change

The physical and cultural landscape change is schematized in Figures 4 and 5. Between 7500 and 5000 yr BP, in the northern sector, the SLR interrupted the connection between Mt. Argentario and the mainland [47], leaving only a narrow and incomplete isthmus. The sea, penetrating the Albegna River paleovalley, gave rise to a bay. Around 6000 yr BP, the decrease in the SLR rate allowed for a greater fluvial sedimentary supply to the coast. The bay filled up and the Albegna River began to wander into the coastal plain that was forming. At the end of the period, the river formed a wave-dominated delta cusp north of the incomplete isthmus. To the south of the Ansedonia promontory, a Holocene beach developed in the first phase, separating the Pleistocene coastal deposits from the sea. Further southeast, a narrow lagoon bordered by a thin and discontinuous sandy barrier occurred. The longshore current coming from the southeast carrying the Fiora, Arrone, and Marta rivers' sediments fed the beach and the barrier. As the SLR progressed, the lagoon extended until Ansedonia, the sandy barrier became less thin and discontinuous, and an organic coastal lake sedimentation occurred.

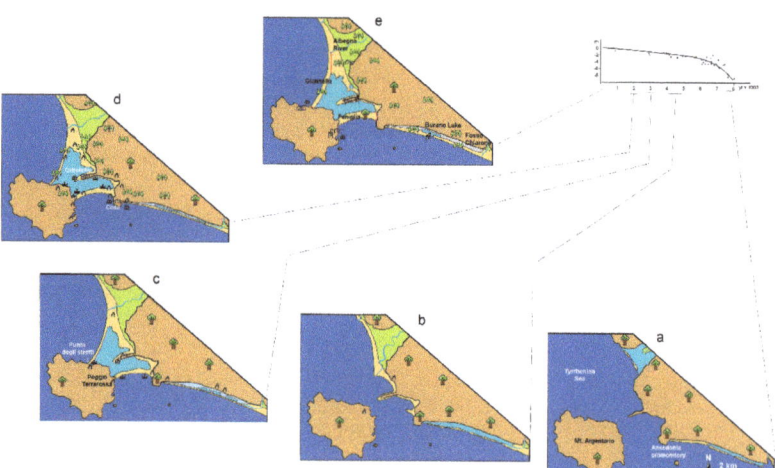

Figure 4. (a–e). Two-dimensional evolutionary patterns of the Tuscany area inferred from underground and historical/archaeological data. The single images illustrate the coastal landscape relating to the period shown on the abscissae of the sea level rise curve. The curve was plotted based on ^{14}C dating from Burano paleolagoon peat levels.

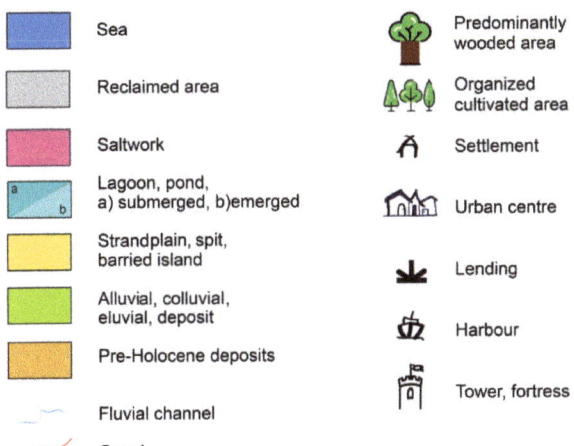

Figure 5. Legend for Figures 4, 8, and 11.

Between 5000 and 3500 yr BP, after the quasi-still stand, the sea level rose with minor rates. In the northern sector, the development of the cusp continued and the Feniglia developed completely while the Orbetello tombolo remained incomplete. Southeast of the Ansedonia, the lagoon reached its maximum width (500–700 m) and, along most of its central axis, the sedimentation of a carbonate coastal lake occurred. About 4000 yr BP, the SRL did not change the planimetric features of the lagoon; however, the carbonate sedimentation showed a more discontinuous distribution and progressively disappeared.

Between 3500 and 1000 yr BP, the Feniglia connected Mt. Argentario with the coastal plain, preventing further development of the Orbetello tombolo. The Albegna River migrated northwards on the trace of the current course. It is not clear whether this event was natural or favored by human work [41,48]. The new position of the mouth favored the development of the Giannella that reached Mt. Argentario, giving the Orbetello lagoon its current morphological setting.

In the southern sector, a large part of the lagoon rapidly dried up, or, at least, turned into a wetland locally and periodically submerged so that part of the organic sediments was subject to pedogenesis, forming a thin brown soil. The area of the Burano Lake and a small area close to the Ansedonia promontory remained submerged.

Between 1000 yr BP and the present, south to Ansedonia, there were no substantial changes until the hydraulic reclamation (first part of the 20th century), which definitively dried up the entire area except for the Burano Lake whose depth is currently not over 1 m.

3.1.5. Diachronic Cultural Landscape Change

The morphological and ecosystemic evolutions of the coast between Talamone and Burano Lake marked it up to the entire pre-protohistoric transition. The landscape, today characterized by the presence of the Giannella and Feniglia tombolos, displays a concentration of functional characteristics for anthropization partially comparable to those of the Tiber and Garigliano deltas.

On the basis of the archaeological research, the population appears to be very fluctuating at least up to the end of Ancient Bronze Age, both due to the demographic and environmental evolutions. Some funerary sites, such as the Grottino di Ansedonia [49–51], and some findings of commonly used ceramics, such as Poggio Terrarossa [52,53], are recorded. On the base of limited documentation, the presence of small settlements, scattered around the lagoon, is hypothesized. These settlements were mainly linked to a subsistence based on the consumption of wild plants and shellfish. Therefore, up to this stage, a low impact on the ecosystem was outlined, due to the presence of small and sparse communities with a predominantly subsistence economy. However, the proliferation and the affirmation of the small perilagunar sites, attributed to the first phase of the Bronze Age, testify to a demographic increase. These sites were probably small, seasonal settlements linked to transhumance [54] and pastoral activities that periodically offered adequate pastures and water.

As in other Italian areas, the Middle Bronze Age for the perilagunar settlements increased and the discovery of new founded settlements represented an important demographic and cultural crossing point. Both the Albegna and Fiora rivers show new, long-lasting settlements located on heights and river terraces, testifying to the first evidence of a territorial system. These will be generalized to characterize the Late Bronze and Iron ages. In this period, the islands of the Tuscan archipelago were also perceived as a territorial segment coherent and functional to the coastal one, beginning maritime activity and routes to Corsica. The exploitation of the islands persisted at least until the first Roman era.

Except for the phase of the Recent Bronze, when a decrease in housing and cultic attestations is evidenced, during the first half of the third millennium BP transition between the Late Bronze and the Iron ages, human presence in the lagoon area was strengthened. A prime example was the settlement of Punta degli Stretti at the junction between the Argentario and Giannella tombolos [51,55–58]. This returned a large amount of evidence of a vital and full-bodied settlement. In this context of cultural change on the Tuscan coast, the first cremation graves characterizing the Villanovan were recorded. The exploitation of the lagoon's food resources and salt and the extraction of metals from the Argentario represented signs of a territorial restructuring persisting until the beginning of the Etruscan era. The Iron Age marked a generalized demographic, cultural, and political change highlighted by the diffusion in most of Italy of Villanovan evidence and a territorial reorganization, which in Etruria was related to the development of *Velch* (Vulci), *Kaisra* (Cerveteri), *Tarcuna* (Tarquinia), and *Vatluna* (Vetulonia) cities. Although the Orbetello coast was also affected by a change in settlements and in the productive–cultural context, it did not suffer a noticeable decline in population. Although the settlement of Punta degli Stretti was ending, the area in question, mainly in the Feniglia tombolo sector [51–53,59–62] of Burano [51], was characterized by various settlements always focused on the exploitation of fish resources, salt, landings, and metals.

From the mid-eighth century BC until the sixth century BC, *Velch*, the hegemonic city of the area, controlled and exploited the territory in a capillary and differentiated way. Moreover, even if some important centers arose, a first network of solitary settlements spread, probably the first agricultural network based on farms.

Among the centers, such as Saturnia or Marsiliana (both along the Albegna River), Orbetello [63] stood out for the exploitation of the natural resources typical of the Argentario area. Orbetello, a fortified town, maintained its importance even beyond the crisis that Etruria went through in the fifth century BC and appeared to be an important reference also after the definitive conquest of the volcanic territory by Rome at the beginning of the third century BC. In the new territorial organization, Rome established *Cosa*, a maritime colony, located near the current Ansedonia, whose *ager* interacted with the Orbetello structures. This condition ended with the Roman conquest of Sardinia that changed the maritime horizons of Rome. *Cosa* played an essential role since the second half of the third century BC: The building of two ports (hydraulically connected by a canal carved into the promontory limestone), south of the Ansedonia promontory and north (*Portus Fenilie*) between the Feniglia and the promontory itself, played a significant role for its logistic and mercantile development. Moreover, its proximity to the Via *Aurelia* [64] and all the local branches made it a reference point for land transport. Alongside the maritime vocation, the agricultural one was increasingly growing. Following the centuriations, a profitable network of *villae rusticae* developed throughout the Republican age, which could exploit both the good coastal soils and the dense and advantageous viability in the mercantile environment. At least until the passage between the Republic and the Empire, the integration between the port structures was widespread also throughout the Feniglia, dedicated to the export of local wine and oil, and the agricultural area, increasingly organized in rich and well-structured funds. Consistent with this economic framework, it is noted that the population, even up to the beginning of the second century, was distributed in small *vici*, even if *Cosa* maintained an administrative role. From demographic and social points of view, it did not have significant growth [65]. Especially after the Trajan age, the scenario changed significantly, and the Empire incorporated wide and varied territories and markets. The amalgamation of various funds led to *latifundia* formation, characterized by high but scarce productivity. At this time, the activities of the *Portus Fenilie* continued, while the patrician villas destined for *otium* populated the coastal strip. *Cosa* definitely lost its attractiveness, as evidenced by some building collapses [66].

During the Middle Empire, therefore, there was a significant change in the landscape due to new economic choices and more favorable social conditions for a small number of patrician families. In addition, at the same time, the arrival of malaria, as reported by Pliny (Ep. V, 6.2), *"gravis et pestifer ora Tuscorum quae per litus extenditur"*, affected the coastal strip. The new agricultural structure with grain estates or pastures was preferred to the previous one, also due to less organization difficulty and less capillary control. All this led to an inefficient management of the waters, which began to invade some areas. The first signs of swamping appeared in this period.

The generalized crisis of the third century AD strengthened the negative economic trend and the demographic decline of the area that was less and less under control. Between the Caracalla and Aureliano periods, it was established *Res Publica Cosanorum* [67], which probably managed the food supply by means of the army.

The fourth and fifth centuries AD marked a definitive decline of the area. The first was characterized by a sharp contraction of the productive and port structures, around which the remaining population gathered, followed by a substantial abandonment. Between the fall of the Western Roman Empire and the Greek-Gothic war, there is no longer any evidence of an institutionalized structure. In fact, only due to the war context of the sixth century AD, the evidence shows an exploitation of some ancient centers for military purposes. The coast at the beginning of the Middle Ages was a sparsely inhabited area, wild and malarial, controlled by the Roman Church, and divided among the main families of princely and bishopric rank. Between the late Middle Ages and the Renaissance, the coast from Talamone to Burano Lake was directly controlled by the Kingdom of Spain, which built a series of fortifications against Ottoman attacks. Subsequently, the war interest for the area decayed, and the zone remained substantially underutilized until the unification of Italy.

3.2. Tiber Delta Area

A NE–SW fault system [23,68] characterized the Latium Tyrrhenian margin. This structural configuration determined a lowered and articulated area between Palo and Lavinio, where the Pliocene deposits outcrop. In this area, 100 meters more of Pleistocene sediments and the Holocene Tiber delta were deposited [69]. The delta is the result of a complex evolution starting after the LGM. The reconstruction of the delta evolution was mainly possible by several cores, located along the emerged sector of the delta, and geophysical surveys carried out mainly in its submerged portion [34,69–76]. The Tiber delta shoreline stretches for about 35 km, but the Tiber River sediments affect completely the central coast of Latium.

3.2.1. Morphological Setting

The Tiber delta is a cuspate, wave-dominated delta with a strand plain (outer delta plain) and extended land–sea up to 4 km, characterized by beach/dune ridges locally 5 m high (Figure 6). In the strand plain at Capo due Rami, the Tiber River splits into two distributaries. The subordinate one (Fiumicino channel), flowing almost perpendicular to the coast, is the evolution of a Roman canal dredged in the 1st–2nd centuries AD. The main distributary (Fiumara Grande) at first is parallel to the coast over about 1.5 km, then turns to the WSW with a sharp, elbow-shaped curve and reaches the sea. Until September 1557, Fiumara Grande formed a wide meander directly landward before reverting seaward. Landward, a flat area (inner delta plain), crossed by the Tiber River channel, develops. Here, on the sides of the Tiber there are two ponds now reclaimed.

Figure 6. Morphological scheme of the Tiber delta area. The yellow lines indicate the trend of the Holocene beach/dune ridges. The red line indicates the outline of the marshes/ponds reclaimed between the 19th and 20th centuries. The labels in the white rectangles indicate the location of the boreholes, shown in Figure 7 (Image Landsat/Copernicus–Google Earth).

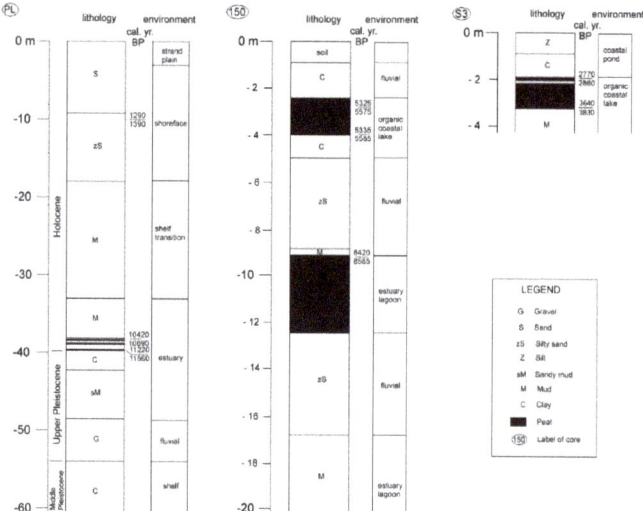

Figure 7. Lithological/environmental scheme of some boreholes executed in the Tiber delta area. PL and 150 are rotating boreholes; S3 is percussion boreholes. PL is modified after [68]; S3 is modified after [34].

3.2.2. Vegetation Frame

Between about 8300 and 5400 yr BP, the regional vegetation was characterized by dense forests dominated by oaks with a scarce presence of *Olea*, *Phillyrea*, *Pistacia*, *Pinus*, and *Juniperus*, testifying to a limited development of beach/dune ridges. Successively in the northern part, a wide diffusion of riparian trees (in particular, *Alnus* and *Salix*) and a significant presence of *Quercus* evergreen, *Quercus* deciduous, and *Fraxinus* occur up to 2900 yr BP [77]. In the southern part [35], up to 2600 yr BP, a mixed oak-dominated woodland with evergreen elements characterizes the regional context. Around 2400 yr BP, a significant amount of cultivated and anthropochore plants is recorded, including a clear increase in *Olea*, *Vitis*, and Cannabaceae, along with a progressive appearance of cereals, *Mercurialis*, *Juglans*, and *Castanea*.

3.2.3. Facies

Five facies characterize the architecture of the sediments deposited in the last 8000 years (Figure 7).

Facies 1. Two subfacies constitute facies 1. The first one occurs in the inner delta plain between 5 and 15 m below the ground surface. It constitutes a ribbon-shaped body elongated from land to sea up to the limit with the strand plain. Medium-fine sands, barren in fauna and muddy intercalations with plant remains, mainly characterize it. This lithofacies is embedded in estuary/lagoon deposits (see below). The second unit crops out throughout the inner delta plain, mainly near the modern river channel, with thicknesses between 1 and 5 m. It constitutes a sheet layer of gray-green to brown mud with scarce freshwater fauna, diatomites, and altered volcanic material. In the lower part, there are peaty intercalations with thin shell levels.

Interpretation: The first subfacies is attributed to a bayhead delta body, while the second one is attributed to alluvial plain deposits.

Facies 2. It is present in the subsoil, both in the inner delta plain, below the river sediments, and under the strand plain sediments. It consists of gray-blue mud with thin layers of fine sand often with bioclastic debris or with *Cerastoderma* and *Ostrea* valves interbedded. Plant remains as well as more or less wide peaty lenses are present at different depths. The herbaceous vegetation was mostly composed of sedges and herbs with the presence of *Myriophyllum spicatum*. After 5400 yr BP, Poaceae dominated followed by Cyperaceae and other herbaceous taxa. A peak of micro-charcoal and the presence of cereal-type pollen characterized the northern zone.

Interpretation: The fauna indicates an environment with variable salinity both in time and in space. However, a general trend towards lower salinity is evident both landwards and upwards. The presence of *Myriophyllum spicatum* suggests a large development of areas with fresh waters certainly related to the influence of the Tiber River. After 5400 yr BP, taxa suggest a lowering of the water table probably induced by changes in the sedimentation dynamics of coastal areas [77]. The pollen record from the northern zone provides clear evidence of human impact. These sediments are attributed to deposition in an estuary/lagoon that developed from the beginning of the Holocene up to 5000–6000 yr BP when a further and progressive decrease in marine influence caused a large coastal lake.

Facies 3. In the inner delta plain, above the estuary–lagoon sediments, lies an almost continuous layer of peat, rich in plant remains and barren in fauna. The peat layer has a thickness ranging from 0.5 to over 4 m with the top lying between 2 and 5 m below the ground level and largely settled between 5000 and 2600 yr BP. Pollen data in the northern part [77] indicate, between 5000 and 2900 yr BP, the scarce presence of herbaceous taxa. An important drop of Arboreal Pollen (AP) percentage values, a progressive expansion of Cyperaceae, and an increase in aquatics, mainly *Sparganium/Typha*, *Typha latifolia*, *Alisma*, and Nymphaeaceae, were recorded after 2900 yr BP. A slight increase in Chenopodiaceae around 2600 yr BP occurred. Few are anthropogenic markers indicating cultivation or pastures. In the southern part [35], up to 2600 yr BP, sedge vegetation characterized the environment.

Interpretation: This facies is attributable to an organic coastal lake where the presence of Nymphaeaceae, Lythraceae, *Callitriche*, and *Myriophyllum* highlight a freshwater lake that locally, around 3100 yr BP, showed drying phases (peaks of Asteroideae, Apiaceae, and ferns). Around 2600 yr BP, the definitive disappearance of freshwater hydrophytes, an increase in Chenopodiaceae, and the appearance of *Ruppia* indicate a salinity increase. Anthropogenic markers are negligible.

Facies 4. This facies, 1–3 m thick, developed in the most depressed areas around 2800–2600 yr BP and lies locally on the coastal lake deposits. It consists of sandy, gray to brown, or locally organic muds [78], containing small valves of *Cerastoderma glaucum*, *Hydrobia ventrosa*, and *Abra segmentum* and some foraminifera (e.g., *Ammonia tepida*, *Ammonia parkinsoniana, Elphidium crispum*). Locally, there are freshwater mollusks (*Armiger crista, Valva cristata, Lymnea peregra*). Among the pollens, the Chenopodiaceae are abundant and *Ruppia* appears.

Interpretation: The sediments, which were deposited up to the end of the 19th century, are attributed to a sea-connected marsh/coastal pond locally and sporadically were subjected to freshwater inputs.

Facies 5. Present in the outer delta plain, it consists of well-sorted, fine to medium sand, rich in femic minerals. It is about 20 m thick along the shoreline and closes wedge-shaped towards the inner delta plain. In the central area of the delta, it extends for about 5 km in a land–sea direction, decreasing a few hundred meters towards the delta wings. Towards the sea, it lies on neritic sediments and, landwards, on the lagoon/estuary sediments, above described. Shallow, benthic foraminifera (*Ammonia tepida, Elphidium crispum, Lobatula, Ammonia beccarii*) associated with rare reworked planktonic taxa (e.g., *Globigerina falconensis, Globigerina pachyderma, Globigerinoides ruber*) represent the faunal content. Some gastropods, such as *Theba pisana* and *Colchicella barbara*, appear at the top.

Interpretation: This facies, developing mostly in the last 6000 years, highlights a strand plain with sandbars, beach ridges, and coastal dunes.

3.2.4. Diachronic Physical Landscape Change

The different physical and cultural landscapes that have occurred over the last 8000 years (Figure 8) can be outlined as follows.

Between 8000 and 7000 yr BP, during the last phase of the postglacial transgression, a wide estuary/lagoon, partially closed seawards by sandy bars, in which the Tiber, slowly prograding, built a bayhead delta, characterized the area. About 8000 years ago, the sea–lagoon limit ran to the east of Ostia Antica, near Capo due Rami, continuing northwards about halfway through the current Fiumicino airport. The described landscape was maintained at least up to 7000 years ago when the SLR rate rapidly dropped, allowing a rapid Tiber River mouth progradation towards the sandbars.

Between 7000 and 5500 yr BP, a significant landscape change took place. The bayhead delta reached the sandbars, and the waves began to rework the river sediment, making the sandbars more continuous. The lagoon progressively was isolated from the sea, giving rise to two lakes separated by the river course.

Between 5500 and 2900 yr BP, in the first 400 years of this interval, the lake's level decreased. This event could be correlated to a change in the sedimentary dynamics, after the closure of the old lagoon, with consequent overflow of the fluvial sediment in the lakes. In this way, organic sedimentation progressively increased in the lakes, until the end of this interval. An evaporitic level, found in Le Cerquete-Fianello and datable to around 4000 yr BP [79], could be correlated with the 4.2 dry event, which would have temporarily reduced the width of these lakes. The Tiber mouth was probably just north of the current position [35,80,81].

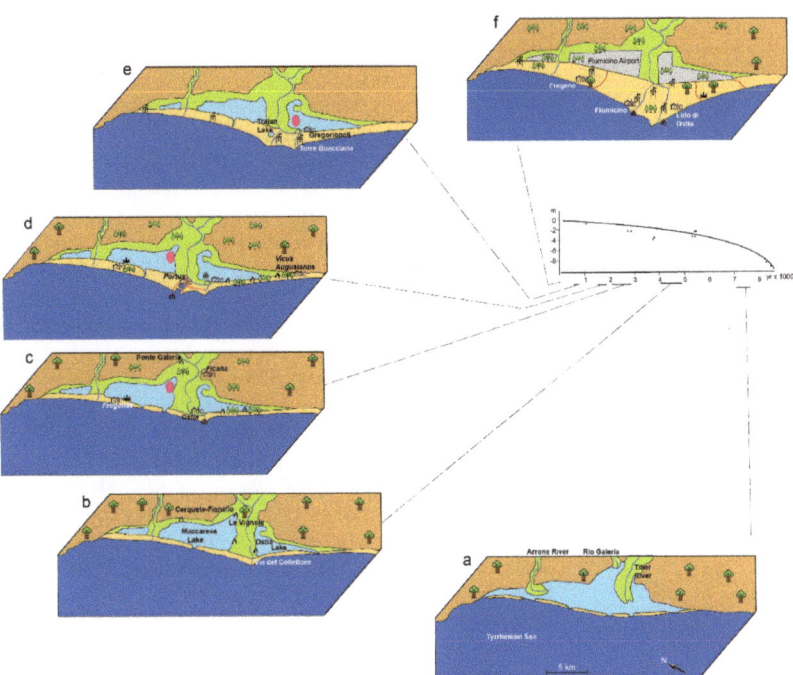

Figure 8. (a–f). Two-dimensional evolutionary patterns of the Tiber delta area inferred from underground and historical/archaeological data. The individual images illustrate the coastal landscape relating to the period shown on the abscissa of the sea level rise curve. The Tiber delta was an area locally subject to repeated changes (excavation of canals and ports and abandonment of meanders) so as to make it difficult to trace a single sea level rise curve [81]. It was, therefore, decided to draw the curve based on only ^{14}C datings coming from the peat levels of Maccarese Lake. However, those of the Ostia Lake, which are positioned below the traced curve, are also shown in red on the graph.

Between 2900 and 2600 yr BP, a significant change occurred in this period. It is still possible today to observe the river coming from the alluvial valley, instead of heading seawards, and bending landwards, in an anomalous way, before pointing towards the Tyrrhenian Sea with a last elbow. Following this, a meander lapping the southern lake (henceforth Ostia Lake) developed from the Roman period until the 16th century. It is assumed that around 2900–2800 yr BP a break of the left bank, due to a flood or the weak but active local tectonics (the red data of the SLR curves in Figure 8 could indicate a lowering of the left wing of the delta) induced the avulsion southwards of the river. The event caused a local destruction of the beach ridges that isolated the lake from the sea, so that the waters became brackish and silts replaced peat. The abandonment of the previous mouth had to cause a partial erosion of the beach ridges that isolated the northern lake (henceforth Maccarese Lake), which also became brackish with clastic sedimentation.

Between 2600 and 1900 yr BP, in this period the human presence became progressively significant. Starting from 2400 yr BP, the increase in cultivated and anthropochore plants testifies to a progressive increase in the human presence. The new Tiber mouth triggered the progradation of a delta cusp in the position of the current branch of Fiumara Grande. The cusp development was very rapid in the initial phase and, in the fourth century BC, on the left riverbank, between the meander and the mouth, as well there occurred the development of the first Roman settlement in the area (the *Ostia* town). The town developed among the sea, the river, and the Ostia Lake, certainly connected with the sea by means of an inlet located about 3 km south of the Tiber channel. The growth of the cusp was almost continuous, albeit at different rates, due both to climatic periods and the Roman river basin management [74].

Between 1900 and 1400 yr BP, the territorial setting changed during the Imperial Period mainly due to anthropic activity. In the third century AD, a moderate erosive phase was triggered especially south of the Tiber mouth. This seems to have been caused both by the end of the Roman Warm Period (first BC—second AD centuries) and the cutting of a canal (*Fossa Traiana*), which opened a second and smaller river mouth, moving part of the river sediment further toward the north. However, the erosion did not affect the distal part of the southern wing, where, near *Laurentum*, a limited progradation occurred [82] during the shoreline rectification phases. In addition to the *Fossa Traiana*, several canals were dug to serve a complex port system (Claudius and Trajan harbor, 1st–2nd centuries AD, respectively), which interacted with the longshore current modifying the shoreline. In the first century AD, the evolution of the meander interrupted near *Ostia* the Ostiense road [83].

Between 1400 and 600 yr BP, this period largely coincides with the Middle Ages during which two different climatic periods followed. The cold–humid period (5th–8th centuries; Dark Age Cold Period) with the progressive depopulation occurring after the fall of Rome produced a little progradation and an expansion of the marshy areas in the delta. Subsequently, during the Medieval Warm Period (10th–12th centuries), the shoreline was not subject to significant changes and periodically the minor mouth was obstructed, making the *Fossa Traiana* impracticable. Significant changes also affected the imperial ports. The docks of the Claudius port were partially destroyed, and the basin was partially infilled. Even the hexagonal Trajan basin, no longer connected to the sea, became unusable. The Maccarese and Ostia lakes changed into coastal ponds.

Between 600 yr BP and the present, a new, marked environmental change occurred in coincidence with the cold–humid phase (Little Ice Age), developed above all in the 15th–17th centuries, and ended in the mid-19th century. The major historical floods of the Tiber River were concentrated between 1495 and 1606 AD. The increased energy of the river made both mouths constantly active and conveyed a lot of sediment to the mouths, causing a significant and rapid delta progradation, which assumed a particularly cuspid shape. Several dune ridges developed, facilitating the change of the two coastal lakes into ponds (known as Stagno di Maccarese and Stagno di Ostia). The flood of 1557 AD cut the Ostia meander, which, after 1562 AD, was isolated from the river, assuming the toponym of Fiume Morto. The Claudius port was completely infilled, and the Trajan port became a hexagonal lake. The whole area became malarial until the second half of the 19th century when the reclamation of the two ponds began. The emerged delta reached the maximum expansion at the beginning of the 20th century. In the second half of the century, an erosive phase, largely determined by the construction of hydroelectric basins in the Tiber River catchment, started.

3.2.5. Diachronic Cultural Landscape Change

The anthropization of the Tiber delta area was deep and widespread in all the historical phases examined, thanks to both the environmental benefits and the geographical setting. Moreover, the Tiber delta represented the main outlet on the Tyrrhenian coast, as well as a point of contact among Etruria, Apennines, and southern Latium. The Eneolithic and Ancient Bronze ages' human activities were centered on the right bank of the river, whereas the left side, the Ostiense one, was deeply re-elaborated, affecting significantly the state of preservation the archaeological data.

The area surrounding the Tiber River mouth and the Maccarese and Ostia lakes was mainly characterized by fertile land and easy access to fishing areas as well as to springs [84]. In this context, the settlement of Le Cerquete-Fianello [85] developed. The site, located on an offshoot of land that extended towards the Maccarese Lake, gave back important traces of housing structures, hotbeds, material of common use, and a ritual burial of a horse. The stratigraphy of this and nearby sites testifies substantially to a stable presence from the Eneolithic to at least the entire Middle Bronze Age. The territory offered a rich and favorable ecosystem for both the inhabited area and productive activities, widely exploited during these centuries. The housing model underwent some changes during the Middle and Final Bronze Ages. The site of Le Vignole dates back to this period [86–91] and it testifies to the development of productive activities with seasonal timing. The housing structures were positioned on small artificial mounds composed of sandy fill soil and wooden intertwining that constituted a drainage substrate of the hut walkway. These bumps were periodically renewed, as can be deduced from the succession of sandy layers and the discharge of materials. The anthropic impact had a dual effect due to the optimal access to resources for gathering, fishing, and hunting during the Eneolithic, and the optimal availability of resources suitable for both crafts and sheep farming during the Bronze Age.

The territorial system of the mouth during the whole Middle and Final Bronze Ages, characterized by mostly seasonal and productive settlements, had to be widespread along the Italian peninsula during the whole period. Other environmental conditions displayed other settlement choices. Going up to the hills closest to the coast is where we have evidence of an important settlement, *Ficana*, on the top of Mt. Cugno, a naturally easily defensible area, positioned near a possible ford of the Tiber. *Ficana* [92–96] was a settlement with a stable presence from the end of the Middle Bronze Age up to the Roman conquest in the royal era and which has returned a substantial series of structural evidence, tombs, ceramics, and other objects. Moreover, the delta plain and the surrounding area were characterized by a set of seasonal sites located near the freshwater stretches, on moist and silty–clayey soils. Coinciding with the beginning of the Iron Age, around the 9th–8th centuries BC, the Maccarese Lake changed into a brackish basin making possible the installation of the Etruscan salt pans, and, at the same time, other important anthropic changes marked the entire area (e.g., the foundation of *Veii*, north to the Rome, in the 10th century BC). The change from a freshwater into brackish basin caused human abandonment along the Maccarese Lake border and the demographic increase along the alluvial fan near the confluence of the Rio Galeria in the Tiber River [97–99]. Here, a small settlement with a rural structure developed. Between the eighth and sixth centuries BC, this characterized the hills near the north bank of the Tiber, between *Veii* and the salt pans, sustaining the economy until the clash with Rome.

The wars, which Rome waged between the eighth and fourth centuries BC, turned out to be a real battle for salt, the main resource of the area. The Roman strategies for territorial control led to the foundation of *Ostia* and the occupation of *Ficana*. According to ancient historiography, both events would be ascribed to Anco Marzio, in the seventh century BC; however, it should be pointed out that the archaeological evidence of a destructive event in *Ficana* dates to the seventh century BC, while the most ancient traces of *Ostia* appear to be those of the castrum related to the fourth century BC.

It is important to note that the area surrounding the Tiber delta began to play a primary role in dominating the salt pans (*Campus Salinarum Romanarum*) and access to the sea, basic elements in the economy of Rome. In fact, starting from the Republican era, *Ostia* played a central role in the logistics and economy of the city, as evidenced by the deep and systematic anthropization of the entire coastal strip that today is included in the Municipality of Fiumicino and Rome (Municipality X). Simultaneously with the enlargement of the urban portion, which already in the Republican era relied on logistics, transport, and commerce, there was a progressive expansion in the southern portion of the countryside, as well as eastern along Via *Ostiense*, which connected *Ostia* to Rome in continuity with the Via *Salaria* [100]. The *ager* that separated the coast from the lake during the first century BC, and a portion of the stretch along the *Ostiense* road, was a very structured rural area with a network of roads and paths and a diversified productive area, including, among others, orchards and farmyard animals (Varrone, De re rustica, Book III) and elements suitable for the drainage of the most humid areas [101]. In addition, in the Republican era context, it is necessary to emphasize that *Ostia* was equipped with landings located along the river stretch, which served the city's *horrea* and served as a hub for Rome.

Moreover, there was a pier located in the wide meander just upstream of today's village of Ostia Antica.

A significant increase in commercial and military traffic characterized the first and second centuries AD. The Tiber delta became one of the main hubs of the Empire and this needed an expansion of the structures toward the north. The port of Claudius was built and later it was expanded and improved with the structure of the Trajan basin. The new harbor center (*Portus*) was surrounded by logistic structures and connected to *Ostia* by canals and roads such as the Via *Flavia*, a northern continuation of the Via *Severiana* [102]. The wide and populous economic pole consisting of *Portus* and *Ostia* changed the area, which was also strongly tested by the clashes of the last phase of the republic age as well as the use of some of its parts. The rustic area, widely structured in the Republican era, was more and more a place of the patrician life. In the strip closest to the coast, some purely peasant spaces were transformed into rich and decorated *villae maritimae*. From the first to the third centuries AD, the territory no longer considered rural was occupied by a necropolis, as well as to the north, today's Isola Sacra.

The generalized crisis of the third century AD caused a contraction of economic activities and a partial abandonment of the areas, evidenced by the substantial closure of the *Ostiense* necropolis. With a progressive economic stagnation during the fourth century AD, the work activities were increasingly limited to the district of the Claudius and Trajan ports, while *Ostia* underwent a further sharp demographic decrease.

The area of the Tiber delta, therefore, was characterized by a slowed economy, with building limited to the reuse of materials and the remodeling of ancient spaces and use starting as a burial area from the fifth century AD (commercial buildings). The few innovations included the building of Christian basilicas either along the peri-urban roads or along Via *Severiana* near the so-called *Villa di Plinio* [103], on a connecting axis with the other streets of the Laurentian quadrant, between fourth and fifth centuries AD.

Germanic raids, the end of the Western Roman Empire, the birth of the Roman-Barbarian kingdoms, and the bloody Greek-Gothic war characterized the troubled period between the fourth and sixth centuries AD. All these factors pushed the inhabitants of the Ostia area to take refuge around the primitive church of S. Aurea. In the ninth century AD, Pope Gregory IV formalized this settlement by defining the new, fortified inhabited area, hence called Gregoriopoli. This new village stood near the last meander of the Tiber, a prominent position on the coast, determining the Saracen attacks in 846 AD. After the success of the Christian league (sea battle of Ostia 849 AD), the Saracens' raids stopped. The new settlement found stability, becoming more and more a reference center both for the management of the salt pans (moved from Maccarese to Ostia Lake) and for the control of river navigation.

Throughout the Middle Ages, the imperial ports became unusable. Archaeological evidence was rather scarce and mainly referred to the remodeling of the fortifications of Gregoriopoli. However, the Tiber delta is reported in different historical sources and archeological proof (i.e., Torre Boacciana) evidenced of stripping of the decorative materials of the ancient city repeated up to at least the 15th century. From the mid-1400s to the mid-1500s, the Ostiense side returned to be the political center. From the archaeological data and, above all, from the study of documents and topographical maps, a first expansion of the village emerged from both the ecclesiastical structures and the defensive ones.

This new impulse also included the reconstruction of the river pier right near the village and the fortress. All these structures represented the papal custom and the transshipment point of goods, which, by mean of boats, were sent toward Rome. In 1557, the Tiber changed course by cutting the meander, which gradually became silted up and swamped; the pier became unusable, and customs was moved first to Torre Boacciana and, successively, to Torre San Michele.

The meander, later called the dead river, and the ancient coastal lakes became malarial areas, and the residual population was directly linked to the salt pans; the reversal trend occurred only after the unification of Italy through the reclamation and elimination of salt pans.

3.3. Campania Area (the Garigliano Coastal Plain)

The Garigliano delta plain developed, between Latium and Campania, within a wide Quaternary extensional basin belt that stretched from the Tyrrhenian margin to the Central–Southern Apennine chain [3,104,105]. The coastal plain was characterized by the terminal stretch of the Garigliano River that built its delta on Plio-Pleistocene sediments and tephra (from Campanian and Roccamonfina volcanic centers). These deposits infilled a graben developed during a time interval ranging from the Miocene to about 125,000 yr BP [106–109].

3.3.1. Morphological Setting

The Garigliano River separates two strand plains, trending NW–SE, characterized by wet and depressed areas (Figure 9). The inner strand plain is referred to as the Eutyrrhenian [110], and the outer one is part of the present (Holocene) Garigliano River delta. The northern triangular depression is locally up to 1.5 m b.s.l. Its shorter side parallels the river through about 800 m, and, in turn, the longer one stretches about 1.4 km along the Holocene strand plain. The trapezoid shape of the southern depression is up to 1.2 km wide, extends about 4 km parallel to the coast, and it is locally deepening to 2.5 m b.s.l. Historical maps show that up to the 18th century both depressions were partially submerged, whereas today, after the land reclamation, they are totally dry. Based on aerial photographs, paleochannel traces, recording the past wandering of the Garigliano River, were recognized in the northern depression between the present river channel and the southern depression and along the eastern side of the Eutyrrhenian strand plain. However, the numerous Roman witnesses along the river prove that the river course has not changed at least since Roman Times. The Holocene strand plain shows beach/dune ridges 2–3 m high to the north of the Garigliano River and up to 9 m high to the south.

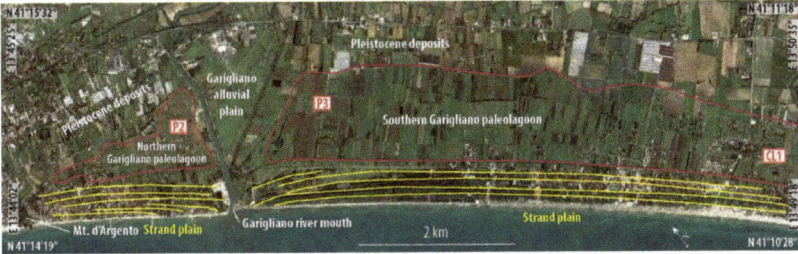

Figure 9. Morphological scheme of the Campania area. The yellow lines indicate the trend of the Holocene beach/dune ridges. The red line indicates the outline of the marshes/ponds, today reclaimed. The labels in the white rectangles indicate the location of the boreholes, shown in Figure 10 (Image Landsat/Copernicus—Google Earth).

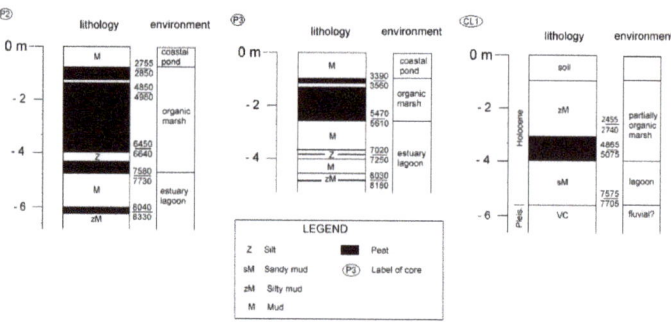

Figure 10. Lithological/environmental diagram of some boreholes carried out in the Campania area. P2 and P3 are percussion boreholes; CL1 is rotating boreholes. P2 and P3 are modified after [36]; CL1 is modified after [111].

3.3.2. Vegetation Frame

The appearance of vegetation and its evolution in the Garigliano area are discussed in detail in [36,111,112]. Between 8200 and 5800 yr BP, in a coastal plain, a large forest cover, mainly consisting of *Quercus* deciduous, *Alnus*, *Salix*, *Populus*, and *Juniperus*, occurred. The presence, in the first phase, of abundant *Alnus* closest to the current river mouth suggests a riparian forest. Subsequently, oak woods expanded, and the Mediterranean taxa increased. Towards the end of this period in the innermost areas, a mosaic of freshwater and brackish water environments coexisted with open environments and mixed woods located far from the site. Between 5800 and 3100 yr BP, forest cover was more present with considerable incidence of hygrophilous trees (*Alnus* and *Salix*), which also extended into the southernmost portion of the coastal plain, and an increasing of the Mediterranean taxa such as *Olea and Castanea* and coprophilous fungal spores occurred. Between 3100 and 2700 yr BP, the forest cover decreased with minor incidence of *Salix and Juniperus* types and the Mediterranean taxa. In contrast, *Populus* as well as several freshwater aquatics' plants spread. The *Olea, Juglans,* and *Castanea* (OJC group) included all the three trees, and cereals occurred.

During the Roman and post-Roman time, *Alnus* and *Myrtus* increased, while *Salix* and *Quercus* deciduous decreased. In the OJC group, *Juglans* was absent and *Castanea* increased. In the modern age, forest cover dropped notably, owing to the decline of riparian trees and the Mediterranean taxa. Broadleaved trees remained steady and the complete OJC group was maximized.

3.3.3. Facies

Five different facies developed in the last 8000 years (Figure 10).

Facies 1 developed between the end of the postglacial sea level rise and the early stage of the subsequent quasi-still stand. Near the Garigliano River, it consists of dark-gray mud levels with brackish and marine fauna (*Ecrobia ventrosa, Cerastoderma glaucum, Abra segmentum, Mytilus galloprovincialis, Parvicardium exiguum, A. tepida, Haynesina germanica, H. depressula,* and *Porosononion granosum*), wood debris, peat, and gray silt intercalations. Farther from the river channel are present mud and silt with local peaty levels or mollusk fragments and sandy bed with parallel lamination with *Bittium reticulatum, Loripes lucinalis, Abra segmentum,* and *C. glaucum,* in addition to *A. beccarii, A. tepida, A. parkinsoniana, L. lobatula, H. germanica,* and Cyperaceae, Poaceae, and Salicornia pollens. Dinoflagellates are in the southernmost part of the coastal plain [112], where this facies lies on the "Campanian Ignimbrite" volcanic deposits [111].

Interpretation: The facies highlights a lagoon/estuary rich in hygrophilous woody vegetation with probable fluvial inputs. In the southern part, fauna and pollens reflect an ongoing lagoon infilling under variable marine and fluvial inputs.

Facies 2. It consists of a homogeneous peat suite, devoid of fauna but with well-preserved vegetal remnants (*Alnus, Salix*) that witnessed the spreading trend in hygrophilous woods. Furthermore, gray-greenish silts, sometimes peaty and locally with pumice grains, and gray medium-fine sand with parallel laminations occur.

Interpretation: The sediments reveal an organic coastal lake in which clastic levels locally break the organic sedimentation and, according to the fauna remnants, witness washover or crevasse events. This facies largely developed between 5500 and 3500 yr BP.

Facies 3. It is characterized by dark-gray or brown mud with local peat levels and rare intercalations of gray-ochre silt levels with occasional laminations and pumice grains. The fauna only includes rare freshwater gastropods such as *Acroloxus lacustris, Bithynia tentaculata, Gyraulus laevis, Planorbis planorbis,* and *Valvata cristata*. The pollen content mainly consists of Callitriche, *Nymphaea alba* type, *Potamogeton,* and *Myriophyllum*.

Interpretation: The facies is attributable to a freshwater basin (marsh/coastal pond) governed by decantation and occasional flooding inputs. It is believed that drying phases alternated with shallow, limpid ephemeral ponds prior to the final drying up. The development of this facies began around 4000 yr BP and tended to shrink after the Roman period until it disappeared following the land reclamation in the 19th century.

Facies 4. It spread mainly near the present Garigliano River. It includes gray-greenish mud levels with thin, reddish silt laminae. Locally, calcrete nodules, small and rounded pebbles, pumice, subfossil tree roots, and rare brick fragments occur.

Interpretation: This is a fluvial facies.

Facies 5. The sediments are characterized by coarse sand with polygenic pebbles and marine bivalves changing upwards in medium-fine sand containing subrounded or bladed carbonate and volcanic clasts. Sand is interbedded with a decimetric, dark mud level with altered plant remains. Locally, marine shells and low-angle parallel lamination occur. At the top, there is well-sorted, fine-medium ochre sand locally exhibiting cross lamination.

Interpretation: The sediments mark the Holocene strand plain.

3.3.4. Diachronic Physical Landscape Change

The stratigraphical data allow us to reconstruct the evolution of the local physical and cultural landscape for the past 8000 years (Figure 11).

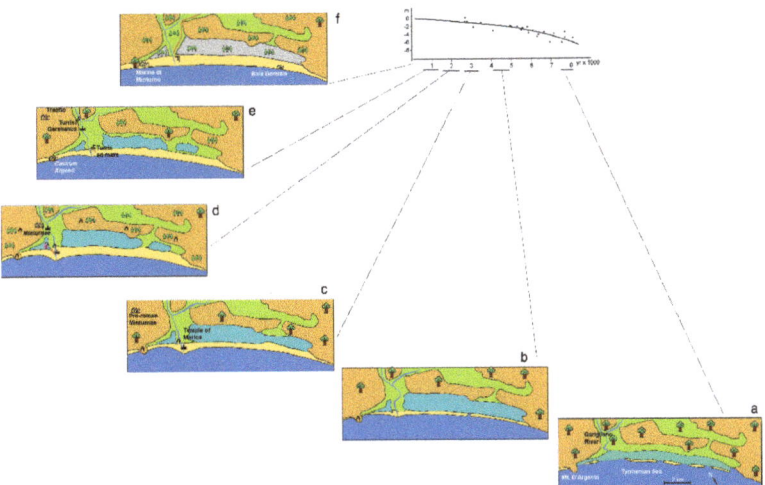

Figure 11. (a–f). Two-dimensional evolutionary schemes of the Campania area inferred from underground and historical/archaeological data. The individual images illustrate the coastal landscape relating to the period reported on the abscissa of the sea level rise curve drawn mainly based on ^{14}C dating of marsh/pond peat levels [36].

Between 8000 and 7000 yr BP, a partially emerged wet zone, with a riparian wood, was present in the northern zone. On the contrary, proceeding south, a sheltered, narrow lagoon, locally well connected with the sea, developed. Only landward, a continental environment influenced by alluvial deposition occurred.

Between 7000 and 5500 yr BP, during the final transgression stage and the beginning of the sea level still stand, the lagoon migrated landward. In the northern zone, a coastal barrier bordered a freshwater coastal lake with organic sedimentation. In the southern zone, the lagoon, less and less connected with the sea for the expansion of the coastal bars, progressively extended, enriching itself in organic sediment and producing a local mosaic of fresh and brackish water environments.

Between 5500 and 2600 yr BP, at the beginning, a delta cusp developed in the northern zone, turning the coastal bars into a strand plain throughout the area, and a freshwater coastal lake developed in the southern part. Brackish environments progressively disappeared, and the river wandered between the twin lakes. Perhaps due to the end of the 4.2 yr BP dry event, a greater input of alluvial sediments progressively changed the coastal lakes in coastal ponds/marshes, with prevailing clastic sedimentation. Woodlands declined, and the pollen records reveal an early anthropic impact.

Between 2600 yr BP and the present, two ponds/marshes separated by the river course and by a partially eroded delta cusp characterized the early landscape. Some hundred meters wide, a strand plain bordered the ponds. Human impact on the territory increased but the pollen and fauna data do not show the use of the ponds as salt works. The significant presence of *Myriophyllum alterniflorum* suggests that clear and oligotrophic water filled the ponds, probably because of the proper water management in Roman times. The subsequent lack of maintenance of the drainage system in the Middle Age resulted in enhanced solid inputs to the ponds, with consequential phases of partial drying. The ponds were never filled in, and the areas now must be kept dry by ongoing reclamation. Forest cover decreased contemporary to the reduction of the wetlands and the expansion of agrarian systems. Similarly to other deltas of the Tyrrhenian coast [113,114] from the Roman period, progradational and erosive phases followed one another, and currently the delta cusp is missing.

3.3.5. Diachronic Cultural Landscape Change

Over the millennia, the area of the plain of the Garigliano delta has maintained a set of natural characteristics suitable for human settlement and different types of exploitation.

Throughout the period from the Neolithic to the Middle Bronze Ages, this area had scarce signs of population but was set in an interesting context, as exemplified by the Neolithic sites in the Mt. Massico area, attributable to the Laterza culture [115], and by the later ceramic material discovered on the marine side of the same mountain.

During the Middle Bronze Age, the inhabited areas on the fertile plain disappeared, whereas those on the slopes were scarce. In some centuries (first half of Middle Bronze), the small settlements located preferentially on the dune ridges or on land with pozzolana, always near water pools, became more numerous. A flora useful for feeding, craft, construction, and potentially phytotherapeutic fields (plants with tannins and salicin are included) characterized this habitat. The availability of terrestrial and lake fauna must also be considered, as well as the possibility of exploiting for housing purposes the highly draining soils, sand and pozzolana, and, for the clay production, the more humid areas with clayey silts. At the end of Middle Bronze Age there was a gradual increase in settlements on elevated zones' ground, among the first were the areas surrounding Mt. Massico. In this period and in the following centuries, the exploitation of the area changed in a stable and widespread manner. In addition, the increasing population led to a cultural and social impulse determining a political, economic, and territorial control of the area. The data for the Recent and Late Bronze Ages show an increase in small sites located in a humid coastal habitat. The accumulation of commonly used materials in small stratifications and without particular traces of structures leads us to consider these sites as productive with periodic attendance. The proto-urban centers exploited the different environmental and geomorphological characteristics of the entire area. The socio-political balance in the Late Bronze Age involved the appearance of small agricultural and pastoral settlements on fertile land without natural defenses. As in other contexts (for example, Punta degli Stretti, see Tuscany area), parallel to the above-described structure, autonomous centers located along the coast arose. This was the case of the town on the top of Mt. d'Argento [31,116,117], which was built close to the sea and, at the same time, in an elevated position. The coastal stretch had an increasingly important role, as a point of contact with the Mycenaean world. Many demographic variations were associated with the constant exploitation of the area, similarly to Mt. Massico, a site that characterized the transition to the first Iron Age. Between the ninth and seventh centuries BC, there was also important cultural evidence [117]. Starting from the eight century BC, the archaeological data show varied typology of settlements, which, nevertheless, seemed to remain in the wake of the territorial management described above. From the seventh century BC, well-defended centers developed on topographically high areas, becoming a reference for the scattered settlements in the productive plain. In the sixth century BC, attendance, already attested to on Mt. d'Argento, was stable and large. The presence of areas fortified by polygonal walls suggests a capillary military control by means of communication (routes and passes). These *oppida*, extending along the entire stretch of the Garigliano River, seem to have had control and defense roles of the population [118]. This function, known since the sixth century BC, was strengthened up to the end of the third century BC, when the area was involved in the Samnite wars. The historical sources report that the population, the Aurunci, settled in the three main *urbes*: *Minturnae*, *Vescia* (south of Mt. Massico), and *Ausona* (near today's Ausonia). These *urbes* were probably the afferent centers of *oppida*, *vici*, and farms, as indicated by Tito Livio. In continuity with previous centuries, the coastal area played an important sacral role, as testified to by the first monumental temple dedicated to the goddess Marica [119–126] that was an important sacral structure for all Mediterranean populations [120]. This temple was close to the dune ridges and oriented towards the lake.

From the sixth century BC, significant changes occurred in Italy. The Aurunca population was exposed to the growing Roman power, which determined the conquest and Romanization of the Garigliano coastal plain by the third century BC. After the suppression of the revolt at 314 BC and the destruction of *Ausona*, *Vescia*, and *Minturnae*, the territorial structure was redesigned on the basis of a new foundation of the city, roads (i.e., *Appia* road, a consular road that partially followed pre-existing paths near the coast), and centuriation.

In 296 BC, the reorganization process was completed by the establishment of two further colonies of a *castrum* type, defined *coloniae maritimae* (Lev. XXVII, 38, 3-5), under Roman law: *Minturnae* and *Sinuessa* (near today's Sessa Aurunca). *Minturnae* was located on the Pleistocene dune on the right bank of the Garigliano River about 2 kilometers from the coastline, and *Sinuessa*, near the coast south of Mt. Massico. Their locations show the clear intent to occupy spaces to acquire the maximum control, and, so, to guarantee stability to the Romanization policies. This led to a widespread population and exploitation of rural potential, as indicated by *Corpus Agrimensorum Romanorum* of Igino (Hyg. De Lim. Const., 178, 6–9) and confirmed by the analysis of aerial photos. Throughout the middle and late republics, for the proliferation of rustic and patrician villas along the coast and the hinterland, the area began a vital and populous territory. The urban areas started an evident monumentalization while the rural one was terraced to facilitate both building and farming. The coast of *Minturnae* became the reference point for all the main agricultural production of the Garigliano basin. Particularly the wine production assumed international significance, as testified to by the *Minturnae* remains along the routes to Gaul, Spain, and the Middle East [127]. The recovery of various logistics and shipbuilding structures and a pier suggests that the fulcrum of this flourishing activity was the harbor. This scenario grew and settled, down to the whole of the first century AD, including even the perilagunar area, as evidenced by the presence of many *villae maritimae*, remains of structures related to fishing, fish farming, and *garum* production. In addition, epigraphs for the *salinatores* members certified the salt production is in the urban area [128]. Similarly to other equally prosperous *agri*, the territory gradually was structured in wide estates managed by the exponents of the main Roman families. The turbulent centuries of the second half of the republic saw *Minturnae* at the center of many events, among which were the servile revolt and the flight of Gaius Marius. Successively, the colony and its territory suffered considerable and repeated damage. Between the first BC and first AD centuries, there was a substantial reorganization, as reported by Igino in the *De Limitibus Constituendis* (Hyg. De Lim. Const., 177, 8–15; 178, 1–9). In the first century AD, a new territorial parceling in the south of the river ensured new prosperity, so much so that many of the *villae* turned into real patrician residences. Until the whole second century AD, *Minturnae* was undergoing a major overhaul. In addition to the improvement of the recreational, cultural, and worship structures, there was also a modernization of the harbor and the mouth. In fact, probably logistical structures from the first century AD were identified. The strong mercantile characterization of the area led us to suppose the existence of a harbor also in the lagoon area [129–134], although there is no archaeological evidence. The second century AD, as in most of Italy, marked a moment of less imperial control, and a progressive impoverishment of economic and building activities ensued.

Additionally attributable to this phase was the new reconstruction of the ancient temple of Marica [119], now oriented towards the river and, therefore, disconnected from the pond. The center of commerce and wealth that moved all this building fervor, and perhaps the arrival of new cults, such as that of Isis [119], had to play a more heartfelt role than the tradition.

In the third century AD, the crisis of the Empire reached its peak and the economy underwent a general contraction; consequently, *Minturnae* suffered a significant impoverishment. The sepulchral reuse of the commercial area was an eloquent testimony of the situation until the fourth and fifth centuries AD. The demographic collapse and the Greek-Gothic war marked the end of *Minturnae*, which, in a few decades, lost all economic or political importance. The last traces are those relating to the lime production centers through the recycling of building material. In 590 AD, a letter from Pope Gregory the Great attested to the passage of the town church to the diocese of Formia, an evident sign of depopulation. Italy, since the sixth century AD, was under Lombard dominion, with the exception almost exclusively of the territory governed by the pope. The Garigliano River was its border [135]. The alluvial plain was depopulated; the few inhabitants looked for safe areas suitable for a subsistence economy. The Minturnese *ager* was dotted with small settlements, often grown around the ancient *villae rusticae* or positioned on the plateaus already exploited in protohistoric times. Until the 10th century, the only vital centers were Gaeta and Sessa Aurunca.

A discontinuity with respect to the set of small towns under the direct influence of the papacy was the settlement witnessed by some historiographical, toponymical, and partly archaeological traces of a Saracen group. In the last two decades of the ninth century AD, historical sources mention *Mons Garelianus* [136] as the town exploited for the control of the coast and piracy carried out by the Saracens. The ancient port structures were exploited, suitable for the seafaring activity that scourged the entire Tyrrhenian coast. The grip of Christian cities allied under the aegis of the Pope, which since the mid-ninth century AD clashed with the Saracen forces, suffocated the base of the Garigliano that was abandoned following the battle of 915 AD [137]. It followed that the coast acquired an important military control role; the first watchtowers were built in the aftermath of the Christian victory at Garigliano. *Turris Gariliani* [137,138], on the right bank of the river, already erected at the end of the ninth century AD reusing Roman structures, was repaired following the battle, and *Turris ad mare* [138] was erected on the left bank between 961 and 981 AD. With a progressively more stable political situation, the towers took on a role of control of the mouth of the river, which, being still navigable, connected the coast with the inland centers; this made them a point of interest for part of the population (i.e., Mt. d'Argento, *Castrum Argenti*) [139,140]. Some fortified structures agglomerated inhabitants around them. From the 12th century AD, for about two centuries, there was a wall, a church, and some houses, as well as a road connecting the fortified plateau and the town of *Traetto*.

These settlements remained isolated throughout the Middle Ages, and the water management was very difficult under the Little Ice Age meteorological worsening. The progressive swamping pushed the resident communities to abandonment. Until the unity of Italy, the area was uninhabited, acquiring agricultural and tourist interests only after the reclamation of the coastal ponds.

4. General Discussion

The evolution of the coastal landscape of the three studied areas shows both common elements and significant diversities. The latter are due to both the peculiar geological and morphological characteristics of each area (presence of islands, headlands, river mouths) and the different anthropogenic activities (socio-political and commercial), which the areas acquired during the historical times.

In the period between 8000 and 6000 yr BP, the SRL rate, although decreased with respect to the previous period, limited the sedimentary supply to the coastal belt. A dense forest dominated by deciduous oaks with a local presence of riparian tree covered the three areas (Figures 4a, 8a and 11a). The population was scarce and the human influence on the environment was negligible. Near the river mouths (Albegna, Tiber, Garigliano), the coastline inflected, generating more or less wide inlets/estuary that were slowly being filled. In the Tuscany area, Mt. Argentario was an island close to the coast, and to the SE of the Ansedonia promontory the shoreline migrated landwards and formed a long and narrow lagoon locally well connected with the sea (Figure 4a). In the Tiber area, the wide estuary, bordered by barrier islands partially migrating to the land, contained the bayhead Tiber delta (Figure 8a). In the Campania area, to the north of the Garigliano, a small freshwater lake was forming; in the southern part, a barrier, partially discontinuous and slowly migrating to the land, bordered a lagoon locally connected to the sea (Figure 11a).

Between 6000 and 4000 years ago, the SLR rate, stabilized at values close to 1 mm/yr, was no longer a limit to the sedimentary supply of the coastal area. The river sediments, reworked by the longshore currents, produced a general progradation of the coastlines that caused changes of the landscape in each three areas. In the Tuscany area, the pollen content indicates that the variations in vegetation cover in the Neolithic are related more to climatic fluctuations than to anthropic activity [33,42,43]. The Feniglia developed and in the SE of the Ansedonia promontory, and the long lagoon become wider (Figure 4b). A carbonate sedimentation, probably linked to the rise of hydrothermal fluids, partly characterized the well protected from the sea lagoon [37]. The territory, still covered by a dense forest with the development of herbaceous plants in the perilagoon areas, was inhabited by small and sparse communities with a mainly subsistence economy, which settled, scattered close to the lagoon areas; in this regard, a comparison was proposed [33] with the site of Le Cerquete-Fianello of Maccarese. In the Tiber area, the bayhead delta reached the barrier islands, causing their coalescence and the transformation of the wide inlet into two coastal lakes (Figure 8b). The river separated the lakes, characterized by fresh water, mainly towards the inner margin.

The progradation of the Tiber mouth gave rise to a delta cusp. The area was vegetated by a dense forest of oaks and riparian trees and subordinately by a dune vegetation. The evolution of the physical landscape brought an environment favorable to human stable settlements. The development of the Eneolithic Le Cerquete-Fianello site on the northern lake is evidence of that for its long stability and for the activities that well exceeded the subsistence economy. In the area of Campania, the sediments of Garigliano allowed the development of an almost continuous barrier that separated a long coastal lake from the sea.

The river wandered between the lakes and progressively produced a weakly cuspate strand plain (Figure 11b). In the oak forest, the hygrophilous plants and Mediterranean taxa increased. The small population produced only minimal settlements in the plain.

Between 4000 and 3000 years ago, the physical evolution of the territory was determined by the sedimentary contribution to the coast. Although some areas changed significantly, a more widespread anthropic presence occurred. In the Tuscany area, the dune systems developed, and Mt. Argentario was permanently connected to the coastal plain by two tombolos that enclosed a wide lagoon. At the SE of the Ansedonia promontory, the long coastal lake progressively filled up, splitting into smaller water bodies (Figure 4c).

The vegetation was enriched with herbaceous plants and dune vegetation. The demographic increase was evidenced by some stable settlements developed on the fluvial terraces and in proximity of the coast for the control of the new marine routes. Considering the Punta degli Stretti site, which displayed settlement similarities with the Puntata di Fonteblanda site near Talamone, we can hypothesize a well-defined management method, in which these centers controlled the main resources and the maritime communication routes. Moreover, small, mostly seasonal, perilagoon settlements proliferated, exploiting the water and meadows' supply. The physical landscape of the Tiber area had no great variations other than an extension of the delta cusp and a limited reduction of the coastal lakes. The vegetation was enriched with riparian trees, especially in the northern part, and dune vegetation. For some centuries, the Le Cerquete-Fianello site was active but, towards the middle of the period, the use of the territory was characterized by small, seasonal peri-lacustrine settlements, not exclusive of the northern lake, which devolved essentially to farming and handicrafts (Figure 8b). At the same time, in a dominant position on the hill between the Tiber and a southern lake, a new stable site began to develop. In the Campania area, the variation of the physical landscape consisted of an increase in the strand plain and a partial burial of the coastal lakes (Figure 11c). Forest cover and water plants decreased while some cereals extended. Mostly seasonal settlements developed in the plain close to the lakes and the river that represented productive sites suitable for grazing [141]. In the second half of the period, stable settlements rose on the elevated areas at the edge of the plain even though they continued to control and use the seasonal productive sites. The proto-urban root of Lazio can be identified in the agglomeration of the population in well-defined and well-defended areas, testifying to a persisting territorial attraction from productive and economic points of view.

The coastal landscape showed both significant physical and anthropic evolutions starting from 3000 yr BP. Social and political changes had the first important effects on human–environmental interaction. In the Tuscany area (Figure 4c,d), the physical landscape began as very similar to the current one: Mt. Argentario, well connected to the coastal plain by the two tombolos and the Albegna River mouth, permanently had a position at the eastern root of the Giannella. At the SE of the Ansedonia promontory, the ancient coastal lake evolved in a wet zone well isolated from the sea by a continuous dune ridge. Starting from the Bronze Age, the anthropogenic effect on the vegetation appears significant, and, in the Roman period, the arboreal cover was characterized, similarly to today, by the Mediterranean scrub mixed with oak wood [44]. The demographic increase involved the growth of small settlements located mainly both around the Burano Lake, for the production of salt, and Feniglia, now dotted with landing places. In this period, Etruscans organized the territory. On the central tombolo, Orbetello, the first fortified urban center, developed. In its surroundings, an organized agricultural system and the use of the lagoon were undertaken. In the second part of the millennium, the Roman control extended on the territory. On the Ansedonia promontory, *Cosa* town was founded. The city provided landings along the entire Feniglia by means of ports located on both sides of the promontory that were connected by a canal carved into the mountain. Centuriation, road networks, and agriculture, characterized by well-structured funds, developed. In the Tiber area (Figure 8c), the river moved the mouth further south where a new delta cusp developed, while the previous one was partially eroded and the two lakes became brackish. At first, the Latin center of *Ficana* and the northern one by the Etruscan center of *Veii*, which implanted the salt pans in the Maccarese Lake, controlled the southern part of the delta. With the expansion of Rome, the salt pans remained, *Veii* and *Ficana* disappeared, and *Ostia* with its river port was born near the Tiber mouth. A road network developed as well as a series of *villae* for agricultural production and breeding. In the Campania area (Figure 11c,d), there were no significant variations in the physical landscape. The northern lake took on an important cultic function and some fortified centers were built in the area. Towards the end of the period, the Roman conquest modified the landscape through the centuriation, road development, and the foundation of *Minturnae*. This urban center provided a river port from which the agricultural products departed.

In the first millennium of the Common Era, two distinct phases of coastal landscape evolution followed one another. In the first half of the period, anthropic action was an important forcing, but it was significantly reduced in the next phase. In the Tuscany area (Figure 4d), at first the coast became home to patrician villas and the agricultural structure evolved towards *latifundium*. However, this pattern did not favor territorial management, and, in the central centuries of the period, a phase of swamping began, causing a decrease in the population and port activity. For the rest of the time the area was marshy, malarial, and sparsely populated. The Tiber area was deeply affected by human impact in the early part of the period (Figure 8d). On the northern part, salt pans were active and ports were built. All these activities involved commercial structures and roads, as well as the building of canals that determined a double-mouth delta, partially modifying the coastal dynamics.

On the southern part, the agricultural organization displayed large villas that followed one another along the coastal road. These villas represented both productive and leisure centers for the rich owners. After the third century, the population slowly decreased, shrinking strongly at the end of the empire. In the early Middle Ages, the great salt pans were no longer productive and the ports were partly buried. The territory, not carefully managed, became marshy, and a small town, close to oldest *Ostia*, gathered a small number of inhabitants. In the Campania area (Figure 11d,e), in the first period, *Pagus* and *vici* created the political pattern to manage the first agricultural production structures, already conceivable as real farms. Commercial, maritime, agricultural, and salt production continued. It is believed that the salt pans were located at the mouth of the Garigliano [142], but in a freshwater context of problematic localization. The trace, detectable in an aerial photo of a narrow channel extending from the northern lake towards the river, as well as for the medieval salt pans of *Ostia*, could suggest the presence of a small salt pan. As with the other areas, the coast hosted the patrician villas that replaced the rustic villas. However, after the third century, as the result of a demographic decline and the fall of the Roman Empire, *Minturnae* disappeared. In the early Middle Ages, a subsistence economy was established again and the few inhabitants settled in small, elevated centers abandoning almost completely the plain, subject to swamping and river floods.

In the first part of the last millennium (late Middle Ages), the natural change in physical landscape was limited and human influence poorly affected the coastal landscape. In the second part of the last millennium, some areas showed a remarkable change in the physical landscape due to a significant human–environment interference occurring mainly in the last two centuries. In the Tuscany area, slight variations were related at the edge of the Orbetello lagoon. The sparsely populated area was dotted with towers and coastal defenses in the middle of the millennium. Only in the last century (Figure 4e) was the Burano area reclaimed, determining an agricultural increase, whereas, in Orbetello, ichthyoculture and tourism were on the rise. The history of the Tiber area was different (Figure 8e,f). Here, already in the first centuries, salt production began and continued until the 18th century, no longer in the Maccarese Lake but in the Ostia Lake. The imperial ports disappeared, but, as with the lake, the Trajan dock remained. Starting from the 15th century, the rapid progradation of the Tiber mouths, associated with a final modification of the main channel, expanded the strand plain that was enriched with towers guarding the mouths. The scarce population used the territory mainly for pasture. At the end of the 19th century, reclamation drained the ponds, the population increased, and agriculture extended; in the 20th century, tourism and urbanization strongly developed. The area that represented, during the Roman period, the connection between Rome and the world through the port structures returned to have the same function in the 20th century by the Fiumicino airport. In the Campania area (Figure 11e,f), the situation occurring *quo ante* persisted until the middle of the millennium, until, because of the worsening climate, a phase of progradation occurred.

This fact produced an extension of the strand plain and a further hydraulic disorder. The whole area was almost abandoned and used for grazing. At the beginning of the 20th century, the reclamation of the coastal ponds [143] and the Garigliano damming allowed agricultural use and a limited tourism development.

5. Conclusions

In the final analysis, different forcings caused the evolution of the coastal landscape in the considered areas over the last 8000 years. They effected differently in time and space. The natural forcing was closely linked to the Milankovian and sub-Milankovian climatic variations to which, during certain periods, anthropic activity overlapped. If the SLR dominated in the first two millennia, the amount of sediment, coming to the coast, dominated over the next two millennia when, locally, humans began to organize settlements and activities using the resources of the territory. Human impact progressively increased, mainly in the Tuscany and partially in the Tiber areas, when the social and political structure of the Etruscans deeply modified the territory mainly through the salt pans and the organization of the ports. During the Roman period, the anthropic influence increased, highlighting firmly the relationship between landscape change and socio-political organization of the population that inhabited it. The landscape of the Tiber area, close to the center of power, resulted as the most impacted. The degradation of that organization caused the renewed dominance of natural forcing. Only the restoration of the socio-political organization of the new Italian State, together with the technological evolution of the last two centuries, has allowed limiting the effect of natural forcing.

Author Contributions: Conceptualization, M.D., P.B., L.D. and L.D.B.; Methodology, P.B., T.B. and L.D.; Supervision, M.D., P.B., T.B., L.D. and L.D.B.; Writing—original draft, M.D., P.B., T.B., L.D. and L.D.B. All the authors participated in writing, reviewing, and editing the manuscript. All authors have read and agreed to the published version of the manuscript.

Funding: This research received no external funding.

Institutional Review Board Statement: Not applicable.

Informed Consent Statement: Not applicable.

Acknowledgments: We would like to thank Silvana Falcetti and Laura Di Pietro for the drawing of figures.

Conflicts of Interest: The authors declare no conflict of interest.

References

1. Nisi, M.F.; Antonioli, F.; Pra, G.; Leoni, G.; Silenzi, S. Coastal deformation between the Versilia and the Garigliano Plains (Italy) since the Last Interglacial stage. *J. Quat. Sci.* **2003**, *18*, 709–721. [CrossRef]
2. Antonioli, F.; Ferranti, L.; Fontana, A.; Amorosi, A.M.; Bondesan, A.; Braitenberg, C.; Dutton, A.; Fontolan, G.; Furlani, S.; Lambeck, K.; et al. Holocene relative sea-level changes and vertical movements along the Italian and Istrian coastlines. *Quat. Int.* **2009**, *206*, 102–133. [CrossRef]
3. Ferranti, L.; Oldow, J.S.; Sacchi, M. Pre-Quaternary orogen-extension in the Southern Apennine belt, Italy. *Tectonophysics* **1996**, *260*, 325–347. [CrossRef]
4. Antonioli, F.; Kershaw, S.; Renda, P.; Rust, D.; Belluomini, G.; Cerasoli, M.; Radtke, U.; Silenzi, S. Elevation of the last interglacial highstand in Sicily (Italy): A benchmark of coastal tectonics. *Quat. Int.* **2006**, *145*, 3–18. [CrossRef]
5. Alley, R.B. Ice-core evidence of abrupt climate changes. *Proc. Natl. Acad. Sci. USA* **2000**, *97*, 1331–1334. [CrossRef]
6. Bond, G.; Kromer, B.; Beer, J.; Muscheler, R.; Evans, M.; Showers, W.; Hoffmann, S.; Lotti-Bond, R.; Hajdas, G.; Bonani, G. Persistent solar influence on North Atlantic climate during the Holocene. *Science* **2001**, *294*, 2130–2136. [CrossRef]
7. Mayewski, P.A.; Rohling, E.E.; Stager, J.C.; Karlend, W.; Maascha, K.A.; Meekere, L.D.; Meyersona, E.A.; Gassef, F.; van Kreveldg, S.; Holmgrend, K.; et al. Holocene climate variability. *Quat. Res.* **2004**, *62*, 243–255. [CrossRef]
8. Ribeiro, S.; Moros, M.; Ellegaard, M.; Kuijpers, A. Climate variability in West Greenland during the past 1500 years: Evidence from a high-resolution marine palynological record from Disko Bay. *Boreas* **2012**, *41*, 68–83. [CrossRef]

9. Nieto-Moreno, V.; Martinez-Ruiz, F.; Giralt, S.; Gallego-Torres, D.; García-Orellana, J.; Masqué, P.; Ortega-Huertas, M. Climate imprints during the 'Medieval Climate Anomaly' and the 'Little Ice Age' in marine records from the Alboran Sea basin. *Holocene* **2013**, *23*, 1227–1237. [CrossRef]
10. Zalasiewicz, J.; Waters, C.N.; Williams, M.; Barnosky, A.D.; Cearreta, A.; Crutzen, P.; Ellis, E.; Ellis, M.A.; Fairchild, I.J.; Grinevald, J.; et al. When did the Anthropocene begin? A mid-twentieth century boundary level is stratigraphically optimal. *Quat. Int.* **2015**, *383*, 196–203. [CrossRef]
11. Sabatier, P.; Dezileau, L.; Colin, C.; Briqueu, L.; Bouchette, F.; Martinez, P.; Siani, G.; Raynal, O.; Von Grafenstein, U. 7000 years of paleostorm activity in the NW Mediterranean Sea in response to Holocene climate events. *Quat. Res.* **2012**, *77*, 1–11. [CrossRef]
12. Kaniewski, D.; Marriner, N.; Morhange, C.; Faivre, S.; Otto, T.; van Campo, E. Solar pacing of storm surges, coastal fooding and agricultural losses in the Central Mediterranean. *Sci. Rep.* **2016**, *6*, 25197. [CrossRef] [PubMed]
13. Ghilardi, M.; Istria, D.; Curras, A.; Vacchi, M.; Contreras, D.; Vella, C.; Dussouillez, P.; Crest, Y.; Guiter, P.; Delanghe, D. Reconstructing the landscape evolution and the human occupation of the Lower Sagone River (Western Corsica, France) from the Bronze Age to the Medieval period. *J. Archaeol. Sci. Rep.* **2017**, *12*, 741–754. [CrossRef]
14. Camuffo, D.; Enzi, S. The Analysis of Two Bi-Millenary Series: Tiber and Po River Floods. In *Climatic Variations and Forcing Mechanisms of the Last 2000 Years*; Jones, P.D., Bradley, R.S., Jouzel, J., Eds.; Springer: Berlin/Heidelberg, Germany, 1995; Volume 41, pp. 433–450.
15. Glaser, R.; Riemann, D.; Schonbein, J.; Barriendos, M.; Bradzil, R.; Bertolin, C.; Camuffo, D.; Deutsch, M.; Dobrovolny, P.; van Engelen, A.; et al. The variability of European floods since AD 1500. *Clim. Change* **2010**, *101*, 235–256. [CrossRef]
16. Benito, G.; Macklin, M.G.; Zielhofer, C.; Jones, A.F.; Machado, M.J. Holocene flooding and climate change in the Mediterranean. *Catena* **2015**, *130*, 13–33. [CrossRef]
17. Bellotti, P.; Caputo, C.; Dall'Aglio, P.L.; Davoli, L.; Ferrari, K. Human settlement in an evolving landscape. Man–environment interaction in the Sibari Plain (Ionian Calabria). *Alp. Mediterr. Quat.* **2009**, *22*, 61–72.
18. Goiran, J.P.; Pavlopoulos, K.P.; Fouache, E.; Triantaphyllou, M.; Etienne, R. Piraeus, the ancient island of Athens: Evidence from Holocene sediments and historical archives. *Geology* **2011**, *39*, 531–534. [CrossRef]
19. Bini, M.; Brückner, H.; Chelli, A.; Pappalardo, M.; Da Prato, S.; Gervasini, L. Palaeogeographies of the Magra Valley coastal plain to constrain the location of the Roman harbour of Luna (NW Italy). *Palaeogeogr. Palaeoclimatol. Palaeoecol.* **2012**, *337–338*, 37–51. [CrossRef]
20. Amorosi, A.; Bini, M.; Giacomelli, S.; Pappalardo, M.; Ribecai, C.; Rossi, V.; Sammartino, I.; Sarti, G. Middle to late Holocene environmental evolution of the Pisa plain (Tuscany, Italy) and early human settlements. *Quat. Int.* **2013**, *303*, 93–106. [CrossRef]
21. Anthony, E.J.; Marriner, N.; Morhange, C. Human influence and the changing geomorphology of Mediterranean deltas and coasts over last 6000 years: From progradation to destruction phase? *Earth Sci. Rev.* **2014**, *139*, 336–361. [CrossRef]
22. De Rita, D.; Faccenna, C.; Funiciello, R.; Rosa, C. Stratigraphy and Volcano-Tectonics. In *The Volcano of Alban Hills*; Trigila, R., Ed.; Università degli Studi di Roma "La Sapienza": Rome, Italy, 1995; Volume 1, pp. 33–47.
23. Bartole, R. Caratteri sismostratigrafici, strutturali e paleogeografici della piattaforma continentale tosco-laziale; suoi rapporti con l'Appennino settentrionale. *Ital. J. Geosci.* **1990**, *109*, 599–622.
24. Bordoni, P.; Valensise, G. Deformation of the 125 ka Marine Terrace in Italy: Tectonic Implications. In *Coastal Tectonics*; Stewart, I.S., Vita-Finzi, C., Eds.; Geological Society London: London, UK, 1999; Volume 146, pp. 71–110.
25. Lambeck, K.; Antonioli, F.; Anzidei, M.; Ferranti, L.; Scicchitano, G.; Silenzi, S. Sea level change along the Italian coast during the Holocene and projections for the future. *Quat. Int.* **2010**, *232*, 250–257. [CrossRef]
26. Benincasa, F.; De Vincenzi, M. Il Monitoraggio Costiero Italiano. In *La Costa d'Italia*; Ginesu, S., Ed.; Carlo Delfino: Sassari, Italy, 2011; Volume 1, pp. 119–134.
27. Mattei, G.; Caporizzo, C.; Corrado, G.; Vacchi, M.; Stocchi, P.; Pappone, G.; Schiattarella, M.; Aucelli, P. On the influence of vertical ground movements on Late-Quaternary sea-level records. A comprehensive assessment along the mid-Tyrrhenian coast of Italy (Mediterranean Sea). *Quat. Sci. Rev.* **2022**, *279*, 107384. [CrossRef]
28. Segre, A.G. Giacimenti pleistocenici con fauna e industria litica a Monte Argentario (Grosseto). *Riv. Sci. Pr.* **1959**, *14*, 1–17.
29. Blanc, A.C. L'uomo fossile del Monte Circeo. Un cranio neandertaliano nella Grotta Guattari a San Felice Circeo. *Riv. Antropol.* **1938**, *32*, 1–16.
30. Carboni, G.; Conati Barbaro, C.; Manfredini, A. The Copper Age Settlement Le Cerquete-Fianello (Maccarese, Rome): A Sedentary Community in the Lagoon Environment of Maccarese (Rome). In Proceedings of the Union Internationale de Sciences Préhistoriques et Protohistoriques, Atti del XII Congresso, Forlì, Italy, 8–14 September 1996; pp. 27–34.
31. Alessandri, L. *L'Occupazione Costiera Protostorica del Lazio Centromeridionale*; BAR International Series 1592; British Archaeological Reports Oxford Ltd.: Oxford, UK, 2007; Volume 1, pp. 1–241.
32. Bietti Sestieri, A.M. L'Età del Bronzo finale nella Penisola Italiana. *Padusa* **2008**, *44*, 7–54.
33. Dolfini, A. Neolitico, Eneolitico ed Età del Bronzo. In *Paesaggi d'Acque La laguna di Orbetello e il Monte Argentario tra Preistoria ed Età Romana*; Negroni-Catacchio, N., Cardosa, M., Dolfini, A., Eds.; Centro Studi di Preistoria e Archeoologia: Milano, Italy, 2017; Volume 1, pp. 310–327.
34. Bellotti, P.; Calderoni, G.; Carboni, M.G.; Di Bella, L.; Tortora, P.; Valeri, P.; Zernitskaya, V. Late Quaternary landscape evolution of the Tiber River delta plain (Central Italy): New evidence from pollen data, biostratigraphy and 14C dating. *Z. Geomorph.* **2007**, *51*, 505–534. [CrossRef]

35. Bellotti, P.; Calderoni, G.; Di Rita, F.; D'Orefice, M.; D'Amico, C.; Esu, D.; Magri, D.; Preite Martinez, M.; Tortora, P.; Valeri, P. The Tiber river delta plain (central Italy): Coastal evolution and implications for the ancient Ostia Roman settlement. *Holocene* **2011**, *21*, 1105–1116. [CrossRef]
36. Bellotti, P.; Calderoni, G.; Dall'Aglio, P.L.; D'Amico, C.; Davoli, L.; Di Bella, L.; D'Orefice, M.; Esu, D.; Ferrari, K.; Bandini Mazzanti, M.; et al. Middle-to Late-Holocene environmental changes in the Garigliano delta plain (Central Italy): Which landscape witnessed the development of the Minturnae Roman colony? *Holocene* **2016**, *26*, 1457–1471. [CrossRef]
37. D'Orefice, M.; Bellotti, P.; Bertini, A.; Calderoni, G.; Neri, P.; Di Bella, L.; Fiorenza, D.; Foresi, L.M.; Louvari, M.A.; Rainone, L.; et al. Holocene Evolution of the Burano Paleo-Lagoon (Southern Tuscany, Italy). *Water* **2020**, *12*, 1007. [CrossRef]
38. Folk, R.L. The Distinction between Grain Size and Mineral Composition in Sedimentary-Rock Nomenclature. *J. Geol.* **1954**, *62*, 344–359. [CrossRef]
39. Stuiver, M.; Polach, H.A. Discussion Reporting of 14C data. *Radiocarbon* **1977**, *19*, 355–363. [CrossRef]
40. Stuiver, M.; Reimer, P.J. Extended 14C Data Base and Revised CALIB 3.0 14C Age Program. *Radiocarbon* **1993**, *35*, 215–230. [CrossRef]
41. Coltorti, M.; Ravani, S. Caratteri Geomorfologici della Fascia Costiera Compresa tra la Foce del Fiume Albegna, la Laguna di Orbetello e Ansedonia. In *Paesaggi d'Acque La laguna di Orbetello e il Monte Argentario tra Preistoria ed Età Romana*; Centro Studi di Preistoria e Archeoologia: Milano, Italy, 2017; Volume 1, pp. 27–34.
42. Giraudi, C. The sediments of the 'Stagno di Maccarese' marsh (Tiber river delta, central Italy): A late-Holocene record of natural and human-induced environmental changes. *Holocene* **2011**, *21*, 1233–1243. [CrossRef]
43. Sadori, L.; Jahns, S.; Peyron, O. Mid-Holocene vegetation history of the central Mediterranean. *Holocene* **2011**, *21*, 117–129. [CrossRef]
44. Dolci, M. L'Età Romana e Tardo-Antica. In *Paesaggi d'Acque La Laguna di Orbetello e il Monte Argentario tra Preistoria ed Età Romana*; Centro Studi di Preistoria e Archeoologia: Milano, Italy, 2017; Volume 1, pp. 344–374.
45. Mazzini, I.; Anadon, P.; Barbieri, M.; Castorina, F.; Ferreli, L.; Gliozzi, E.; Mola, M.; Vittori, E. Late Quaternary sea-level changes along the Tyrrhenian coast near Orbetello (Tuscany, central Italy): Palaeoenvironmental reconstruction using ostracods. *Mar. Micropaleontol.* **1999**, *37*, 289–311. [CrossRef]
46. Bartolini, C.; Corda, L.; D'Alessandro, L.; La Monica, G.B.; Regini, E. Studi di Geomorfologia costiera; III, Il tombolo di Feniglia. *Boll. Soc. Geol. Ital.* **1977**, *96*, 117–157.
47. Bellotti, P. La Laguna e I Suoi Tomboli: Meccanismi di Formazione e Caratteri Sedimentologici. In *Paesaggi d'Acque La laguna di Orbetello e il Monte Argentario tra Preistoria ed Età Romana*; Centro Studi di Preistoria e Archeoologia: Milano, Italy, 2017; Volume 1, pp. 296–301.
48. Mazzanti, R. Il punto sul Quaternario della fascia costiera e dell'Arcipelago della Toscana. *Boll. Soc. Geol. Ital.* **1983**, *102*, 419–556.
49. Cardini, L.; Rittatore, F. Ansedonia. *Riv. Sc. Preist.* **1953**, *8*, 210–211.
50. Revedin, A.; Mella, A. Materiali ceramici dal "Grottino" di Ansedonia. *Riv. Sc. Preist.* **1990**, *42*, 155–170.
51. Negroni Catacchio, N.; Cardosa, M. Dalle Sorgenti al Mare. Rapporti tra l'Area Interna e le Lagune Costiere nel Territorio tra Fiora e Albegna. In Proceedings of the Atti del V incontro di studi Preistoria e Protostoria in Etruria Paesaggi d'acque, Sorano, Italy, 12–14 May 2000; Centro Studi di Preistoria e Archeoologia: Milano, Italy, 2002; pp. 157–177.
52. Bronson, R.C.; Uggeri, G. Isola del Giglio, Isola di Giannutri, Monte Argentario, Laguna di Orbetello. *Studi Etruschi* **1970**, *38*, 201–214.
53. Cardosa, M. Paesaggi d'Acque al Monte Argentario. In Proceedings of the Atti del VI incontro di studi Preistoria e Protostoria—Etruria Paesaggi d'acque, Valentano-Pitigliano, Italy, 13–15 September 2002; Centro Studi di Preistoria e Archeoologia: Milano, Italy, 2004; pp. 405–415.
54. Barker, G. Landscape and Society: Prehistoric Central Italy. *Acad. Press* **1981**, *73*, 1–281.
55. Poggesi, G. Punta Degli Stretti. In *Memorie Sommerse. Archeologia Subacquea in Toscana*; Poggesi, G., Rendini, P., Eds.; Pitigliano, Italy, 1998; Volume 1, pp. 216–222.
56. Arcangeli, L.; Pellegrini, E.; Poggesi, G. L'Insediamento Sommerso dell'età del Bronzo finale di Punta Degli Stretti Nella Laguna di Orbetello (GR). In Proceedings of the Atti del V incontro di Studi Preistoria e Protostoria in Etruria Paesaggi d'Acque, Sorano, Italy, 12–14 May 2000; Centro Studi di Preistoria e Archeoologia: Milano, Italy, 2002; pp. 133–143.
57. Negroni Catacchio, N. Orbetello e Monte Argentario (GR). Il progetto "Paesaggi d'Acque". *Not. Soprintend. Beni Archeol. Toscana* **2005**, *1*, 448–451.
58. Poesini, S. La Produzione Ceramica di Punta Degli Stretti (Orbetello, GR): Aggiornamento Degli Studi. In Proceedings of the Atti del X Incontro di Studi Preistoria e Protostoria in Etruria Paesaggi d'Acque, Valentano-Pitigliano, Italy, 10–12 September 2010; Centro Studi di Preistoria e Archeoologia: Milano, Italy, 2012; pp. 553–566.
59. Tolomei, C.; Cattaneo, P. *Della Edificazione di Una Città sul Monte Argentario. Ragionamenti di Claudio Tolomei e Pietro Cattaneo (1544–1547)*; Milanesi, G., Ed.; Tipografia dell'Arte della Stampa: Firenze, Italy, 1885; pp. 1–19.
60. Santangelo, M. *L'Antiquarium di Orbetello*; Generico: Roma, Italy, 1954; pp. 1–28.
61. Benedetti, L.; Capuzzo, P.; Fontana, L.; Rossi, F. Paesaggi d'Acque. Duna Feniglia, loc. Ansedonia. Scavo di un Insediamento del Primo Ferro: Risultati e Prospettive. In Proceedings of the Atti del VIII Incontro di Studi Preistoria e Protostoria in Etruria Paesaggi d'Acque, Valentano-Pitigliano, Italy, 15–17 September 2006; Centro Studi di Preistoria e Archeoologia: Milano, Italy, 2008; pp. 261–284.

62. Benedetti, L.; Capuzzo, P.; Fontana, L.; Rossi, F. Nuovi Dati Dallo Scavo di Duna Feniglia (Orbetello, GR). In Proceedings of the Atti del IX Incontro di Studi Preistoria e Protostoria in Etruria Paesaggi d'Acque, Valentano-Pitigliano, Italy, 12–14 September 2008; Centro Studi di Preistoria e Archeoologia: Milano, Italy, 2010; pp. 157–167.
63. Negroni Catacchio, N.; Cardosa, M.; Domanico, L.; Miari, M. New Information on the Late Bronze Age Settlements of Sorgenti della Nova (Viterbo) and Sovana (Grosseto) within the Framework of Etruscan and Italian Protovillanovan. In Proceedings of the XIII International Congress of Prehistoric and Protohistoric Sciences, Forlì, Italy, 8–14 September 1996; ABACO: Forlì, Italy, 1998; Volume 4, pp. 401–408.
64. Coarelli, F. Colonizzazione romana e viabilità. *Dialoghi Archeol.* **1988**, *6*, 35–48.
65. Fentress, E. Peopling the Countryside: Roman Demography in the Albegna Valley and Jerba. In *Quantyfing the Roman Economy: Methods and Problems*; Oxford Scholarship Online: Oxford, UK, 2009; Volume 1, pp. 127–161.
66. Fentress, E. Cosa in the Empire: The unmaking of a Roman town. *J. Rom. Archaeol.* **1994**, *7*, 208–222. [CrossRef]
67. Scott, R.T. A New Inscription of the Emperor Maximinus at Cosa. *Chiron* **1981**, *11*, 309–314.
68. Faccenna, C.; Funiciello, R.; Bruni, A.; Mattei, M.; Sagnotti, L. Evolution of a transfer related basin: The Ardea basin (Latium, Central Italy). *Basin Res.* **1994**, *6*, 35–46. [CrossRef]
69. Milli, S.; D'Ambrogi, C.; Bellotti, P.; Calderoni, G.; Carboni, M.G.; Celant, A.; Di Bella, L.; Di Rita, F.; Frezza, V.; Magri, D.; et al. The transition from wave-dominated estuary to wave-dominated delta: The Late Quaternary stratigraphic architecture of Tiber River deltaic succession (Italy). *Sediment. Geol.* **2013**, *284–285*, 159–180. [CrossRef]
70. Belluomini, G.; Iuzzolini, P.; Manfra, L.; Mortari, R.; Zalaffi, M. Evoluzione recente del delta del Tevere. *Geol. Romana* **1986**, *25*, 213–234.
71. Bellotti, P.; Carboni, M.G.; Milli, S.; Tortora, P.; Valeri, P. La piana deltizia del Fiume Tevere: Analisi di facies ed ipotesi evolutiva dall'ultimo 'low stand' glaciale all'attuale. *G. Geol.* **1989**, *51*, 71–91.
72. Bellotti, P.; Chiocci, F.L.; Milli, S.; Tortora, P.; Valeri, P. Sequence stratigraphy and depositional setting of the Tiber delta: Integration of high resolution seismics, well logs and archaeological data. *J. Sediment. Res.* **1994**, *64*, 416–432.
73. Bellotti, P.; Milli, S.; Tortora, P.; Valeri, P. Physical stratigraphy and sedimentology of the late Pleistocene–Holocene Tiber Delta depositional sequence. *Sedimentology* **1995**, *42*, 617–634. [CrossRef]
74. Bellotti, P.; Davoli, L.; Sadori, L. Landscape diachronic reconstruction in the Tiber delta during historical time: A holistic approach. *Geogr. Fis. Dinam. Quat.* **2018**, *41*, 3–21.
75. Amorosi, A.; Milli, S. Late Quaternary depositional architecture of Po and Tevere river deltas (Italy) and worldwide comparison with coeval deltaic successions. *Sediment. Geol.* **2001**, *144*, 357–375. [CrossRef]
76. Milli, S.; Mancini, M.; Moscatelli, M.; Stigliano, F.; Marini, M.; Cavinato, G. From river to shelf, anatomy of a high-frequency depositional sequence: The Late Pleistocene to Holocene Tiber depositional sequence. *Sedimentology* **2016**, *63*, 1886–1928. [CrossRef]
77. Di Rita, F.; Celant, A.; Magri, D. Holocene environmental instability in the wetland north of the Tiber delta (Rome, Italy): Sea–lake–man interactions. *J. Paleolimnol.* **2010**, *44*, 51–67. [CrossRef]
78. De Angelis d'Ossat, G. Geo-pedogenesi delle terre sul delta del Tevere. *Réch. Sol.* **1938**, *6*, 138–168.
79. Giraudi, C. Evoluzione tardo-olocenica del delta del Tevere. *Quaternario* **2004**, *17*, 477–492.
80. Arnoldus-Huyzendveld, A. The Lower Tiber valley, environmental change and resources in historical time. *Eur. J. Post Class. Archaeol.* **2017**, *7*, 173–198.
81. Salomon, F. Les origines d'Ostie: Quelles interactions avec la dynamique d'embouchure ? (Delta du Tibre, Italie). *Arch. Archéol. Hist. Anc.* **2020**, *7*, 129–140. [CrossRef]
82. Bicket, A.R.; Rendell, H.M.; Claridge, A.; Rose, P.; Andrews, J.; Brown, F.S.J. A multiscale geoarchaeological approach from the Laurentine shore (Castelporziano, Lazio, Italy). *Géomorphologie* **2009**, *15*, 241–256. [CrossRef]
83. Arnoldus-Huyzendveld, A.; Paroli, L. Alcune considerazioni sullo sviluppo storico dell'ansa del Tevere presso Ostia e sul porto-canale. *Archeol. Laz.* **1995**, *12*, 383–392.
84. Alessandri, L. Il Lazio Centromeridionale Nelle età del Bronzo e del Ferro. Ph.D. Thesis, Rijksuniversiteit Groningen, Groningen, The Netherlands, October 2009; pp. 1–620.
85. Manfredini, A. *Le Dune, il Lago, il Mare. Una Comunità di Villaggio dell'Età del Rame a Maccarese*; Istituto italiano di preistoria e protostoria: Firenze, Italy, 2002; pp. 1–268.
86. Castelli, R.; Facciolo, A.; Gala, M.; Grossi, M.C.; Rinaldi, M.; Ruggeri, D.L.; Sivilli, S. Scavi e Ritrovamenti. In *Interporto Roma-Fiumicino. Prove di Dialogo tra Archeologia, Architettura e Paesaggio*; Alinea: Firenze, Italy, 2008; Volume 1, pp. 69–86.
87. Morelli, C.; Carbonara, A.; Forte, V.; Giudice, R.; Manacorda, P. The Landscape of the Ager Portuensis, Rome: Some New Discoveries, 2000–2002. In *Archaeology and Landscape in Central Italy (Papers in Memory of John A. Lloyd)*; Lock, G., Faustoferri, A., Eds.; Oxford University School of Archaeology: Oxford, UK, 2008; Volume 69, pp. 213–232.
88. Ruggeri, D.; Gala, M.; Facciolo, A.; Grossi, M.C.; Morelli, C.; Rinaldi, M.L.; Sivilli, S.; Carrisi, E.; Citro, D.; De Castro, F.R. Località le Vignole–Maccarese (Fiumicino–Roma): Risultati Preliminari dello Scavo Protostorico. In *L'alba dell'Etruria. Fenomeni di Continuità e Trasformazioni nei Secoli XII-VIII a.C. Ricerche e Scavi, Proceedings of the Atti del IX Incontro di Studi Preistoria e Protostoria in Etruria Paesaggi d'Acque, Valentano-Pitigliano, Italy, 12–14 September 2008*; Centro Studi di Preistoria e Archeoologia: Milano, Italy, 2010; pp. 327–337.

89. Morelli, C.; Carbonara, A.; Forte, V.; Grossi, M.C.; Arnoldus-Huyzendveld, A. La Topografia Romana dell'Agro Portuense alla Luce delle Nuove Indagini. In *Portus and its Hinterlands: Recent Archaeological Research, Archaeological Monographs*; The British School at Rome: London, UK, 2011; Volume 18, pp. 261–285.
90. Morelli, C.; Forte, V. Il Campus Salinarum Romanarum e l'epigrafe dei conductores. *Melanges Ec. Fr.* **2014**, *126*, 9–21. [CrossRef]
91. Grossi, M.C.; Sivilli, S.; Arnoldus-Huyzendveld, A.; Facciolo, A.; Rinaldi, M.L.; Ruggeri, D.; Morelli, C. A Complex Relationship between Human and Natural Landscape: A Multidisciplinary Approach to the Study of the Roman Saltworks in "Le Vignole-Interporto" (Maccarese, Fiumicino—Roma). In *Archaeology of Salt. Approaching an Invisible Past*; Sidestone Press: Leiden, The Netherlands, 2015; Volume 1, pp. 83–101.
92. Malmgren, C. Ficana (Acilia). In *Enea nel Lazio: Archeologia e Mito*; catalogo della mostra; Palombi eds: Roma, Italy, 1981; pp. 102–104.
93. Fischer Hansen, T. *Scavi di Ficana I. Topografia Generale*; Istituto Poligrafico dello Stato: Roma, Italy, 1990; pp. 1–21.
94. Malmgren, C. Early Settlement at Ficana. In Proceedings of the Munuscula Romana Conference, Lund, Sweden, 1–2 October 1988; Svenska Institutet I Rom: Stokholm, Sweden, 1991; pp. 17–28.
95. Brandt, J.R. *Scavi di Ficana II,1. Il Periodo Protostorico ed Arcaico*; Istituto Poligrafico dello Stato: Roma, Italy, 1996; pp. 1–27.
96. Malmgren, C. *Ficana, the Final Bronze and Early Iron Age*; Lund University: Lund, Sweden, 1997; pp. 1–144.
97. Arnoldus-Huyzendveld, A.; Pellegrino, A. Development of the Lower Tiber Valley in Historical Times. *Mem. descr. Carta Geol. Ital.* **2000**, *54*, 219–226.
98. Arnoldus-Huyzendveld, A. The Natural Environment of the Agro Portuense. In *Portus. An Archaeological Survey of the Port of Imperial Rome*; The British School at Rome: Rome, Italy, 2005; pp. 14–30.
99. Arnoldus-Huyzendveld, A. Le Dinamiche Evolutive dell'Ambiente Costiero e del Tevere. In *Paesaggi dell'Archeologia Invisibile. Il Caso del Distretto Portuense*; Quodlibet: Recanati, Italy, 2014; pp. 72–89.
100. Pannuzi, S. Viabilità e Utilizzo del Territorio. Il Suburbio Sud-Orientale di Ostia alla Luce dei Recenti Rinvenimenti Archeologici. In Proceedings of the Atti del Terzo Seminario Ostiense, Ricerche su Ostia e il Suo Territorio, Roma, Italy, 21–22 October 2015; L'Ecole Francaise de Rome: Roma, Italy, 2015; pp. 181–212.
101. Pannuzi, S. Il Suburbio Sud-Orientale di Ostia dall'Età Pre-Protostorica all'Età Moderna. In *Alle foci del Tevere: Territorio, Storia, Attualità*; Società Italiana di Geologia Ambientale: Roma, Italy, 2019; Volume 1, pp. 12–25.
102. Germoni, P.; Keay, S.; Millett, M.; Strutt, K. Ostia beyond the Tiber: Recent Archaeological Discoveries in the Isola Sacra. In Proceedings of the Atti del Terzo Seminario Ostiense, Ricerche su Ostia e il Suo Territorio, Roma, Italy, 21–22 October 2015; L'Ecole Francaise de Rome: Roma, Italy, 2015; pp. 127–143.
103. Buonaguro, S. La Basilica Paleocristiana Anonima di Castelfusano (Ostia). In *Marmoribus Vestita: Miscellanea in Onore di Federico Guidobaldi. Studi di Antichità Cristiana*; Pontificio Istituto di Archeologia Cristiana: Rome, Italy, 2011; Volume 63, pp. 287–303.
104. Casciello, E.; Cesarano, E.; Pappone, G. Extensional detachment faulting on the Tyrrhenian margin of the southern Apennines contractional belt (Italy). *J. Geol. Soc.* **2006**, *163*, 617–629. [CrossRef]
105. Malinverno, A.; Ryan, W.B.F. Extension in the Tyrrhenian Sea and shortening in the Apennines as result of arc migration driven by sinking of the lithosphere. *Tectonics* **1986**, *5*, 227–245. [CrossRef]
106. Ippolito, F.; Ortolani, F.; Russo, M. Struttura marginale tirrenica dell'Appennino campano: Reinterpretazione di dati di antiche ricerche di idrocarburi. *Mem. Soc. Geol. It.* **1973**, *12*, 227–250.
107. Abate, D.; De Pippo, D.; Ilardi, M.; Pennetta, M. Studio delle caratteristiche morfoevolutive quaternarie della piana del Garigliano. *Quaternario* **1998**, *11*, 149–158.
108. Cosentino, D.; Federici, I.; Cipollari, P.; Gliozzi, E. Environments and tectonic instability in Central Italy (Garigliano Basin) during the late Messinian Lago-Mare episode: New data from the onshore Mondragone 1 well. *Sediment. Geol.* **2006**, *188*, 297–317. [CrossRef]
109. Milia, A.; Iannace, P.; Tesauro, M.; Torrente, M.M. Upper plate deformation as marker for the Northern STEP fault of the Ionian slab (Tyrrhenian Sea, Central Mediterranean). *Tectonophysics* **2017**, *710*, 127–148. [CrossRef]
110. Pennetta, M.; Corbelli, V.; Esposito, P.; Gattullo, V.; Nappi, R. Environmental Impact of Coastal Dunes in the Area Located to the Left of the Garigliano River Mouth (Campany, Italy). *J. Coast. Res.* **2011**, *61*, 421–427. [CrossRef]
111. Aiello, G.; Amato, V.; Aucelli, P.; Barra, D.; Corrado, G.; Di Leo, P.; Di Lorenzo, H.; Jicha, B.; Pappone, G.; Parisi, R.; et al. Multiproxy study of cores from the Garigliano Plain: An insight into the Late Quaternary coastal evolution of Central-Southern Italy. *Palaeogeogr. Palaeoclimatol. Palaeoecol.* **2021**, *567*, 110–298. [CrossRef]
112. Di Lorenzo, H.; Aucelli, P.; Corrado, G.; De Iorio, M.; Schiattarella, M.; Ermolli, E. Environmental evolution and anthropogenic forcing in the Garigliano coastal plain (Italy) during the Holocene. *Holocene* **2021**, *31*, 1089–1099. [CrossRef]
113. Tarragoni, C.; Bellotti, P.; Davoli, L.; Petronio, B.M.; Pietroletti, M. Historical and recent environmental changes of the Ombrone Delta (Central Italy). *J. Coast. Res.* **2011**, *61*, 344–352. [CrossRef]
114. Tarragoni, C.; Bellotti, P.; Davoli, L.; Raffi, R.; Lupia Palmieri, E. Assessment of coastal vulnerability to erosion: The case of Tiber River Delta (Tyrrhenian Sea, central Italy). *Ital. J. Eng. Geol. Environ.* **2015**, *2*, 1–15.
115. Guidi, A. Alcune osservazioni sul popolamento protostorico tra il golfo di Gaeta e gli Aurunci. *Latium* **1991**, *8*, 5–31.

116. Guidi, A. Il Popolamento del Territorio di Mondragone tra il Neolitico e la Prima età del Ferro. In Proceedings of the Atti della 40° Riunione Scientifica dell'Istituto Italiano di Preistoria e Protostoria: Strategie di Insediamento fra Lazio e Campania in età Preistorica e Protostorica, Roma, Napoli, Pompei, Italy, 30 November–3 December 2005; Istituto Italiano di Preistoria e Protostoria: Firenze, Italy, 2007; pp. 671–682.
117. Morandini, A. Gli insediamenti costieri di età protostorica nel Lazio Meridionale. *Latium* **1999**, *16*, 5–47.
118. Conta Haller, G. *Ricerche su Alcuni Centri Fortificati in Opera Poligonale in Area Campano-Sannitica (Valle del Volturno. Territorio tra Liri e Volturno)*; Accademia di Archeologia, Lettere e Belle Arti di Napoli: Napoli, Italy, 1978; Volume 1, pp. 31–40.
119. Mingazzini, P. *Il Santuario della dea Marica alle Foci del Garigliano*; Monumenti Antichi dell'Accademia Nazionale dei Lincei, Hoepli: Milano, Italy, 1938; Volume 37, pp. 693–957.
120. Talamo, P. *L'Area Aurunca nel Quadro dell'Italia Centromeridionale. Testimonianze Archeologiche di Età Arcaica*; British Archaeological Reports: Oxford, UK, 1987; pp. 1–191.
121. D'Urso, M.T. Il tempio della dea Marica alle foci del Garigliano. *Storia* **1985**, *1*, 1–143.
122. Trotta, F. Minturnae Preromana e il Culto di Marica. In *Minturnae*; Fillipo Coarelli: Roma, Italy, 1989; Volume 1, pp. 11–28.
123. Laforgia, E. Nuove osservazioni sul tempio di Marica. *Ann. Archeol. Stor. Antica* **1992**, *14*, 69–76.
124. Rescigno, C. L'edificio arcaico del santuario di Marica alle foci del Garigliano: Le terrecotte architettoniche. *Ann. Archeol. Stor. Antica* **1993**, *15*, 85–108.
125. Andreani, M. Sul Santuario di Marica alla Foce del Garigliano. In *Santuari e Luoghi di Culto nell'Italia Antica*; Quilici, L., Quilici, S., Eds.; L'Erma di Bretschneider: Roma, Italy, 2003; Volume 1, pp. 177–207.
126. Livi, V. Religious Locales in the Territory of Minturnae: Aspects of Romanization. In *Religion in Republica Italy*; Schultz, C.E., Harvey, P.B., Eds.; Yale Classical Studies: New Haven, CT, USA, 2006; Volume 33, pp. 90–116.
127. Bellini, G.R.; Trigona, S.L. Minturnae e il Garigliano. In *Centre and Periphery in the Ancient World, Proceedings of the XVIIIth International Congress of Archeology, Merida, Spain, 13–17 May 2013*; Alvarez Martinez, J.M., Nogales, T., Rodà, I., Eds.; Museo Nacional de Arte Romano: Merida, Spain, 2014; pp. 721–724.
128. Pompilio, F. Fonti Epigrafiche e Letterarie. In *Monete dal Garigliano. IV. Monete Romane (Caesar-Nero): Fonti Epigrafiche e Letterarie*; Bellini, G.R., Ed.; Edizioni Ennerre: Milano, Italy, 1999; Volume 1, pp. 85–110.
129. Remmelzwaal, A. Soil genesis and Quaternary landscape development in the Tyrrhenian coastal area of south-central Italy. *Publ. Fys. Geogr. Bodemkd. Lab.* **1978**, *28*, 1–310.
130. Ruegg, S.D. Minturnae: A Roman River Seaport on the Garigliano River, Italy. In *Archaeology of Coastal Changes, Proceedings of the First International Symposium Cities on the Sea. Past and Present, Haifa, Israel, 22–29 September 1986*; Raban, A., Ed.; British Archaeological Association: Oxford, UK, 1988; pp. 209–228.
131. Ruegg, S.D. *Underwater Investigations at Roman Minturnae. Part I: Report (147 pp.); Part 2: Catalogue of Artifacts (221 pp.)*; Förlag Astroms, P., Ed.; Göttemborg: Sävedalen, Sweden, 1995; Volume 1, pp. 1–221.
132. Bellini, G.R. Minturnae: 296 a.C.–44 a.C. Dalla Deduzione della Colonia alla Morte di Cesare. In *Minturnae Antiquarium. Monete dal Garigliano. II. Monete Greche, Provinciali, Romane e Tessere Romane (di Bronzo e di Piombo)*; Bellini, G.R., Ed.; Edizioni Ennerre: Milano, Italy, 1998; Volume 1, pp. 9–15.
133. Bellini, G.R. Minturnae porto del Mediterraneo. *Romula* **2007**, *6*, 7–28.
134. Bellini, G.R.; Trigona, S.L.; Matullo, G. Minturnae. Il Garigliano. In *Lazio e Sabina, Proceedings of the Settimo Incontro di Studi sul Lazio e la Sabina, Roma, Italy, 9–11 Marzo 2010*; Ghini, G., Ed.; Quasar: Roma, Italy, 2011; pp. 563–574.
135. Arthur, P. Assetto Territoriale ed Insediamento fra Tardo Antico ed Alto Medioevo nel Bacino del Garigliano. In *Minturnae*; Coarelli: Roma, Italy, 1989; pp. 183–192.
136. Torre, P. Monte d'Argento: Indagini preliminari. *Archeol. Laz.* **1988**, *9*, 432–440.
137. Rossillo, O. La colonia musulmana del Garigliano. *Studi Stor. Archeol. Archeoclub Minturnae Scauri* **1985**, *1*, 1–67.
138. Di Biasio, A. *Il Passo del Garigliano Nella Storia d'Italia. Il Ponte di Luigi Giura*; Caramanica: Marina di Minturno, Italy, 1994; pp. 1–295.
139. Torre, P. Il rinvenimento di Ceramiche Invetriate e Smaltate con Motivi Decorativi nell'Insediamento di Monte d'Argento. In *Le Ceramiche di Roma e del Lazio Meridionale in età Medievale e Moderna, Proceedings of the Atti del III Convegno di Studi, Roma, Italy, 19–20 Aprile 1996*; de Minicis, E., Ed.; Kappa Bologna: Italy, 1998; pp. 183–296.
140. Ciarrocchi, B.; Torre, P. Reperti Vitrei dello Scavo di Monte d'Argento (Minturno, LT): Un Panorama Tipologico dal Tardoantico al Basso Medioevo. In *Il Vetro in Italia Meridionale e Insulare: Settime Giornate Nazionali di Studio Comitato Nazionale AIHV, Proceedings of the Atti del Secondo Convegno Multidisciplinare, Napoli, Italy, 5–7 December 2001*; Piccoli, C., Sogliani, F., Eds.; De Freda: Napoli, Italy, 2003; pp. 185–196.
141. Ferrari, K. *Ad Ostium Liris Fluvii. Storia del Paesaggio Costiero alla Foce del Garigliano*; Bononia University Press: Bologna, Italy, 2016; pp. 1–189.
142. Gregori, G.L.; Nonnis, D. Dal Liris al Mediterraneo: L'Apporto dell'Epigrafia Repubblicana alla Storia del Porto di Minturnae. In Proceedings of the Atti del Convegno Immensa Aequora—Ricerche Archeologiche, Archeometriche e Informatiche per la Ricostruzione dell'Economia e dei Commerci del Bacino Occidentale del Mediterraneo, Roma, Italy, 24–26 January 2011; Olcese, G., Ed.; pp. 163–177.
143. Ferrari, K.; Dall'Aglio, P.L.; Bellotti, P.; Davoli, L.; Di Bella, L.; Torri, P.; Bandini Mazzanti, M. Holocene landscape evolution at the Garigliano River mouth. *Ann. Bot.* **2013**, *3*, 191–198.

Article

Weakening of Coastlines and Coastal Erosion in the Gulf of Guinea: The Case of the Kribi Coast in Cameroon

Philippes Mbevo Fendoung [1], Mesmin Tchindjang [2,*] and Aurélia Hubert-Ferrari [3]

[1] Department of Forest Engineering, Advanced Teachers' Training College for Technical Education, University of Douala, Douala P.O. Box 1872, Cameroon
[2] Department of Geography, Faculty of Arts, Letters and Social Sciences, University of Yaoundé 1, Yaoundé P.O. Box 30464, Cameroon
[3] SPHERES Research Unit, Department of Geography, Faculty of Science, University of Liege-Belgium, Quarter Village 4, Clos Mercator 3, Building 11 B, 4000 Liege, Belgium
* Correspondence: mtchind@yahoo.fr

Citation: Mbevo Fendoung, P.; Tchindjang, M.; Hubert-Ferrari, A. Weakening of Coastlines and Coastal Erosion in the Gulf of Guinea: The Case of the Kribi Coast in Cameroon. *Land* **2022**, *11*, 1557. https://doi.org/10.3390/land11091557

Academic Editors: Pietro Aucelli, Angela Rizzo, Rodolfo Silva Casarín and Giorgio Anfuso

Received: 11 July 2022
Accepted: 5 September 2022
Published: 13 September 2022

Publisher's Note: MDPI stays neutral with regard to jurisdictional claims in published maps and institutional affiliations.

Copyright: © 2022 by the authors. Licensee MDPI, Basel, Switzerland. This article is an open access article distributed under the terms and conditions of the Creative Commons Attribution (CC BY) license (https://creativecommons.org/licenses/by/4.0/).

Abstract: For more than four decades, the Gulf of Guinea's coasts have been undergoing a significant phenomenon of erosion, resulting from the pressures of both anthropogenic and marine weather forcings. From the coasts of West Africa (Senegal, Ivory Coast, Ghana, Benin, Togo, and Nigeria) to those of Central Africa (Gabon, Equatorial Guinea, and Cameroon), the phenomenon has been growing for more than four decades. The southern Cameroonian coastline from Kribi to Campo has become the scene of significant environmental dynamics that render it vulnerable to coastal erosion, which appears to be the major hazard of this coastal territory and causes a gradual degradation of the vegetative cover, thereby leading to the degradation of the coast's land/ground cover and human-made infrastructure. The objective of this work is to analyze the kinematics of the Kribian coastline between 1973 and 2020; to quantify the levels of retreat, accretion, and stability; and finally, to discuss the factors influencing the evolution of the coastline. The methodological approach is based on the large-scale processing of Landsat images with a spatial resolution of 30 m. Then, small-scale processing is carried out around the autonomous port of Kribi using Pléiades and Google Earth images from the years 2013, 2018, and 2020 with a 0.5 m spatial resolution. The Digital Shoreline Analysis System (DSAS) version 5 and ArcMap 10.5® tool are used to model coastal kinematics. In addition, the dynamics of the agro-industrial plantations are assessed via satellite images and landscape perception. Environmental degradation is measured with respect to the entire Cameroonian coastline through the supervised classification of Landsat images (1986–2020). The results show that erosion is in its initial phase in Kribi because significant retreats of the coastline are noticeable over the period from 2015–2020. Thus, between 1973 and 2020, the linear data present a certain stability. In total, +72.32% of the line remained stable, with values of +1.3% for accretion and +26.33% for erosion—obtained from Landsat images of 30 m resolution—with an average retreat of +1.3 m/year and an average accretion of 0.9 m/year between 1973–2020. Based on high-resolution images, between 2013 and 2019, the average retreat of the coastline on the Kribian coast was −8.5 m/year and the average accretion was about 7 m/year. Agro-industrial plantations are responsible for environmental degradation. Thus, at SOCAPALM in Apouh, there has been a clear growth in plantations, which has fallen from 53% in 1990 to 78% in 2020, i.e., an increase of 25% of its baseline area. This is linked to the fact that plantations are growing significantly, with increases of 16% in 1990, 28% in 2000, and 29% in 2020, for old plantations.

Keywords: coastal erosion; coastline; Gulf of Guinea; Kribi; fragilization

1. Introduction and Background of the Study

The African continent is currently experiencing galloping demographic growth. Coastal areas, especially urban centers, concentrate most of this population. The increase in the urban population in West Africa grew from 16% of the total population in 1961 to 47% in 2014 and it is expected to reach 66% in 2050 [1]. This significant growth of the urban population in Africa is linked to the migratory phenomenon as its main driver [2]. Population projections in Sub-Saharan Africa predict 1.8 billion inhabitants in 2050 and 2.6 billion in 2100 and do not envisage any stabilization before 2100 [3,4]. This population growth is dazzling in coastal areas, and it is accompanied by developments that contribute to the weakening (fragilization) of these areas [5–7]. In most cases, this worrying situation results in hazards including floods [8], pollution [9], and coastal erosion [10,11] with various consequences. Coastal erosion is the primary concern of the present study. It affects coasts around the world, with disproportionate repercussions, often dependent on the level of consideration of its management in the public policies of the related countries. Thus, it is shown that 70% of the coasts are experiencing coastal erosion, 20% are stable, and only 10% are enriching [12]. All the coasts of the world are concerned, and the entire international scientific community is being challenged.

The coasts of the Gulf of Guinea also experience various natural hazards, among which coastal erosion is the most important on African coasts [13–17]. From Mauritania to Nigeria, West African coasts are threatened [18,19]. In Central Africa, rapid erosion has been observed for the Pointe-Noire (Congo) and Libreville (Gabon) shorelines [20–22]. Issues directly affecting the African coastal zone include population growth and poverty and the loss of habitat and land through coastal erosion [22].

Throughout the world, several authors have dealt with the issue of coastal erosion using approaches specific to the geographical areas concerned. Thus, some of them have addressed this issue in the context of climate change [23]. In a mangrove context, authors have characterized coastal erosion and its impacts on this ecosystem in southern Thailand [24]. As an example of the role of human activities in exacerbating coastal erosion, one study reported the construction of the Akosombo dam in Ghana in West Africa, which led to the erosion of the Togolese coast [25]. The situation is similar on the coasts of Senegal and Benin [26]. The management of this hazard is complex. For over eight decades, the National Research Council [27] has presented some management strategies based on integrated approaches.

On the Cameroonian coasts, erosion significantly destroys the coast by depositing mudflats and sand along the coastline [28,29]. This mechanism is orchestrated by the rip currents or flood currents that strike the coastline, undermining the loose substrate. The deposited sediments (sand) are collected and marketed by the coastal populations. The effects of interior erosion on roads, houses, agricultural fields in ruins, and the destruction of landing sites for boats—and for canoes in the northern section of the Cameroonian coast, particularly in Isangele, Bamusso, Kombo Itindi, and Kombo Abedimo—can be observed. On the southern Cameroonian coast, from the Wouri estuary [11] to Campo, there are visible signs of coastal erosion around Cap Cameroun, Yoyo I and II, Kribi, and Campo. The markers of this erosion are, among others, the presence of falling trees, destroyed houses, and damaged dykes.

Additionally, the Kribi-Campo coastal section is currently experiencing a recent shoreline adjustment, resulting in a loss of sediment, fattening, and stability. Indeed, of the 135 km of the south Cameroon shoreline, 88% of this land is experiencing erosion and 22% is experiencing accretion, while 0.24% remains stable [30]. This explains the vulnerability of the Kribi coastline to natural hazards (erosion), and particularly this coastal area around the Kribi Sea Port and above Campo [31]. Recent data [8] show that these coastal areas around Kribi experienced sediment enrichment between 1973 and 2007 with an average surface area of 4,081,400 m^2 representing 45% of the total dynamics, while the recorded regressive dynamic corresponding to the linear coastal retreat of the concave sectors of the coast represents 55% (i.e., 4,949,550 m^2) of the total dynamics [8].

The causes mentioned above are likely related to a sediment deficit caused by both anthropogenic and meteo-marine forcings [32]. The breaking of the waves feeding the tides, the rise in the sea level, and the action of marine currents and waves [19] can also be conducive factors. The noted anthropogenic forcings involve the multiple impacts of man on the coastal environment. Indeed, the strong anthropogenic pressure on the mangrove ecosystems of Campo is the main factor of coastal erosion [33]. The construction of the Kribi Sea Port (between 2007 and 2018) has orchestrated a notorious thinning of this section of the southern Cameroonian coast that is likely to aggravate erosion. The population growth of Kribi city and the resulting needs are also the basis of the weakening of the coastline. Other anthropogenic factors are significant, such as the construction of dams as in the case of the Memve'ele hydroelectric dam (15 km from Kribi) built on the Ntem River, which traps a large part of the sediments in transit towards the Kribi coasts. However, additional factors include the port facilities, the cutting of mangroves, and the extraction of sand.

These observations allow us to address the main objective of this paper, which is to analyze the kinematics of the Kribian coastline between 1973 and 2020. This analysis includes the quantification of the levels of retreat, accretion, and stability and finally a discussion of the factors influencing the evolution of the coastline. Thus, the dynamic observed creates environmental upheavals at the coastal level, with repercussions on ecosystems and populations. To better study this phenomenon, the coastlines of the years 1973, 2000, 2015, and 2020 were analyzed using Landsat sensors since the objective is to quantify the rate of retreat of the coastline.

2. Geographical and Geomorphological Settings

2.1. Geographical and Administrative Situation of the Kribian Coast

The Kribi coast is located in the South Administrative Region and at the Head of the Ocean Division. The coastal section observed in this study extends over three municipalities: Kribi 1, Kribi 2, and Campo (which is the largest). It extends between 2°10′3″ N and 9°10′10″ E (Figure 1). Kribi is the second economic pole in Cameroon and is therefore booming. It is a seaside town with a cosmopolitan population of approximately 104,000 inhabitants [34]. Originally, Kribi was inhabited by the Pygmies from the Bagyeli tribe, later followed by the Batanga and Mabi tribes [35].

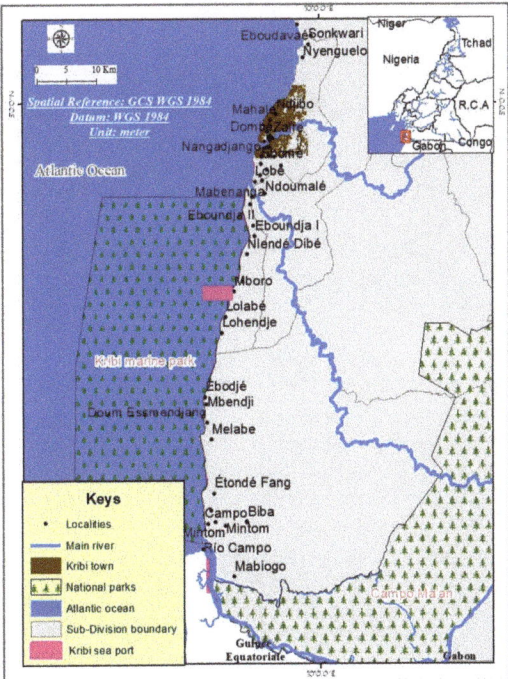

Figure 1. Location map of the study area.

2.2. Geological Context of the Kribian Coast

Geologically, the Kribian coast is not only composed of sedimentary rocks but also of crystallophyllian rocks from the basic metamorphic complex composed of gneiss, micaschists, quartzites, etc. [36]. The Kribian coastline spreads over a low-altitude coastal plain that is less than 100 m in elevation. This coast is sandy or rocky in places, with soils resulting from the weathering of rocks and that are suitable for agriculture.

Sandstone formations have also been recorded for this coastal section, which are linked to the phenomenon of progression both toward the north and the south of the ocean. These sandstones range from the Upper Aptian at Campo to the Albian with a thickness that varies from 500 to 600 m [37]. The presence of an Archean and Paleoproterozoic greenstone belt, Archean charnockites and granitoids, gneiss, a mobile granitic zone, a sedimentary basin, and an overlap zone can also be noticed (Figure 2) [38].

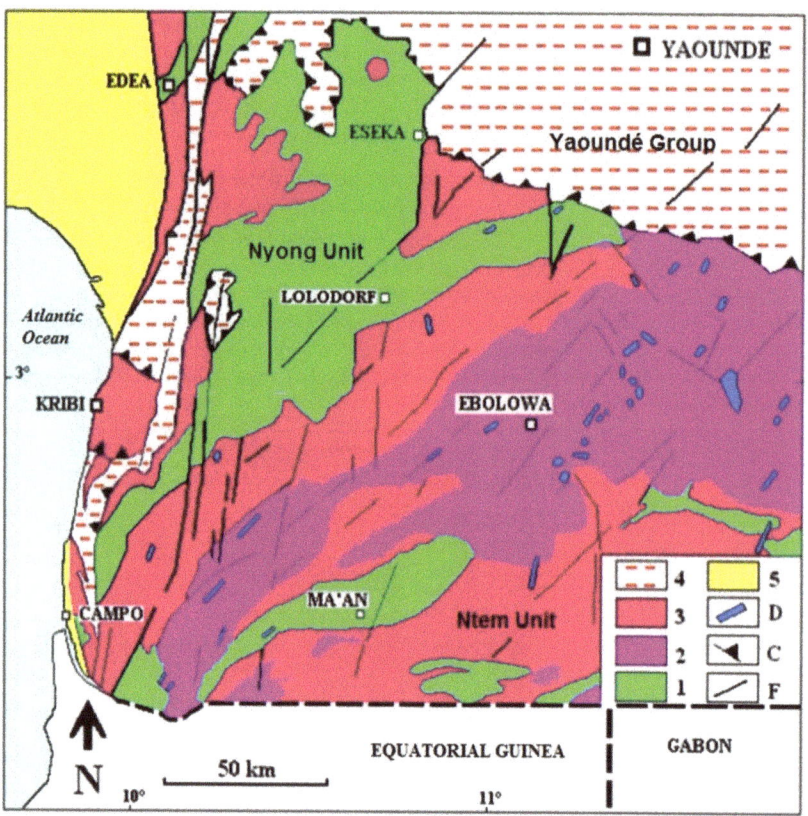

Figure 2. Geological structure of the Kribian coast (Source: adapted from Oslisly [38]). Caption: (1) yellow ferralitic soils showing a high level of acidity, with a degree of saturation varying from 15 to 20% at the surface and less than 10% at depth. (2) Soils on gneiss and lateritic schists, composed of highly leached humus sand and red sandy clay on lateritic schists. (3) Hydromorphic soils rich in organic matter, which generally lie alongside the banks of rivers and marshy and mangrove areas (mouth of the Ntem). (4) Mobile granite zones; (5) sedimentary basin; (C) overlaps; (D) dolerites; (F) main faults.

From a pedological point of view, the Kribian coast mainly presents: (1) yellow ferralitic soils showing a high level of acidity, with a degree of saturation that varies from 15 to 20% on the surface and less than 10% in depth [39]; (2) soils on gneiss and lateritic schists, composed of highly leached humus sand and red sandy clay on lateritic schists; (3) and hydromorphic soils rich in organic matter generally located near rivers, area swamps, and mangroves (mouth of the Ntem) (Figure 2) [40].

The 2014 Global Baseline for Soil Resources updated in 2015 ([41]) classifies tropical African countries, such as Kenya, the DR Congo, and Cameroon, as containing Nitisols, Plinthosols (soils with a high pisolite content up to 80%), and Andosols (derived from parent materials other than glass-rich volcanic products that are located in humid regions).

This physical setting influences the erosion of the Kribi coasts and needs to be analyzed to better understand the processes underway in this area.

2.3. Hydrogeomorphological Characterization of the Kribi-Campo Area

2.3.1. Geomorphology of the Kribian Coast

The Cameroonian coastal sedimentary basin, located on the edge of the Gulf of Guinea, covers an area of 7000 km^2 [42]. It corresponds to a subsiding trench formed during the Cretaceous that is gradually deepening towards the ocean where it reaches thicknesses of 4000 m at 40 km and 8000 m further offshore. Globally, the Kribian coast is less than 5 m in altitude with respect to areas that are very close to the sea; however, after the coastline, the altitude can reach 15 to 50 m in some areas (Figure 3).

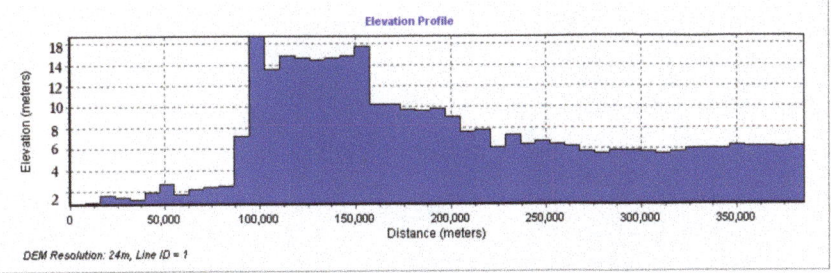

Figure 3. Kribi elevation profile, generated from the Digital Terrain Model (DTM).

The Kribian coasts' main geological formations are black marls and clays as well as sandstone sands. These formations develop on a varied typology of the coast, including—among other things—low loose coasts, rocky coasts, concave coasts, etc.

(1) A low coast

This is the most developed type of coastline, spanning from the Wouri estuary and extending to Kribi for more than 150 km, with an average altitude varying from 0 to 30 m. Its slopes are gently in contact with the ocean and vary from 0 to 6%. The low coast in the Campo area, for example, has a flat morphotype typical of large beaches [42], with a generally uniform profile with no break in the slope. These profiles are characterized by the area's flattest swash and surf zones. In the area of the town of Kribi, there are some marine terraces corresponding to the moderately concave morphotype, presenting a moderate slope.

(2) The rocky coasts

These coasts are rough and made up of rocky areas with narrow beaches, which run from Kribi to Campo over nearly 80 km. These beaches were shaped on crystalline schist fragmenting into large blocks. These shales give a particular aspect to the coastal landscape of Kribi, which is renowned for its fine sandy beaches. According to Hegge's [43] classification of coastal morphotypes, this rocky coast has a very narrow beach and a relatively steep intertidal zone, sometimes not exceeding 2 m. It is characterized by a strong step-like break in the subtidal zone. From north to south, the first break in the slope is located towards the Lobé Falls. It increases from almost 2% to almost 5%. Another break is present to the north of the Kribi seaport. The coastline profiles shown in Figure 4 illustrate the complex configuration of the Kribi coast.

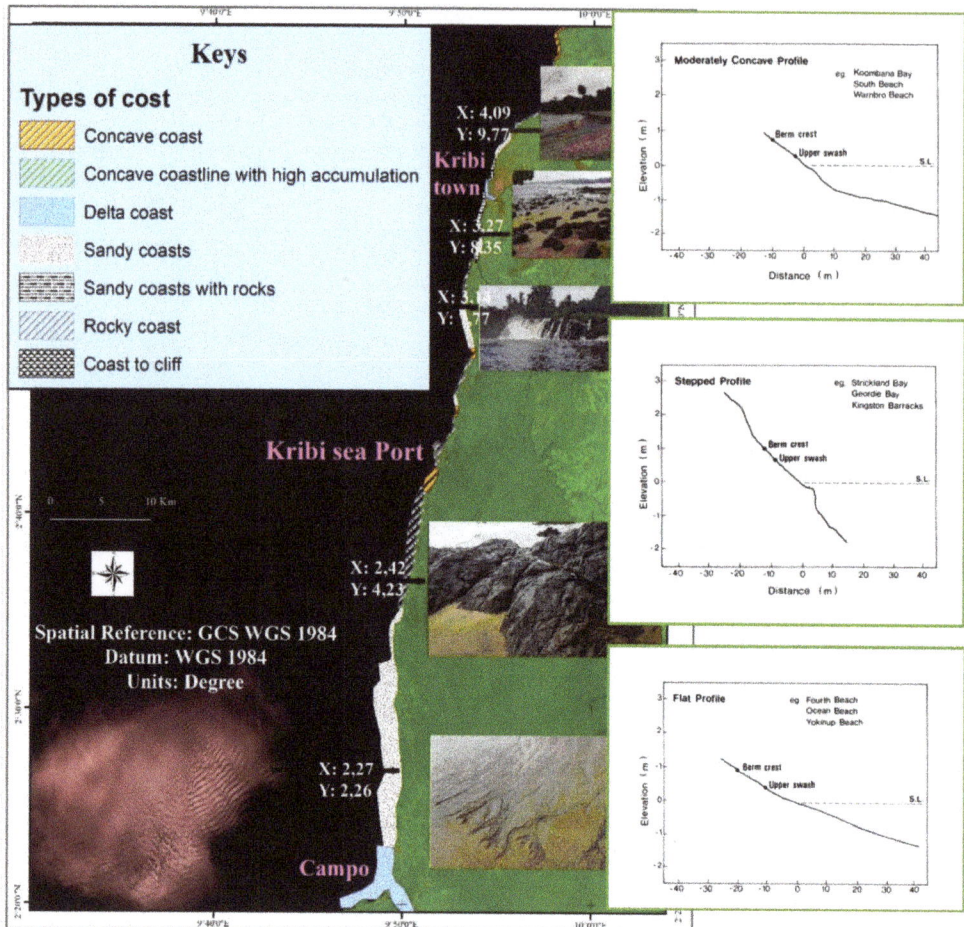

Figure 4. Kribian types and morphotypes of the coast. These figures show an alternation between sandy low coast (flat and concave morphotype profiles) and rocky coast (stepped profile).

Profile 1 characterizes these relatively small beaches, with a forebeach of less than 10 m and a swash zone of less than 5 m, and moderate slope breaks along the profile. Profile 2 corresponds to the staircase morphotype. This is a very narrow beach with a relatively steep intertidal zone. It is characterized by a strong break in the slope in the subtidal zone. This type of profile can be found at the Lobé Falls in Kribi. Finally, the third profile presents a flat morphotype characteristic of wide beaches, with a uniform profile with no break in the slope. These profiles are characterized by the flattest swash and surf zones. The flattening extends into the submerged area.

2.3.2. The Kienké River and Its Hydrogeomorphological Characteristics

The Kienké is a small river in southern Cameroon that flows into the Atlantic Ocean near Kribi. With a 190 km length, the Kienké River occupies a catchment area of more than 1435 km^2 in Kribi, with a flow rate of around 40.5 m^3/s [44]. A report [45] characterized the basins of several rivers on the Kribian coast (with respect to land use, slope, geology, etc.). It appears that the Kienké is divided into four geomorphological units: altitudes >176 m, between 177–361 m and 362–595 m, and a last high unit situated between 596–1172 m. Upstream, the waters of the Kienké flow over a unit with a fairly high average slope (1‰)

and 1.6‰ upstream, which gradually decreases as the river approaches the ocean. The geology mainly consists of upper and lower gneiss. It is a watershed dominated by a vast Atlantic Forest, which degrades as one approaches the town of Kribi due to the presence of the rubber and palm agro-industries. The dominant tree species are Sterculiaceae and Ulmaceae [46]. The waters of the Kienké, as in the other catchments, have a diffuse flow upstream on the Kienké massif, with several tributaries, before forming a single channel at the outlet to the Atlantic Ocean. The flow of this river has a strong seasonal variation. The former and first Kribi Sea Port was built at its mouth. The city of Kribi spread towards its two eastern and western banks. There is no element to assess the solid inflows of sediments by the Kienké River, but the examination of the sounding plans of the Kribi Sea Port seems to show that the inflows of sand must be relatively low. By analogy with other rivers in Cameroon, notably the Wouri River in particular, the sand inputs should not exceed 10,000 to 20,000 t/year, i.e., 5000 to 10,000 m^3 of the materials in place. On the other hand, the silt inputs could be much greater without having values in kind to quantify them (10 times more, perhaps). However, the current Kienké transport capacity has been estimated at 1200 m^3/month [47], and this value is only reached during the rainy season.

2.3.3. The Lobé River and Its Hydrogeomorphological Characteristics

More than 130 km long, the Lobé River has its source in the Ntem massif in the center of the Campo-Ma'an National Park and joins the ocean through the equatorial forest. Its mouth, a few kilometers south of Kribi, forms a kind of delta with many small arms that spread out over 1 km before emerging into the Atlantic Ocean through a series of waterfalls, the highest of which measures around 15 m (Figure 5). Its watershed is 1940 km^2 wide with an estimated flow of 105 m^3/s [48], which varies from 20 m^3/s (February) to 300 m^3/s (October) [47]. Sediment transport for the Lobe River is estimated at 11.1 m^3/year, with a strong seasonal variation [47]. The Lobé runs its course on a granite-gneissic base, which is more or less lateralized, with a maximum altitude of the catchment area of 500 m and regular forest cover [47].

Figure 5. The Lobé waterfalls, Kribi coast, and the Kienké flowing to the sea by small waterfalls under the bridge at the Kribi landing stage (Credit: Mbevo, August 2018 and Tchindjang, July 2017).

2.3.4. The Ntem River and Its Hydrogeomorphological Characteristic

The Ntem is a river that serves as a natural border between three states of Central Africa, namely, Gabon, Equatorial Guinea, and Cameroon. It has its source in the Gabonese Province of Woleu-Ntem, and it then flows into the Atlantic Ocean in the extreme south of Cameroon at Campo. Its catchment area is 31,000 km^2, with a length of 460 km. Its average flow rate is 195.3 m^3/s [48]. The Ntem watershed has an altimetry that varies between 524–613 m, and the waters flow on an average slope unit of 0.5‰, with very small areas of breakage [49]. The Ntem sub-basin covers nearly 70% of the waters of the region that feeds the Atlantic basin, while that of the Lobé covers just 30% [50]. The Ntem River runs through an ancient crystalline basement whose formation is marked by a significant folding phase (Liberian orogeny), which led to the creation of the Ntem Complex. Today, its extension is limited to southern Cameroon and constitutes the northern borders of the Congo craton (Archean) [40]. There are granite deposits whose age refers to Precambrian D (2.7–3 billion years) or Precambrian C (2 to 2.3 Ma.) [50].

2.4. Vegetation on the Kribi Coast

The Kribi coast's vegetation essentially consists of dense forest and is highly dependent on paleoclimatic variations. The botanist Letouzey [46] reports the omnipresence of the *Lophira alata* species that he linked to past anthropic clearings. There is a certain number of woody species of great importance such as Okoumé (*Aukoumea klaineana*) or Azobé (*Lophira alata*), which are highly dependent on ancient human occupations. The most beautiful stands of these two species are once again located around cleared areas or former occupations [48]. Finally, there are also small pockets of mangroves located in the mouths of the Kienké, Lobé, and Ntem rivers.

A supervised classification of Landsat images from the years 2000, 2015, and 2020 allowed us to see the evolution of the vegetation cover on the Kribian coast (Figure 6).

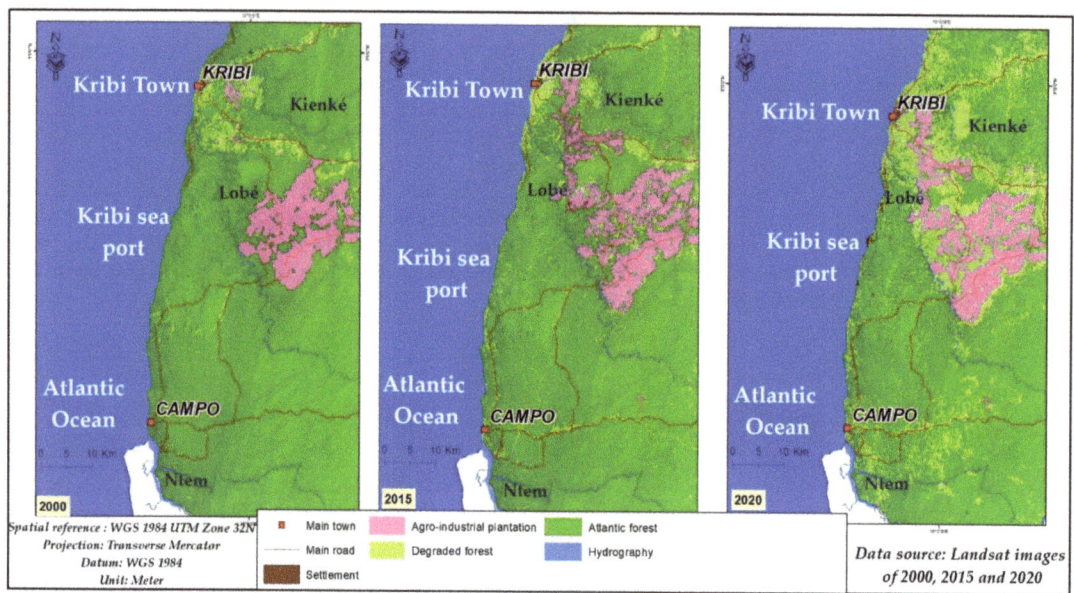

Figure 6. Evolution of land use on the Kribi-Campo coast between 2000 and 2020.

There is a clear degradation of the vegetation cover and an increase in agro-industrial plantations and built-up areas. Between 2000 and 2010, the Atlantic Forest declined. This decline will be consolidated between 2010 and 2020. Urbanization and agro-industrial activity are therefore the main factors of degradation affecting the Kribian coast (Figure 7).

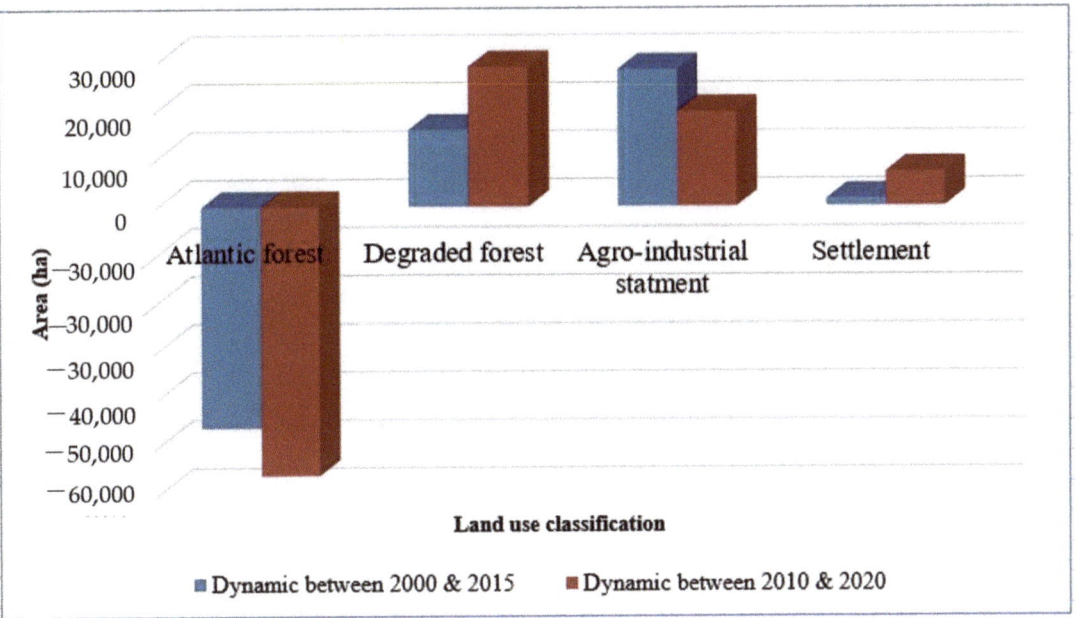

Figure 7. Dynamics of the different land use classes between 2000–2020.

2.5. Climate of the Kribi Coast

The Kribian climate is equatorial, humid, and subject to the typical monsoons of coastal areas [51]. The annual rainfall is about 2900 mm in Kribi. December, January, and February constitute the long dry season, with an average rainfall of 262 mm, while September, October, and November belong to the long rainy season, with a more than 1183 mm average rainfall [52]. The area's temperatures are relatively high: the diurnal thermal amplitude is 2.42 in Kribi, the maximum temperatures are between 26 to 31 °C, and the peaks are generally observed in the dry season (December, January, February, and March). Minimum temperatures fluctuate between 23 and 25 °C) during the rainy season (April, May, September, and October) (Figure 8). On the Kribian coast, rainfall has little effect on soil erosion. Indeed, it is a forested area, and the soils are sufficiently protected. The only places where rainfall has an effect are the areas that have suffered from degradation/deforestation. With approximately 2900 mm of rain per year, the Kribi coastline remains one of the least watered, compared to Douala, which receives between 3500 and 4000 mm of rain per year, and Debunscha with approximately 13,000 mm per year. For this reason, the main factors that amplify coastal erosion on the Kribi coast are anthropogenic forcings.

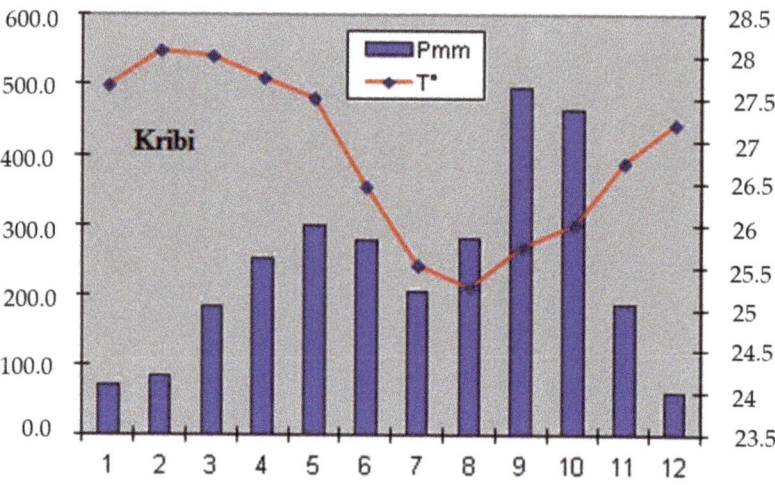

Figure 8. Onbrothermal diagram of Kribi.

2.6. Tide and Sediment Transit on the Kribi Coast

There is evidence that waves and tides are significant factors with respect to the modeling and reshaping of the coastline. However, the Kribi coastline has one of the lowest tides compared to other areas such as the Wouri estuary. The average height is 1.2 m [52], with some variation between the Kribi Sea Port (Mboro) and Lolabé.

The sediment transit for the Kribi coast, according to the Kribi Sea Port data on four sections, presents a maximum of 140,000 m^3/year north of the current Kribi Sea Port and a minimum of 80,000 m^3/year to the south from this port. At the port site, the sediment transport is estimated at 50,000 m^3/year (Figure 5).

2.7. Wind on the Kribian Coast

The analysis of the annual and seasonal boreal wind rose for the continental shelf region of the Cameroon coast (2–3° N & 5–10° E) from 1960 to 2001 [53], which made it possible to understand the directions and wind speeds of the Kribi coast. Old data (until 1978) on the winds (month, speed, and direction) around Kribi showed that the highest speeds are recorded in December, with a value of 20 m/s (Figure 9). The sediment transport is more influenced by winds blowing from the ocean to the coast, since winds blowing from the interior to the coast are blocked by the Atlantic Forest. These winds give speed to the littoral drift and the swell that affect the coast. In the soft coast, the erosion is more important than in the rocky one.

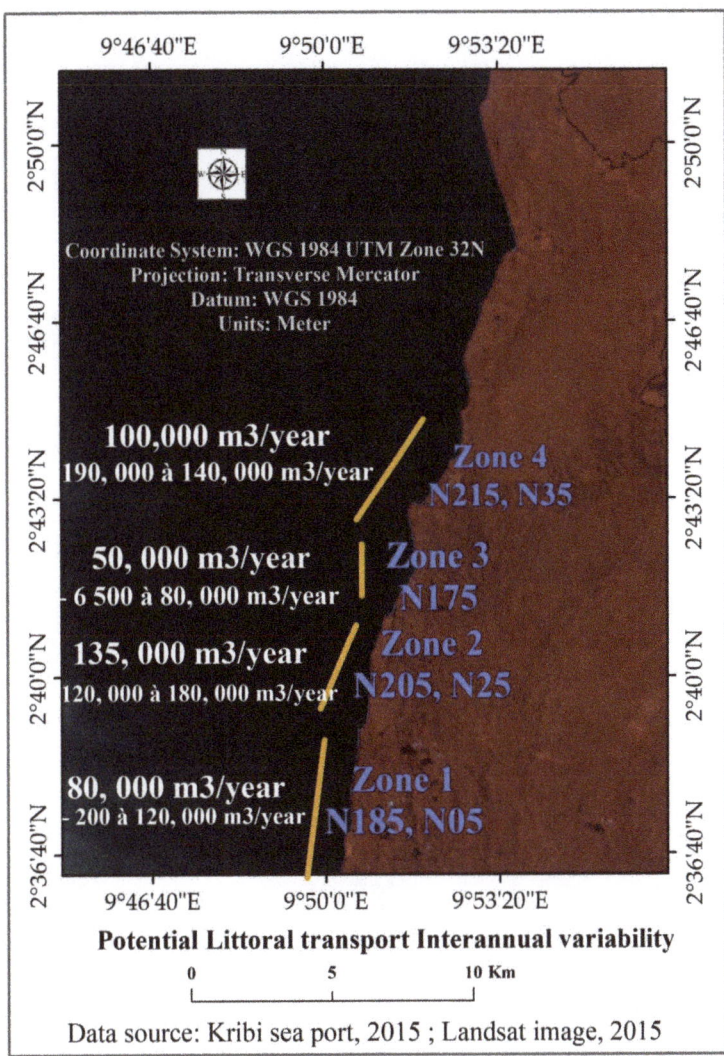

Figure 9. Coastal drift and sediment transport on a portion of the Kribi coast (Source: [54]).

Recent wind data around Kribi [55] confirm this influence of seasonality. Between December, January, and February, the prevailing winds were from the SW direction (24 km/h), S (16 km/h), and SSW (14 km/h), respectively. The same trend continued in March, April, and May, with the only difference that the W winds (16 km/h) appeared and those from the S decreased in intensity (8 km/h). Between June and August, SW (24 km/h) and SSW (24 km/h) winds were dominant, followed by SWS and S winds. In September, October, and November, SW winds regained the upper hand, with a greater force of 24 km/h, followed by WSW (16 km/h) and SSW (16 km/h) winds. An annual summary shows that the SW winds are dominant [56].

These winds correspond to the trade winds (regular winds in the intertropical regions between 23–27° N and 23–27° S). They regularly blow from east to west, from the subtropical high pressures (subtropical ridge) to the equatorial low pressures (convergence zone intertropical), and they are very active in the Gulf of Guinea [55]. They mainly take place from December to March (dry season).

2.8. Marine Currents around the Kribi Coastline

The main current at work on the Cameroonian coast is the coastal drift. Its orientation is South–North for the entire Cameroonian coast and it is essentially conditioned by the waves generated by the local SW-oriented wind and by the swells of long periods, which affect the whole West African coast and originate in southern latitudes (45–60°) [56,57]. Two types of swells are identified [58]. The one oriented NNW is generated by the easterly winds in the region of western Namibia. The other is generated by the westerly winds in the southern Atlantic Ocean around latitudes 50° S [56]. The height of the swell varies significantly, with a maximum from June–August and a minimum from December–February [56]. On the Kribi coast specifically, these currents are relatively weak, with variations depending on the site. A series of three measurements was made around the Kribi port site (1st, 2nd, 3rd). The results obtained show little variation in the speed of the currents: from 0.3 m/s to 0.6 m/s. (Figure 10). It can be concluded from the third measurement that the shape of the Kribi coastline influences the direction and speed of the currents.

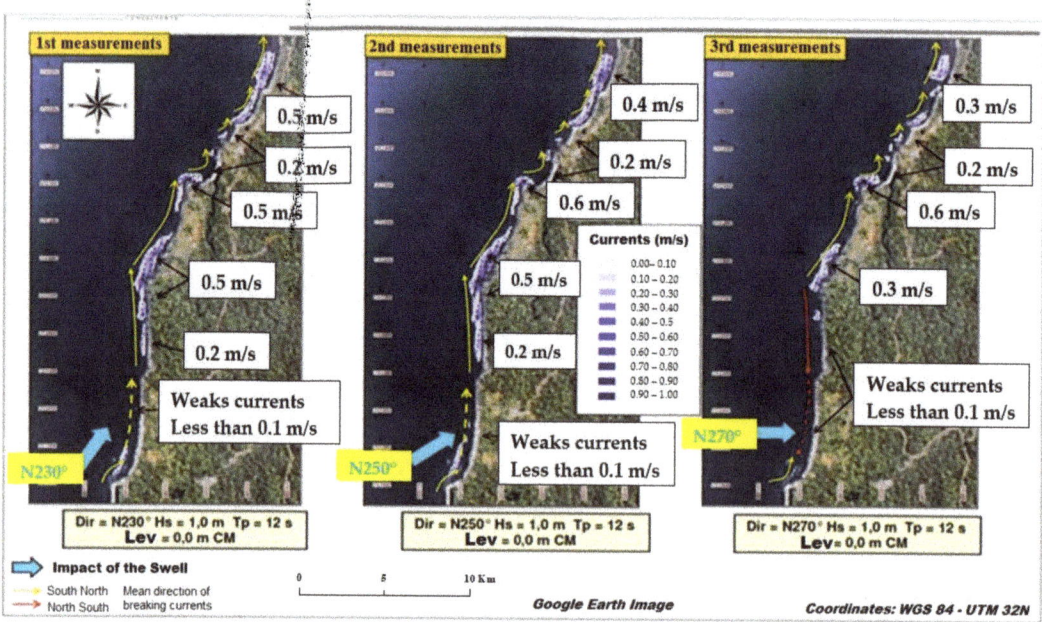

Figure 10. Speed of currents affecting the Kribi coastline (Source: [54]).

All in all, this coastal area is subject to the action of large swells, and the materials brought by the rivers (sands and kaolin clays, etc.) are taken up by the south–north coastal drift and by a west–west coastal current (swell) [55]. (Data from the Kribi Sea Port [54].) Sediment transport on the coastal section of Kribi was significant between 1992 and 2008 (Figure 11). From 1992–1993, the rate of sediment transit was around 100,000 m^3/year, which fell slightly to about 90,000 m^3/year between 1994–1996. A peak was observed in 1997, with a sediment transit of 140,000 m^3/year (Figure 11).

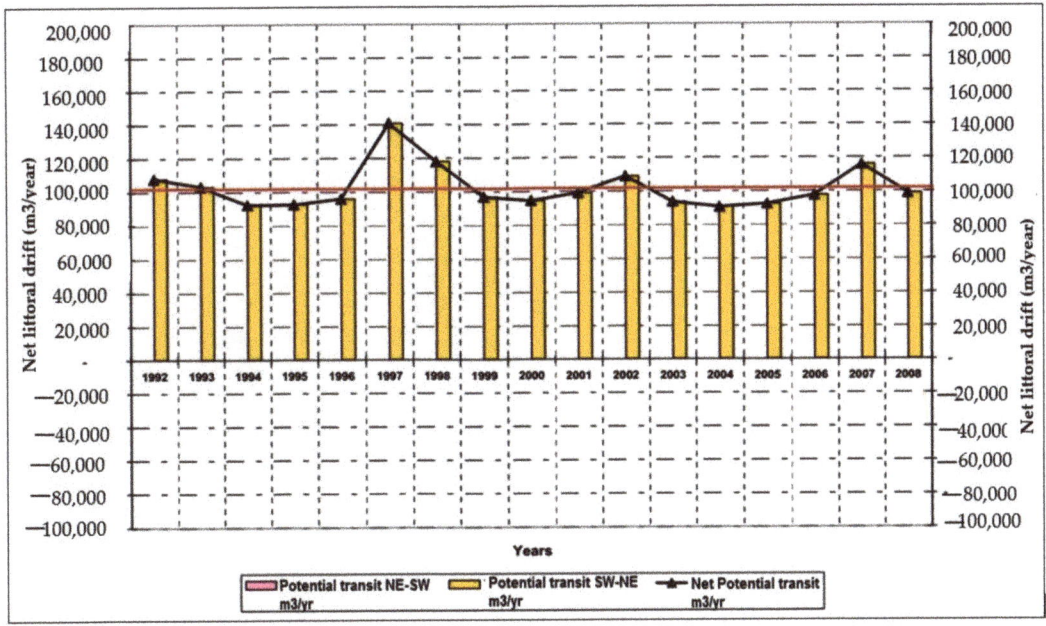

Figure 11. Sediment transport by littoral drift on the Kribi coast (Source: Kribi sea port [54]).

The above data reveal that the sedimentation of the Kribi coast is taking place and is exacerbated by the implementation of the Kribi Sea Port (we will use the French acronym PAK). Indeed, sedimentation was generally present on the Kribian and Cameroonian coastline [53]. However, since the beginning of the port's construction in 2007, this sedimentation seems to have increased, given the hydro-sedimentary disturbances observed in situ. Thus, there is a high accumulation of sediment to the south of the harbor due to the presence of the harbor's protection dam. This should be confirmed through the analysis of the bathymetric configuration of the Cameroonian coast to understand its influence on marine processes.

2.9. Bathymetric Configuration in the Kribi Area

Characteristic of coasts with a straight profile and rocky structures, the Kribi coastline presents a consistent bathymetric variation from the open sea towards the coast. The lowest values can be recorded at the mouth of the Kienké River (Figure 12), which is more than 8 m in depth and about 3 km in width. The depths on the shore increase when moving longitudinally to the south. From the mouth of the Lobé to the PAK site, the dominant bathymetric values are between 10, 14, and 16 m, with a more than 2.5 km width at the mouth of the Lobé, 5 km at Mahale, 9 km at Eboundja II, and 3 km at the PAK site. These values increase when moving offshore to more than 30 m depth (Figure 12).

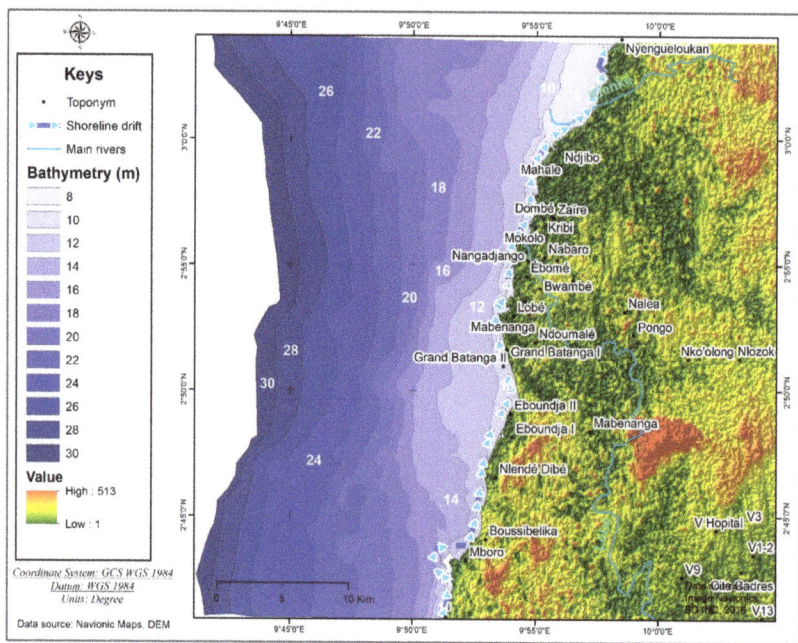

Figure 12. Bathymetric configuration of the Kribi coastline (Source: DEM and digitized navionic bathymetric map).

These natural forcing factors are determinants of the current observed dynamics of the Cameroonian coast and have a great influence on the sedimentary load and coastal erosion.

3. Materials and Methods

3.1. Land Cover Mapping

The land cover mapping was based on supervised classification of Landsat images from 2000, 2010, and 2020. After importing and assembling the bands in the Erdas Imagine® 2014 software (sold by Intergraph (Madison, AL, USA) in the United States), radiometric corrections were applied to the different images. Supervised classification was performed following the maximum likelihood algorithm. Table 1 shows the characteristics of the images used.

Table 1. Characteristics of processed Landsat images.

Date	Path and Row	Resolution	Radiometry	Sensor	Season	Purpose
1973/11/27	LM03_L1TP_201 057	30 m	8 bits	Mss	Dry	Extraction of the coastline
2000/11/6	LE7 186 057	30 m	8 bits	ETM	Dry	Extraction of the coastline and Land use classification
2015/02/01	LE7 186 056	30 m	8 bits	ETM	Dry	Extraction of the coastline and Land use classification
2020/02/21	LC08 187 057	30/15 m	16 bits	L8/OLI	Dry	Extraction of the coastline and Land use classification

3.2. Mapping of the Coastal Kinematics of Kribi

The mapping of coastal kinematics was performed at large and small scales. The first mapping procedure one considered the entire Kribian coastline, from the north of the city of Kribi to the south at Campo, using Landsat images (Table 1). The second focused on the northern and southern parts of the Kribi seaport. Therefore, two sources of remote-sensing data were used. The first source consisted of Landsat images with an average spatial resolution of 30 m.

The latter used Google Earth aerial photos (0.5 m) and Pleiades very high-resolution images (0.5 m) (Table 2). Both data sources aided in detecting and mapping coastline variations and erosion on the Kribi coast—at different spatial resolutions—in order to more accurately compare and draw conclusions. The types of software used for these processing operations were Erdas Imagine v14, ARCMAP v10.5 under ARCGIS v10.5, and Digital Shoreline Analysis Systems (DSAS v5), made by ESRI (Redlands, CA, USA).

Table 2. Characteristics of the processed Pleiades images.

Properties	Characteristics
Area	Kribi Pleiades Images 1
Image capture date	15 June 2013
Spatial Resolution	0.5 m
Sensor	Pleiades
ID	DS_PHR1B_201306150956399_SE1_PX_E009N02_1118_01728
Area	Kribi Pleiades Images 2
Image capture date	12 March 2016
Spatial Resolution	0.5 m
Sensor	Pleiades
ID	DS_PHR1A_201603120955019_FR1_PX_E009N02_1120_03618
Area	Kribi Pleiades Images 3
Image capture date	23 February 2019
Spatial Resolution	0.5 m
Sensor	Pleiades
ID	DS_PHR1A_201902231003214_FR1_PX_E009N02_1015_03204
Kribi	Google Earth Images 1 and 2
Image capture date	23/11/2015 and 03/05/2017
Spatial Resolution	0.5 m

3.2.1. Pretreatments under Erdas Imagine

Preprocessing under Erdas Imagine v14

The preprocessing scheme under Erdas consisted in making radiometric corrections of the selected Landsat images. These corrections had several purposes: resampling, haze reduction, and periodic noise elimination.

Resampling (Resolution Merge)

This process was applied to the 1973 Landsat images to reduce the spatial resolution from 60 m to 30 m in order to make this image compatible with other images from different sensors.

Correction of Atmospheric Noise (Haze Reduction)

This operation made it possible to reduce atmospheric errors of Landsat images as much as possible, without degrading the pixels. Such errors include mist, clouds, etc. For multispectral images, this method was based on a transformation that produces a component correlated with the haze.

Periodic Noise Elimination

It is this form of correction that eliminates tide-related errors in an image. The input image is first divided into overlapping blocks of 128 by 128 pixels. The Fourier transform of each block is calculated and the logarithms of each block are averaged.

3.2.2. Processing under ArcMap/DSAS

Coastline Extraction under ArcMap 10.5

Coastline definition is not an easy task because confusion appears between this notion and the shoreline [59]. The coastline is simply defined as the dividing line between sea and land [60]. Depending on the type of coast considered and as soon as we seek to draw this "limit", the notion of coastline becomes more complex and can be characterized in different ways, using several markers, and depending on the data available such as: (1) the limit of vegetation and (2) intersection line of the topographic surface with the level of the highest astronomical seas [61,62].

In the context of this study, it appears that the Kribi coast has a contrasting morphology as illustrated above (Figure 4). It alternates between rocky and soft coasts, with a straight profile; therefore, the coastline has varied in the linear analysis (Figure 4). We opted for the wet sand boundary, as did a similar study based on Landsat images by the authors of [63]. The rectilinear nature of the Kribi coastal profile allows for this boundary line to be remarkably visualized on the images, whether at high or low spatial resolutions.

Treatment Operations under DSAS v5

There are four main operations under DSAS v5 (made by United States Geological Survey, Reston, VA, USA) [64,65]: (1) Parameterization of the baseline and the shoreline, (2) Definition of transects, (3) Statistical analysis, and (4) Calculation of uncertainties and errors.

To succeed in these operations, we first defined transects' length (1000 m) and spacing between them (200 m) and set the shoreline [64,65], which were generated automatically. These transects are equidistant and perpendicular to the baseline. A pixel error corresponding to the resolution of each image was defined (30 m for Landsat images and 0.5 for Pleiades and Google Earth images). Regarding statistical analysis, DSAS uses several statistical techniques known as end point rate (EPR) method to compare the positions of the coastline over time to estimate the evolution of the coastline [64,65]. As shown by Equation (1), it is the distance on the transect between the two most recent and oldest coastlines divided by the number of years separating these coastlines [64].

$$R = D/T_e \qquad (1)$$

R is the speed in meters per year (m/year), D is the distance in meters, and T_e is the time elapsed between the oldest and the most recent coastline (years). The EPR still works even though only two coastlines are used to analyze the evolution [64].

Uncertainties and errors calculated can be grouped into four types: pixel errors (Ep), those related to the extraction of coastlines, digitizing (Ed), and planimetric (EP) [11]. These types of errors seem random and cannot appear automatically on all images. Their sum is determined by the total value of the errors (Et), which is equal to the square root of the different errors (Equation (2) from [64]).

$$ET = \pm\sqrt{Ep^2 + ERMs^2 + Ed2 + EP^2} \qquad (2)$$

The total error was estimated from three sources (Table 3): (i) the total shoreline position error was calculated for three periods; (ii) the measured transect error (Em) and annualized error (Ea) associated with the rate of shoreline change at a given transect, which was calculated over three time periods (1973–2000; 2000–2016; 2016–2020); and over a period, the annualized error was calculated using the following equation [66–68]:

$$Ea = \frac{\sqrt{Et1^2 + Et2^2 + Et3^2}}{\text{Total period (years)}} \qquad (3)$$

Table 3. Errors related to the erosion model on the Kribi coast.

Years	1973–2000	2000–2015	2015–2020	1973–2020
Pixel error (*Ep*)	2	5	/	/
RMS ortho-rectification (*ERMs*)	15	22	12	11
Digitalization error (*Ed*)	7	15	/	/
Planimetric error (*EP*)	/	23	23	/
Total error (*Et*)	15	38	10	14
Measured errors (*Em*) (m)	42	56	30	29
Annual error (*Ea*) (m/45 years)	/	0.65	/	0.65
Uncertainties in Calculations (*ECI*) (m/year)	2	7.2	0.52	0.43

Source: Statistical calculations in DSAS.

Figure 13 combines all the technical tools and methodological steps used to achieve the results that will be the subject of the following paragraphs.

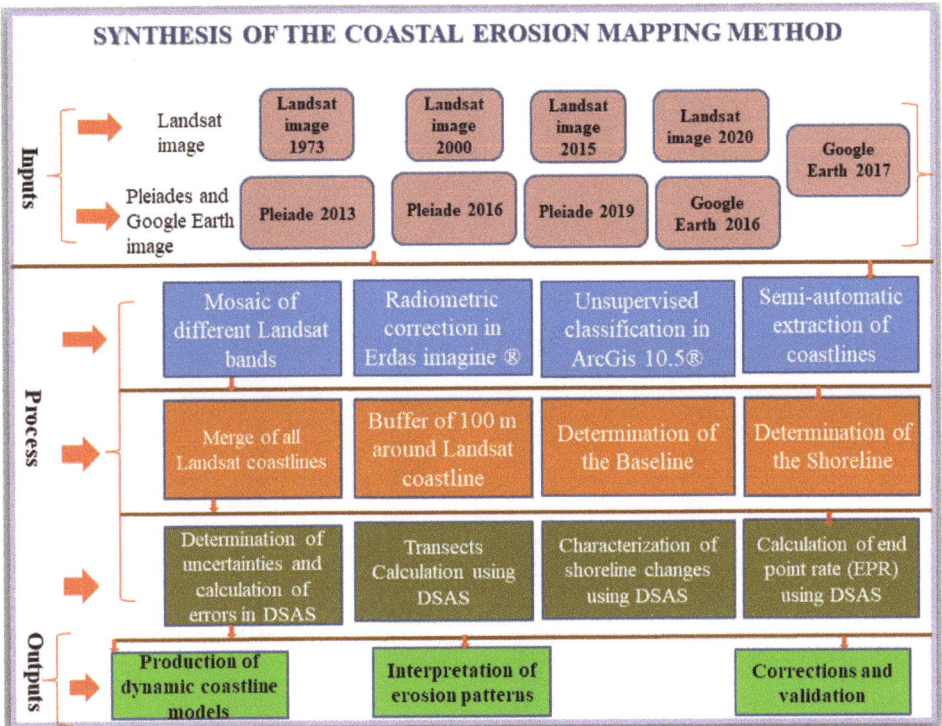

Figure 13. Flow Chart of tools and techniques used in assessing the Kribian coastal dynamic.

4. Main Findings and Interpretations

4.1. Modeling Coastal Erosion on the Kribi Coast

4.1.1. Large-Scale Modeling Using Landsat Images from 1973–2020

This modeling revolves around three major moments: before, during, and after the installation of the Kribi Sea Port (KSP). Retreat and accretion values between −5 m and 5 m are considered stable for observing erosion [65]. Diagrams showing the coastline dynamics were designed and are associated with the different figures.

Coastline Dynamics before the Construction of the PAK (1973–2000, i.e., 27 Years)

Prior to the establishment of the PAK, the analysis of the Landsat images from the period from 1973 to 2000 shows that the Kribi coastline had virtually no erosion sectors. It recorded an accretion on a significant slice of its linear dimension. Even the current PAK site was stable. The coastline is largely marked by stability (Figure 14).

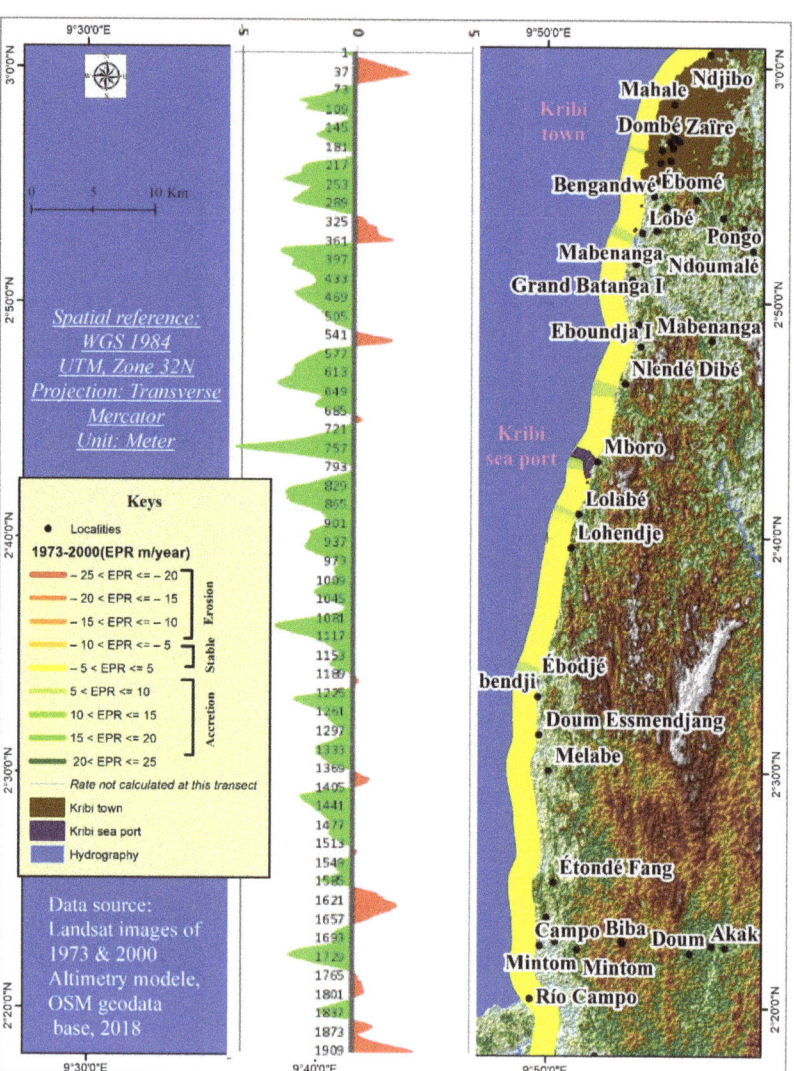

Figure 14. Rate of change (meters per year) in Kribi coastal erosion between 1973–2000. The figure was obtained with DSAS, from the linear extracted from Landsat images. The coastlines are divided into 1909 transects separated by 50 m. The cartographic representation is on the left. The graphic representation is on the right, with the number of transects on the ordinate axis, and the EPR in m/year on the abscissa axis.

An average accretion rate of +0.4 m/year and an accretion rate of 1.5 was determined. The accretion peak is situated in the area where the PAK is located, at about 5 m/year, followed by the Lolabé area (3.5 m/year) and Ebodjé (2.6 m/year). The erosion marks are disparate. At Etondéfon, we observe a decline rate of 2.1 m/year. In Eboundja, the retreat speeds are 1.8 m/year. Conclusively, all these values are less than −5, which implies the absence of real erosion. This coastline's stability during this 27-year period may be due to several factors. Firstly, the lack of urbanization allows the area to maintain its natural character. Secondly, the presence of the forest constitutes a natural barrier to any erosion. Thirdly, the main rivers (Kienké and Lobé) feeding the littoral drift (from the South to the North) do not experience wide sedimentary hindrances such as dams.

Coastline Dynamics during the Construction of the PAK (2000–2015, i.e., 15 Years)

During the period from 2000–2015, there was a regressive coastline dynamic, unlike in previous years. This took place at three essential points: towards the Lobé falls, around the PAK, and towards the locality of Etondé Fang (Figure 15).

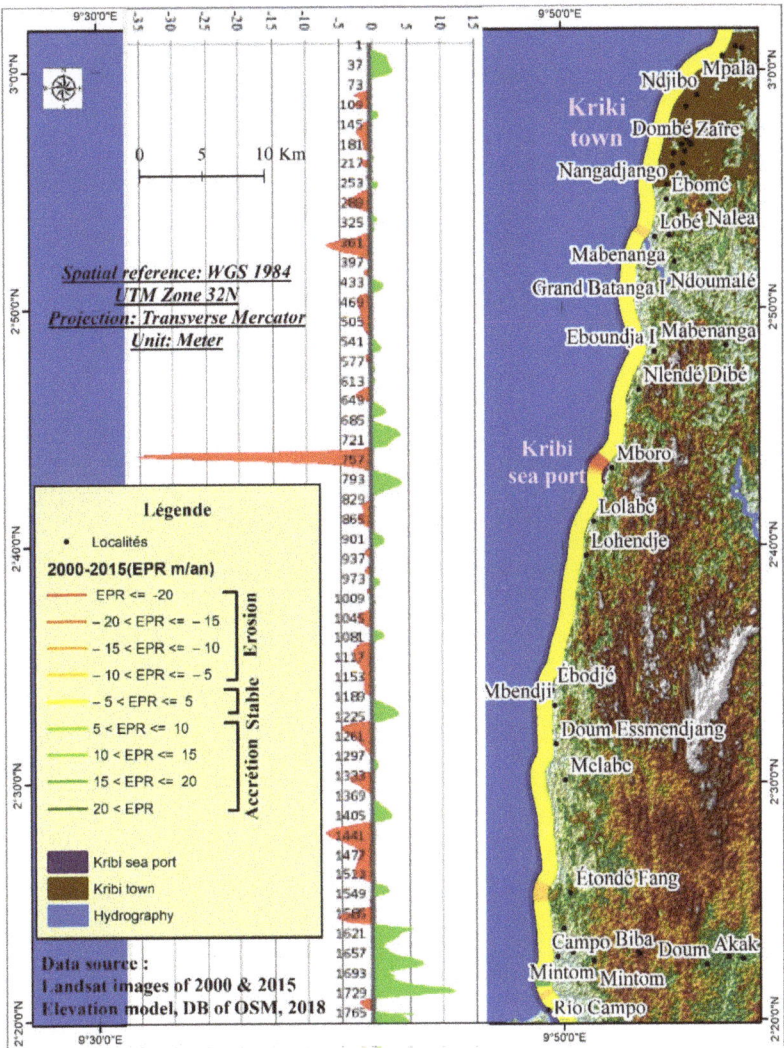

Figure 15. Coastal erosion on the Kribi coast in EPR (m/year) between 2000 and 2015 obtained with DSAS, through linear data extracted from Landsat images.

An average retreat of 0.9 m/year was calculated for this period with an accretion rate of 0.2. There was a decline of about 7 m/year at the level of the Lobé falls and a significant retreat of the coastline at the PAK site, with a regression of more than 35 m. Finally, the locality of Etondé Fang experienced a decline of −6 m/year (Figure 15). The other declines seem less significant: Doum Essimendjang (−3 m/year), Ebodjé (−1 m/year), south of Mboro (−4 m/year), and north of the urban center (−2 m/year). Accretion marks were also observed from the linear data, from North to South. A slight accretion was observed near Campo (13 m/year) and at Tondéfon (6 m/year). Other, less significant accretions were observed towards the South of Campo (approximately +2 m/year), Malabe (+2 m/year), Mboro (+4 m/year), and further north towards the town of Kribi (+3 m/year) and Ebomé (+4 m/year).

Coastline Dynamics under the PAK Implantation and Management (2015–2020)

An almost generalized erosion on the Kribi coast also marks this period, characterized by the beginning of the activities of the Kribi Sea Port. It is accentuated upstream and downstream of the PAK because of the second phase of Kribi Sea Port's developments marked by deforestation in many areas such as Lolabé, Lohendje, and up to Ebodjé (Figure 16).

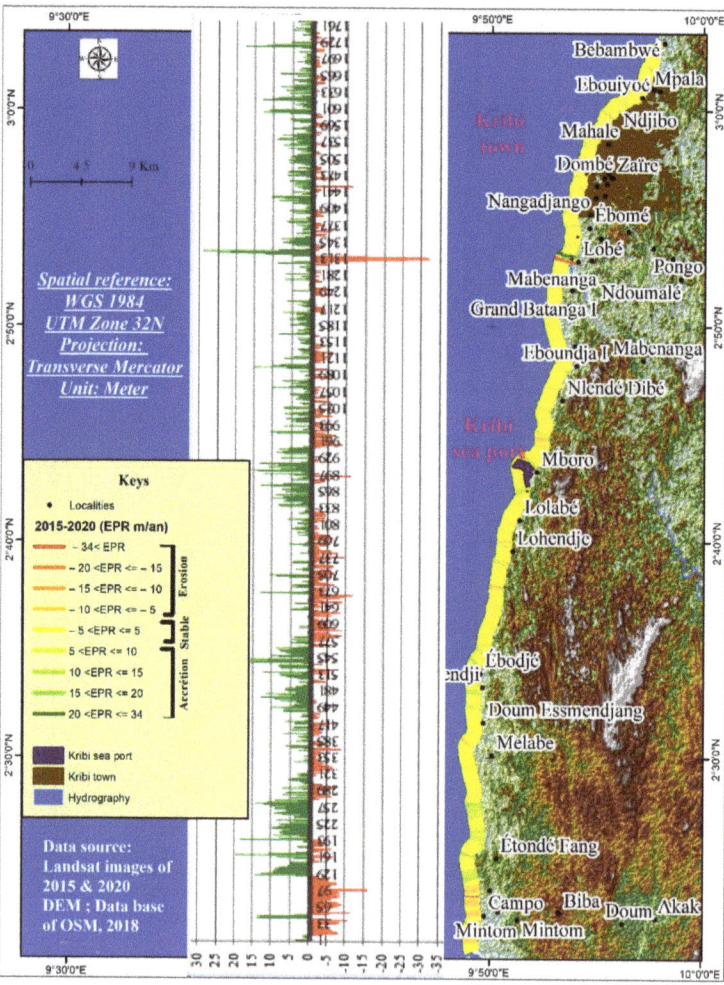

Figure 16. Coastal erosion on the Kribi coast with respect to speed (meters per year) between 2015 and 2020 obtained with DSAS, from linear data extracted from Landsat images.

This output image shows that coastal erosion has intensified on the Kribi coastline. An average retreat of −2 m/year was quantified, against an average accretion of 1 m/year. The Campo area in the South records a decline of 15 m/year. The peak of regression is around the Lobé falls, with a decline of about −32 m/year. Indeed, in this area, there is a spit before the Lobé falls, which would trap sediments coming from the south. This probably explains why there is an accretion peak of +30 m/year in the same place. In addition, a significant accretion was visible south of Kribi city, with about +15 m/year. Between the PAK and Lolabé village, an accretion of more than 15 m/year was measured, corresponding to the coastal dyke protecting the seaport. This zone constitutes a place for the accumulation of the sediments trapped by the aforementioned dyke. Conclusively, the Kribian coastline evolved in a saw-toothed manner between 2015 and 2020. The sections of erosion succeed those of accretion and stability. This is due to the very nature of the Kribi coastline, which alternates between soft and rocky coasts.

These results obtained from Landsat images with a 30 m resolution show the effectiveness of coastal erosion on the Kribian coastline, with an average retreat of −1.3 m/year and an average accretion of +0.9 m/year between 1973 and 2020. These retreat and accretion values (+1.3 m/year and +0.9 m/year) obtained from the Landsat images' processing are reliable insofar as they are above the total error value which is +0.65 (Table 3). A previous study [11] has used the same methods. Globally, 72.32% of the coastline remained stable, +1.3% is experiencing accretion, and −26.33% is experiencing erosion. However, the spatial resolution of the Landsat images is low; for this reason, we used very high-resolution images (the Pleiades and Google Earth) to refine these results. Given that these images are not available for the entire Kribi coastline, these in-depth analyses will focus on the northern and southern parts of the Kribi seaport.

4.1.2. Modeling of Small-Scale Coastal Erosion in Kribi, Using Pleiades Images and Google Earth Aerial Photos

Pleiades images and Google Earth aerial photos with a 0.5 m resolution have made it possible to refine the mapping of coastal erosion on the Kribi coast, specifically in the areas upstream and downstream of the PAK.

Coastline Dynamics in EPR (m/year) from Google Earth Aerial Photos (2015) and Pleiades Images (2013–2016, i.e., 3 Years)

This small-scale and very high-resolution processing confirms the ongoing erosion of the Kribi coast, previously highlighted by Landsat images. Between 2013 and 2016, the pockets of erosion remained concentrated in the southern part of the PAK and its northern end, towards the Lobé falls. Accreting and stable sections dominate the northern part of the PAK and a weak southern part. Overall, 61% of the linear data are experiencing accretion, 28% are experiencing erosion, and only 11% is stable (Figure 17).

In this three-year period, the erosion peak was about 6 km north of the PAK site, with a decline of −25 m/year. The average decline was estimated at ±4 m/year over the entire area, with an accretion rate of +3.16 m/year. In terms of accretion, the peak was south of the Lolabé village, with a value of about +17 m/year.

Thus, coastal erosion is active on the Kribi coast. It has intensified since the implementation of the Kribi industrial-port complex (CIPK) project, of which the PAK is a major component. High-spatial resolution mapping has made it possible to better visualize this erosion and to correct as much as possible the imperfections of the mapping carried out with low-resolution Landsat images.

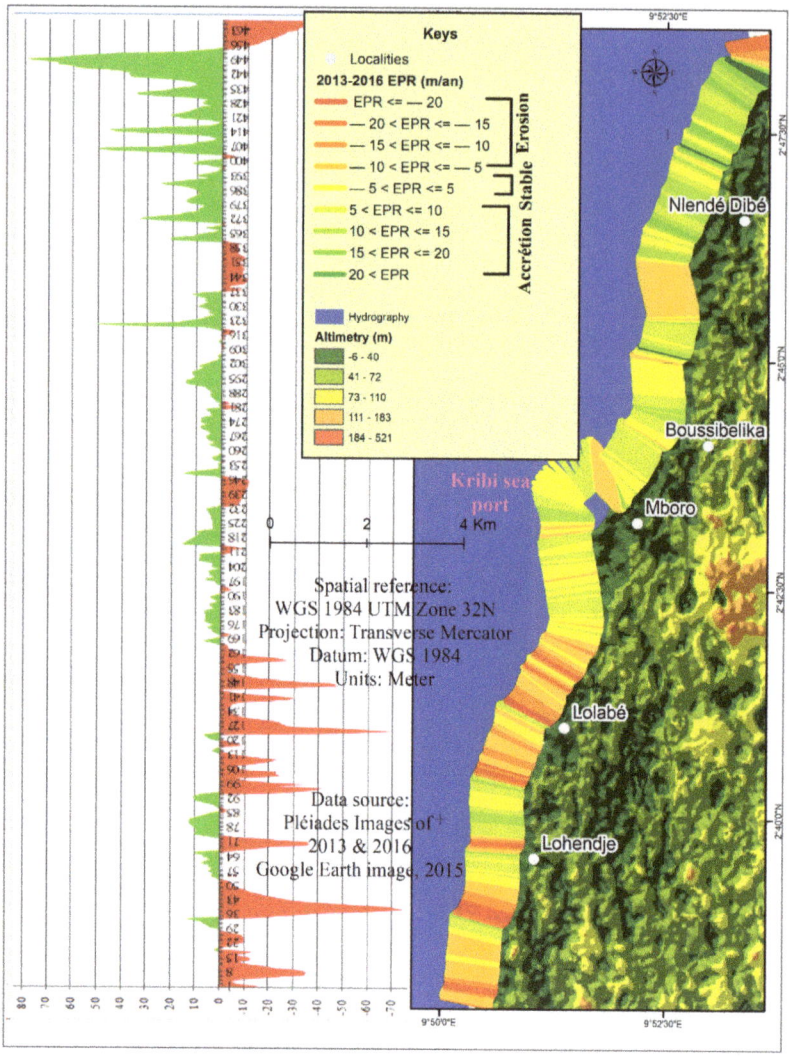

Figure 17. Coastline dynamics in Kribi between 2013 and 2016, obtained with DSAS from linear extracts of Pleiades images. Coastlines were divided into 463 transects separated by 50 m each; on the **left**, there is a cartographic representation, and a graphic representation is on the **right**, with the number of transects on the ordinate and the EPR in m/year on the abscissa.

Coastline Dynamics in EPR (m/year) from Google Earth Image (2017) and Pleiades Images (2016–2019, i.e.)

This temporal space, which saw the entry into the function of the PAK, is marked by a stabilization of the coastline in the southern part of the PAK, and an accentuation of coastal erosion in the northern part. From north to south, erosion is present but at a low rate of retreat. The accretion situations are isolated. They are located to the north before the Lobé falls and to the south around Lolabé and Lohendje. On the other hand, the stable zones are dominant (Figure 18). The output image shows that more than 71% of the Kribian coastline is stable, 25% is experiencing erosion, and only 4% is experiencing accretion.

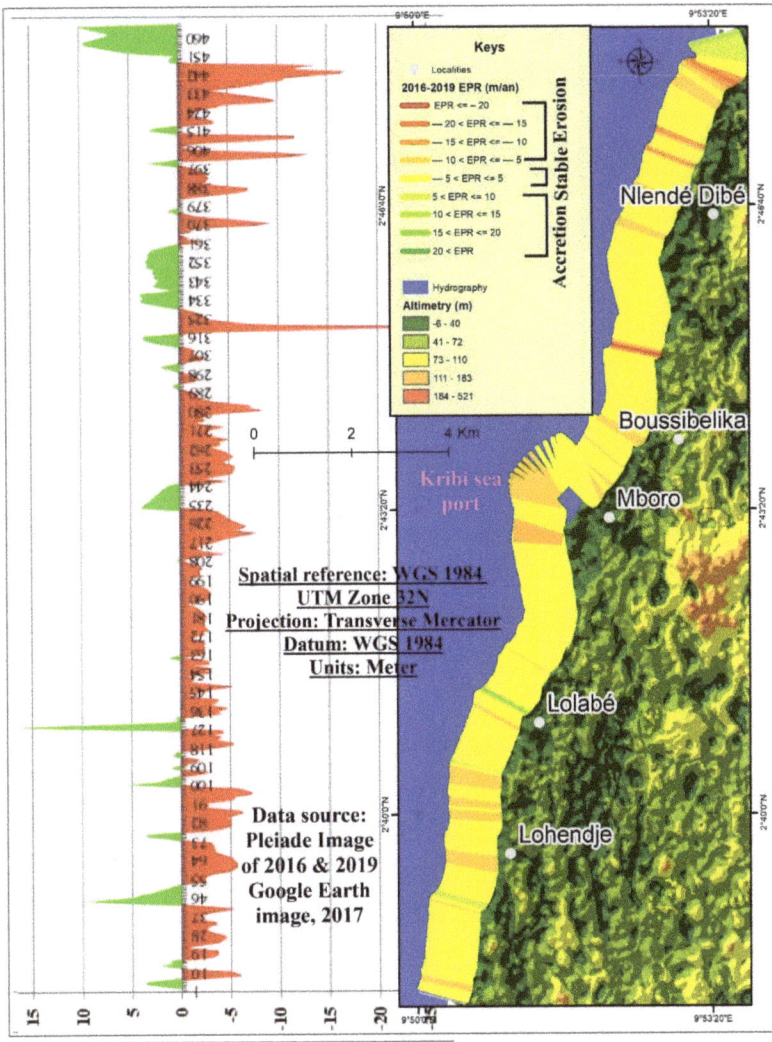

Figure 18. Coastline dynamics in Kribi between 2016 and 2019 from Pleiades images (2016–2019) and a Google Earth aerial photo (2017) (the Southern part corresponds to transect 1).

From south to north, the greatest decline was estimated at −35 m/year and −75 m/year south of Lohendje village. An average retreat of −13.8 m/year was calculated for this period with an accretion rate of +11 m/an. To the south and north of Lolabé, the regression reaches −70 m/year and −45 m/year, respectively. This area corresponds to the first PAK extension site currently being developed. Close to the current site of the PAK, the erosion of a small spatial extent was evaluated at −10 m/year. This also corresponds to the site being developed for the phase two extension of the Kribi seaport. In the South of Nlendé Dibé, the decline is estimated at −10 m/year.

The accretion peak is located in the northern part of Nlendé Dibé village, just before the Lobé falls, with a record value of around +80 m/year. Such accretion is due to the presence of a small spit that has trapped sediments in transit towards the north. The Boussibélika village is experiencing a +10 m/year accretion corresponding to the accumulation site of rock fragments and other sediments linked to the development of the current PAK since 2007. The southern part of the PAK site also presents an accretion of +15 m/year downstream of the protection dyke and this can be explained by the deposit of sediments trapped by the protection of the PAK.

Based on very high-resolution images, between 2013 and 2019, the average retreat of the Kribian coastline is −8.5 m/year, and the average accretion is about 7 m/year. However, there is a wide disparity in terms of recession and accretion. The Landsat images, due to their spatial resolution, tend to minimize the values of recession and accretion although the spatial repair of accretion, erosion, and stability areas are almost similar between the two data sources.

For the entire coastline, three main areas present worrying levels of erosion and require priority intervention due to anthropogenic forcings.

- Londji beach, because of coastal tourism and a high human population density up to the seafront, which is itself reduced, thereby implying a great exposure to human, societal, economic, and environmental issues.
- The Kribi Plaza bridge facing the Kribi Bilingual High School, where erosion is linked to high-density housing and equipment (high school) as well as infrastructure (national road N°7). With a poor management system, the collapse of the national N°7 will impact the mobility of goods and people.
- Bongadoué and Tara Beach are areas with extensive hotel facilities and an uncontrolled exploitation of sand.

4.2. Sustainable Management of Coastal Erosion on the Kribi Coast: Current Status and Prospects

The coastal erosion data show the weakening of the Kribian coastline since the construction and implementation of the Kribi Sea Port. The observations made raise the problems of the sustainable management of the coastal space for which there have been various initiatives taken by different actors to mitigate or solve this worrying situation.

4.2.1. Current Status of Coastal Erosion Management on the Kribi Coastline

The Cameroonian government, non-governmental organizations (NGOs), and local populations have taken initiatives to combat coastal erosion on the Kribi coast.

Measures Undertaken by the State: Construction of Buildings to Protect Against Coastal Erosion in Kribi

Sustainably managing natural risks in Kribi city certainly entails considering measures likely to curb their harmful effects. These measures range from tested and approved accommodations to the operationalization of existing laws in the field. In 2012, a dyke was built to protect the Lobé Falls against coastal erosion (Figure 19). Indeed, the erosion at the level of the soft shore of the Lobé falls originates from the upwelling of sea currents resulting from the construction of the PAK. This initiative was salutary although limited given the interest in the Lobé Waterfalls coveted by UNESCO as a World Heritage site.

Figure 19. Construction of a protective dyke at the Lobé waterfalls in Kribi. (Credit: Mbevo, 2017; Tchindjang, 2013).

Unfortunately, the dyke did not resist erosion by upwelling currents. The dyke was dismantled in 2017 by the phenomenon it was supposed to fight against. Such a situation raised a question: why do protective measures and structures that are so expensive have such a short lifespan in Cameroon?

Pursue the Ngoye-Kribi Beach Development and Electrification Project

Kribi would benefit from promoting the sustainable city initiative. Tourist development will simultaneously preserve the natural assets of this city while promoting economic profitability. It could also increase the competitiveness of the tourism sector (seaside) in Cameroon through international-type attractions, which are very popular with Western tourists. In order to preserve the environment and the rights of local populations, a Regional Environmental assessment (REA) was developed in 2008 by Royal Haskoning and the Environmental and Social Impact Assessment (ESIA). Many more projects were realized later (between 2012 and 2016).

Local Response Strategies

The local populations of the Kribi coast, faced with the aggressive nature of the sea, have taken action to deal with the phenomenon. These initiatives, far from being the most appropriate and effective, at least make it possible to maintain security around their wealth and assets. They can be summarized by the arrangement of a stony cordon (Figure 20).

Protection of the banks against marine erosion

The spectacular advance of the sea is Mboamanga

Fighting against marine erosion

Construction of a protective stone bund in Mboamanga

Figure 20. Coastal Erosion at work on the Kribi coastline and ineffective local response measures (Source: CUK 2015).

Non-Governmental Organizations (NGOs) Measures

In the Kribi-Campo area, the Organization for the Environment and Sustainable Development (OPED), funded by the African Development Bank (AfDB), has conducted studies on reducing the rate of mangrove degradation by improving fish-smoking technologies.

This action led to a reduction of around 60% of cut mangrove wood, with the construction of 185 improved smokehouses [33]. In the Rio Ntem estuary (South Cameroon), the FAO launched a project called "sustainable community management and conservation of mangrove ecosystems". This project has allowed for the reforestation of more than 1000 hectares of mangroves. Overall, these initiatives seem to be ineffective, hence the need to consider other more appropriate measures capable of significantly reducing the harmful impacts of coastal erosion on the Kribi coast.

4.2.2. What Measures Should Be Considered on the Coastline?

The Study Mission for Ocean Planning (MEAO) organized a brainstorming workshop on sustainable coastal erosion management strategies in Kribi, bringing together Cameroonian experts in the field, and the following measures have been adopted.

- Limit the sand extraction from the beach. For this measure, it is necessary (i) to sensitize the mining actors and the consumers of sand with respect to the question of coastal erosion (its harmful effects); (ii) and to inform such actors and consumers of the nature of the sea sand, which is not suitable for construction. This sensitization needs to be accompanied by alternative activities, either through association or through technical, financial, and material support.
- Promote a policy on the regeneration of coastal mangroves. There is already an experimental phase in Londji and Ebodje that has succeeded. It would be interesting to replicate it elsewhere in a participatory and integrated grassroots approach. Such an approach seems appropriate for areas of low vulnerability.
- Execute the territorial restructuring and relocation of populations occupying the seafront.
- Implement the rockfill approach, which requires the availability of supply points for rock material. This will promote the establishment of sediment accumulation zones.

5. Discussion

5.1. Coastline Dynamic

Finally, it should be noted that the recoil values put forward in this work remain to be put into perspective. The coastline extraction, which was based on a semi-automatic and manual method, based on Landsat images, took wet sand as the limit [66]. That assumes that the action of the tide constantly influences the coastal line. Consequently, to obtain a more plausible idea, the use of Pleiades images with a finer spatial resolution was beneficial, as confirmed by a recent report [69].

In Kribi, in addition to the results of the 2007 research [8], the measurements from studies commissioned by the PAK [70] also show very strong coastal dynamics, with migrations of the coastline of the order of several tens of meters (10–30 m on average) in 5–7 years (2003–2010), with accretion dominating the period from 2003–2010 and erosion dominating the period from 2010–2015/16. These studies, based on the analysis of Quicbird, Ikonos, Worldview, and GeoEye images (0.35 to 1 m spatial resolution), confirm our observations between 2003 and 2010. These erosions lead to significant sedimentary reworking, which can rapidly remodel (in less than 5 years) sandy forms in the estuaries.

In some African coastal countries, there are works dealing with the kinematics of the cost but performed with field measurements or at least low-resolution images.

Thus, in the southwest of the Ivory Coast, the coastal perimeter of San-Pédro and the beach segment of Assouindé Valtour are experiencing critical natural regressive evolutions of the shoreline and are subject to the impact of developments. Thus, they were subjected to a morpho-sedimentary study [71]. The assessment of the morpho-dynamic follow-up in the medium and long term (2008–2012 and 1985–2012) showed an estimated decline of 0.5–1 m/year over the long term and more than 4 m/year over the short term. The same phenomenon was observed in Senegal at the De Guédiawaye and Malika beaches in 2021, with an estimate of −0.15 m/year from 1942 to 2011 from Landsat images using DSAS [72].

Aerial photos (1979–1989) and field data (2007–2009) also show that the rate of erosion of the bay of Port-Bouet and the beaches of Assinie, which generally does not exceed

1.5 m/year [73], is exacerbated for short periods by violent swells that originate in the South Atlantic Ocean (2.3–18 m in a tidal cycle). In Mauritania, the coastline varied from −20.91 m/year to 22.95 m/year between 1978–1988 and 1988–1999 according to Landsat image processing [74].

In Libreville, the measurement of erosion phenomena on the northern coast and the monitoring of the morpho-sedimentary dynamics of intertidal sandy beaches has been the subject of a thesis carried out with in situ measurements [75]. The results show that between 1977 and 2013, the shoreline motion reveals variation rates from −2.07 m/year to +1.78 m/year. These rates correspond to 83% eroded beaches against 17% that are accreted. The results show declines between −2 and −12 m between 2012 and 2015, evidence of accelerated erosion due to anthropogenic forcing over short periods, as in San Pedro and Kribi.

Other authors [76] have analyzed the evolution of the Kribi shoreline and have been able to establish that this coastline is undergoing undeniable changes due to the phenomena of the fattening and thinning of the beach. This dynamic is manifested by changes in the coastal morphology, with the main consequence being "the shrinking of tourist recreational areas and the erosion of the coconut grove at the top of the beach". The rate of erosion at these locations is estimated at 13.9 cm per year. The average displacement of the shoreline over all the zones is around 17 cm/year on the Kribi shore.

Unlike the Wouri estuary, which has benefited from earlier more detailed work, the Kribi coastline has not yet been the subject of an in-depth study on coastal erosion. Therefore, the results provided by this study constitute the current basis for observing erosion on this coastal strip in the midst of spatial change.

The erosion phenomenon on the Cameroonian coast can be explained by several factors. Each section of the coastline has its explanatory factors. Indeed, the hydro-sedimentary transit is disturbed by human installations and some physical parameters. The findings that emerge from the previous analyses concern the distribution of erosion on almost the entire Kribi coastline in 2020, although the rate of decline is low. Field observations made in 2014, 2015, and very recently in May 2019 during the internship carried out at the Study Mission for Ocean Planning (MEAO)—the testimony of people who have lived for a long time in the intertidal space—further support these findings.

5.2. The Construction of Kribi Sea Port (PAK) and the Rise of Coastal Erosion

The Kribi Sea Port (PAK) project consists of the development of an Industrial-Port complex in Kribi (CIPK), including the construction of a seaport, the fitting out and development of industrial and logistics areas, and the new town, which could house more than 100,000 people when the commercial port has reached full activity. In this gigantic project, the portion of coastal land to be stripped is 26,000 ha. Similarly, the previous channel, which was 600 m long and 200 m wide with a coast 15 m deep, experienced an extension of the south dyke to the north with new characteristics: it was 1200 m long, 200 m wide, and 15 m deep, while maintaining the same orientation of 125° at the entrance and 305° at the exit [54]. In addition to this, the construction of the main protective dyke with a length of 1850 m constitutes a real sediment trap and is therefore an amplifying factor of coastal erosion.

The building of this dyke thus influences sediment transit. Indeed, the sediments mobilized by the coastal drift (in a south–north orientation) are trapped downstream of the PAK by its dyke. The visualization of aerial photos posted on Google Earth makes it possible to observe the phenomenon described (Figure 21).

Figure 21. Google Earth aerial photo of the PAK in 2017 showing fattening south of the port and thinning to the north, where the beach has disappeared, and the forest has begun to be attacked by waves at high tide. The littoral drift runs from south to north and is interrupted by the dyke, generating this situation.

Finally, the construction of the port of Kribi also generated changes in land use. Two Landsat images from 2004 and 2020 reflecting the state of the environment and the territorial dynamics induced by the construction of the PAK (Figure 22) have been shown. The destruction of the vegetation cover that serves as a natural barrier weakens the coastline and exposes it to coastal erosion.

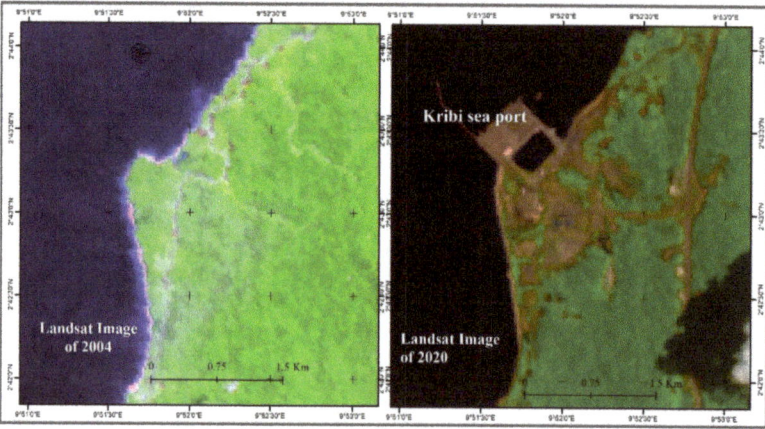

Figure 22. Landsat satellite images from 2004 on the left and from 2020 on the right show the deforestation induced by the construction of the PAK on the Kribi coastal environment.

5.3. Sand Mining and Coastal Erosion

Despite the denunciations and catastrophic situations observed in the Caribbean [77]; in West Africa in Togo [13,78], Sierra Leone [79], and Senegal [80]; and in Central Africa/Cameroon [81,82], the extraction of sea sand continues to weigh on African and Cameroonian beaches in particular. The extraction of sand induces a thinning of the beaches, a deepening of the water column, and therefore the greater erosive action of the waves. This is the case on the Kribi coasts where the exploitation of sand beaches for economic purposes (Figure 23) contributes to the aggravation of erosion despite the prohibition of this activity [8,80].

Even if the erosion of the Cameroonian shore is largely due to the activities of the PAK, elsewhere in Senegal, on the other hand, the observations made in situ in recent years (1990–2003) by the owners of tourist camps show a withdrawal of the line of coast of the order of 15 m/year [83] (linked to sand mining), which was calculated in relation to the property limits or the position of the well. The establishment of sand mining at Pointe Sarène beach has therefore significantly modified the sediment balance of the area, causing an increase in erosion throughout the sector.

Figure 23. Artisanal and industrial sand extraction on the Kribi coast. (Credit: Mbevo, May 2019, Tchindjang, March 2013). This photo board confirms the significant extraction of sand on the Kribi coast.

5.4. Deforestation and Coastal Erosion

Traditionally, it is thought that only mangroves protect these coasts [84,85]. However, the coastal forest also plays a decisive role [86,87]. On the Kribian coast, the reduction of small pockets of mangroves on the littoral zone has affected the stabilization of this zone. Similarly, the destruction of the littoral forest, linked to development projects, significantly exposes the Kribi littoral zone to erosion. This is particularly well demonstrated on the first site chosen for the establishment of the Douala Sea Port, where complete deforestation was carried out. Significant erosion is underway at this site. At the current site of the Kribi Sea Port, despite the construction of a dyke, erosion is still locally threatening. These inappropriate human interventions currently constitute a worrying and significant threat to the environment [82,88].

5.5. Urbanization, Population Growth, and Coastal Erosion

Kribi city is experiencing significant spatial and demographic growth, which impacts the dynamics of the environment. The high demand for housing leads to the extraction of sand, while the supply of river sediment is low on this coastal portion. This removal of sand from the beaches will induce an increase in their slope and therefore an increase in the erosive action of the waves during breaking, which leads to a reduction in the beaches' area. This is amplified by the construction on the beach; the erosion by the waves increases, and the sediment transit between the back beach and the beach is reduced. Authors documenting the coast of Algiers [89,90] also mention that erosion related to urbanization will have consequences on the retreat of the coastline.

5.6. Hotel Activity Impact on the Sea and Coastal Erosion

Douala Metropolis, the largest coastal city in Cameroon, is experiencing accelerated urbanization. As Kribi (a medium-sized city) is concerned, its spatial extension has been developed to the detriment of the mangroves and another Atlantic Forest that constitutes a form of natural protection against erosion [91]. This urban development correlates with tourist activity, which is booming in this area as evidenced by the arrival of many national and international tourists. The walls and dams built to protect hotels—far from stopping it—accentuate erosion [85]. This situation will be amplified with the implementation of the project to electrify and develop Ngoye beach for tourist purposes [86]. Moreover, Kribi already occupies the second place on the chessboard regarding the stay of the customers [92]. The Lobé falls are the main point of attraction [93].

Hotel activities also lead to strong land speculation on the Kribi coast. Such a situation explains why the land belonging to the private domain of the state is sold and transformed into a hotel establishment. Foundation pillars can be observed hugging the shore and disturbing the hydro-sedimentary dynamics in situ (Figure 24).

Figure 24. Land pressure on the coast in Kribi. The photo on the left shows the foundation of a private hotel in Kribi, directly built on the coastline (titled land—feet in the water—for sale). The photo on the right was taken −10 m from the shoreline. Indeed, since the entry into office of the PAK, there has been a major land race in this city, which has become the new growth pole of Cameroon. Land sales are multiplying, even in the private domain of the state, and advertisements are multiplying on the internet. The 2008 state law prohibiting the occupation of land 50 m from the coastline has thus been flouted.

5.7. Changes in Fluvial Inputs and Coastal Erosion

As said before, The Ntem is the largest river on the Kribian coast. Its watershed is 31,000 km^2 wide and 460 km long. It has a relatively low sediment transport (10.6 t/km^2) because its watershed is covered by forest [44,94–96]. The construction of the Memve'ele dam (2°24′09″ N, 10°23′37″ E) on the Ntem has a significant impact on the Kribi coast. This building traps a significant part of the sediment transport of this river, thereby reducing its sediment supply to the coast. This is the main explanatory factor. In addition, recent climate changes that have impacted the Sanaga River must also play a role and participate in the reduction of inputs on the coast.

5.8. Climate Change and Coastal Erosion

In both developed and developing countries, climate change is having an impact on coastal erosion [97]. Sea level rise, increased swell speeds, and rising sea surface temperatures are all climate drivers that exacerbate coastal erosion [98,99]. In Kribi, the relative effects of climate change are exacerbated by anthropogenic forcings because the climate seems more stable in Kribi than in Douala or the West Cameroonian coast (i.e., Limbe, Idenau, and Debunscha). In certain regions of the world, a radical change in the morphology of coastlines has been underlined by some authors [100].

6. Conclusions

The analysis and description of the physical framework in order to assess the coastal kinematics or even the weakening of the Kribi coast between 1973 and 2020 were the objectives of this work. The evaluation of the dynamics of the coastline was performed with Landsat images over three periods: 1973–2000, 2000–2015, and 2015–2020. Regarding the Pleiades images, two three-year periods, namely, 2013–2016 and 2016–2019, were analyzed using geostatistical methods implemented by the Digital Shoreline Analysis System (DSAS) software. The results obtained show an increase in time and space of the eroding areas on the Kribi coast. The period from 1973–2000 shows the clear stability of the coastline, with an average accretion rate of +0.4 m/year. The period from 2000–2015, which corresponds to the construction period of the Kribi Sea Port, coincided with limited erosion, with an average decline of −0.9 m/year. Finally, the period from 2015–2020, which marks the entry into service of the seaport, is also the period when erosion increased considerably, i.e., at an average decline of −2 m/year. The explanatory factors are linked—to the greatest extent possible—to marine weather forcings and to anthropogenic forcings, which appear to be more significant. Therefore, everything suggests that the installation of seaport equipment in Kribi constitutes the amplificatory element of coastal erosion in the area. As in most coasts in the Gulf of Guinea, unsuitable human interventions constitute internal weakening factors. Global climate change is not an exception. The kinematics dynamic that has been analyzed is influenced by the combined effects of the natural forcings and anthropogenic factors mainly linked to the Kribi seaport's establishment (constituting great human influences on coastal processes) on the studied area. Finally, an improved knowledge of coastal dynamics provides policy makers with tools for integrated coastal management.

Author Contributions: P.M.F. was involved in the conceptualization, methodology, mapping, analysis, and drafting of the paper after field investigations. M.T. was involved in field investigation, the conceptualization of the figures and tables, validation, and drafting the paper. A.H.-F.'s role involved constructing the methodology, review, editing, and analysis. All authors have read and agreed to the published version of the manuscript.

Funding: This research received no external funding.

Data Availability Statement: The data supporting reported results can be provided by request.

Conflicts of Interest: The authors declare no conflict of interest.

References

1. Gemenne, F.; Blocher, J.M.D.; De Longueville, F.; Vigil Diaz Telenti, S.; Zickgraf, C.; Gharbaoui, D.; Ozer, P. Changement climatique, catastrophes naturelles et déplacements de populations en Afrique de L'ouest. *Geo Eco Trop Rev. Int. Géol. Géogr. D'écol. Trop.* **2017**, *41*, 22.
2. Bocquier, P.; Traoré, S. Urbanisation et Dynamique Migratoire en Afrique de L'ouest; La Croissance Urbaine en Panne. 2000. Available online: https://dial.uclouvain.be/pr/boreal/object/boreal:78555 (accessed on 2 June 2022).
3. Dubresson, A.; Moreau, S.; Raison, J.P.; Steck, J.F. *L'Afrique Subsaharienne: Une Géographie Du Changement*; Armand Colin: Paris, France, 2011.
4. Janssens, M. Lomé, 100 ans de Croissance Démographique. Presses de L'UB, Lomé. 1998. Available online: https://www.oceandocs.org/bitstream/handle/1834/1193/lome100.pdf?sequence=1 (accessed on 7 June 2022).

5. Hénaff, A. Les Aménagements des Littoraux de la Région Bretagne en vue de Leur Défense Contre L'érosion Depuis 1949 (Protection of the Shoreline from Coastal Erosion Since 1949 in Britanny). *Bull. L'assoc. Géogr. Fr.* **2004**, *81*, 346–359. Available online: https://www.persee.fr/docAsPDF/bagf_0004-5322_2004_num_81_3_2397.pdf (accessed on 2 June 2022). [CrossRef]
6. Tafani, C. Littoral Corse: Entre Préservation de la Nature et Urbanisation, Quelle Place Pour les Terres Agricoles ? *Méditerr. Rev. Géogr Pays Méditerr. J. Mediterr. Geogr.* **2010**, *115*, 79–91. Available online: https://journals.openedition.org/mediterranee/5216 (accessed on 23 May 2022). [CrossRef]
7. Tiafack, O.; Chrétien, N.; Emmanuel, N.N. Development Polarisation in Limbe and Kribi (Littoral Cameroon): Growth Challenges, Lessons from Douala and Options. *Curr. Urban Stud.* **2014**, *2*, 361. Available online: https://www.scirp.org/html/6-1150105_52736.htm (accessed on 17 June 2022). [CrossRef]
8. Tchindjang, M.; Mouliom, A.; Nzieyo Nombo, G. Evolution du rivage kribien depuis 1973. In *Construire La Ville Portuaire De Demain En Afrique Atlantique*; Tchindjang, M., Steck, B., Bopda, A., Eds.; Editions EMS: Paris, France, 2019; pp. 538–568. Available online: https://www.editionsems.fr/livres/collections/afrique-atlantique/ouvrage/525-construire-la-ville-portuaire-de-demain-enafrique-atlantique.html (accessed on 26 July 2022).
9. Laimé, M. *Le Dossier De L'eau. Pénurie, Pollution, Corruption*; Média Diffusion: Paris, France, 2015.
10. Ozer, P.; Hountondji, Y.C.; De Longueville, F. Évolution récente du trait de côte dans le golfe du Bénin; Exemples du Togo et du Bénin. *Geo Eco Trop Rev. Int. Géol. Géogr. D'écol. Trop.* **2017**, *41*.
11. Fossi Fotsi, Y.; Pouvreau, N.; Brenon, I.; Onguene, R.; Etame, J. Temporal (1948–2012) and Dynamic Evolution of the Wouri Estuary Coastline within the Gulf of Guinea. *J. Mar. Sci. Eng.* **2020**, *7*, 343. [CrossRef]
12. UGI. *The Coastline Change: Annual Report*; International Geographical Union: Washington, DC, USA, 2012.
13. Blivi, A. Géomorphologie Et Dynamique Actuelle Du Littoral Du Golfe du Bénin (Afrique de L'ouest). Ph.D. Thesis, Université Michel de Montaigne, Bordeaux, France, 1993.
14. Ndior, V. Les organisations internationales et l'érosion côtière. *Rev. Jurid. L'environ.* **2019**, *44*, 71–78. Available online: https://www.cairn.info/revue-revue-juridique-de-l-environnement-2019-1-page-71.htm (accessed on 25 March 2022).
15. Valère, D.E.M.; Fatoumata, B.; Jeanne, K.M.; Jean-Baptiste, K.A.; Hervé, M.A.B.; Kouamé, A.K.; André, T.J.; Patricia, Y. Cartographie De La Dynamique Du Trait De Côte À Grand-Lahou: Utilisation De L'outil Digital Shoreline Analysis System (Dsas). *Eur. Sci. J.* **2016**, *12*, 327. [CrossRef]
16. Aubié, S.; Mugica, J.; Mallet, C. *Caractérisation De L'aléa Recul Du Trait De Côte Sur Le Littoral De La Côte Aquitaine Aux Horizons 2025 et 2050*; Observatoire Côte Aquitaine: Paris, France, 2011.
17. Degbe, C.G.E. Evolution du trait de côte du littoral béninois de 2011 à 2014. *Sci. Vie Terre Agron.* **2017**, *5*.
18. Tomety, S.F. Analyse des statistiques de vagues au Nord du Golfe de Guinée (Côte d'Ivoire, Ghana, Bénin, Nigéria) dans le cadre du suivi de l'érosion côtière; Mémoire de. Master of Science En Océanographie Physique Et Applications (CIPMA-Chaire UNESCO). Master's Thesis, Université d'Abomey-Calavi, Abomey, Benin, 2013.
19. Rey, T.; Defossez, S. Retours D'expériences Post-Catastrophes Naturelles. 2019. Available online: https://www.researchgate.net/profile/Pottier_Nathalie/publication/336882202.pdf (accessed on 12 July 2021).
20. Moungaga, M.-D. Érosion côtière et risques littoraux face aux changements climatiques; essai d'analyses comparatives des indicateurs de vulnérabilité à Libreville (Gabon) et Pointe-Noire (Congo). In Proceedings of the Communication Du Colloque Du SIFEE, Niamey, Niger, 26–29 May 2009.
21. Ovono, Z.M. Effet des changements climatiques en Afrique Centrale: Le cas de l'érosion côtière sur le littoral du Gabon. In *Revue Territoires D'Afrique*; N°9; Les Impacts Du Changement Climatique Sur Les Littoraux D'Afrique; Territoire D'Afrique: Paris, France, 2017; pp. 17–27.
22. Tweneboah, E. The Role of Environmental Values and Attitudes of Ghanaian Coastal Women in Natural Resource Management. Ph.D. Thesis, Faculty of Environmental Sciences and Process Engineering, Brandenburg University of Technology, Cottbus, Germany, 2009.
23. Zhang, K.; Douglas, B.C.; Leatherman, S.P. Global warming and coastal erosion. *Clim. Chang.* **2004**, *64*, 41–58. [CrossRef]
24. Thampanya, U.; Vermaat, J.E.; Sinsakul, S.; Panapitukkul, N. Coastal erosion and mangrove progradation of Southern Thailand. *Estuar. Coast. Shelf Sci.* **2006**, *68*, 75–85. [CrossRef]
25. Ly, C.K. The role of the Akosombo Dam on the Volta River in causing coastal erosion in central and eastern Ghana (West Africa). *Mar. Geol.* **1980**, *37*, 323–332. [CrossRef]
26. Ndour, A.; Laïbi, R.A.; Sadio, M.; Degbe, C.G.E.; Diaw, A.T.; Oyédé, L.M.; Anthony, E.J.; Dussouillez, P.; Sambou, H. Management strategies for coastal erosion problems in West Africa: Analysis, issues, and constraints drawn from the examples of Senegal and Benin. *Ocean. Coast. Manag.* **2018**, *156*, 92–106. [CrossRef]
27. National Research Council. *Managing Coastal Erosion*; National Academy of Science: Washington, DC, USA, 1930.
28. Kengapet Kouekam, A.; Fowe, P.G.; Togue Kamga, F.; Ngueguim, J.R.; Tsague, J.S. Modélisation de la Dynamique du Trait de Côte sur une Portion de la côte Ouest Cameroun allant de Batoke à Seme Beach par Imagerie Landsat de 1979 à 2018. *Eur. Sci. J.* **2019**, *15*, 1857–7881.
29. ENVIREP CAMEROUN, Plan D'action National de Gestion des Zones Marine et Côtière Valide. 2010. Available online: https://aquadocs.org/bitstream/handle/1834/5227/VERSION%20DU%20PAN%20VALIDEE.pdf?sequence=1&isAllowed=y (accessed on 22 June 2022).

30. Ouabo, R.E.; Tchoffo, R.S.; Ngatcha, B.R. Utilisation de la Géomatique Pour L'analyse des Risques Liés à la Dynamique du Trait de Côte à Kribi. *J. Cameroon Acad. Sci.* **2018**, *14*, 121–136. Available online: https://www.ajol.info/index.php/jcas/article/view/174682/164073 (accessed on 20 September 2021). [CrossRef]
31. Mfombam Nsangou, G.C. *Contribution De La Modélisation A L'évaluation Du Trait De Côte, Mémoire De Master Professionnel En Cartographie, Télédétection Et SIG Appliqués A La Gestion Durable Des Territoires*; Université de Yaoundé 1: Yaoundé, Cameroon, 2016.
32. Mangor, K.; Drønen, N.K.; Kærgaard, K.H.; Kristensen, S.E. *Shoreline Management Guidelines*; DHI Water and Environment: Copehnagen, Denmark, 2004.
33. Ajanino, G.; Tchikangwa, B.; Chuyong, G.; Tchamba, M. Les défis et perspectives de la formulation d'une méthodologie communautaire généralisable pour évaluer la vulnérabilité et l'adaptation des écosystèmes de mangrove aux impacts du changement climatique: Expérience du Cameroun. *Nat. Faune* **1999**, *24*.
34. Mbaha, J.P.; Etoundi, M.L.B.A. Et demain Kribi: Construire une Ville Portuaire Stratégique et Émergente à L'horizon 2035. *Espace Géograph. Soc. Maroc.* **2021**, 43–44. Available online: https://revues.imist.ma/index.php/EGSM/article/download/24572/13011 (accessed on 12 November 2021).
35. MINHDU. *Plan Directeur D'Urbanisme De La Ville De Kribi. Rapport Justificatif*; Communauté Urbaine de Kribi: Kribi, Cameroon, 2014; 152p.
36. MINEPDED/PNUD. *Révision /Opérationnalisation Du PNGE Vers Un Programme Environnemental (PE), Volume I: Diagnostic De La Situation De L'Environnement Au Cameroun*; Ministère de L'environnement et de la Protection de la Nature: Yaoundé, Cameroon, 2009; 203p.
37. Vicat, J.P.; Pouclet, A.; Nsifa, E. Les Dolérites du Groupe du Ntem (Sud Cameroun) et des Régions Voisines (Centrafrique, Gabon, Congo, Bas Zaïre): Caractéristiques Géochimiques et Place dans L'évolution du Craton du Congo au Protérozoïque. *Géologie Environ. Cameroun Collect. GEOCAM* **1998**, 305–324.
38. Oslisly, R. The history of human settlement in the middle Ogooué valley (Gabon): Implications for the environment. *Afr. Rain For. Ecol. Conserv.* **2001**, 101–118.
39. Segalen, P. Les Sols et la Géomorphologie du Cameroun. Cahiers ORSTOM, Série Pédologie. 1967. pp. 137–187. Available online: https://www.thesisonafrica.com/wp-content/uploads/2018/06/Sols_geomorphologie_Cameroun_Segalen.pdf (accessed on 12 April 2022).
40. Nlend Nlend, P. Les Traditions Céramiques dans leur Contexte Archéologique sur le Littoral Camerounais (Kribi-Campo) de 3000 à 500 BP. Ph.D. Thesis, Université Libre De Bruxelles, Brussels, Belgium, 2013. Available online: https://dipot.ulb.ac.be/dspace/bitstream/2013/209563/3/3698de3f-8bfe-4eab-8c30-6bcf7eef2042.txt (accessed on 11 October 2021).
41. IUSS Working Group WRB. *World Reference Base for Soil Resources 2014, Update 2015 International Soil Classification System for Naming Soils and Creating Legends for Soil Maps*; World Soil Resources Reports, No. 106; FAO: Rome, Italy, 2015.
42. Sighomnou, D. Analyse Et Redéfinition Des Régimes Climatiques Et Hydrologiques Du Cameroun: Perspectives D'Evolution Des Ressources En Eau. Ph.D. Thesis, University of Yaoundé I, Yaoundé, Cameroon, 2004.
43. Hegge, B.J. Low Energy Sandy Beaches of Southwestern Australia: Twodimensional Morphology, Sediments and Dynamics. Ph.D. Thesis, Department of Geography, Western Australia University, Perth, Australia, 1994.
44. Liénou, G.; Mahé, G.; Paturel, J.E.; Servat, E.; Sighomnou, D.; Sigha-Nkamdjou, L.; Dieulin, C. Impact de la variabilité climatique en zone équatoriale: Exemple de modification de cycle hydrologique des rivières du sud-Cameroun. *Hydrol. Sci. J. Sci. Hydrol.* **2008**, *53*, 789–801. Available online: https://www.iwra.org/member/congress/resource/abs127_article.pdf (accessed on 25 July 2022). [CrossRef]
45. SOGREAH. *Recensement, Analyse Et Evaluation Des Données Existantes Eventuelles Et Acquises (Lot n°5) Pour Le Projet De Port En Eau Profonde De Kribi, Rapport n°2 Final*; Port Autonome de Kribi: Kribi, Cameroon, 2010.
46. Letouzey, R. *Étude Phytogéographique du Cameroun*; Paris, P., Ed.; Le Chevalier: Paris, France, 1968.
47. Olivry, J.C. Fleuves Et Rivières Du Cameroun, Monographies Hydrologiques ORSTOM. 1986. Available online: https://horizon.documentation.ird.fr/exl-doc/pleins_textes_6/Mon_hydr/25393.pdf (accessed on 19 November 2021).
48. De Chautagne, L. Fiche Descriptive Sur Les Zones Humides Ramsar (FDR). 2007. Available online: https://rsis.ramsar.org/RISapp/files/RISrep/TN1707RIS.pdf (accessed on 2 June 2022).
49. Ebodé, V.B. Variabilité Hydropluviométrique en Afrique Centrale Occidentale Forestière: Entre Analyse des Fluctuations Observées, Modélisation Hydrologique Prédictive et Recherche des Facteurs Explicatifs. Ph.D. Thesis, University of Yaounde 1, Yaoundé, Cameroon, 2020.
50. Oslisly, R. Les traditions culturelles de l'Holocène sur le littoral du Cameroun entre Kribi et Campo. *Grundlegungen. Beiträge Zur Eur. Afr. Arch. Manfred KH Eggert* **2006**, 303–317. Available online: https://horizon.documentation.ird.fr/exl-doc/pleins_textes/divers17-01/010059679.pdf (accessed on 10 April 2022).
51. Suchel, J.B. Rainfall patterns and regimes rainfall in Cameroon. *Doc. Geogr. Trop.* **1987**, *5*, 287.
52. Mena, M.S.; Tchawa, P.; Amougou, J.A.; Tchotsoua, M. Les changements climatiques à travers la modification du régime pluviométrique dans la région de Kribi (1935–2006). *Rev. Ivoire. Sci. Technol.* **2016**, *28*, 389–407. Available online: https://revist.net/REVIST_28/REVIST_28_23.pdf (accessed on 24 May 2022).
53. Lienou, G. Impacts de la Variabilité Climatique sur les Ressources en eau et les Transports de Matières en Suspension de Quelques Bassins Versants Représentatifs au Cameroun. Ph.D. Thesis, Université De Yaoundé I, Yaoundé, Cameroon, 2007.

54. PAK. *Projet De Port En Eau Profonde De Kribi: Études Hydro-Sédimentaires (Lot n°7), Rapport N°4–V1*; Kribi Sea Port: Kribi, Cameroon, 2015.
55. Keugne Signe, E.R. Gestion de L'érosion Côtière sur le Littoral Sud du Cameroun. Master's Thesis, Département De Géographie, Université De Liège, Liege, Belgium, 2018.
56. BIRD, E.C.F. *Coastal Geomorphology: An Introduction*; John Wiley and Sons Ltd: Chichester, UK, 2007.
57. Morin, S.; Kuété, M. Le littoral camerounais: Problèmes morphologiques. In *Travaux De Laboratoire De Géographie Physique Appliquée*; Institut de Géographie–Bordeaux III: Bordeaux, France, 1988.
58. Almar, R.; Kestenare, E.; Reyns, J.; Jouanno, J.; Anthony, E.J.; Laibi, R.; Ranasinghe, R. Response of the Bight of Benin (Gulf of Guinea, West Africa) coastline to anthropogenic and natural forcing, Part1: Wave climate variability and impacts on the longshore sediment transport. *Cont. Shelf Res.* **2015**, *110*, 48–59. [CrossRef]
59. Ondoa, G.A.; Onguéné, R.; Eyango, M.T.; Duhaut, T.; Mama, C.; Angnuureng, D.B.; Almar, R. Assessment of the Evolution of Cameroon Coastline: An Overview from 1986 to 2015. *J. Coast. Res.* **2018**, *81*, 122–129. [CrossRef]
60. Parker, B. Where is the shoreline? The answer is not as simple as one might expect. *Hydrointernational* **2001**, *5*, 6–9.
61. Baulig, H. *Vocabulaire Franco-Anglo-Allemand De Géomorphologie*; Fascicule 130; Société D'édition Les Belles Lettres: Paris, France, 1956.
62. Shom. Available online: http://observatoires-littoral.developpement-durable.gouv.fr/qu-est-ce-que-le-trait-de-cote-r25.html (accessed on 25 July 2021).
63. Bird, E.C.F.; Schwartz, M.L. (Eds.) *The World's Coastline*; Van Nostrand Reinhold: New York, NY, USA, 1985; p. 1071.
64. Thieler, E.R.; Himmelstoss, E.A.; Miller, T. *Digital Shoreline Analysis System (DSAS) Version 3.0: An ArcGIS Extension for Calculating Shoreline Change*; U.S. Geological Survey Open-File Report; USGS: Reston, VA, USA, 2005; p. 1304.
65. Himmelstoss, E.A.; Rachel, E.H.; Meredith, G.K.; Amy, S.F. *Digital Shoreline Analysis System (DSAS) Version 5.0 User Guide*; No. 2018–1179; US Geological Survey: Reston, VA, USA, 2018.
66. Addo, K.; Appeaning, P.N.; Quashigah, J.; Kufogbe, K.S. Quantitative analysis of shorelinechange using medium resolution satellite imagery in Keta, Ghana. *Mar. Sci.* **2011**, *1*, 1–9. [CrossRef]
67. Bird, E.C. Coastal erosion and rising sea-level. In *Sea-Level Rise and Coastal Subsidence*; Springer: Dordrecht, The Netherlands, 1996; pp. 87–103.
68. Jamont, M.F. Etude Des Aléas Naturels Sur Le Sud Vendée Et Marais Poitevin; Rapport De Phase 2 Caractérisation Des Aléas De Référence. 2014. Available online: www.vendee.gouv.fr/IMG/pdf/Rapport_Phase_2.pdf (accessed on 10 September 2021).
69. Fletcher, C.H.; Romine, B.M.; Genz, A.S.; Barbee, M.M.; Dyer, M.; Anderson, T.R.; Lim, S.C.; Vitousek, S.; Bochicchio, C.; Richmond, B.M. *National Assessment of Shoreline Change: Historical Shoreline Change in the Hawaiian Islands*; U.S. Geological Survey OpenFile Report; USGS: Reston, VA, USA, 2011; p. 1051.
70. Port Autonome de Kribi. *Etudes Environnementales (Dont Étude Hydro-Sédimentaire) et Élaboration des Plans de Préservation du Port Autonome de Kribi*; R2–Scénarios D'évolution des Écosystèmes en Fonction des Facteurs Externes N° 47267, Activité 2, Rapport D'étude; Port Autonome de Kribi: Kribi, Cameroon, 2021.
71. Koffi Koffi, P.; Yao Kouadio, S.; Abe, J.; Hauhouot, C.; Bamba Siaka, B. Quelles perspectives face à la dynamique préoccupante des plages d'assouindé valtour et du club nautique, respectivement au sud-est et au sud-ouest de la Côte d'Ivoire. In Proceedings of the XIIIèmes Journées Nationales Génie Côtier–Génie Civil, Dunkerque, France, 2–4 July 2014; Available online: http://www.paralia.fr (accessed on 20 July 2022). [CrossRef]
72. Sagne, P.; Ba, K.; Fall, B.; Youm, J.P.M.; Faye, G.; Sarr, J.P.G.; Sow, E.H. Cartographie De La Dynamique Historique Du Trait De Côte Des Plages De Guédiawaye Et Malika (Dakar, Sénégal). *Eur. Sci. J.* **2021**, *17*, 214. [CrossRef]
73. Konan, E.K.; Jacques, A.; Kouamé, A.; Urs, N.; Jan, N.; André, O. Impacts des houles exceptionnelles sur le littoral ivoirien du Golfe de Guinée. *Géomorphol. Relief Process. Environ.* **2016**, *22*, 11241. [CrossRef]
74. Faye, I. Dynamique Du Trait De Côte Sur Les Littoraux Sableux De La Mauritanie A La Guinée-Bissau (Afrique De L'Ouest): Approches Régionale Et Locale Par Photo-Interprétation, Traitement D'images Et Analyse De Cartes Anciennes. Ph.D. Thesis, Université De Bretagne Occidentale, Brest, France, 2010.
75. Mouyalou, V.M.T. Morphosedimentary Dynamic of a Sandy Beach: From Lycee Léon Mba (Libreville) to La Sablière (Akouango Bay). Ph.D. Thesis, Université Omar Bongo, Libreville, Gabon, 2017.
76. Fangue, N.H.; Tonye, E.; Akono, A.; Ozer, A. Estimation de la vitesse de recul de la ligne du rivage par télédétection sur le rivage Kribien. In Proceedings of the IXèmes Journées Scientifiques du Réseau Télédétection de l'AUF, Yaoundé, Cameroun, 29 November–2 December 2001.
77. Cambers, G. Caribbean beach changes and climate change adaptation. *Aquat. Ecosyst. Health Manag.* **2009**, *12*, 168–176. [CrossRef]
78. Kwassi, A.L. Contribution A L'étude Des Populations Rurales De La Zone Côtière Du Togo. Master's Thesis, University of Lomé, Togo, 2000.
79. PNUE. *Érosion Cor/Ere En Afrique De Ouest Et Du Centre*; Collaboration des Nations Unies et de UNESCO: Paris, France, 1985.
80. Diop, E.; Soumare, A.; Diallo, N.; Guissé, A. *Dynamique De La Mangrove Des Iles Du Gandoul Occidental (Du Nord De L'île De Guissanor Au Sud Du Bôlon De Niodior, Saloum/Sénégal)*; Rapport Final De l'EPEEC-8; Cyber Geo: Brussels, Belgium, 1996.
81. Kouekam Kengap, A.; Fowe Kwetche, P.G.; Togue Kamga, F.; Ngueguim, J.R.; Dingong Atoukoh, T.G. Importance Des Paramètres Hydrodynamiques Dans La Répartition Spatiale Des Sédiments Superficiels Des Plages De Limbe (Sud-Ouest Cameroun). 2019. Available online: https://core.ac.uk/reader/236411710 (accessed on 5 March 2022).

82. Kuété, M.; Assongmo, T. Développement contre Environnement sous les Tropiques: L'exemple du littoral de la région de Kribi (Cameroun). *Cah. D'outre Mer.* **2002**, *55*, 279–306. [CrossRef]
83. Cesaraccio, M.; Thomas, Y.-F.; Diaw, A.-T.; Ouegnimaoua, L. Impact des activités humaines sur la dynamique littorale: Prélèvements de sables sur le site de Pointe Sarène, Sénégal/Impact of sand extractions on coastal dynamics (Sarène Point beach, Senegal). *Géomorph. Relief Process. Environ.* **2004**, *10*, 55–63. [CrossRef]
84. Ndour, N.; Dieng, S.; Fall, M. Rôles des mangroves, modes et perspectives de gestion au Delta du Saloum (Sénégal). *Vertigo Rev. Électron. Sci. L'environ.* **2012**, *11*. [CrossRef]
85. Ackermann, G.; Alexandre, F.; Andrieu, J.; Mering, C.; Ollivier, C. Dynamique des paysages et perspectives de développement durable sur la petite côte et dans le delta du Sine–Saloum (Sénégal). *Vertigo Rev. Électron. Sci. L'environ.* **2006**, *7*.
86. Sangne, C.; Barima, Y.; Bamba, I.; N'Doumé, C.T. Dynamique forestière post-conflits armés de la Forêt classée du Haut-Sassandra (Côte d'Ivoire). *Vertigo Rev. Électron. Sci. L'environ.* **2015**, *15*. [CrossRef]
87. Merckelbagh, A. *Et Si Le Littoral Allait Jusqu'à La Mer: La Politique Du Littoral Sous La Ve République*; Éditions Quae: Paris, France, 2009.
88. Fongnzossié, F.E.; Sonwa, D.J.; Kemeuze, V.; Mengelt, C.; Nkongmeneck, B. *Assessing Climate Change Vulnerability And Local Adaptation Strategies in the Kribi-Campo Coastal Ecosystems, South Cameroon*; Urban Climate: Amsterdam, The Netherlands, 2018.
89. Otmani, H.; Belkessa, R.; Rabehi, W.; Guerfi, M.; Boukhdiche, W. Dégradation des dunes côtières algéroises entre pression de l'urbanisation et conséquences sur l'évolution de la ligne de rivage. *GeoEcoMarina* **2019**, *25*, 131–145.
90. Tarik, G.; Bouziane, S. Urbanisation côtière en Algérie, Processus et impacts sur l'environnement: Le cas de la baie d'Aïn el Turck. *Études Caribéennes* **2010**, *15*, 4431. [CrossRef]
91. Saha, F.; Tchindjang, M. Dynamique spatiale de la ville de Kribi: Facteurs et conséquences. In *Construire La Ville Portuaire De Demain En Afrique Atlantique*; Tchindjang, M., Steck, B., Bouopda, A., Eds.; Edition EM: Paris, France, 2019; pp. 148–164. Available online: https://www.researchgate.net/profile/Frederic_Saha/publication/332445874_Chapitre_6_Dynamique_spatiale_de_la_ville_de_Kribi_facteurs_et_consequences/links/5f34e53ea6fdcccc43c5b101/Chapitre-6-Dynamique-spatiale-de-la-ville-de-Kribi-facteurs-et-consequences.pdf (accessed on 10 May 2019).
92. Guyomarc'h, J.P.; Le Foll, F. *Milieux Côtiers, Ressources Marines Et Société*; Ceser Bretagne. 2011. Available online: https://www.bretagne.bzh/upload/docs/application/pdf/201112/milieux_cotiers_ressources_marines_et_societe_internet_light_2011-12-07_10-15-49_709.pdf (accessed on 14 July 2022).
93. MINEPAT; PCFC. *Rapport Final De L'étude D'impact Environnemental Et Social Sommaire Du Projet D'aménagement Et D'électrification De La Plage De Ngoye-Kribi A Des Fins Touristiques*; Ministère de la L'économie, de la Planification et de L'aménagement du Territoire: Yaoundé, Cameroon, 2016.
94. Tchindjang, M.; Éwolo Onana, Z.; Mahend, E.; Mbohou, S. Tourisme et création d'emplois au Cameroun: Mythe, utopie, panacée ou réalité. In *Repenser La Promotion Du Tourisme Au Cameroun: Approches Pour Une Redynamisation Stratégique*; Kamdem, P., Tchindjang, M., Eds.; Linx: Paris, France, 2011; pp. 315–358.
95. Tchindjang, M.; Etoga, M.H. The Lobé Waterfall, an exceptional geocultural heritage on the coast of Cameroon between sustainable tourism and the conservation of cultural identities. *Via. Tour. Rev.* **2014**, *5*, 964. Available online: https://journals.openedition.org/viatourism/951 (accessed on 29 March 2022). [CrossRef]
96. Sigha-Nkamdjou, L.; Sighomnou, D.; Liénou, G.; Ndam Ngoupayou, J.R.; Bello, M.; Kamgang, G.R.; Servat, E. Impacts des modifications climatiques et anthropiques sur les flux de matières de quelques bassins fluviaux du Cameroun. *Sediment Budg.* **2005**, *2*, 291–298.
97. Masselink, G.; Russell, P. Impacts of climate change on coastal erosion. *MCCIP Sci. Rev.* **2013**, 71–86.
98. Toimil, A.; Camus, P.; Losada, I.J.; Cozannet, G.L.; Nicholls, R.J.; Idier, D.; Maspataud, A. Climate change-driven coastal erosion modelling in temperate sandy beaches: Methods and uncertainty treatment. *Earth Sci. Rev.* **2020**, *202*, 103110. [CrossRef]
99. Cai, F.; Su, X.; Liu, J.; Li, B.; Lei, G. Coastal erosion in China under the condition of global climate change and measures for its prevention. *Prog. Nat. Sci.* **2009**, *19*, 415–426. [CrossRef]
100. Masselink, G.; Russell, P.; Rennie, A.; Brooks, S.; Spencer, T. Impacts of climate change on coastal geomorphology and coastal erosion relevant to the coastal and marine environment around the UK. *MCCIP Sci. Rev.* **2020**, *2020*, 158–189.

Article

Evolution of Sediment Parameters after a Beach Nourishment

Juan J. Santos-Vendoiro [1], Juan J. Muñoz-Perez [1,*], Patricia Lopez-García [1], Jose Manuel Jodar [1], Javier Mera [1], Antonio Contreras [2], Francisco Contreras [2] and Bismarck Jigena [1]

[1] Facultad de Ciencias del Mar y Ambientales, Universidad de Cadiz, 11510 Puerto Real, Cadiz, Spain; juanjo.santosvendoiro@alum.uca.es (J.J.S.-V.); patricia.lopezgarcia@uca.es (P.L.-G.); josemanuel.jodar@uca.es (J.M.J.); javier.merabaston@alum.uca.es (J.M.); bismarck.jigena@gm.uca.es (B.J.)

[2] Escuela Politecnica, Universidad de Cadiz, 11202 Algeciras, Spain; antonio.contreras@uca.es (A.C.); francisco.contreras@uca.es (F.C.)

* Correspondence: juanjose.munoz@uca.es

Abstract: A methodology for monitoring the behaviour and size of sand after a beach nourishment process is presented herein. Four sampling campaigns (before and just after the nourishment, after six months and one year later) were performed on four beaches of the Gulf of Cadiz (Spain). D_{50} and sorting size parameters were analysed. Among the results, it should be noted that differences of up to 20% between native and nourished sand values disappear only one year after the nourishment.

Keywords: beach nourishment; coastal zone management; beach erosion

1. Introduction

Beach nourishment is a common process conducted when the coastline is being subjected to erosion because of natural or anthropogenic causes, as it replaces sediments within a littoral system and allows natural forces to continue their operation [1,2]. This process involves the placement of large volumes of sand along the beach profile [3]. Nourishments are also often associated with exposed coasts with intensive levels of development or great recreational value [4]. Due to their response to different forms of energy, such as wind, storms, waves, or modifications of the sea level, beaches are undergo a flux of erosion-accretion [5]. This is not different in the Gulf of Cadiz zone, where this study was centred. For example, during the 2000s, more than 47 restoration operations were carried out on several beaches in the area [6,7]. The importance of the nourishments is rising, mostly because of the growth of the population living in coastal areas, which is increasing the budget destined to these proceedings [8]. For example, in the U.S., 53% of the population live in coastal states (having increased by 33 million between 1980–2003). In fact, houses built in non-nourishing zones tend to be significantly smaller than those located in nourishing zones [9]. In areas where there is potential for tourist or urban development and erosion problems are detected, scientifically based engineering solutions are expected to control or mitigate these phenomena [10]. Additionally, a great number of residents of coastal areas are aware of coastal erosion/beach loss, which is important to raise a better understanding of coastal risks and hazards [11]. On the contrary, some studies have even suggested that beach nourishment and other hazard mitigation measures could encourage coastal development, thus increasing risk [12].

Although the effects of erosion in populated areas present a major problem, it is worth noting that these nourishments are not only performed on urban beaches, but also in places where there is a lower economic impact but a significant ecological value (for more information about these processes, see nourishments of that kind in [13]). Additionally, there have been reports suggesting that not every nourishment has a positive environmental effect if performed incorrectly, or in the wrong place. For instance, the formation of rip currents attributable to sandbars caused by the modification of the original state of the beach have been described [14]. The addition of new sediment to beaches requires a sound understanding of form–process continuums in order to achieve the desired response of morphodynamic systems [15]. Moreover, as nourished sand does not last forever, the periodicity of maintenance work must be established or at least foreseen. That is why the Shore Protection Manual [16] featured in 1984 the James' renourishment factor (R_J) [17] is trying to answer the basic question of how often renourishment is required when the borrow source is different from the native beach sand. Unfortunately, due to the lack of accuracy in the prediction, this abacus was removed from the new version, the Coastal Engineering Manual [18]. Thus, afterwards, new attempts to address the problem have been presented by other researchers, e.g., [19,20]. Moreover, some statistical studies have been performed where renourishment rates for U.S. projects are typically in the range of 5% per year (or less) of the initial nourishment volume [21].

Due to these maintenance cost and safety problems, it is important to investigate the evolution over time of the borrow sand dumped on the beach. A crucial question appears: is it possible that the granulometric characteristics of the borrow sand evolve towards the original ones of the native sand and, if so, when?

Thus, the objective of this study was to analyse the behaviour of the sand after beach nourishment. A methodology to study the evolution of the two most representative values, D_{50} (mean diameter of the grain) and sorting (standard deviation with respect to the mean grain), generalisable to any other site, was applied.

2. Study Area

Samples were taken from four beaches located in the Gulf of Cadiz: Santa María del Mar (SMM), Victoria (VB), Camposoto (CB), and La Barrosa (BB) (Figure 1). SMM has the particularity that it is embedded between two lateral groynes that confine it.

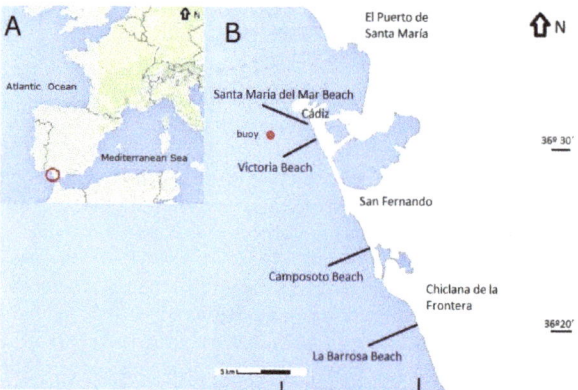

Figure 1. Location of the study area (**A**), the four beaches analysed in the Gulf of Cadiz and the wave buoy (**B**).

An aerial view of these four beaches and their monitored profiles, as well as an indicative wave rose diagram, is shown in Figure 2.

SMM (Figure 2A) is a pocket beach enclosed between two groynes [22]: the northern one has dimensions of about 240 linear metres, while the southern is about 212 linear metres long. The length of the beach measured between the starting points of the two groynes is 600 m and about 400 m between the heads. It is influenced by rip currents and undertows, corresponding to the characteristic outflow pulses of the incoming water mass. This phenomenon causes a loss of sand that is impossible to recover, due to the bathymetric conditions consisting of the existence of a small rocky step [23].

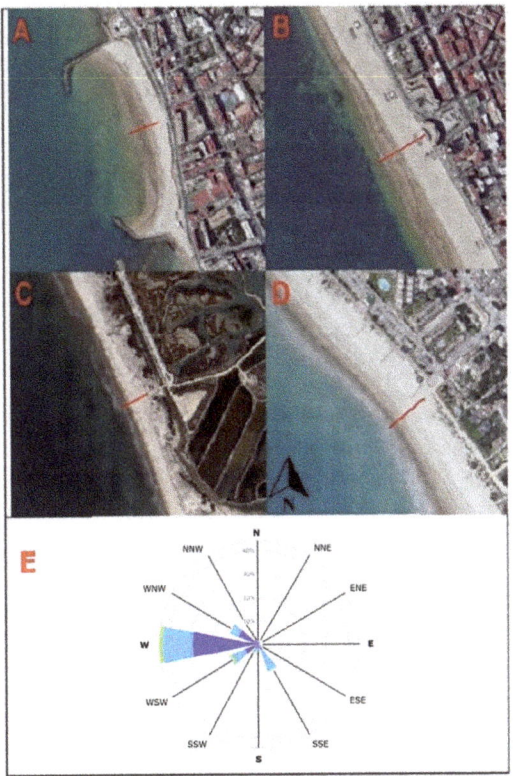

Figure 2. Location of the profiles monitored in the four different beaches: SMM (**A**), Victoria (**B**), Camposoto (**C**), and Barrosa (**D**). A wind rose diagram (**E**) is also presented. The orientation of the beaches is approximately NNW in all of them.

Victoria Beach (Figure 2B) is a three-km-long beach located, like SMM, in the city of Cadiz. Some submerged rocky shoals in front of its shoreline furnish it with a certain amount of heterogeneity [24].

Camposoto Beach (Figure 2C) is located in the south, also facing the Atlantic Ocean, in a littoral spit which consists of quartz-rich sand beaches, dune ridges (locally showing washover fans), and salt marshes [25].

Barrosa Beach (Figure 2D) is the southernmost beach in this study, with a general northwest/southeast orientation. Its total length is 3 km and it has both a promenade with a high urban development (northward) and a dune ridge with lower human occupation (southward) [26]. Nourishment was performed in the urbanised northern sector.

Victoria and Camposoto are part of the same physiographic unit. They are also large beaches that naturally recover most of the sediment. Therefore, annual nourishments for tourist purposes are not necessary.

On the other hand, Santa María and Barrosa are urban beaches with larger sand grain sizes and, subsequently, a shorter intertidal zone. Anthropic actions like scrapping accelerate the natural process of recovery in those beaches.

Climate and Morphologic Characteristics

The climate in the study area is Mediterranean, with a regime of sea surface temperatures of a semi-warm subtropical type (mean value of 16.6 °C) [27]. Rains are within a

Humid Mediterranean regime, with October/November and March/April being the most intense months, but not surpassing 600 mm of water annually [28].

Wave regimes in the area are highly seasonally dependent, with a mean significant height (Hs) of 0.84 m and a mean period (Tz) of 7 s [29]. Sea waves are responsible for 28.5% of the wave energy while swell waves comprise remaining 71.5% [30,31]. For a more detailed description of the wave data, www.puertos.es (15 August 2021) can be consulted. The position of the local buoy (6.33° W, 36.50° N) is located in Figure 1.

Wave runup is important to coastal managers, nearshore oceanographers, and coastal engineers because it delivers much of the energy responsible for beach erosion [32,33].

The tidal range has a mean amplitude of 2.2 m (meso-tidal) [30,31], with the highest amplitude being of about 4 m. The effect of wind and atmospheric pressure on sea level variations is not negligible on this stretch of coast [34]. The sand from the four beaches studied here consists of fine-medium sediment. The average D_{50} is about 0.25 mm, consisting of 90–95% quartz and 5–10% bioclasts [35].

For a proper comprehension, visual data of Hs, peak period (Tp), and wave direction are shown in Figure 3.

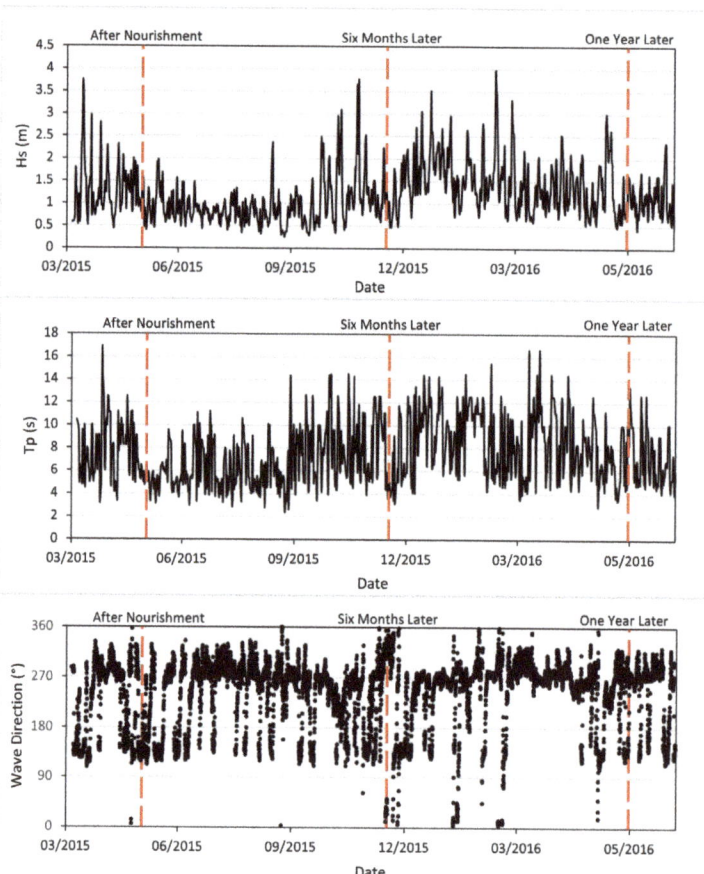

Figure 3. Wave data during the monitoring time between March 2015 and June 2016: Hs (in meters, (**top**)), Tp (in seconds, (**centre**)), and wave direction (being 0° and 360° in the north, 90° in the east, 180° in the south, and 270° in the west, (**bottom**)). The survey dates used in the analysis are marked vertically. Source: own elaboration with data from the Spanish port administration www.puertos.es (15 August 2021).

3. Materials and Methods

This analysis was carried out with samples taken from the SMM, Victoria, Camposoto, and Barrosa beaches during the 2015–2016 period. Dredging and beach nourishment works began and were finished in May–June 2015. A sampling of sand before nourishment, at the end of April, was performed to determine the natural configuration of every beach and, therefore, its native sand parameters. Afterward, three additional samplings of sand were performed semi-annually after the nourishment to monitor their evolution. D_{50} was measured in mm, while sorting was calculated in phi units, a logarithmic scale, following the research of Damveld et al. [36]. The dates of these four campaigns are shown in Table 1.

Table 1. Dates of the monitoring campaigns when samples were taken.

State	Date
Native	April 2015
Just after nourishment	May–June 2015
six months later	November–December 2015
one year later	May–June 2016

This section is separated in two parts: first, a description of how these samples were taken from the beaches, and after that the procedure followed at the laboratory.

3.1. Beach Sand Sampling

Samples were taken from a profile located at the centre of each beach. These profiles, as shown in Figure 2, are perpendicular to the coastline.

Sampling was carried out for four different levels in each transverse profile (Figure 4). The elevations of these levels were: −1 m (submerged), 0 m (the Lowest Low Water Level, LLWL or datum), 2 m (intertidal zone), and 4 m (over Highest High Water Level, HHWL or dry beach).

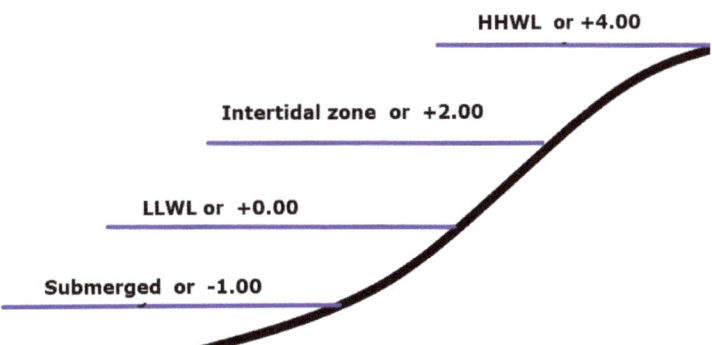

Figure 4. Different levels surveyed in each beach profile.

Once these profiles were defined, the next step involved taking the samples following the procedure of [37], established according to the recommendations proposed in [38].

Subsequently, the process of extracting the sand from the different points (carried out manually) began. This whole procedure is shown in Figure 5, offering a visual description of the different steps of the sampling at the different levels (Figure 5A).

The first step involved the elimination of a first layer, the most superficial one where bioclasts (pieces of shells) were found (Figure 5B). Then, a 20 × 20-centimetre square was made (Figure 5C). Excavation began with a small shovel to a depth of 20 centimetres to collect the sand taken in a small mound (Figure 5D). Next, one part of that (about 300 g) collected sand was deposited in small plastic bags (Figure 5E). Finally, the identifying data (place and date) of each of the points sampled was labelled on each of the plastic bags using a permanent ink marker (Figure 5F)

A Van Veen grab can typically be used to collect submerged samples by throwing it into the sea at a depth of −1 m. Sand accumulates inside it, and the rope is later hoisted to lift the bucket. This procedure can also be done by hand if the conditions are favourable, i.e., at a shallow depth (~1 m) and under slow-current conditions. In our case, samples were collected directly by hand, because fines tend to escape from the Van Veen grab.

Figure 5. Methodology proposed by [37] for beach sand sampling. (**A**): profile with different levels; (**B**): elimination of the first layer because of bioclasts; (**C**): drawing of the 20 cm square for extraction; (**D**): excavation and production of a pile adjacent to the sampling pit; (**E**): mixing and sampling; (**F**): labelling.

3.2. Laboratory Analysis

Grain size is one of the most important sediment particle properties. Sieve analysis is known to be an essential technique for classifying materials and sedimentary environments. It is, therefore, a widely used method in fields such as marine geology or coastal engineering [39].

Grain size is related to the tendency of the sediments to remain in suspension [40]. That is, the greater the D_{50}, the more difficult its removal is. In this case, the laboratory procedures for the granulometry and statistical studies followed the methodologies of [37,39].

According to the usual requirements of the Spanish Coastal Department, the sieves used in the laboratory had the following mesh openings:

- 2 mm
- 1 mm
- 500 µm
- 355 µm
- 250 µm
- 125 µm
- 75 µm

The process was the identical for every sample:

The sample (about 100 gr) was placed on top of the sieve column (i.e., the 2 mm mesh) and, for the filtration process, we placed the sieves into a machine which shook them for 10 min. The fractions for each mash were determined by weighting with a digital scale which was accurate to within 0.1 gr. Two different 100 g samples from each bag were tested and the rest reserved. The differences found were never greater than 0.1 gr (0.1%).

Subsequently, the results were analysed with specific software (Gradistat, a Microsoft Excel add-on [41], was used here).

4. Results and Discussion

4.1. D_{50}

The D_{50} data results were compiled for each of the beaches (see Table 2). To facilitate the observation of temporal patterns, these data are also plotted in Figure 6. Moreover, total variations in D_{50} (between the sand one year later and the native sand) were also compiled and a percentage difference (Dif %) was calculated according to Equation (1) and shown in Table 3 to facilitate the understanding of the sediment's size evolution.

$$Dif \% = \frac{D_{50Native} - D_{50Final}}{D_{50Native}} * 100 \qquad (1)$$

Table 2. Values of D_{50} at SMM, VB, CB, and BB for each monitoring campaign.

Beach	Time	D_{50} (mm) in Each Zone			
		Submerged	LLWL	Intertidal	HHWL
Santa Maria (SMM)	Native (N)	0.2	0.18	0.18	0.23
	After nourishment (AN)	0.21	0.21	0.25	0.27
	six months later (6M)	0.27	0.28	0.29	0.23
	one year later	0.2	0.2	0.2	0.23
Victoria (VB)	Native	0.16	0.18	0.28	0.26
	After nourishment	0.25	0.22	0.24	0.24
	six months later	0.3	0.28	0.29	0.21
	one year later	0.21	0.2	0.28	0.26
Camposoto (CB)	Native	0.18	0.22	0.24	0.23
	After nourishment	0.33	0.28	0.28	0.24
	six months later	0.34	0.32	0.33	0.22
	one year later	0.21	0.22	0.24	0.23
Barrosa (BB)	Native	0.23	0.26	0.38	0.21
	After nourishment	0.27	0.25	0.33	0.24
	six months later	0.32	0.31	0.33	0.21
	one year later	0.22	0.26	0.34	0.21

Evolution of D_{50} values during the different stages of the study:

In general, sands of medium-fine size with a characteristic golden colour and with little variability along the entire beach were observed [42]. However, Figure 6 shows D_{50} values equal to or greater than native sand after the nourishment of the beach. Values between 0.25 and 0.30 mm were maintained in all the analysis, with this being of vital importance for the stability of the beach. This larger size resulted from the larger grain size of the sand used for nourishment relative to the native sand (see the D_{50} values reached just after nourishment in Table 2). For instance, we can see how the grain size of the native sand was less than that of the borrowed sand at SMM. The largest D_{50} values within the nourished sand were found in the upper area of the beach, given that sand was dumped almost entirely in the upper area of the beach due to the pumping system, to reach a settling of the sand [43] as shown in Figure 7.

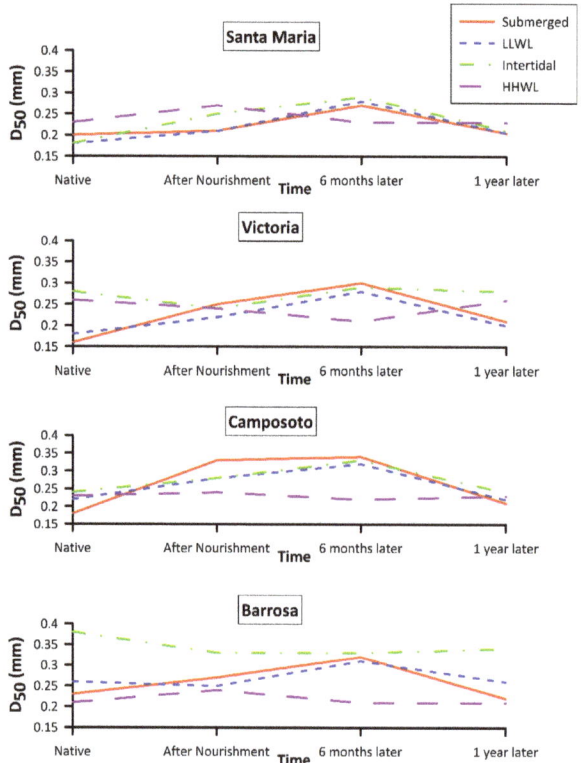

Figure 6. Results of D_{50} (in mm) obtained at the four beaches (Santa Maria del Mar, Victoria, Camposoto, and Barrosa) during one year of monitoring.

Table 3. Total variation of D_{50} (between native sand and nourished sand after one year) at SMM, VB, CB, and BB at the different levels of the profile (Submerged, LLWL, Intertidal, and HHWL).

		Total Variation in mm and Dif% of D_{50} in Each Zone			
	Beach	Submerged	LLWL	Intertidal	HHWL
Variation (in mm and in %)	Santa María	0.00 mm 0.00%	−0.02 mm −10.10%	−0.02 mm −10.10%	0.00 mm 0.00%
	Victoria	−0.05 mm −31.30%	−0.02 mm −10.10%	0.00 mm 0.00%	0.00 mm 0.00%
	Camposoto	−0.03 mm −16.70%	0.00 mm 0.00%)	0.00 mm 0.00%	0.00 mm 0.00%
	Barrosa	0.01 mm 4.30%	0.00 mm 0.00%	0.04 mm 10.50%	0.00 mm 0.00%

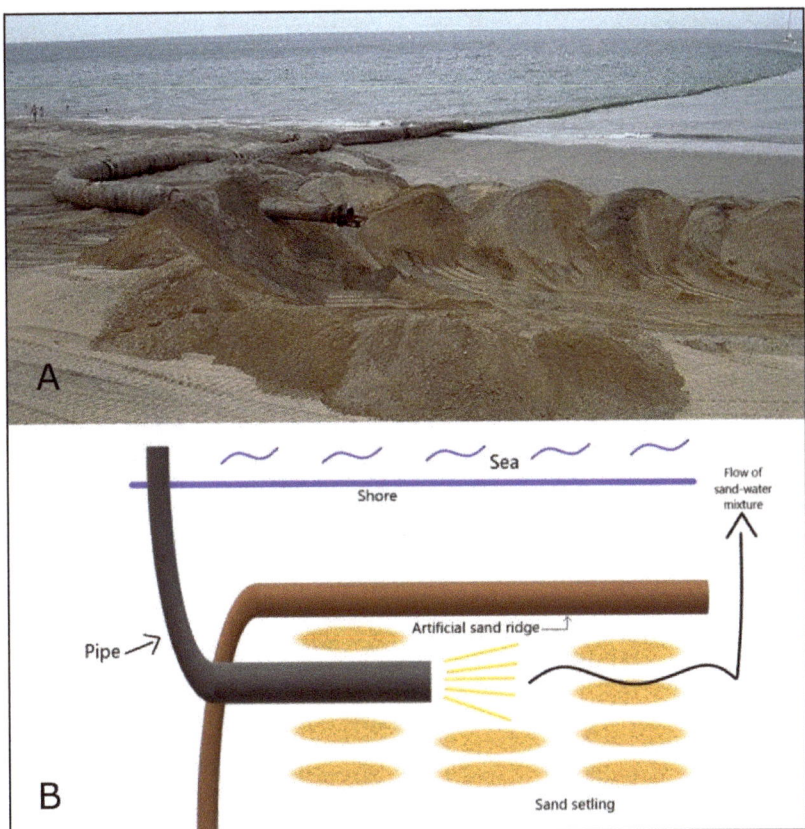

Figure 7. Sand pumping system in a photo (**A**) and schematised diagram (**B**). The pipe is brought to the beach and placed within an artificial ridge. This pipe releases a mixture of sand and water (approximately in a 1/5 proportion) that, thanks to the ridge, settles most of the sediment before reaching the shore.

To understand the behaviour over time of the data shown in Figure 6, it is worth noting that the erosion season (sea waves) ends by May while the sand accretion season (swell waves) lasts until October (see Hs in Figure 3). Summer conditions or a warm climatic season started just after nourishment helping to maintain this structure of the profile. Thus, the migration of fines from the submerged to the emerged zone explains why the D_{50} values decreased slightly during the months immediately after nourishment on the emerged beach but increased on the submerged beach (Figure 6). This is consistent with the variable morphological impacts (depending on the forcing factors, such as storm surge intensity, magnitude, and duration) observed by other authors such as Monteruil et al. [44]. Reeve and Spevack [45] also stated that storm events can lead to large but often transient deformations of the beach, with the effects being smoothed over a period of months.

In the transit between both moments, i.e., six months after the nourishment, great differences (up to 0.07 mm) could be observed between the different points (levels) of the profile: the D_{50} decreased in HHWL (e.g., from 0.27 to 0.23 in SMM) while D_{50} increased in the rest of the profile (e.g., D_{50} increased by 20% in the submerged zone in VB, Table 3). This change is produced by the summer waves (swell) that the beach receives during this period, transporting fine material from the submerged area and low tide to the upper part of the beach [22]. Obviously, the process is reversed during the stormy season [46].

In summary, nourishment (anthropic process) ended in May, at the moment when swell waves (a natural process) started transporting fines from submerged zones to the dry beach, as shown in Figure 8.

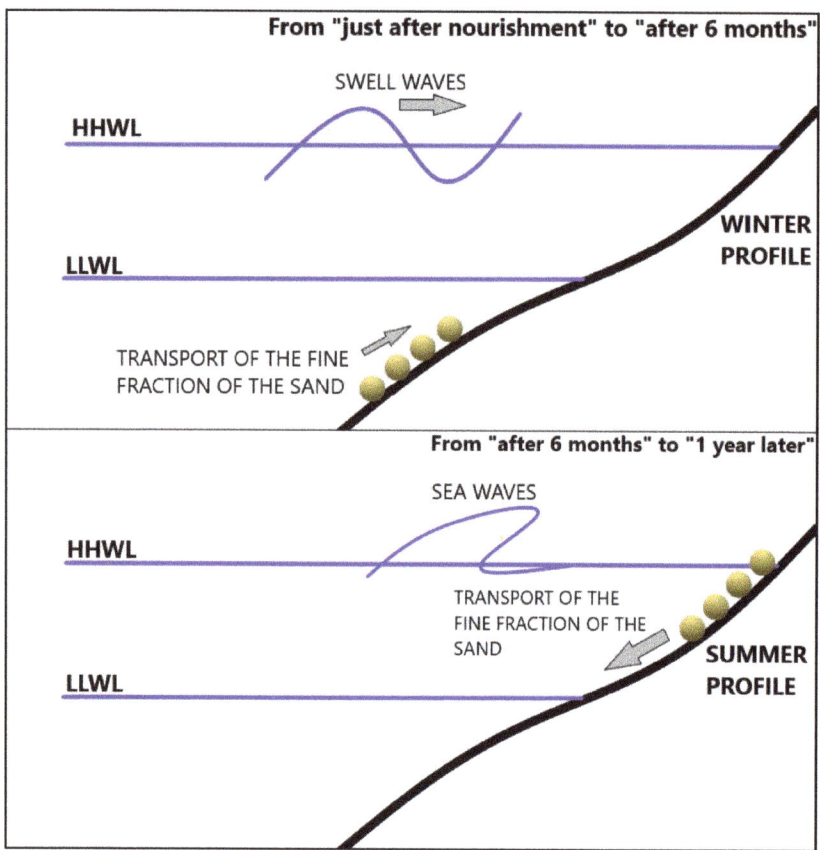

Figure 8. Sketch showing the upwards movement of the fine fraction of the sand by swell waves during the summer season (**top**) and downwards due to sea waves during the winter season (**bottom**).

After one year, the new D_{50} values matched those of the native sand D_{50} in HWWL on all the beaches (Table 3); variations in D_{50} (between native and nourished sand) ranged from 0% to 10% at most in the rest of the points on the profile. Thus, sand size distribution was practically the same as before the nourishment except for the case of the submerged zones of VB and CB where D_{50} was still larger than before (0.05 and 0.03 mm, respectively). These beaches only differ in that their slope is milder than that of SMM and BB. Thus, a clear explanation for this anomaly has not been found and, therefore, further investigations are needed. However, a worthy conclusion is that the old fashioned renourishment factor R_J becomes 1 just one year after the nourishment work and, thus, the erosion rate becomes the same that previously.

The success of the beach nourishment has been verified through the homogeneity of D_{50} values, which were very similar to the native values, achieved just one year after the nourishment works. However, the irreversible loss of sand is also still the same because the hydrodynamics have not changed, with estimates of erosion in some areas of the Gulf of Cadiz of more than 1 m/year [47].

4.2. Sorting

In the same way as was carried out for the D_{50} values, the sorting results were compiled for each of the beaches (see Table 4). To facilitate the observation of temporal patterns, the evolution over time of sorting values is shown in Figure 9. Moreover, since the sorting values are expressed in phi units (not as easy to interpret as mm), the sorting values were interpreted by using the indications given in Table 5 adapted from Roman-Sierra et al. [37]. We must remember that the mathematical expression of the phi scale (φ) is given in Equation (2) by:

$$D(\varphi) = -\log_2 D(mm) \qquad (2)$$

where D is the grain diameter in phi units and d is the grain diameter in millimetres.

Table 4. Sorting values for every beach and surveying campaign.

Beach	Time	Sorting (φ Units) in Each Zone			
		Submerged	LLWL	Intertidal	HHWL
Santa María	Native	0.68	0.6	0.73	0.57
	After nourishment	0.47	0.46	0.98	1.2
	six months later	1.62	1.15	1	0.64
	one year later	0.72	1.09	0.63	0.7
Victoria	Native	0.61	0.5	0.63	0.58
	After nourishment	0.66	0.51	0.58	0.52
	six months later	1.26	1.06	0.62	0.47
	one year later	0.61	0.59	0.74	0.55
Camposoto	Native	0.66	0.82	1.03	0.57
	After nourishment	0.81	0.45	0.58	0.61
	six months later	1.31	0.98	0.64	0.56
	one year later	0.72	0.93	0.84	0.67
Barrosa	Native	0.88	0.55	0.62	0.59
	After nourishment	1.08	0.46	0.61	0.55
	six months later	1.17	1.01	0.65	0.5
	one year later	0.73	0.64	0.67	0.57

Table 5. Classification of sorting values, adapted from Roman-Sierra et al. [37].

Phi Range	Standard Deviation (Sorting)
<0.35	Very well-sorted
0.35–0.50	Well-sorted
0.50–0.71	Moderately well-sorted
0.71–1.00	Moderately sorted
1.00–2.00	Poorly sorted
2.00–4.00	Very poorly sorted
>4.00	Extremely poorly sorted

Native sorting values ranged from 0.50 to 1.03 for the four beaches, therefore classifying these as having moderately well-sorted sands with some scarce locations of moderately sorted material (Table 5). Some of the beaches, such as SMM and VB, seemed to have almost negligible variations in their grain sizes of less than 0.2 ø.

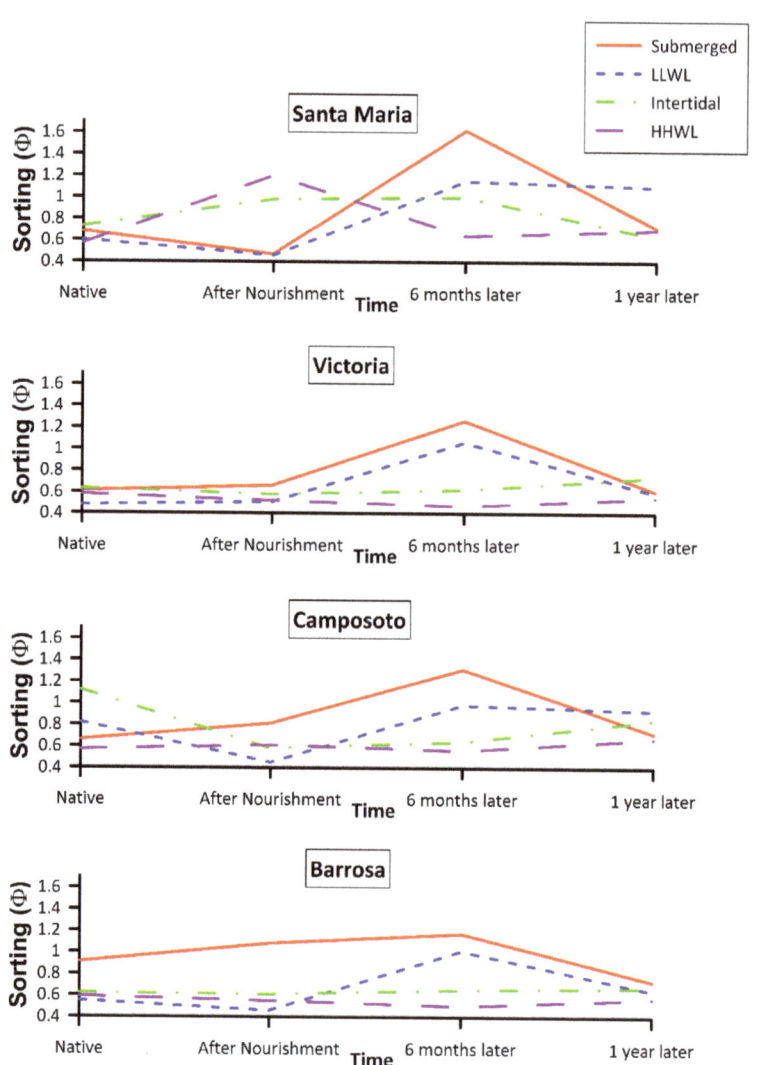

Figure 9. Results of sorting (φ units) obtained for the established elevations (submerged, LLWL, intertidal and HHWL) in the four monitored beaches.

Because nourishment was carried out with sand whose sorting is similar to the native sand's values and because of the sand being poured on the upper beach (Figure 7), there was not a big difference in the sorting at any level except for the intertidal and HHWL at SMM (Figure 9) just after the nourishment. Some changes were observed six months later in the lower (LLWL) and submerged parts of the beach profile, probably due to the movement of the finer sand grains. After one year, the sorting values decreased, becoming very close to the previous native values, being less than 0.75 ø (moderately well sorted) at all levels except at the LLWL at SMM and CB, where slightly higher values were observed (1.09 and 0.93 ø, respectively). Thus, we can state that there was not a significant difference between the sorting of native and nourished sand after one year.

5. Conclusions

The study aimed to identify and explain the behaviour of sand size after a beach nourishment. A generalisable methodology to study the evolution of the D_{50} and the sorting was presented. For this purpose, samples were taken from four different beaches, at different levels (submerged, LLWL, intertidal zone, and HHWL) over time. These campaigns were carried out before and just after the nourishment (in May at the end of the eroding season), six months later (after the swell or accretion season), and finally, one year after the nourishment.

Thus, a decrease of D_{50} (up to 20%) was observed in HHWL six months after the nourishment, whereas D_{50} increased in the rest of the profile. This phenomenon coincided with the summer season during which swell waves moved upwards the fine fraction of the sand from the submerged area. The process was reversed during the winter or stormy season and, eventually, D_{50} became almost identical to the native values one year after the nourishment.

Regarding sorting, the four beaches had native sorting values ranging between 0.50 to 1.03 and, therefore, were classified as having moderately well-sorted sands with some scarce locations of moderately sorted sands. No sorting differences were detected just after the nourishment except in the upper part of the beach where the sand was poured. Six months later, some changes were observed in the LLWL and the submerged zone, probably due to the upward movement of the finer fraction of the sand. Nevertheless, one year after the nourishment and similarly to the D_{50}, sorting values became very close to the original native sand's ones.

Author Contributions: Conceptualization, J.M.J., J.M. and J.J.M.-P.; methodology, J.J.S.-V., P.L.-G., J.M.J., J.M., B.J., A.C., F.C. and J.J.M.-P.; software, J.J.S.-V., P.L.-G., J.M.J., J.M.; validation, J.M.J., B.J., A.C., F.C. and J.J.M.-P.; formal analysis, J.M.J., B.J., A.C., F.C. and J.J.M.-P.; investigation, J.J.S.-V., P.L.-G., J.M.J., J.M., B.J., A.C., F.C. and J.J.M.-P.; original draft preparation, J.J.S.-V., P.L.-G., J.M.J., J.M.; review and editing, J.M.J., B.J., A.C., F.C. and J.J.M.-P. All authors have read and agreed to the published version of the manuscript.

Funding: Funds for APC payment were provided by Coastal Engineering Research Group from University of Cadiz.

Institutional Review Board Statement: Not applicable.

Informed Consent Statement: Not applicable.

Data Availability Statement: Not applicable.

Conflicts of Interest: The authors declare no conflict of interest.

References

1. Silva, R.; Martínez, M.L.; Hesp, P.A.; Catalan, P.; Osorio, A.F.; Martell, R.; Fossati, M.; Miot da Silva, G.; Mariño-Tapia, I.; Pereira, P.; et al. Present and Future Challenges of Coastal Erosion in Latin America. *J. Coast. Res.* **2014**, *71*, 1–16. [CrossRef]
2. Elko, N.; Briggs, T.R.; Benedet, L.; Robertson, Q.; Thomson, G.; Webb, B.M.; Garvey, K. A century of U.S. beach nourishment. *Ocean Coast. Manag.* **2021**, *199*, 105406. [CrossRef]
3. Liu, G.; Qi, H.; Cai, F.; Zhu, J.; Lei, G.; Liu, J.; Zhao, S.; Cao, C. Morphodynamic Evolution of Post-Nourishment Beach Scarps in Low-Energy and Micro-Tidal Environment. *J. Mar. Sci. Eng.* **2021**, *9*, 303. [CrossRef]
4. Asensio-Montesinos, F.; Pranzini, E.; Martínez-Martínez, J.; Cinelli, I.; Anfuso, G.; Corbí, H. The Origin of Sand and Its Colour on the South-Eastern Coast of Spain: Implications for Erosion Management. *Water* **2020**, *12*, 377. [CrossRef]
5. Jóia Santos, C.; Andriolo, U.; Ferreira, J.C. Shoreline Response to a Sandy Nourishment in a Wave-Dominated Coast Using Video Monitoring. *Water* **2020**, *12*, 1632. [CrossRef]
6. Muñoz-Perez, J.J.; Roman-Sierra, J.; Navarro-Pons, M.; da Graça Neves, M.; del Campo, J.M. Comments on "Confirmation of beach accretion by grain-size trend analysis: Camposoto beach, Cádiz, SW Spain" by E. Poizot et al. (2013) Geo-Marine Letters 33(4). *Geo-Mar. Lett.* **2014**, *34*, 75–78. [CrossRef]
7. Poizot, E.; Anfuso, G.; Méar, Y.; Bellido, C. Confirmation of beach accretion by grain-size trend analysis: Camposoto beach, Cádiz, SW Spain. *Geo-Mar. Lett.* **2014**, *33*, 263–272. [CrossRef]
8. Gopalakrishnan, S.; Smith, M.D.; Slott, J.M.; Murray, A.B. The value of disappearing beaches: A hedonic pricing model with endogenous beach width. *J. Environ. Econ. Manag.* **2011**, *61*, 297–310. [CrossRef]

9. Armstrong, S.B.; Lazarus, E.D.; Limber, P.W.; Goldstein, E.B.; Thorpe, C.; Ballinger, R.C. Indications of a positive feedback between coastal development and beach nourishment. *Earth's Future* **2016**, *4*, 626–635. [CrossRef]
10. Martell, R.; Mendoza, E.; Mariño-Tapia, I.; Odériz, I.; Silva, R. How Effective Were the Beach Nourishments at Cancun? *J. Mar. Sci. Eng.* **2020**, *8*, 388. [CrossRef]
11. Alves, B.; Rigall-I-Torrent, R.; Ballester, R.; Benavente, J.; Ferreira, Ó. Coastal erosion perception and willingness to pay for beach management (Cadiz, Spain). *J. Coast. Conserv.* **2015**, *19*, 269–280. [CrossRef]
12. Cutler, E.M.; Albert, M.R.; White, K.D. Tradeoffs between beach nourishment and managed retreat: Insights from dynamic programming for climate adaptation decisions. *Environ. Model. Softw.* **2020**, *125*, 104603. [CrossRef]
13. Herrera, A.; Gomez-Pina, G.; Fages, L.; de la Casa, A.; Munoz-Perez, J.J. Environmental Impact of Beach Nourishment: A Case Study of the Rio San Pedro Beach (SW Spain). *Open Oceanogr. J.* **2010**, *4*, 32–41. [CrossRef]
14. Fletemeyer, J.; Hearin, J.; Haus, B.; Sullivan, A. The impact of sand nourishment on beach safety. *J. Coast. Res.* **2018**, *34*, 1–5. [CrossRef]
15. Benedet, L.; Finkl, C.W.; Campbell, T.; Klein, A. Predicting the effect of beach nourishment and cross-shore sediment variation on beach morphodynamic assessment. *Coast. Eng.* **2004**, *51*, 839–861. [CrossRef]
16. USACE, US Army Corps of Engineers. *Shore Protection Manual*; USACE: Washington, DC, USA, 1984.
17. James, W.R. *Techniques in Evaluating Suitability of Borrow Material for Beach Nourishment (No. 60)*; US Coastal Engineering Research Center, 1975; Available online: https://erdc-library.erdc.dren.mil/jspui/bitstream/11681/2871/1/TM-CERC-No-60.pd (accessed on 15 August 2021).
18. USACE, US Army Corps of Engineers. Coastal Engineering Manual. 2002. Available online: https://www.publications.usace.army.mil/USACE-Publications/Engi (accessed on 15 August 2021).
19. Chu, M.L.; Guzman, J.A.; Muñoz-Carpena, R.; Kiker, G.A.; Linkov, I. A simplified approach for simulating changes in beach habitat due to the combined effects of long-term sea level rise, storm erosion, and nourishment. *Environ. Model. Softw.* **2014**, *52*, 111–120. [CrossRef]
20. Anthony, E.J.; Cohen, O.; Sabatier, F. Chronic offshore loss of nourishment on Nice beach, French Riviera: A case of over-nourishment of a steep beach. *Coast. Eng.* **2011**, *58*, 374–383. [CrossRef]
21. Campbell, T.; Benedet, L. Beach nourishment magnitudes and trends in the US. *J. Coast. Res.* **2006**, *39*, 57–64.
22. Bernabeu Tello, A.M.; Muñoz Pérez, J.J.; Medina Santamaría, R. Influence of a rocky platform in the profile morphology: Victoria Beach, Cádiz (Spain). *Ciencias Mar.* **2002**, *28*, 181–192. [CrossRef]
23. Muñoz-Perez, J.J.; Gutierrez-Mas, J.M.; Parrado, J.M.; Moreno, L. Sediment Transport Velocity by Tracer Experiment at Regla Beach (Spain). *J. Waterw. Port Coast. Ocean Eng.* **1999**, *125*, 332–335. [CrossRef]
24. Muñoz-Pérez, J.J.; Medina, R.; Tejedor, B. Evolution of longshore beach contour lines determined by the E.O.F. method. *Sci. Mar.* **2001**, *65*, 393–402. [CrossRef]
25. Bellido, C.; Anfuso, G.; Plomaritis, T.A.; Rangel-Buitrago, N. Morphodynamic behaviour, disturbance depth and longshore transport at Camposoto Beach (Cadiz, SW Spain). *J. Coast. Res.* **2011**, 35–39.
26. Benavente, J.; Reyes, J.L. The application of morphodynamic indices to exposed beaches of Cadiz Bay. *Bol. Inst. Esp. Ocean.* **1999**, *15*, 213–220.
27. Vargas, J.M.; García-Lafuente, J.; Delgado, J.; Criado, F. Seasonal and wind-induced variability of Sea Surface Temperature patterns in the Gulf of Cádiz. *J. Mar. Syst.* **2003**, *38*, 205–219. [CrossRef]
28. Montero de Burgos, J.L.; González Rebollar, J.L. *Diagramas Bioclimaticos*; Ministerio de Agricultura, Pesca y Alimentación: Madrid, Spain, 1974; ISBN 9788474792058.
29. Reyes, J.L.; Martins, J.T.; Benavente, J.; Ferreira, Ó.; Gracia, F.J.; Alveirinho-Dias, J.M.; López-Aguayo, F. Gulf of Cadiz beaches: A comparative response to storm events. *Bol. Inst. Esp. Ocean.* **1999**, *15*, 221–228.
30. Ministry of Public Works. *Maritime Works Recommendations. Anex I: Wave Climate on the Spanish Coast*; Ministerio Obras Publicas: Madrid, Spain, 1992; p. 76.
31. Jódar Tenor, J.M. *Estudio de la Evolución de los Sedimentos tras la Regeneración de la Playa de Santa María del Mar (Cádiz)*; University of Cadiz: Cadiz, Spain, 2001.
32. Stockdon, H.F.; Holman, R.A.; Howd, P.A.; Sallenger, A.H. Empirical parameterization of setup, swash, and runup. *Coast. Eng.* **2006**, *53*, 573–588. [CrossRef]
33. Davis, R.A. Beach and Nearshore Zone. In *Coastal Sedimentary Environments*; Davis, R.A., Ed.; Springer: New York, NY, USA, 1985; pp. 379–444. ISBN 978-1-4612-5078-4.
34. Aboitiz, A.; Tejedor Álvarez, M.B.; Muñoz Pérez, J.J.; Abarca, J.M. Relation between daily variations in sea level and meteorological forcing in Sancti Petri Channel (SW Spain). *Ciencias Mar.* **2008**, *34*, 491–501. [CrossRef]
35. Poullet, P.; Muñoz-Perez, J.J.; Poortvliet, G.; Mera, J.; Contreras, A.; Lopez, P. Influence of different sieving methods on estimation of sand size parameters. *Water* **2019**, *11*, 879. [CrossRef]
36. Damveld, J.H.; Borsje, B.W.; Roos, P.C.; Hulscher, S.J.M.H. Horizontal and Vertical Sediment Sorting in Tidal Sand Waves: Modeling the Finite-Amplitude Stage. *J. Geophys. Res. Earth Surf.* **2020**, *125*, e2019JF005430. [CrossRef]
37. Román-Sierra, J.; Muñoz-perez, J.J.; Navarro-Pons, M. Influence of sieving time on the efficiency and accuracy of grain-size analysis of beach and dune sands. *Sedimentology* **2013**, *60*, 1484–1497. [CrossRef]

38. Syvitski, J.P.M. *Principles, Methods and Application of Particle Size Analysis*; Syvitski, J.P.M., Ed.; Cambridge University Press: Cambridge, UK, 1991; ISBN 9780521364720.
39. Roman-Sierra, J.; Navarro, M.; Muñoz-Perez, J.J.; Gomez-Pina, G. Turbidity and Other Effects Resulting from Trafalgar Sandbank Dredging and Palmar Beach Nourishment. *J. Waterw. Port Coast. Ocean Eng.* **2011**, *137*, 332–343. [CrossRef]
40. Black, K.P.; Parry, G.D. Entrainment, dispersal, and settlement of scallop dredge sediment plumes: Field measurements and numerical modelling. *Can. J. Fish. Aquat. Sci.* **1999**, *56*, 2271–2281. [CrossRef]
41. Muzambiq, S. Sedimentation Model Area of Lau Kawar Lake from Volkanic Eruption of Sinabung Mountain in Karo District, North Sumatra Province. *Int. J. Adv. Eng. Manag. Sci.* **2019**, *5*, 269–274. [CrossRef]
42. Edwards, A.C. Grain size and sorting in modern beach sands. *J. Coast. Res.* **2001**, *17*, 38–52.
43. Muñoz-Perez, J.J.; Gutiérrez-Mas, J.M.; Moreno, J.; Español, L.; Moreno, L.; Bernabeu, A. Portable Meter System for Dry Weight Control in Dredging Hoppers. *J. Waterw. Port Coast. Ocean Eng.* **2003**, *129*, 79–85. [CrossRef]
44. Montreuil, A.L.; Chen, M.; Brand, E.; Verwaest, T.; Houthuys, R. Post-storm recovery assessment of urbanized versus natural sandy macro-tidal beaches and their geomorphic variability. *Geomorphology* **2020**, *356*, 107096. [CrossRef]
45. Reeve, D.E.; Spivack, M. Evolution of shoreline position moments. *Coast. Eng.* **2004**, *51*, 661–673. [CrossRef]
46. Payo, A.; Kobayashi, N.; Muñoz-Pérez, J.; Yamada, F. Scarping predictability of sandy beaches in a multidirectional wave basin. *Cienc. Mar.* **2008**, *34*, 45–54. [CrossRef]
47. Anfuso, G.; Benavente, J.; Gracia, F.J. Morphodynamic responses of nourished beaches in SW Spain. *J. Coast. Conserv.* **2001**, *7*, 71–80. [CrossRef]

Article

Interaction between Tourism Carrying Capacity and Coastal Squeeze in Mazatlan, Mexico

Pedro Aguilar, Edgar Mendoza * and Rodolfo Silva

Instituto de Ingeniería, Universidad Nacional Autónoma de México, Ciudad Universitaria, Circuito Exterior S/N, Coyoacán, Mexico City 04510, Mexico; paguilarc@iingen.unam.mx (P.A.); rsilvac@iingen.unam.mx (R.S.)
* Correspondence: emendozab@iingen.unam.mx

Abstract: While many coastal areas are affected by coastal squeeze, quantitative estimations of this phenomenon are still limited. Ambiguity concerning the degree of coastal squeeze, combined with a lack of knowledge on its interaction with human activities may lead to inadequate and unsuccessful management responses. The objective of the present research was to quantify the degree of coastal squeeze on the highly urbanized coast of Mazatlan, Mexico, and to investigate the relationship between the development of tourism and coastal squeeze from various time perspectives. The Drivers, Exchanges, States of the environment, Consequences, and Responses (DESCR) framework was applied to identify the chronic, negative consequences of dense tourism in the area, together with the assessment of coastal squeeze. A Tourism Load Capacity (TLC) estimation was made and correlated with the DESCR results, showing that coastal squeeze is inversely correlated with tourism load in Mazatlan. The medium-intensity coastal squeeze currently experienced in Mazatlan requires interventions to avoid severe degradation of the ecosystem on which the local tourism industry relies, for which immediate, long-term, and administrative recommendations are given.

Keywords: coastal squeeze; tourism carrying capacity; coastal management; DESCR; urbanized coasts

1. Introduction

Loss of coastal territory as a result of natural or human activities is a situation faced by most countries with ocean boundaries (Silva et al.) [1]. This threat is the subject of much ongoing research, since it affects both ecosystems and human activities. The intrinsic vulnerability of any coastal zone is undeniable, as these are the most dynamic environments on Earth—the only places where the terrestrial environment, atmosphere, seawater, and freshwater all interact (Silva et al.) [2]. The coastal zone can be delimited as the area between the oceanic boundary of the continental shelf and the first significant topographic change above the maximum storm surge elevation (USACE) [3]. The adaptability of coastal areas, together with the dynamics of the ecosystems they host (wetlands, dunes, and beaches), allow them to control the energy of marine hydrometeorological events and thus, one of the primary services they provide is protection (Silva et al.) [1].

Coastal areas in Europe have always struggled against the loss of territory, particularly in England, where the term coastal squeeze was born (Doody, Tros de Ilarduya) [4–6]. This term was initially used to describe the loss of coastline and habitats to sea defences (Pontee et al.) [7]. In general, coastal squeeze has been understood as the process in which hydrometeorological hazards threaten coastal ecosystems through the combination of sea-level rise (SLR) and the presence of rigid barriers which prevent the ecosystems' adaptation, such as human infrastructure. This situation impedes the terrestrial migration of ecosystems and species as the coast moves inland, and thus they are exposed to local extinction (Martinez et al.) [8]. Natural processes that trigger coastal squeeze include the natural variability in sea level, extreme cyclical events (e.g., storm surges and flooding), and inland landscape morphology, which can function as a static ecological barrier to species migration (Doody) [5]. Known factors contributing to coastal squeeze include global and local climate change and local effects of poorly planned coastal infrastructure (Doody, Pontee et al.) [4,5,7]. The present research considers coastal squeeze as a process in which rising sea levels and other factors, such as hard infrastructure, cause a loss of space in land and sea, and where the ecosystems no longer have the necessary conditions to maintain their essential functions (Silva et al.) [2].

Given its importance and impact on the coasts of many countries, the evaluation of coastal squeeze has been the subject of numerous research efforts. Notable works include those by Jackson et al., Mazaris et al., and Schleupner et al. [9–11], who developed methodologies and spatial models to quantify habitat loss. They explained the responses of coastal ecosystems in different study sites (wetlands, mangroves, sea turtle populations, and intertidal organisms). Another example is Torio et al. [12], who developed a coastal squeeze index from a spatial model that can be used along the boundaries of a single wetland and ranks threats faced by multiple wetlands. Coastal squeeze in the state of Veracruz, Mexico, was investigated by Martinez et al. [8]. They considered urban expansion along the coast, an analysis of coastal geodynamics, and a projection of the potential effects of sea-level rise and the distribution of two focal plant species that are endemic to the coastal dunes in Mexico. Using systematic spatial planning, Mills et al. [13] assessed the optimal configuration and the trade-offs involved in SLR adaptation, incorporating spatial models of inundation, urban growth, and ecosystem migration. None of these works included efforts to forecast coastal squeeze, apart from considering some sea-level-rise scenarios. Hildinger and Braun [14] proposed a methodology that considers three main aspects determining the dynamics of coastal squeeze: the geosphere, the biosphere, and the anthropogenic impact. They conducted a small- and large-scale risk analysis to regulate land use, but did not present any quantification or forecast. Luo et al. [15] combined the coastal squeeze index (CSI) and the assessment method proposed by Torio et al. [12] to evaluate the coastal squeeze potential of the Yellow River Delta coastal wetlands in future SLR scenarios. They focused on the effects of slope and impervious surfaces on adjacent uplands with regard to potential wetland migration. This study was applied to a wetland area but not to urbanized coasts. Luo et al. and Ramirez-Vargas et al. [15,16] developed fuzzy-logic-based coastal squeeze indexes to quantify coastal squeeze intensity. They included ecological, geomorphological, and socioeconomic variables. Silva et al. [2] developed the DESCR (Drivers, Exchanges, States, Consequences, and Responses) framework, which examines the relationships between drivers, exchanges, and environmental states to subsequently assess chronic and negative consequences and determine potential responses to combat coastal squeeze. A recurrent gap in all of the cited work is the assessment of coastal squeeze along with possible response actions, including an evaluation of tourism carrying capacity (Cifuentes) [17].

Three main aspects have increased the rise in consciousness regarding coastal erosion: (a) the continuous growth of human coastal settlements, (b) the lack of knowledge on coastline behavior in the short and medium term and (c) the inefficient regulation of coastal urbanization (Silva et al.) [1]. In Mexico, the federal government has expressed interest in developing a strategy to manage all coastal zones and solve the problems faced there. Unfortunately, they have not yet successfully implemented integrated management, so the issues have been addressed individually: that is, in response to the specific needs, or emergencies, of owners or concessionaires (Cortes-Macías et al., Escofet) [18,19]. For the last 50 years, the Mexican coast, particularly along the Caribbean and the Central Pacific, has hosted tourism, residential, and industrial developments. In this time, countless structures have been built with inappropriate designs and with severe impacts on coastal dynamics. The lack of specific regulatory criteria has meant it is impossible to stop or improve poorly planned developments (Silva et al.) [1]. Unfortunately, most anthropic activities affect the coast, directly or indirectly, causing negative consequences on it (Petrișor et al., Senouci and Taibi)) [20,21]. The transcendence of the phenomenon means it is a critical issue; it is induced by the lack of public policies and specific programs to protect and sustain the coastal zone (Huang et al.) [22]. There is an urgent need to establish programs to control, monitor, and predict the behavior of coasts, in order to minimize the physical, ecological, and socioeconomic consequences of their deterioration under different scenarios (natural and anthropogenic).

Given predictions of current and future climate change scenarios and the continuous modification of coastal ecosystems (urbanization), there has been a growing interest in the study of coastal squeeze assessment (Lithgow et al.) [23]. This paper evaluates and quantifies the degree of coastal squeeze occurring at Mazatlan, Mexico. The initial approach to a coastal squeeze intensity forecast was conducted using available historical data and linear models. Next, the tourist carrying capacity was estimated, and a forecast is provided to correlate its evolution with that of the intensity of coastal squeeze. The study site was modeled under different scenarios (past, present, and future) to seek alternatives to reverse the negative trends found. Recommendations for coastal tourism management to avoid increasing coastal squeeze in Mazatlan are proposed.

In Section 2 the methods used are presented, including the study site description and the data sources; Section 3 shows the results of coastline evolution, estimation of maximum water elevation, and the assessment of coastal squeeze and tourism carrying capacity in Mazatlan. The discussion and conclusions of our findings are given in Section 4.

2. Methods

2.1. Study Site Description

The coastal unit of Mazatlan, a municipality in Sinaloa, Mexico, is bordered to the north by San Ignacio Municipality and the state of Durango, to the east by Concordia Municipality, to the south by Rosario Municipality, and to the west by the Pacific Ocean (Figure 1). Mazatlan is located between the coordinates 105°46′23″, and 106°30′51″ W and 23°04′25″ and 23°50′22″ N. According to the 2010 census by INEGI [24], Mazatlan has a population of 500,000 and an area of 3068 km^2 [25].

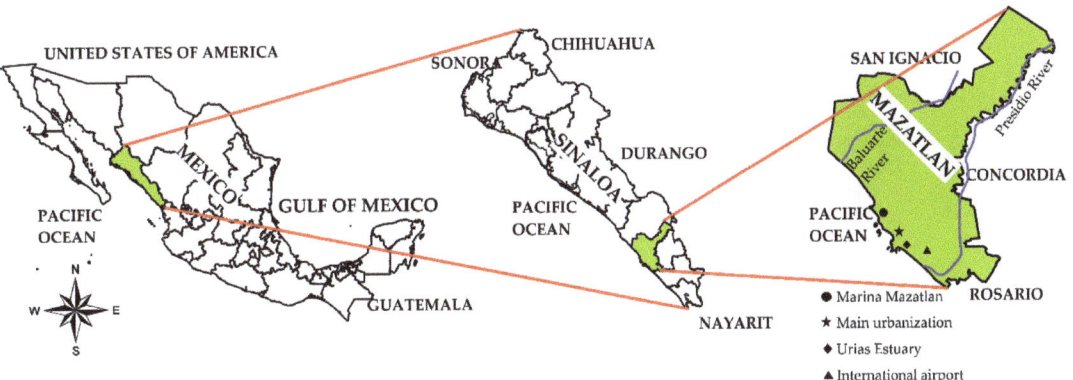

Figure 1. Geographic location of Mazatlan in Sinaloa, Mexico.

Mazatlan is one of the chief ports on the Mexican Pacific due to its maritime conditions, celebrated fishing activity, and tourist activities (it was the sixth national "sun and sea" destination in 2014 according to DATATUR [26]). In 2015, Mazatlan received more than 2,177,000 visitors, an 8% growth compared to 2014. The economic revenue from tourism increased 9% from 2014 to 2015, with over 1 billion USD in revenue from domestic and international tourism and the cruise-ship sector [26]. Apart from the direct economic benefits of the Mazatlan coast (i.e., recreational and tourism activities), port activities and transportation, resource extraction (fishing and aquaculture), education, and scientific research can also be considered local economic drivers.

The Mazatlan coastline is 80 km long and features coastal lagoons, rivers, and streams, with gently rolling hills (formed by wind and marine deposits) with elevations scarcely higher than 50 m above mean sea level. There are no deltas or alluvial plains, but there are sand dunes on the backshore area. According to Fredrickson [27], the coastal strip of Mazatlan is formed by a combination of igneous and volcanic rocks from the Miocene. These rocks underlie a wide layer of alluvial fine and coarse sand. The climate is warm, ranging from 10 to 40 °C (50 to 104 °F) with an annual precipitation average of 722 mm. The monthly average wind speed ranges from 1.4 m/s to 6.6 m/s, with the overall average being 3.5 m/s. The prevailing wind directions along the year are WNW, N, and NNW (Mexican CONAGUA) [28].

The main problems found on the coast of Mazatlan are related to human interventions. Arguably, the most detrimental actions were expanding the boardwalk on the seafront and the modernization of some sections of the area known as the Golden Zone. The former, carried out between Rafael Buelna Avenue and Gutierrez Najera Avenue (see Figure 2), destabilized the beach and destroyed 80% of the coastal dunes. The modernization consisted of dredging and the construction of breakwaters to keep the mouth of the marina open and navigable all year round (see Figure 2). This interrupted the sedimentary longshore balance, producing deficits in some areas and accumulations in others (Oyedutun et al.) [29].

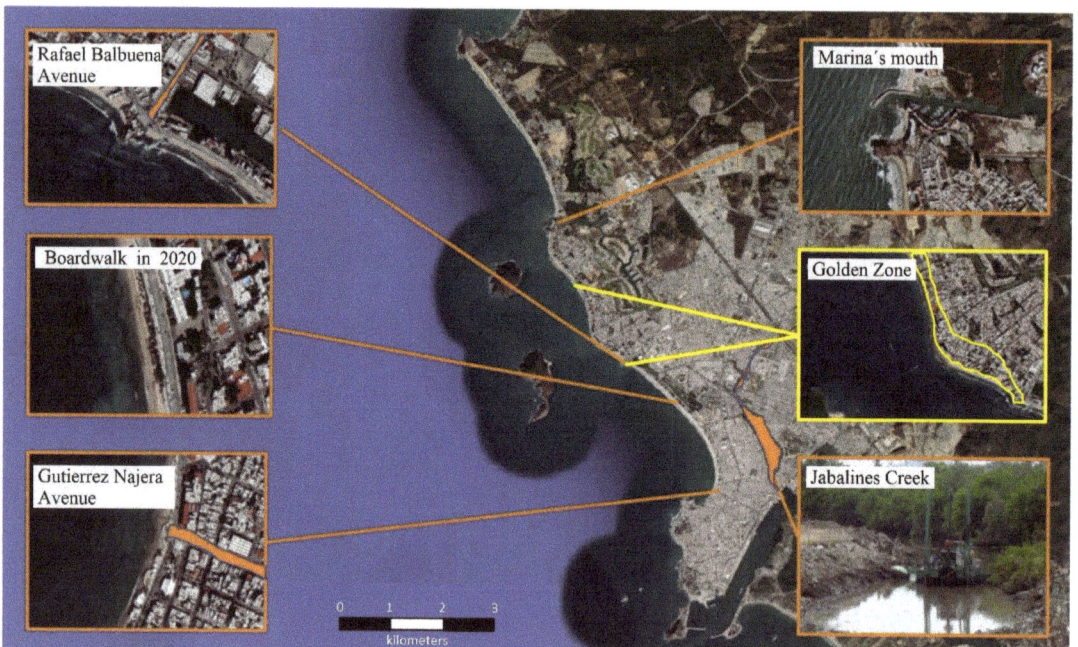

Figure 2. Locations with severe erosion problems on the coast of Mazatlan.

Other actions with negative consequences were the construction of apartments and a shopping center on the shrimp lagoon (a wetland) and the dredging and expansion of Jabalines Creek, for which many mangroves were cut down. In addition, in some parts of the Golden Zone, the construction of buildings very close to the sea has also altered the natural dynamics of the beaches, in most cases narrowing them. To recover the beach width, the owners of the buildings constructed breakwaters and walls without carrying out sufficient studies. The result was inadequate designs and construction work that affected the landscape and caused erosion in nearby areas (Oyedutun et al.) [30]. Figure 2 contains images of some areas in a critical state along the coast of Mazatlan.

2.1.1. Characterization of Marine Climate

Wave data were obtained from the ERA5 reanalysis [31]. The model output contains hourly data of significant wave height (Hs), mean period (Tm), and wave direction for 1979–2020. The data were downloaded for the point located at 23° N 106.5° W (approximately 120 m depth). Figure 3 shows the annual average rose diagrams for significant wave height and mean period. Waves from the SSW clearly govern the marine climate.

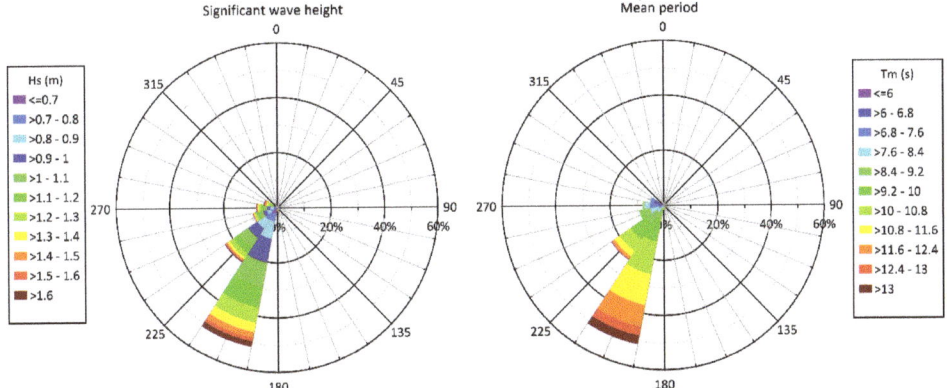

Figure 3. Significant wave height and mean period annual rose diagrams for Mazatlan (1979–2020).

The tidal regime at Mazatlan is microtidal, with a range of 0.44 m at neap tide and 0.55 m at spring tide. These values and the tide levels shown in Table 1 were taken from the Mexican Servicio Mareográfico Nacional [32].

Table 1. Tide levels at Mazatlan [32].

Elevation	Meters above the Mean Sea Level
Maximum registered height	1.462
Highest astronomical tide	1.127
Mean higher high water	0.528
Mean high water	0.455
Mean sea level	0.000
Mean low water	−0.444
Mean lower low water	−0.616
Lowest astronomical tide	−1.250
Minimum registered height	−1.342

Together with human activity, the Mazatlan coast is threatened by sea-level rise (SLR). In this work, SLR was characterized for four scenarios: 1999, 2019, 2059, and 2100. The past and future mean sea levels were estimated by linearly adjusting the IPCC 2014 [33] prediction of a 0.98 m rise for 2100. To estimate the highest water elevations in front of the coast, the Delft 3D model was run for each SLR scenario. The numerical domain was generated by combining a Digital Elevation Model (DEM) of 20 m resolution, the Nautical Chart 363.3 from the Mexican Secretaria de Marina and a topo-bathymetric survey from 2019. The digitizing was carried out with Autodesk Civil 3D software, and the processing and interpolation were performed with Surfer®software to a maximum depth of 85 m. A regular mesh of 272 × 227 nodes in the X and Y directions was set. The mesh contained

61,744 squared cells of 36 m in length. Figure 4 shows the bathymetry obtained for Mazatlan Bay, where a very regular seafloor can be seen, except for some islands.

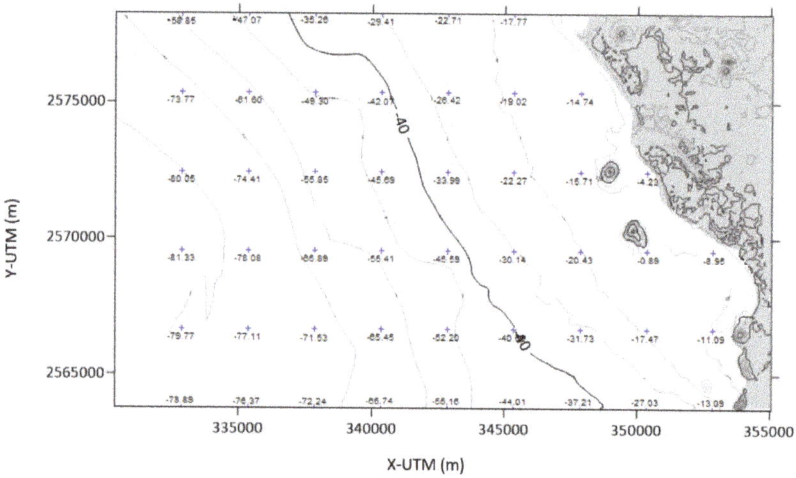

Figure 4. Bathymetry of the study area, soundings in meters.

2.1.2. Coastline Evolution

The Mazatlan coastline has been reported to be rapidly retreating landward in recent years [29]. The analysis presented here was conducted using the Digital Shoreline Analysis System (DSAS), version 4.3 [34] and satellite images available on Google Earth PRO. The shoreline studied comprises the upper limit of the swash zone seen in the available satellite images. The area is delimited by Punta Cerritos and Punta Tiburon (see Figure 5).

The digitized shorelines (the wet/dry lines of the images were manually extracted) and a fixed landward baseline were the inputs to the DSAS. The domain was divided into 278 transects, 50 m apart. The DSAS outputs considered in this research were the Net Shoreline Movement (NSM) and the End Point Rate (EPR). The NSM gives the distance between the oldest and most recent shorelines, regardless of whether they coincide with the positions of the most erosional or cumulative shorelines. The EPR is the NSM value divided by the number of years in each period, giving an annual rate of movement in m/year.

2.1.3. Urban Growth

Urbanization of the Mazatlan coast has increased rapidly since 1999. To characterize this and quantify its growth over time, the urban area of Mazatlan was extracted from Google Earth PRO images for 1999, 2004, 2010, 2015, and 2019. The area was obtained using the semi-automatic classification plugin for QGIS following Chapa et al. [35]

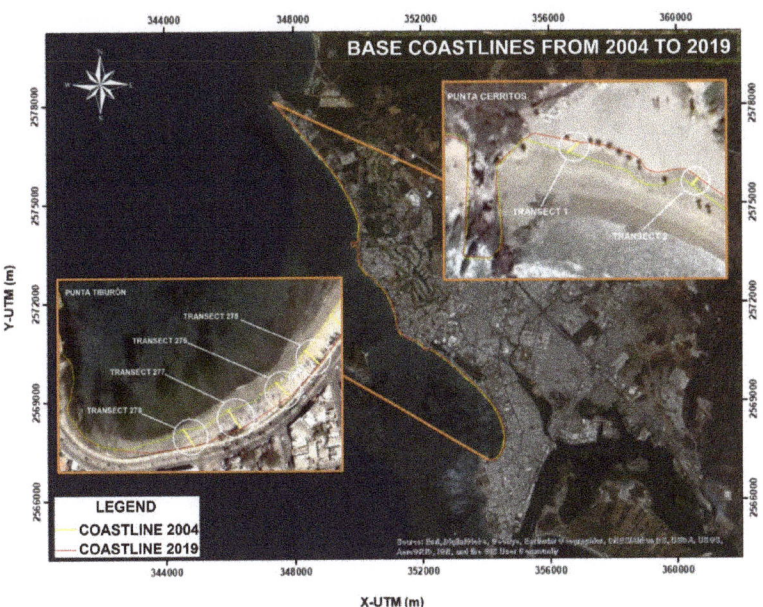

Figure 5. Aerial views of the shoreline in 2004 and 2019, showing the locations of transects 1–2 and 275–278.

2.1.4. Extreme Events

For the present work, the extreme events considered included all categories of tropical storms. According to Hernández et al. [36], 21 tropical storms made landfall close to the Mazatlan area between 1921 and 1999. Hurricanes induce human and material losses, while the rains accompanying these natural phenomena generally cause flooding and the strong winds, intense waves, and storm surge can produce temporal or permanent coastal erosion. The information on extreme weather events was obtained by combining data from a variety of sources: IMPLAN [37], the Mexican National Centre for Disaster Prevention [38] and the National Weather Service and NOAA [39]. On average, 1.7 events occur near Mazatlan per year, and the year with most events was 1981, with 5 (Hernández et al.) [36]. The full list of hurricanes that affected Mazatlan from 1950 to 2019 is shown in Appendix A.

2.1.5. Storm Surge

The water elevation due to storm surge was computed following Villatoro et al. [40], who developed a parametric model of sea surface elevation in front of the coast as a function of wind fields (intensity and direction), given by Equation (1).

$$\eta_s = \alpha V + \beta V$$
$$\text{with}$$
$$\alpha = a + b\theta + c\theta^2$$
$$\beta = d + e\theta + f\theta^2$$
(1)

where η_s is the maximum storm surge, V is the wind velocity in km/h, θ is the wind direction (0° coming from the west and increasing positively, counterclockwise), and α and β are best fit parameters, for which the coefficients a, b, c, d, e and f were obtained and validated by Villatoro et al. [40] from hydrodynamic modeling with MATO [41]. The values of coefficients a–f for Mazatlan are shown in Table 2.

Table 2. Values of the coefficients for storm surge estimation in Mazatlan.

a	b	c	d	e	f
−0.027	0.00025	−0.00000049	−0.00037	0.0000035	−0.000000007

Wind data for the calculation of the storm surge was obtained from the NCEP/NCAR [42] database. The wind intensity time series covered 1949 to 2009. Figure 6 shows the annual rose diagram of wind velocity and direction (incoming) for Mazatlán.

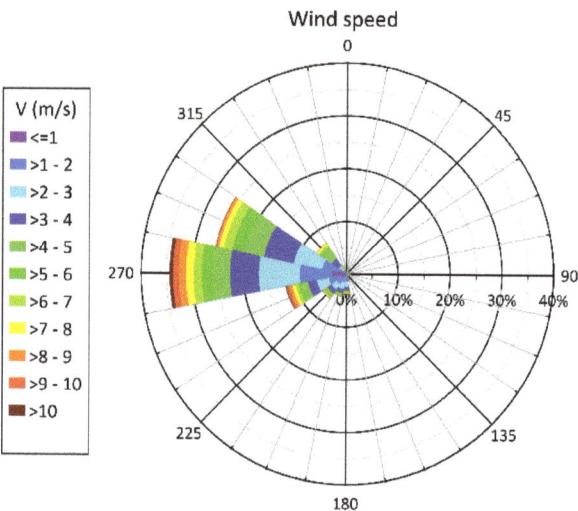

Figure 6. Wind rose diagram for Mazatlan from 1949 to 2009 with data from NCEP/NCAR [42].

Figure 7 shows the yearly maximum values for storm surge (water elevation in front of the coast) obtained from Equation (1).

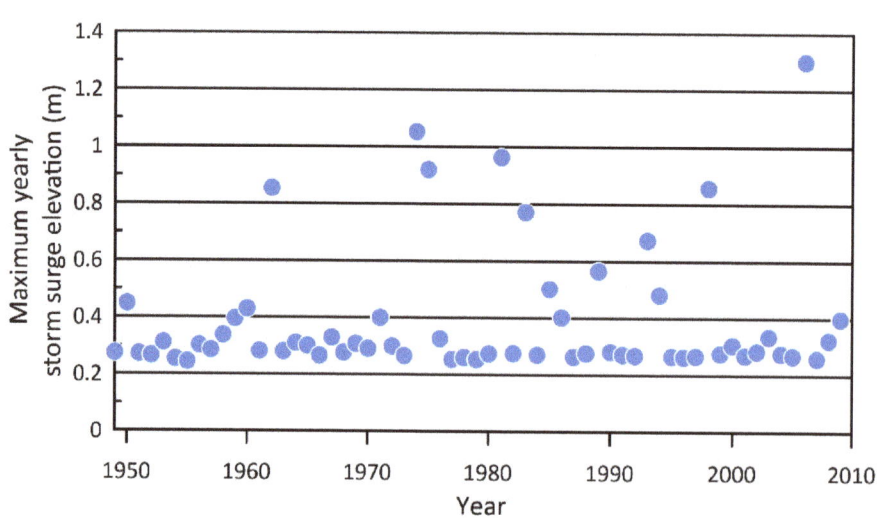

Figure 7. Yearly maximum values of storm surge obtained from Equation (1).

As shown by the data in Figure 7, the mean of the yearly maximum values was 0.39 m and the standard deviation was 0.23 m. The Gumbel probability distribution for a 61-point data sample is

$$x = -\ln(-\ln(f(x))) \times \alpha + \mu$$
$$\alpha = S_x/1.1759 \qquad (2)$$
$$\mu = \bar{x} - (0.5524 \times \alpha)$$

where x is the value to be exceeded, $f(x)$ the probability (inverse of the return period), S_x the standard deviation, and $\bar{x}$ the mean of the sample. From Equation (2), the storm surge values for return periods of 50 and 91 years (i.e., the storm surge elevations for 2059 and 2100) yielded 1.04 m and 1.16 m, respectively.

2.2. Coastal Squeeze Assessment

The coastal squeeze assessment presented by Schleupner et al. [11] was used in this work. The variables were ranked and organized following Saaty [43]. This methodology, known as AHP, solves complex problems with multiple criteria in the following way: the problem is classified according to its causes and solved separately; the solutions are evaluated and ordered in a hierarchical model; and finally, the values are homogenized in order to compute a single value for each scenario (past, present, and future).

The normalization and hierarchization process begins with the construction of a parallel comparison matrix, A. This is a squared matrix in which the variables inducing coastal squeeze are placed in a row and in a column. The elements of A correspond to a comparative relevance value between variables, following the scale shown in Table 3.

Table 3. Comparative relevance scale [43].

Comparative Relevance	Value
Equal importance	1
Moderate importance	3
Strong importance	5
Very strong, demonstrated importance	7
Extremely strong importance	9

The comparative relevance values are placed in the upper diagonals of A, and the bottom diagonals are filled with the multiplicative inverses of the comparative values, yielding

$$A = \begin{bmatrix} 1 & a_{1,2} & \cdots & a_{1,n} \\ 1/a_{1,2} & 1 & \cdots & a_{2,n} \\ \vdots & \vdots & \vdots & \vdots \\ 1/a_{1,n} & 1/a_{2,n} & \cdots & 1 \end{bmatrix} \qquad (3)$$

The second step is the construction of a normalized matrix M, for which the elements of A are added, column by column, and then each element is divided by the sum of its corresponding column, that is

$$M = \begin{bmatrix} \frac{1}{S_1} & \frac{a_{1,2}}{S_2} & \cdots & \frac{a_{1,n}}{S_n} \\ \frac{1/a_{1,2}}{S_1} & \frac{1}{S_2} & \cdots & \frac{a_{2,n}}{S_n} \\ \vdots & \vdots & \vdots & \vdots \\ \frac{1/a_{1,n}}{S_1} & \frac{1/a_{2,n}}{S_2} & \cdots & \frac{1}{S_n} \end{bmatrix} \qquad (4)$$

$$S_j = \sum_{i=1}^{n} a_{i,j}$$

The AHP ends with computation of the weight vector, W, the elements of which are the average of the rows of matrix M as seen in Equation (5).

$$W = \begin{bmatrix} \frac{1}{n}\sum_{j=1}^{n} m_{1,j} \\ \frac{1}{n}\sum_{j=1}^{n} m_{2,j} \\ \vdots \\ \frac{1}{n}\sum_{j=1}^{n} mN_{n,j} \end{bmatrix} \quad (5)$$

Given that the weights obtained from AHP are already hierarchized and homogenized, the intensity of coastal squeeze (ICS) can be obtained as the sum of the value of each particular variable (in its native units), multiplied by its corresponding AHP weight, that is

$$ICS = \sum_{i=1}^{n} w_i C_i \quad (6)$$

where w is the weight and C the characteristic value of each variable inducing coastal squeeze. The result of Equation (4) can be divided by 100 to provide the ICS as a percentage.

2.3. Tourist Load Capacity

Tourism is the main economic activity in Mazatlan; thus, its environmental impact should be systematically monitored. As stated by Fisher et al. [44], a method to quantify whether a tourist resort is negatively impacting the coast is by estimating the Tourist Load Capacity (TLC). The TLC is the maximum number of visitors that an area can accommodate without exceeding the maximum environmental stress, while at the same time maintaining the quality of their experience (Dias et al.) [45]. If the TLC is consistently exceeded, both the tourism industry and the environment begin to degrade.

The methodology used in this work to assess the TLC was that of Cifuentes [17], which consists of calculating the physical carrying capacity (PCC) as

$$PPC = \frac{Sv}{A} \quad (7)$$

where S is the beach area available, v is the average time a tourist stays on the beach, and A is the beach area occupied by a visitor (~2 to 4 m²).

The current carrying capacity (CCC) is calculated by applying a local factor (TCF) to the PPC. The correction coefficients include environmental, social, and economic aspects that may prevent tourists from staying on the beach for the expected time or even from visiting it.

$$CCC = PPC \times TCF$$
$$TCF = \prod_{i=1}^{n} FC_i \quad (8)$$

where FC is the correction coefficient, expressed as a percentage, due to aspect i.

The TLC of effective carrying capacity is obtained by multiplying the CCC by a management capacity coefficient (MC). The management capacity is defined as the best state or conditions that the administration of a protected area can maintain, if it is to carry out its activities and achieve its objectives. Personnel, infrastructure, and equipment variables are used to measure management capacity.

$$TLC = CCC \times MC \quad (9)$$

MC is expressed as a percentage of functionality, with the value being set according to the experience of the administrators of the tourist resort.

3. Results

In this section, the characterization of drivers (SLR, extreme events, and urban growth) is presented. In turn, the exchanges through hydrodynamics and coastline evolution are assessed and the consequences, understood as the intensity of coastal squeeze and impact of tourism, are evaluated.

3.1. Highest Water Elevation in Front of the Coast

3.1.1. Astronomical Tide Level

The input data for each numerical scenario are summarized in Table 4. The wave data taken from ERA5 did not show any increasing trend, so the same conditions were used for all the scenarios. Given that the only bathymetric survey available was for 2019, the mean sea level for this year was taken as a reference, i.e., MSL = 0. The conditions selected coincided with those producing the highest water levels (spring tide and large wave periods) throughout the year.

Table 4. Inputs per numerical simulation scenario.

	1999 MSL = −0.064 m	2019 MSL = 0.0 m	2059 MSL = 0.45 m	2100 MSL = 0.98 m
Astronomical tide	October 18 to 25	October 18 to 25	October 18 to 25	October 18 to 25
Wind	Speed: 4.72 m/s Direction: 315°	Speed: 4.72 m/s Direction: 315°	Speed: 4.72 m/s Direction: 315°	Speed: 4.72 m/s Direction: 315°
Wave	Hs: 0.92 m Tp: 14.3 s Direction: 225°	Hs: 0.92 m Tp: 14.3 s Direction: 225°	Hs: 0.92 m Tp: 14.3 s Direction: 225°	Hs: 0.92 m Tp: 14.3 s Direction: 225°

Figure 8 shows the free surface elevation results of the two-dimensional (vertically averaged) numerical modeling. The modeled time for the coupled tide–wave–wind simulation was 7 days for each scenario, with a calculation time step of 1 min and results recorded every 15 min. The moments of maximum water elevations for the years 1999, 2019, 2059, and 2100 are shown in Figure 6.

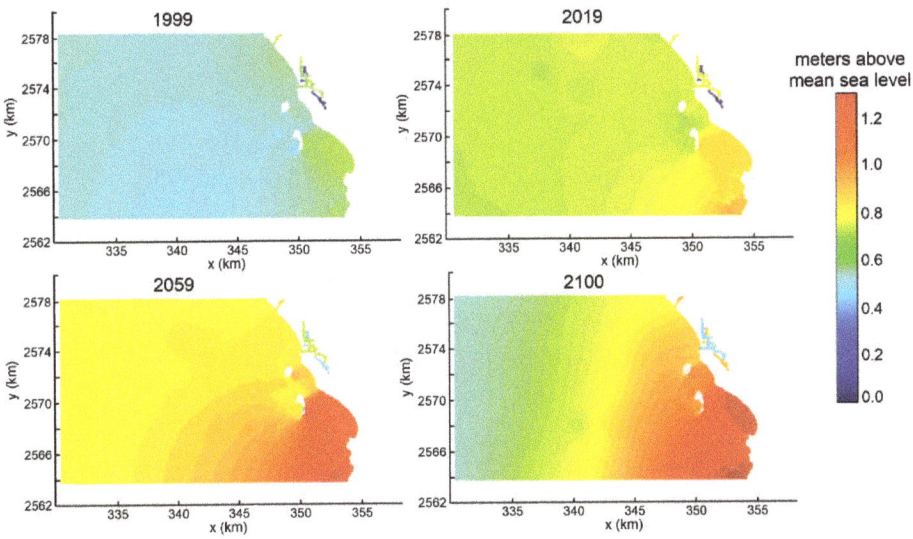

Figure 8. Numerical results for free surface elevations at high tide for 1999 (**top left**), 2019 (**top right**), 2059 (**bottom left**), and 2100 (**bottom right**).

Table 5 summarizes the elevations at high tide and the maximum water elevations for each scenario, obtained from the numerical modeling (Figure 8).

Table 5. Maximum elevation in front of the coast for the predicted SLR scenarios.

	1999	2004	2010	2015	2019	2059	2100
MSL (m)	−0.064	−0.048	−0.029	−0.013	0.000	0.450	0.980
High tide (m)	0.97	1.09	1.05	1.01	0.82	1.07	1.19
Max elevation (m)	1.034	1.138	1.079	1.022	0.820	1.520	2.170

3.1.2. Storm Surge Water Level

The storm surge elevations for the years of interest in this study are summarized in Table 6. The years after 2009 were estimated by Gumbel fit to maximum yearly storm surge values, which gives a worst-case prediction.

Table 6. Storm surges for the years of interest.

	1999	2004	2010	2015	2019	2059	2100
Storm surge (m)	0.28	0.28	0.39	0.61	0.72	1.04	1.15

The total maximum water elevation considers the worst case possible: that is, the simultaneous occurrence of a storm and high tide during the highest spring tides of the year. This is summarized in Table 7.

Table 7. Total maximum water elevation.

	1999	2004	2010	2015	2019	2059	2100
Total maximum water elevation (m)	1.314	1.418	1.469	1.632	1.54	2.56	3.32

3.2. Coastline Evolution

Transects 177 and 179 had the greatest beach retreats, as shown in Figure 9.

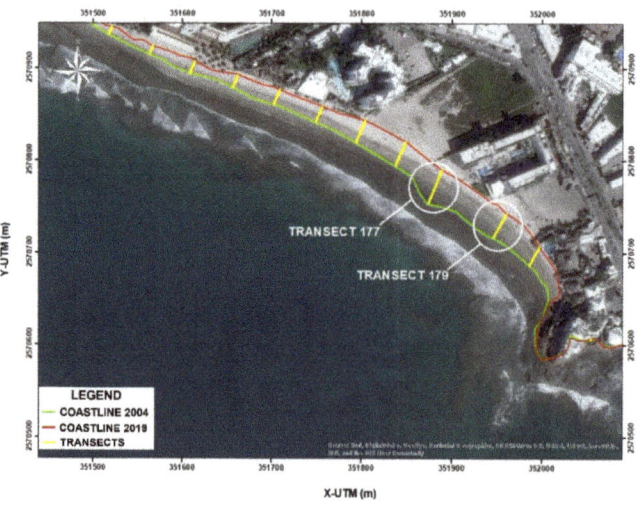

Figure 9. Locations of the transects with greatest retreats on the beach in the Golden Zone.

Figure 10 shows the results of the NSM and the EPR in the left and right panels, respectively. Positive values represent coastline displacement seawards and negative values represent displacement landwards. It can be seen that although the general trend is one of erosion, a small area, near the central part of the beach, was found to be accumulative. Analysis of its evolution from 2004 to 2019 showed an average retreat (erosion) of 23.5 m, with the greatest loss being 2.5 m per year.

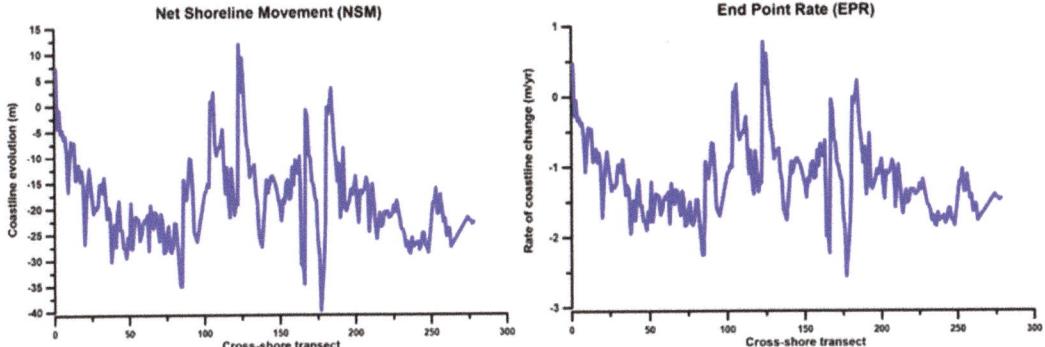

Figure 10. Evolution of the coastline between 2004 and 2019, showing the net shoreline movement (NSM) and the end point rate (EPR) (**left** and **right** panels, respectively).

In this work, the average NSM along the Mazatlan coast was considered as the variable representing the coastal squeeze process. From the DSAS results and fitting of a linear model for the future scenarios, Table 8 shows the values of shore movement for the years considered.

Table 8. Evolution of the Mazatlan coastline.

	1999	2004	2010	2015	2019	2059	2100
Average net shoreline movement (m)	1.1	6.7	13.4	19.0	23.5	68.3	114.2

3.3. Urban Growth

Figure 11 shows the urban area obtained from the satellite images available. With the four available areas, a linear model was fitted to get the areas for 1999, 2059, and 2100. The results are shown in Table 9, where it can be seen that the average urban growth rate is 1.82 km^2 per year.

Table 9. Evolution of the Mazatlan urban area.

	1999	2004	2010	2015	2019	2059	2100
Urban area (km^2)	40.4	48.6	58.4	65.1	73.2	133.1	195.5

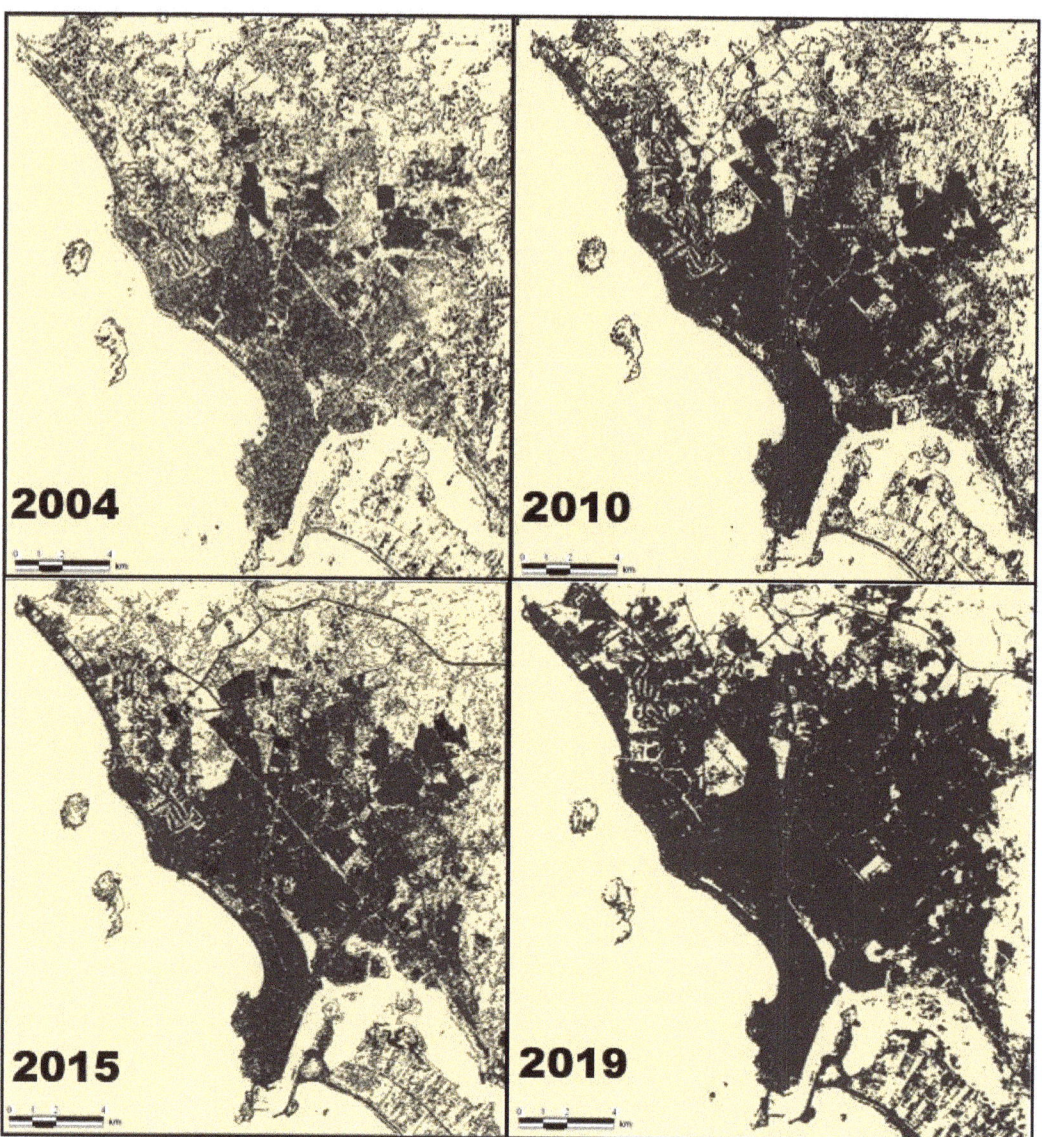

Figure 11. Digitization of satellite images to quantify urbanization in Mazatlan, 2004–2019. The dark gray indicates urban area.

3.4. Coastal Squeeze Assessment

Table 10 shows the hierarchical ranking and weighting of the elements that cause coastal squeeze in Mazatlan. The comparative relevance values were obtained using the authors' experience and personal communication with other experts. The normalized matrix was obtained from Equation (4) and the weight vector from Equation (5).

Table 10. Results of hierarchical analysis and weighting of drivers causing coastal squeeze.

Drivers	Urban Area	Water Elevation	NSM	Extreme Events	Normalized Matrix				W
Urban area	1	7	5	9	0.69	0.84	0.45	0.38	0.59
Water elevation	1/7	1	5	7	0.10	0.12	0.45	0.29	0.24
NSM	1/5	1/5	1	7	0.14	0.02	0.09	0.29	0.13
Extreme events	1/9	1/7	1/7	1	0.08	0.02	0.01	0.04	0.04
Total	1.45	8.34	11.14	24					

The result was a total coastal squeeze value for each scenario. Table 11 shows the results obtained for each driver and the final coastal squeeze value for 1999, 2004, 2010, 2015, 2019, 2059, and 2100.

Table 11. Degree of coastal squeeze for each scenario.

	Scenario						
	1999	2004	2010	2015	2019	2059	2100
Urbanization of the coast	40.4	48.6	58.4	65.1	73.2	133.1	195.5
Maximum water elevation	1.314	1.418	1.469	1.632	1.54	2.56	3.32
Coastline evolution	1.1	6.7	13.4	19.0	23.5	68.3	114.2
Extreme events	1	0	0	1	2	1	1
Coastal squeeze intensity	0.24	0.30	0.37	0.41	0.47	0.88	1.31

The values shown in Table 11 for the urbanization of the coast were taken from Table 9. The maximum water elevation is that reported in Table 7, and the coastline evolution is that from Table 8. The number of extreme events was set using the average number in the period 1950–2019, one event every two years. We used this average as there was no evidence of an increase or decrease in the local historical data. The coastal squeeze intensity was obtained by substituting the values of the upper rows into Equation (6).

The degree of coastal squeeze was determined using the scale by Ramírez-Vargas et al. [17], modified by the authors. The final scale is shown in Table 12.

Table 12. Degree of coastal squeeze, taken and modified from [17].

Degree of Coastal Squeeze	Value
No coastal squeeze	0
Very low	0.00–0.20
Low	0.20–0.40
Medium	0.40–0.60
High	0.60–0.80
Very high	0.80–1.00

As can be seen in Tables 11 and 12, up to 2010 Mazatlan had low-intensity coastal squeeze; since then, up to the current scenario for 2019, the degree of coastal squeeze has been medium, and for 2059 and beyond, the intensity is forecast to be very high.

3.5. Tourist Load Capacity

Table 13 shows the results obtained for Mazatlan for each of the years considered in this study. The beach area was obtained by digitizing the satellite images available in Google Earth PRO and fitting a linear model as done for the urban area. Using survey data from other Mexican beaches, the time per visitor was set to 1.66 h and the area used per person to 4 m^2 (Quijano) [46]; using these values and Equation (7), the PCC was obtained. TLC was determined using Equation (9).

Table 13. The tourist load capacity for Mazatlan.

	Scenario (year)						
	1999	2004	2010	2015	2019	2059	2100
Beach area (m^2)	644,722	602,103	550,960	478,871	442,008	5418	0
Physical carrying capacity (PCC)	267,560	249,873	228,648	198,731	183,433	2248	0
Correction coefficient (TCF)	0.16	0.16	0.16	0.16	0.16	0.16	0.16
Current carrying capacity (CCC)	42,810	39,980	36,584	31,797	29,349	360	0
Management capacity (MC)	0.59	0.59	0.59	0.59	0.59	0.59	0.59
Tourist load capacity (TLC)	25,258	23,588	21,584	18,760	17,316	212	0

The linear model for the beach area gave null values for 2100 as this reflects the "do-nothing" scenario. The present research seeks to motivate actions that will avoid this scenario. The values of *TCF* and *MC* were 0.143 and 0.6, respectively, as detailed in Appendix B. In the absence of historical information and data to perform any forecast, *TCF* and *MC* were considered constant in time.

It was observed that the *TLC* in Mazatlan decreased, to the extent that by 2019 it had a value of 9841 people, compared to 13,829 people in 1999. For the 2059 scenario, there was a *TLC* forecast of only 1573 people, and for 2100, the tourist load capacity was null. This does not mean that no people will visit the resort, but that offering a high-quality tourist experience will not be possible.

Table 14 compares coastal squeeze and *TLC*, showing that as coastal squeeze increases, the *TLC* decreases. The main trigger for the fall in *TLC* is the loss of beach area.

Table 14. Comparison of coastal squeeze and tourist load capacity in different scenarios.

	Scenario (year)						
	19999	2004	2010	2015	2019	2059	2100
Coastal squeeze	0.24	0.30	0.37	0.41	0.47	0.88	1.31
Tourist load capacity	22,957	21,439	19,618	17,051	15,739	193	0

In order to avoid the gloomy future predicted for Mazatlan and shown in Table 10, immediate, long-term actions are needed. These are detailed in the following section.

4. Discussion

The methodology to assess coastal squeeze applied in this work was based on the works by Martinez et al., Schleupner et al., and Ramírez-Vargas et al. [8,11,16], who considered qualitative and quantitative variables. The main differences between this research and the above are the multicriteria evaluation and decision method, and the Hierarchical Analysis Process developed by Saaty [43], which was used to hierarchize and homogenize the values of the drivers that induce coastal squeeze. The DESCR framework was used (Drivers, Exchanges, and States of the environment to subsequently evaluate the chronic, negative Consequences and determine possible Responses), following Silva et al. [2]. Finally, a Tourist Load Capacity (TLC) assessment (Cifuentes) [17] was carried out to verify the results. The Tourist Load Capacity evaluation took into account environmental, social, and economic aspects; therefore, the work presented here is a step closer to the development of an integrated management framework, rather than only an assessment of coastal squeeze.

In the last decade, it has been recognized that some coasts are subject to the phenomenon of coastal squeeze (Schlacher et al.) [47]. With increasing urbanization and human-induced modifications of the coastal zone, the capacity of beaches to change shape and extent in response to storms and SLR is hindered (Nordstrom) [48]. Although studies on coastal squeeze abound, they have rarely been related to increased coastal tourism activity (Lithgow et al.) [23].

In the municipality of Mazatlan, there is no legal framework to regulate and protect the use of the beaches, nor is there any program for their recovery and restoration. For this reason, technical elements are needed to develop criteria for the regulation and sustainable management of new developments. As the oceanographic and topographic data obtained in the field were not available in sufficient quality or quantity, this research also highlights the need for systematic and permanent monitoring of the Mazatlan coast.

The results indicate that this coast is experiencing a coastal squeeze process of 0.47 (medium degree). This means that the sea and land space are already being reduced, and so the beaches and associated ecosystems may disappear if no action is taken. While it is true that the growth in urbanization and the intense expansion of the tourism industry have had positive impacts on the Mazatlan economy, the cost in terms of environmental degradation may be unacceptably high. These developments pose a risk to many coastal and marine recreational activities, and may reduce the attractiveness of the area for tourists.

It was also seen that there is a close link between tourism development and coastal squeeze along the Mazatlan coast, as shown in Table 14, where tourism load capacity and coastal squeeze were found to be inversely correlated.

4.1. Possible Responses

Natural processes that affect coastal stress in Mazatlán include the natural variability in sea level, extreme cyclical events (storm surges and flooding), and inland landscape morphology. Anthropogenic factors causing coastal squeeze include the effects of global and local climate change (sea-level rise and increased frequency and intensity of storms) and the local effects of poorly planned coastal infrastructure. The combination of all the drivers mentioned above has caused a loss of sea and land space (coastal squeeze) to a medium degree, but worsening with time, degrading both the tourism industry and the remaining coastal ecosystems. However, there is still time to halt this trend. From the knowledge gained in this research and following Martínez-López et al. and Chávez et al. [49,50], we offer some recommendations to tackle the coastal squeeze identified in Mazatlan.

4.1.1. Immediate Actions

- Design a permanent coastal unit surveillance program. Continuous monitoring of the coastal zone will produce reliable information, reduce uncertainties, and ensure appropriate actions are taken.
- Make an inventory of urban and coastal areas that are suitable as territorial reserves for coastal protection and, most importantly, stop constructions being built on the beach.
- Carry out an immediate urban densification plan to promote the reuse of lost or forgotten spaces. Given that urbanization is expanding towards the periphery, or the coast, appropriate urban rearrangement may mean Mazatlan is better prepared to face climate change and the challenges of coastal squeeze that will very soon affect the tourism industry there.
- Maintain an updated municipal risk atlas with the detailed information needed for a vulnerable coastal zone.

4.1.2. Long-Term Actions

- Plan new tourist developments and regulations, taking into consideration the dynamics expected, based on the results obtained in the present research regarding SLR.
- Alter or remove infrastructure where necessary; some buildings, roads, etc. were designed without taking environmental sustainability into account. In other cases, infrastructure can be altered, for example by lifting structures off the ground on stilts or pillars.

- Control human migration into the area to reduce urban growth. Proper, long-term planning will enable the authorities to provide tourist services of good quality.

4.1.3. Management and Administration Actions

- Any intervention or construction in the coastal area must demonstrate how it synchronizes with local natural cycles.
- Update land use regulations, to include climate change and SLR information, and review them periodically.
- Implement mitigation plans to address coastal squeeze and establish strategies that respond quickly and effectively to these emerging issues.
- Develop building regulations for the municipality and adapt building codes and urban and coastal infrastructure for safety and sustainability.

5. Conclusions

The methodology developed in this work for measuring the degree of coastal squeeze can be easily applied on a large scale and in other sites. For the case of Mazatlan, it will support municipal, state, and national coastal resource managers in making decisions. The main goal was to provide sufficient technical elements for the development of regulations and sustainable management criteria to design and implement regeneration and restoration projects. It is essential to mention that it is a valuable tool in implementing integrated coastal zone management at a low cost. Likewise, recommendations for coastal tourism management were conceptualized in order to mitigate coastal squeeze, as it has been found to be a direct risk for the tourism industry.

The definition of coastal squeeze has permeated coastal managers' decision-making and disaster risk reduction strategies. Therefore, further work is needed to improve methodologies and disseminate them on a large scale. Policy and decision makers at all levels of government should act together to promote the improvement and care of coastal zones worldwide.

Author Contributions: Conceptualization, P.A. and E.M.; methodology, P.A.; validation, P.A., E.M. and R.S.; formal analysis, E.M. and R.S.; investigation, P.A.; data curation, P.A. and E.M.; writing—original draft preparation, P.A. and E.M.; writing—review and editing, R.S. and E.M.; visualization, P.A.; supervision, R.S.; funding acquisition, R.S. All authors have read and agreed to the published version of the manuscript.

Funding: This research was funded by the CONACYT-SENER-Sustentabilidad Energética project: FSE-2014-06-249795 Centro Mexicano de Innovación en Energía del Océano (CEMIE-Océano).

Acknowledgments: P.A. thanks the support from the Mexican Secretaría de Educación Pública (SEP) through the Programa para el Desarrollo Profesional Docente (PRODEP) throughout his doctoral research.

Conflicts of Interest: The authors declare no conflict of interest. The funders had no role in the design of the study; in the collection, analyses, or interpretation of data; in the writing of the manuscript, or in the decision to publish the results.

Appendix A

Table A1 presents the most extreme meteorological events that impacted Mazatlan from 1950 to 2019, giving the name, date, and the maximum sustained wind speed for each event.

Table A1. Extreme weather events from 1950 to 2019.

Name	Date	Max Wind Speed (km/h)	Name	Date	Max Wind Speed (km/h)
(No name)	19 June 1950	139	Naomi	29 October 1976	56
(No name)	4 July 1950	139	Paul	26 September 1978	61
(No name)	13 September 1951	83	Irwin	29 August 1981	65
(No name)	30 November 1951	83	Knut	21 September 1981	69
(No name)	16 September 1953	139	Lidia	7 October 1981	69
(No name)	2 October 1955	83	Norma	12 October 1981	167
(No name)	21 September 1957	83	Otis	30 October 1981	83
(No name)	20 October 1957	139	Paul	29 September 1982	176
(No name)	22 October 1957	154	Adolph	28 May 1983	65
(No name)	15 June 1958	54	Tico	19 October 1983	111
(No name)	11 September 1958	83	Waldo	9 October 1985	56
(No name)	30 October 1958	46	Newton	22 September 1986	120
(No name)	12 June 1959	83	Paine	2 October 1986	148
(No name)	9 September 1959	139	Roslyn	22 October 1986	78
(No name)	21 October 1959	83	Eugene	26 July 1987	37
Bonny	25 June 1960	83	Kiko	25 August 1989	80
Diana	19 August 1960	139	Douglas	23 June 1990	65
Hyacinth	23 October 1960	120	Rachel	2 October 1990	94
Valerie	25 June 1962	124	Calvin	8 July 1993	89
Doreen	4 October 1962	139	Lidia	13 September 1993	157
Lillian	28 September 1963	83	Rosa	14 October 1994	144
Mona	18 October 1963	70	Henrietta	4 September 1995	139
Natalie	7 July 1964	83	Ismael	14 September 1995	130
Hazel	26 September 1965	83	Isis	2 September 1998	119
Annette	22 June 1968	46	Madeline	19 October 1998	83
Hyacinth	18 August 1968	83	Greg	8 September 1999	102
Naomi	13 September 1968	139	Norman	22 August 2000	46
Emily	23 August 1969	93	Nora	9 October 2003	46
Glenda	10 September 1969	117	Lane	16 September 2006	185
Jennifer	11 October 1969	102	Rick	21 October 2009	91
Eileen	29 June 1970	59	Norman	28 September 2012	95
Helga	19 July 1970	56	Manuel	17 September 2013	75
Ione1	25 July 1970	65	Vance	5 November 2014	65
Katrina	11 August 1971	91	Sandra	28 November 2015	75
Nanette	7 September 1971	137	Javier	8 August 2016	100
Priscilla	12 October 1971	93	Pilar	25 September 2017	75
Orlene	23 September 1974	135	Willa	23 October 2018	240
Olivia	25 October 1975	185	Lorena	19 September 2019	140
Liza	30 September 1976	222	Narda	30 September 2019	95

Appendix B

This section shows the procedure for the estimation of the *TCF* and *MC* factors used in Table 9.

Appendix B.1. Correction Coefficient (TCF)

Following Cifuentes [17], *TCF* is a limiting coefficient which includes three categories of possible limitations to optimal tourist operation regarding environmental, social, and economic issues. *TCF* is defined as

$$TCF = EF \times SF \times EC \qquad (A1)$$

where *EF*, *SF*, and *EC* are the environmental, social, and economic factors, respectively. Each factor is formed by the product of the local limiting subfactors and each subfactor is defined as

$$SF_i = 1 - \frac{Ml_i}{Mt_i} \tag{A2}$$

where *Sf* is the subfactor, *Ml* the limited value, and *Mt* the unconstrained value of element *i*. The second term of the left-hand side of Equation (A2) is called the comparative ratio. Each factor in *TCF* is expressed as a percentage that reduces the physical tourism capacity.

Appendix B.1.1. Environmental Factors

Erodibility

This is the limitation for tourism activities due to beach erosion. Erodibility can be assessed as a function of beach slope and soil texture. Table A2 summarizes the erodibility evaluation in terms of risk levels.

Table A2. Erodibility risk levels.

Sediment	Beach Slope		
	<10%	10–20%	>20%
Sand or gravel	Low	Medium	High
Silt	Low	High	High
Clay	Low	Medium	High

To assess erodibility, the comparative ratio is the sum of medium- and high-risk beach areas by the total available area. In the case of Mazatlan, given that the mean beach slope is less than 10%, the erodibility subfactor is 1.0.

Accessibility

This is a measure of the ease of transit along the beach. Beach slopes of less than 10% are classified as low difficulty, 10–20% as medium difficulty, and above 20% as high difficulty. Given that Mazatlan beaches have a slope of 5%, the accessibility subfactor value is 1.0.

Rain

Rain may prevent visitors spending the average length of time on the beach or even from visiting it at all. In Mazatlan, the rainy season is from July to September (90 days). Considering 6 h of rain as limiting beach visits, the rain subfactor is

$$SF_r = 1 - \frac{540 \; (limiting \; hours)}{4380 \; (available \; hours)} = 0.88 \tag{A3}$$

Note that the total available hours consider only 12 h a day that visitors will go to the beach.

Perturbation to Fauna

Under optimal beach management for Mazatlan, the reproductive season of the Golfina turtle (*Lepidochelys olivacea*) should be considered for a low-tourism season. This season is from July to November; so, the corresponding subfactor is

$$SF_f = 1 - \frac{5 \; (limiting \; months)}{12 \; (total \; months)} = 0.58 \tag{A4}$$

From Equations (A3) and (A4), the environmental factor is

$$EF = 0.88 \times 0.58 = 0.51 \tag{A5}$$

Appendix B.1.2. Social Factors

Visitor Satisfaction

This is addressed via survey results and is a measure of the overall quality of the tourist resort. The information used in this work was obtained from [51,52]. In this survey, the visitors were asked to evaluate several characteristics of the tourist experience, using the following scale: 1 = Bad, 2 = Regular, 3 = Good, 4 = Excellent. The survey data are shown in Table A3.

Table A3. Visitor perception survey in Mazatlan [51,52].

	Survey											
	1	2	3	4	5	6	7	8	9	10	Avg.	%
Transport in general	4	3	2	4	4	1	4	3	4	2	3.1	77
Attention from personnel	3	4	3	3	2	1	4	3	3	2	2.8	70
Quality of food and drink	4	3	2	4	4	1	4	3	4	3	3.2	80
Quality of service	4	3	2	4	2	1	4	3	3	2	2.8	70
Quality of facilities and infrastructure	3	2	2	3	3	2	4	3	4	2	2.8	70
Services and activities offered	3	4	3	3	2	1	4	3	4	2	2.9	73
												73

The visitor satisfaction subfactor is 0.73.

Resident Satisfaction

This was also obtained via surveys, and measures the perception of local inhabitants with respect to the tourist resort. In this work, the data was taken from [51,52]. The inhabitants of Mazatlan were asked to evaluate using the following scale: 1 = Deficient, 2 = Bad, 3 = Regular, 4 = Good, 5 = Excellent. The survey data are shown in Table A4.

Table A4. Residents' perception survey of Mazatlan [51,52].

	Survey											
	1	2	3	4	5	6	7	8	9	10	Avg.	%
General opinion of the resort	5	4	3	1	5	5	3	5	3	1	3.5	87
The tourism industry helps preserve local culture?	5	3	1	1	5	3	3	4	4	2	3.1	78
The arrival of tourists for your community is:	5	5	3	1	5	1	3	5	4	5	3.7	92
The tourist industry provides jobs and wellbeing?	5	5	4	1	5	2	3	5	2	3	3.5	88
How is the relationship between tourists and local inhabitants?	5	3	1	2	5	2	3	5	3	3	3.2	80
												85

The resident satisfaction subfactor is 0.85.
From the above:

$$SF = 0.73 \times 0.85 = 0.62 \tag{A6}$$

Appendix B.1.3. Economic Factors

Tourist Expense Perception

This is a measure of the perceived cost–benefit of the resort. For this work, we took the data from [51,52], where the tourists were asked to evaluate using the following scale: 1 = prices are too high or too low; 2 = prices are high or low; 3 = prices are fair. The survey data are shown in Table A5.

Table A5. Tourists' price perception survey in Mazatlan [51,52].

	Survey											
	1	2	3	4	5	6	7	8	9	10	Avg.	%
Evaluate prices according to service quality	3	3	3	1	2	1	3	3	3	3	2.5	63

The tourist expense perception subfactor is 0.63.

Resident Income Perception

This is the perception of the local inhabitants regarding the income they receive from the tourist industry. In this case, [51,52] used the following scale: 1 = bad income; 2 = moderate income; 3 = good income; 4 = very good income. The survey data are shown in Table A6.

Table A6. Residents' income perception survey in Mazatlan [51,52].

	Survey											
	1	2	3	4	5	6	7	8	9	10	Avg.	%
Evaluate income related to tourism	4	3	4	2	2	3	3	3	4	4	3.2	80

The residents' income perception subfactor is 0.80.
From the above:
$$EC = 0.63 \times 0.80 = 0.50 \tag{A7}$$

Substituting the factors (Equations (A5)–(A7)) into Equation (A1):
$$TCF = 0.51 \times 0.62 \times 0.50 = 0.16 \tag{A8}$$

Appendix B.2. Management Capacity (MC)

The management capacity is the ability of the administration of the resort to offer visitors an optimal experience. Management capacity is measured by the availability of staff, equipment, and infrastructure. For this work, a survey by [51,52] was used and the percentages of functionality regarding the most relevant facilities were gathered. The survey data are shown in Table A7.

Table A7. Functionality of facilities in Mazatlan [51,52].

Equipment	Functionality (%)
Hotel and restaurant facilities available for all economic levels	80
Efficient distribution of services and travel agencies	60
Coordinated intervention of civil protection institutions	68
Existence of a tourist police corporation	55
Lifesavers and aquatic security	45
Dressing rooms and showers on beaches	20
Zoning, signs and regulations	40
Sufficient normative tools for proper administration	40
International airport	70
Maritime terminal	55
Marinas and leisure ports	70
Terrestrial communication	80
Urban transport	80
Health services	68
Average	59

The management capacity factor is 0.59.

References

1. Silva, R.; Villatoro, M.; Ramos, F.; Pedroza, D.; Ortiz, M.; Mendoza, E.; Cid, A. *Caracterización de la Zona Costera y Planteamiento de Elementos Técnicos para la Elaboración de Criterios de Regulación y Manejo Sustentable*; Instituto de Ingeniería Universidad Nacional Autónoma de México: Mexico City, Mexico, 2014; 118p.
2. Silva, R.; Martínez, M.L.; van Tussenbroek, B.; Guzmán-Rodríguez, L.; Mendoza, E.; López-Portillo, J. A Framework to Manage Coastal Squeeze. *Sustainability* **2020**, *12*, 610. [CrossRef]
3. USACE. *Coastal Engineering Manual. Part I*; U.S. Army Corps of Engineers: Washington, DC, USA, 2006; Chapter 1; 5p.
4. Doody, J.P. Coastal squeeze—An historical perspective. *J. Coast. Conserv.* **2004**, *10*, 129–138. [CrossRef]
5. Doody, J.P. Coastal squeeze and managed realignment in southeast England, does it tell us anything about the future? *Ocean. Coast. Manag.* **2013**, *79*, 34–41. [CrossRef]
6. Tros-de-Ilarduya, M. El reto de la Gestión Integrada de las Zonas Costeras (GIZC) en la Unión Europea. *Boletín Asoc. Geógrafos Españoles* **2008**, *47*, 143–156.
7. Pontee, N. Defining coastal squeeze: A discussion. *Ocean. Coast. Manag.* **2013**, *84*, 204–207. [CrossRef]
8. Martínez, M.L.; Mendoza-Gonzalez, G.; Silva, R.; Mendoza, E. Land use changes and sea level rise may induce a "coastal squeeze" on the coasts of Veracruz, Mexico. *Glob. Environ. Chang.* **2014**, *29*, 180–188. [CrossRef]
9. Jackson, A.C.; McIlvenny, J. Coastal squeeze on rocky shores in northern Scotland and some possible ecological impacts. *J. Exp. Mar. Biol. Ecol.* **2011**, *400*, 314–321. [CrossRef]
10. Mazaris, A.D.; Matsinos, G.; Pantis, J.D. Evaluating the impacts of coastal squeeze on sea turtle nesting. *Ocean. Coast. Manag.* **2009**, *52*, 139–145. [CrossRef]
11. Schleupner, C. Evaluation of coastal squeeze and its consequences for the Caribbean Island Martinique. *Ocean. Coast. Manag.* **2008**, *51*, 383–390. [CrossRef]
12. Torio, D.D.; Chmura, G.L. Assessing Coastal Squeeze of Tidal Wetlands. *J. Coast. Res.* **2013**, *29*, 1049–1061. [CrossRef]
13. Mills, M.; Leon, J.X.; Saunders, M.I.; Bell, J.; Liu, Y.; O'Mara, J.; Lovelock, C.E.; Mumby, P.J.; Phinn, S.; Possingham, H.P.; et al. Reconciling Development and Conservation under Coastal Squeeze from Rising Sea Level. *Conserv. Lett.* **2016**, *9*, 361–368. [CrossRef]
14. Hildinger, A.; Braun, A. Outlining an approach to address Geospherical and Biospherical aspects of Coastal Squeeze in the Mediterranean. *J. Coast. Res.* **2016**, *75*, 997–1001. [CrossRef]
15. Luo, S.; Shao, D.; Long, W.; Liu, Y.; Sun, T.; Cui, B. Assessing coastal squeeze of wetlands at the Yellow River Delta in China: A case study. *Ocean. Coast. Manag.* **2018**, *153*, 193–202. [CrossRef]
16. Ramírez-Vargas, D.L.; Mendoza, E.; Lithgow, D.; Silva, R. A Quantitative Methodology for Evaluating Coastal Squeeze Based on a Fuzzy Logic Approach: Case Study of Campeche, Mexico. *J. Coast. Res.* **2019**, *92*, 101–111. [CrossRef]
17. Cifuentes, M. *Determinación de Capacidad de Carga Turística en Áreas Protegidas*; Centro Agronómico tropical de Investigación y Enseñanza (CATIE): Turrialba, Costa Rica, 1992; 23p.
18. Cortes-Macias, R.; Navarro-Jurado, E.; Ruiz-Sinoga, J.D.; Delgado-Peña, J.J.; Noa, R.R.; Salinas-Chávez, E.; Fernández, J.M. Manejo integrado costero en Cuba, la ensenada Sibarimar. *Baetica. Estud. Arte Geogr. Hist.* **2010**, *32*, 45–65.
19. Escofet, A. Marco operativo de macro y mesoescala para estudios de planeación de zona costera en el Pacífico mexicano. In *El Manejo Costero en México*. *México*; Arriaga, E., Azuz, I., Villalobos, G., Eds.; Centro EPOMEX, Universidad Autónoma de Campeche: Campeche, Mexico, 2004; pp. 223–233.
20. Petrișor, A.-I.; Hamma, W.; Nguyen, H.D.; Randazzo, G.; Muzirafuti, A.; Stan, M.-I.; Tran, V.T.; Așteřănoaiei, R.; Bui, Q.-T.; Vintilă, D.-F.; et al. Degradation of Coastlines under the Pressure of Urbanization and Tourism: Evidence on the Change of Land Systems from Europe, Asia and Africa. *Land* **2020**, *9*, 275. [CrossRef]
21. Senouci, R.; Taibi, N.E. Impact of the urbanization on coastal dune: Case of Kharrouba, west of Algeria. *J. Sediment. Environ.* **2019**, *4*, 90–98. [CrossRef]
22. Huang, F.; Huang, B.; Huang, J.; Li, S. Measuring Land Change in Coastal Zone around a Rapidly Urbanized Bay. *Int. J. Environ. Res. Public Health* **2018**, *15*, 1059. [CrossRef]
23. Lithgow, D.; Martínez, M.L.; Gallego-Fernández, J.B.; Silva, R.; Ramírez-Vargas, D.L. Exploring the co-occurrence between coastal squeeze and coastal tourism in a changing climate and its consequences. *Tour. Manag.* **2019**, *74*, 43–54. [CrossRef]
24. INEGI. Censo de Población y Vivienda 2010. Instituto Nacional de Estadística y Geografía. Dirección General de Estadísticas Sociodemográficas. Dirección General Adjunta del Censo de Población y Vivienda. 2010. Available online: https://www.inegi.org.mx/rnm/index.php/catalog/71/related_materials?idPro= (accessed on 15 April 2021).
25. INEGI. Directorio Estadístico Nacional de Unidades Económicas (DENUE). 2015. Available online: http://www.beta.inegi.org.mx/temas/turismo/ (accessed on 15 April 2021).
26. DATATUR. Análisis Integral del Turismo, Secretaría de Turismo. (2015–2016). 2016. Available online: http://www.datatur.sectur.gob.mx (accessed on 15 April 2021).
27. Fredrikson, G. *Geology of Mazatlan Area, Sinaloa Western Mexico*; University of Texas: Austin, TX, USA, 1974; 418p.
28. CONAGUA. Mazatlan-Station. 2020. Available online: https://smn.conagua.gob.mx/es/informacion-climatologica-por-estado?estado=sin (accessed on 15 April 2021).
29. Oyedotun, T.D.T.; Ruiz-Luna, A.; Navarro-Hernández, A.G. Coastline morphodynamics and defences in Mazatlán, Mexico. *Interdiscip. Environ. Rev.* **2018**, *19*, 168–183. [CrossRef]

30. Oyedotun, T.D.T.; Ruiz-Luna, A.; Navarro-Hernández, A.G. Contemporary shoreline changes and consequences at a tropical coastal domain. *Geol. Ecol. Landsc.* **2018**, *2*, 104–114. [CrossRef]
31. ECMCF. ERA5 Hourly Data on Single Levels from 1979 to Present. 2021. Available online: https://doi.org/10.24381/cds.adbb2d47 (accessed on 15 April 2021).
32. SMN. Servico Mareográfico. Instituto de Geofisica, UNAM. 2020. Available online: http://www.mareografico.unam.mx/portal/index.php?page=Estaciones&id=16 (accessed on 15 April 2021).
33. Pachauri, R.K.; Allen, M.R.; Barros, V.R.; Broome, J.; Cramer, W.; Christ, R.; Church, J.A.; Clarke, L.; Dahe, Q.; Dasgupta, P.; et al. *Climate change 2014: Synthesis Report. Contribution of Working Groups I, II and III to the Fifth Assessment Report of the Intergovernmental Panel on Climate Change*; IPCC: Geneva, Switzerland, 2014; 151p.
34. Thieler, E.; Himmelstoss, E.; Miller, T. *User Guide and Tutorial for the Digital Shoreline Analysis System (DSAS) Version 3.2. Extension for ArcGIS*; USGS: Reston, VA, USA, 2005; 33p.
35. Chapa, F.; Hariharan, S.; Hack, J. A New Approach to High-Resolution Urban Land Use Classification Using Open Access Software and True Color Satellite Images. *Sustainability* **2019**, *11*, 5266. [CrossRef]
36. Hernández, M.; Azpra, E.; Carrasco, G.; Delgado, O.; Villicaña, F. *Los Ciclones Tropicales de México I.6.1*; Plaza y Valdes, Instituto de Geografía, UNAM: Mexico City, Mexico, 2001; pp. 117–120.
37. IMPLAN-Mazatlán. Programa Municipal de Desarrollo Urbano. 2020. Available online: http://www.implanmazatlan.mx/programas/ (accessed on 15 April 2021).
38. CENAPRED. Centro Nacional de Prevención de Desastres. 2019. Available online: https://www.gob.mx/cenapred (accessed on 15 April 2021).
39. NHC/NOAA. Eastern Pacific Hurricane Archive. Available online: https://www.nhc.noaa.gov/data/tcr/index.php?season=2021&basin=epac (accessed on 15 January 2021).
40. Villatoro, M.; Silva, R.; Méndez, F.; Zanuttigh, B.; Pan, S.; Trifonova, E.; Losada, I.J.; Izaguirre, C.; Simmonds, D.; Reeve, D.E.; et al. An approach to assess flooding and erosion risk for open beaches in a changing climate. *Coast. Eng.* **2014**, *87*, 50–76. [CrossRef]
41. Posada, G.; Simmonds, D.; Silva, R.; Pedrozo, A. A 2D hydrodynamic model with multi quadtree mesh. In *Ocean Engineering Research Advances*; Alan, I., Ed.; Nova Publishers: Prescot, UK, 2008; pp. 205–241.
42. NCEP/NCAR 40-Year Reanalysis Project. *Bull. Am. Meteorol. Soc.* **1996**, *77*, 437–472. [CrossRef]
43. Saaty, T. Método analítico jerárquico (AHP)-principios básicos. In *Evaluación y Decisión Multicriterio. Reflexiones y Experiencias*; Martínez, E., Escudey, M., Eds.; USACH, UNESCO: Santiago, Chile, 1998; pp. 17–46.
44. Fisher, A.C.; Krutilla, J.V. Determination of optimal capacity of resource-based recreation facilities. *Nat. Resour. J.* **1972**, *12*, 417–444.
45. Dias, I.; Körössy, N.; Selva, V.F. Determinación de la capacidad de carga turística: El caso de Playa de Tamandaré-Pernambuco-Brasil. *Estud. Perspect. Tur.* **2012**, *21*, 1630–1645.
46. Quijano, I. Capacidad de Carga Turística en Tres Playas del Norte de Tuxpan, Veracruz. Specialty in Environmental Mangement and Impact Thesis, Universidad Veracruzana, Xalapa, Mexico, 2019.
47. Schlacher, T.A.; Dugan, J.; Schoeman, D.S.; Lastra, M.; Jones, A.; Scapini, F.; McLachlan, A.; Defeo, O. Sandy beaches at the brink. *Divers. Distrib.* **2007**, *13*, 556–560. [CrossRef]
48. Nordstrom, K.F. *Beaches and Dunes of Developed Coasts*; Cambridge University Press: Cambridge, UK, 2000; 340p.
49. Chávez, V.; Lithgow, D.; Losada, M.; Silva-Casarin, R. Coastal green infrastructure to mitigate coastal squeeze. *J. Infrastruct. Presero. Resil.* **2021**, *2*, 1–12. [CrossRef]
50. Martínez-López, J.; Teixeira, H.; Morgado, M.; Almagro, M.; Sousa, A.I.; Villa, F.; Balbi, S.; Genua-Olmedo, A.; Nogueira, A.J.A.; Lillebø, A.I. Participatory coastal management through elicitation of ecosystem service preferences and modelling driven by "coastal squeeze". *Sci. Total Environ.* **2019**, *652*, 1113–1128. [CrossRef]
51. SECTUR. Agendas de Competitividad de los Destinos Turísticos de México: Mazatlán Sinaloa. Universidad del Occidente. 2014. Available online: http://www.sectur.gob.mx/wp-content/uploads/2015/02/PDF-Mazatlan.pdf (accessed on 15 April 2021).
52. SECTUR/CONACYT. Estudio de la Vulnerabilidad y Programa de Adaptación ante la Variabilidad Climática y el Cambio Climático en diez Destinos Turísticos Estratégicos, así como Propuesta de un Sistema de Alerta Temprana a Eventos Hidrometeorológicos Extremos. Sección VIII Vulnerabilidad del Destino Turístico Mazatlán, Sinaloa. 2014. Available online: http://www.sectur.gob.mx/wp-content/uploads/2014/09/SECCION-VIII.-MAZATLAN.pdf (accessed on 15 April 2021).

Article

History, Current Situation and Challenges of Gold Mining in Ecuador's Litoral Region

Carlos Mestanza-Ramón [1,2,*], Selene Paz-Mena [3], Carlos López-Paredes [4], Mirian Jimenez-Gutierrez [4], Greys Herrera-Morales [4], Giovanni D'Orio [5] and Salvatore Straface [1]

1. Department of Environmental Engineering, University of Calabria, 87036 Rende, Italy; salvatore.straface@unical.it
2. Research Group YASUNI-SDC, Escuela Superior Politécnica de Chimborazo, Sede Orellana, El Coca 220001, Ecuador
3. Green Amazon, Research Center, Nueva Loja 210150, Ecuador; yennifer.paz@espoch.edu.ec
4. Escuela Superior Politécnica de Chimborazo, Sede Orellana, El Coca 220001, Ecuador; carlosr.lopez@espoch.edu.ec (C.L.-P.); mirian.jimenez@espoch.edu.ec (M.J.-G.); greys.herrera@espoch.edu.ec (G.H.-M.)
5. Department of Economics, Statistics and Finance, University of Calabria, Arcavacata, 87036 Rende, Italy; giovanni.dorio@unical.it
* Correspondence: cmestanza@ug.uchile.cl

Citation: Mestanza-Ramón, C.; Paz-Mena, S.; López-Paredes, C.; Jimenez-Gutierrez, M.; Herrera-Morales, G.; D'Orio, G.; Straface, S. History, Current Situation and Challenges of Gold Mining in Ecuador's Litoral Region. *Land* **2021**, *10*, 1220. https://doi.org/10.3390/land10111220

Academic Editors: Pietro Aucelli, Angela Rizzo, Rodolfo Silva Casarín and Giorgio Anfuso

Received: 5 October 2021
Accepted: 8 November 2021
Published: 11 November 2021

Publisher's Note: MDPI stays neutral with regard to jurisdictional claims in published maps and institutional affiliations.

Copyright: © 2021 by the authors. Licensee MDPI, Basel, Switzerland. This article is an open access article distributed under the terms and conditions of the Creative Commons Attribution (CC BY) license (https://creativecommons.org/licenses/by/4.0/).

Abstract: Gold mining in Ecuador has been present in the country since Inca times; over the years interest in the mineral has increased, leading to the creation of legislation to control the mining sector in a safe manner. The Litoral region consists of seven provinces, six of which have registered gold concessions; the most affected provinces are El Oro and Esmeraldas. The objective of this study was to analyze the historical and current situation of artisanal and industrial gold mining in the Litoral region of Ecuador. Different methodologies were used for the elaboration of this study, including bibliographic review, grey literature, field interviews and a validation of expert judgment. The main results indicate that El Oro and Esmeraldas are essentially the most conflictive areas in the region, as they have sometimes had to establish precautionary measures due to the risks caused by illegal mining. In addition, in both areas there is a great socioeconomic impact ranging from lack of opportunities, forgetfulness, migration, emigration, and violation of rights, among others. With respect to environmental impacts, the study highlights the contamination of water sources (which leads to a lack of drinking water for people), and damage to aquatic and terrestrial ecosystems. Finally, the study concludes that the authorities should control the mining sector more by implementing more laws and carrying out inspections to put an end to illegal gold mining, in order to improve the situation in the areas.

Keywords: mercury; gold; socioeconomic impacts; political management; environmental management

1. Introduction

Gold mining and extraction dates back to ancient times, during the primitive era approximately 4000 years ago [1]. This economic activity has been transformed globally over the last two decades to such an extent that investment in developing countries has increased [2]. This event is of great significance, as millions of people depend on industrial, artisanal and small-scale gold mining for their survival [3]. In the context of world gold production, it is estimated that there will be a 5% reduction to a five-year low of 3359 t; an event related to the interruption of mining activities during the COVID-19 pandemic. On the other hand, the countries that have positioned themselves as the largest gold producers in recent years are China, Russia, Australia, United States, Canada, Peru, Ghana, South Africa, Mexico, and Brazil, among others [4].

In Ecuador, gold mining makes a great contribution to the country's economy, and in recent years laws and policies have been implemented to maintain a lower impact on the environment. In Ecuador there are four types of mining according to the Mining Law: artisanal mining, small-scale mining, medium-scale mining and large-scale mining. This categorization has made it possible to maintain better control over mining, as these activities increase over the years [5]. Likewise, there has been an increase in illegality and informality in the gold mining sector, which is considered an obstacle to the development of an accurate record of all the gold produced and exported in the territory [6]. However, the Agency for Regulation and Control of Energy and Non-Renewable Natural Resources (ARCERNNR, for its acronym in Spanish) keeps a record of the economic aspects in the context of the mining activity of companies that exploit the mineral [7].

Mining activities can have a transformative effect on socio-economic development in the areas where it is carried out. It can produce compliance with high social, environmental and safety standards by offering employment opportunities. However, it can also give rise to false hopes of wealth and cause armed conflict [8]. In addition, it may include social conflicts such as lack of gender inclusion, labor exploitation and discrimination due to migratory status, lack of organization, child exploitation, alcoholism, citizen insecurity, and violence associated with illicit activity groups, among other things [9]. In recent years, measures have been proposed to promote mining investment, as it is believed that it can support the dollarization policy and move towards promoting a larger and more responsible mining sector [10].

Gold mining can cause serious environmental impacts. This is mainly related to illegal mining, since no technified methods are used to extract the mineral [5]. As a result, chemical pollution is produced on the banks of the water bodies that flow through the mining areas by processes that release substances such as mercury, cyanide, nitric acid, zinc, lead, arsenic, cadmium and manganese [11–14]. They can also affect aquatic diversity, terrestrial ecosystems and even people [15–18].

Five strategic mining projects and six second-generation projects are currently underway in Ecuadorian territory [19]. However, it is essential to bear in mind that mining exploitation always has repercussions in the socioeconomic and environmental spheres, generating serious impacts. In this sense, few studies have focused on analyzing these areas from a historical and current perspective with regard to gold mining in the coastal region of Ecuador. This has prevented a better understanding of the real situation of the sector, generating the urgent need to offer a real and updated vision. Based on the above, this study was based on the hypothesis that Ecuador does not currently have studies that provide decision-makers, local government administrators and scientists with updated information related to gold mining in the coastal region, which has prevented adequate socialization, technical and scientific assistance on the processes of formalization, regulation and environmental management of these activities.

Based on this hypothesis, the purpose of this research was to analyze the historical and current situation and the challenges of artisanal and industrial gold mining in the coastal region of Ecuador. The study focused mainly on the political sphere (laws applied), socioeconomic aspects (referring to population displacement, loss of livelihoods, changes in population dynamics, cost of living, water scarcity and health impacts) and finally environmental impacts (biotic and abiotic environments). For the development of this research, different methodologies were used: with respect to the historical context and the current situation, bibliographic and grey literature review techniques were used. In addition, field interviews were conducted for the current situation and, finally, expert judgment was used to establish the challenges of gold mining in the coastal zone.

2. Materials and Methods

2.1. Study Area

The Republic of Ecuador has a geographical area of 283,561 km^2 and is located on the equator in South America [5]. Its strategic location provides it with a privileged climate that benefits the development of economic activities in general, including mining [20]. Ecuador is divided into four regions: Insular, Litoral, Andean and Amazonian [21]. However, the present investigation focuses on the Litoral region of the country (Figure 1).

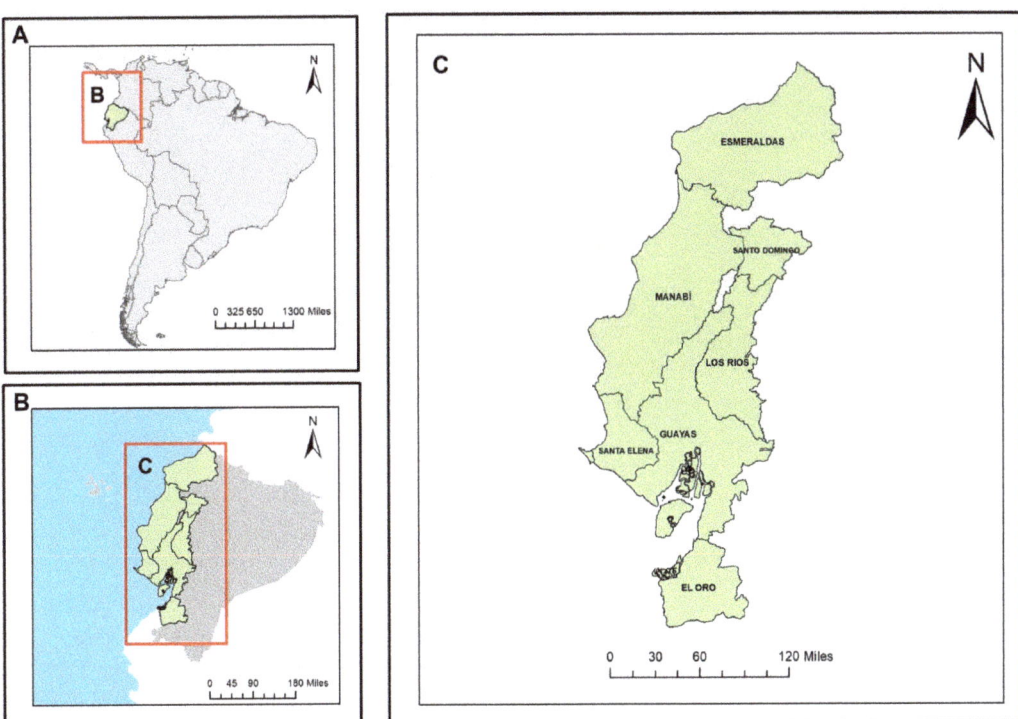

Figure 1. Study area. (**A**) Geographic location of Ecuador; (**B**) Litoral region of Ecuador; (**C**) provinces belonging to the Litoral region.

The Litoral or Coastal region is located between the coastal profile and the Andes mountain range [22]. The study area corresponds to a total of 70,647 km^2, covering the provinces of Guayas, Santa Elena, Manabí, El Oro, Los Ríos, Santo Domingo de los Tsáchilas and Esmeraldas (Figure 1) [23]. From a geomorphological point of view, the coastal region is characterized by a variety of reliefs, plains and large types of coasts (high cliffs with small bays; medium and low cliffs with small cliffs and large beaches; low, adeltaic type with an anastomosing network) [24]. On the other hand, it also has a privileged diversity in soil structure, which can vary depending on the province. The soils that can be observed are: vertisols, alfisols, entisols, aridisols, molisols, andisols, inceptisols and inceptisols [25].

There are currently a large number of gold mines spread along the coastal region, both legal and illegal. However, Table 1 only shows the gold mining concessions that are registered, according to the Mining Cadastre Web Geoportal of the Mining Regulation and Control Agency (ARCOM).

Table 1. Gold mining concessions in the Litoral region.

N°	Province	Canton	Concession Regime					Total Concessions by Canton
			Artisanal Mining	Small Mining	Medium Mining	Large Mining	General Regime	
1	Esmeraldas	Esmeraldas	—	—	—	1	—	1
		Atacames	1	—	—	—	—	1
		Eloy Alfaro	2	1	—	—	—	3
		Muisne	—	—	—	—	1	1
		Quinindé	—	—	4	2	—	6
		San Lorenzo	1	—	—	3	—	4
	Total concessions according to regime in Esmeraldas		4	1	4	6	1	—
2	Santo Domingo de los Tsáchilas	Santo Domingo	4	—	—	—	—	4
	Total concessions according to regime in Santo Domingo		4	—	—	—	—	—
3	Manabí	Jama	1	—	—	—	—	1
		Pedernales	1	—	—	—	—	1
	Total concessions according to regime in Manabí		2	—	—	—	—	—
4	Los Ríos	Babahoyo	—	—	—	—	2	2
		Montalvo	—	—	—	—	1	1
		Urdaneta	—	—	—	2	—	2
	Total concessions according to regime in Los Ríos		—	—	—	2	3	—
5	Guayas	Guayaquil	15	—	—	—	—	15
		Balzar	—	1	—	—	—	1
		General Antonio Elizalde	—	—	—	—	2	2
		Naranjal	—	3	—	—	—	3
		Balao	3	—	—	—	—	3
	Total concessions according to regime in Guayas		18	4	—	—	2	—
6	Santa Elena		—	—	—	—	—	—
	Total concessions according to regime in Santa Elena		—	—	—	—	—	—
7	El Oro	Pasaje	29	3	—	—	3	35
		Chilla	3	—	—	—	3	6
		El Guabo	4	7	—	—	—	11
		Santa Rosa	62	16	—	—	2	80
		Atahualpa	36	2	—	—	4	42
		Zaruma	6	10	—	—	5	21
		Marcabeli	3	—	—	—	—	3
		Portovelo	8	2	—	—	2	12
		Piñas	—	2	—	—	1	3
	Total concessions according to regime in El Oro		151	42	—	—	20	—
	Total mining concessions in the Litoral region							264

2.2. Methods

For a better understanding of this study, the methodology was divided into three sections. The first was based on the analysis and interpretation of the historical situation of artisanal and industrial gold mining in the Litoral region of Ecuador. The second section consisted of analyzing and describing the current situation of artisanal and industrial gold mining. Finally, the political, socioeconomic and environmental challenges of gold mining in the Ecuadorian Litoral territory were established.

The development of the first and second sections was carried out using two methods. On the one hand, a systematic literature review focused on the analysis of different documents published in high impact scientific databases such as Scopus and Web of Science. In the search process, filters were applied to establish the years of the publications (pre-Inca era–2018 for history and 2019–2021 for the current situation) and certain search parameters, also known as keywords, were also established (Table 2). Once the search results were obtained with their respective filters, the titles, related terms and abstracts of each publication were analyzed in order to carry out a selection process that would allow the extraction of the information necessary for the research. Twelve documents were obtained on the history of gold mining with respect to the political, socioeconomic and environmental situation, and five documents on the current situation in the area; see Table 2.

Table 2. Methodological process for the bibliographic review.

Subject	Keywords	Period	Scopus/Web of Science
History of gold mining in the Litoral Region	"Gold" and "mining" and "Ecuador".	Inca period–2018	12 [26–37]
Current situation of gold mining in the Litoral Region		2019–2021	5 [14,38–41]

The second method consisted of a review and analysis of grey literature (Table 3). Grey literature refers to a set of documents of various kinds, which have not undergone review or editing processes and which are not usually disseminated through publication but rather through limited (non-conventional) channels.

Table 3. Grey literature methodological process.

Grey Literature	Registration
Foreign investment and mining policy in Ecuador	June 2017
Ecuador's current mining legislation, including the Mining Code	1986
Mining Code Reform Law	1982
National Mining Sector Development Plan	July 2016
Organic Reformatory Law to the Mining Law, the Reformatory Law for Tax Equity in Ecuador and the Organic Law of the Internal Tax Regime.	Official Gazette 037, 16-VII-2013
Integrated environmental management in the Puyango river basin	2013

In relation to the second section, field visits were made to the different provinces, cantons and parishes belonging to the coastal region, where gold mining concessions were present. During the visits to the gold mines, interviews were conducted using semi-structured questions, that is, open-ended questions directed to local miners, environmental directors of the Ministry of Environment, Water and Ecological Transition of Ecuador (MAATE, for its acronym in Spanish), environmental directors of the Autonomous Decentralized Provincial Governments (GADP, for its acronym in Spanish), Autonomous Decentralized Municipal Governments (GADM, for its acronym in Spanish), presidents of the parish councils (PJP), and inhabitants where the mining concessions are located

(Table 4). The issues addressed in the interviews were: displacement of people, loss of livelihoods, changes in population dynamics, cost of living, water scarcity and health impacts.

Table 4. Questions established to analyze the current situation of gold mining.

Participant	Questions
Local miner (26 interviewees, mining concession owner)	What type of mining is developed? Has a permit to carry out mining activities? What type of technique is used for gold extraction, amalgamation or cyanidation? Wastewater in the extraction process is subjected to some treatment process, prior to its environmental discharge? Do you as a miner use any procedures to mitigate the impacts of gold mining? State three elements/components of the environment that are most affected by pollution?
Local authority (26 interviewees, political leaders in the mining area)	In its jurisdiction, ordinances have been created to control and monitor mining activities? Do you know if there is illegal gold mining in your canton and/or parishes? Do you know if the inhabitants of your canton and/or parishes have had health problems associated with gold mining?
MAAE (6 interviewees, government representatives per province)	Do you know if gold mining is developed in your province? Do you know if illegal gold mining is taking place in your province? There have been reports of contamination from gold mining? How has the Environmental Authority developed audits of gold mining concessions (rights)? How has the Environmental Authority developed water monitoring in the water bodies in the mining influence zone? During the visits and/or audits, has the accumulation of mining waste been evidenced? Do you consider that leaching occurs in the residual accumulations?
Mining town residents (26 interviewees, community representatives per canton)	Do you consider that gold mining has improved the quality of life in the area? Your monthly income covers all monthly expenses? Has gold mining caused population displacement for any reason? Has gold mining resulted in the loss of livelihoods? Has the cost of living changed since the advent of gold mining? Do you consider that with gold mining activity there is a shortage of water for daily activities? Have any family members or acquaintances experienced health problems or death due to gold mining contamination?

Finally, in the third section, we proceeded to establish the political, socioeconomic and environmental challenges present in gold mining in the Litoral region. For this, an expert judgment validation method was used, which consists of a set of opinions provided by professional experts on a topic, in order to verify the reliability of a research. Likewise, a brainstorming session and a round table were carried out with the participation of researchers who gave different perspectives to integrate the challenges of gold mining in the coastal region.

3. Results

The results that respond to the objectives of this study are presented below. The first section describes the history of gold mining in the coastal region of Ecuador in relation to mining policy, the socioeconomic sector and the environmental aspect, based on an exhaustive bibliographic review. The second section analyzes the current situation of gold mining in the coastal region with respect to the displacement of people, loss of livelihoods, changes in population dynamics, cost of living, water scarcity and health impacts, based on interviews conducted during field visits to the mining areas. Finally, the challenges present in gold mining in the coastal region in terms of mining policy, the socioeconomic sector and the environmental aspect are presented.

3.1. History

Pre-Hispanic cultures were the ones that developed the most mining activities in the Republic of Ecuador. For example, the people belonging to the Tolita Culture worked with gold and platinum between 500 B.C. and 500 A.D.; while the Cañaris worked with gold and silver [26–31]. There is no certainty about the tools used in pre-Columbian times to exploit gold in shallow mines and in the arid material of the Pindo-Puyango rivers. However, it is known that tools such as the wedge, the combo, the pick, the barreta, the shovel and the wheelbarrow were used in colonial times, causing a low yield in the exploitation of gold quartz [32].

Later, in the mid-sixteenth century, the discovery of gold particles motivated prospectors to go through rivers and mountains until they finally found the "golden nuggets", which belonged to the areas of Zaruma and Portovelo. The South American Development Company (SADCO) was in charge of exploiting the Portovelo mining area at the end of the 19th century until the middle of the 20th century. Once SADCO's activities concluded, the Industrial Mining Associated Company (CIMA) was established and worked until the 1970s, giving way to exploitation by small miners and artisans [26].

3.1.1. Mining Policy

At the beginning, mining did not have regulations or legal bases to control the activities carried out, which caused repudiation and misinformation among the population due to the lack of mining policies [33–41]. Over the years, the mining industry expanded, and was contemporary with the processes of formation of the Ecuadorian State. Around the year 1830, a great interest was generated in the consecration of property rights and the attraction of foreign investment, which led to the development of the Mining Promotion Law. The purpose of this law was to reactivate the productive sector of the country with the help of the alternating mining activity in the southern zone of Ecuador [27]. In 1886, the Mining Code of Ecuador was enacted; this law was intended to awaken the interest of foreign investors in the discovery, prospecting and exploitation of mines, in addition to providing legal certainty for mining concessionaires. This legislation turned around the forms and conditions of appropriation and exploitation of the deposits; it also influenced the transformation of the norms that regulated the subsoil domain [27]. It is important to note that this legislation recognized precious stones (such as gold, silver, etc.) as the object of the same and, upon a reform in 1982 (Reformatory Law of the Mining Code), petroleum was added to the list. This reform also allowed the owners of the properties to dispose of the minerals as owners of the property [42,43].

In 1937, the Mining Law was enacted; this law was in charge of determining that the minerals found in the subsoil belonged to the State's domain. One year later, the

Codification of the Mining Law was issued [41]. By 1991, a new Mining Law, also known as Law 126, was presented. This law incorporated the principles of State ownership of both mines and deposits and categorically involved the rights of miners. It also incorporated the perception that the Ecuadorian State should legalize mining activities that had complied with the requirements set forth in the law [34]. In 2001, some reforms were incorporated into the Mining Law, which were responsible for qualifying mining management as a national interest, a priority and fundamental for development, including regulations on environmental protection [41]. By 2008, the Constitution of the Republic established a range of environmental principles providing rights to native and indigenous peoples as new values of the State [10]. A year later, in January 2009, the Mining Law and its Regulations were approved in order to generate an effective model of economic progress focused on responsible and organized mining. This law was in charge of administering, regulating, controlling and managing the strategic mining sector, in accordance with the principles of sustainability, precaution, prevention and efficiency [9]. Finally, in 2013, an amendment to the Mining Law was made, called the Organic Reformatory Law to the Mining Law, the Reformatory Law for Equity in Ecuador and the Organic Law of the Internal Tax Regime [44].

It is essential to highlight that there are unique legislative measures that are applied in some areas of the country depending on the mining situation in order to avoid catastrophes. A clear example of this, applied in the province of Esmeraldas on 24 March 2011, is an injunction issued by a judge prohibiting mining exploitation, whether legal or illegal, in the San Lorenzo and Eloy Alfaro cantons. This injunction was ratified in 2018; however, in consecutive years the established measure was ignored. In the past, ancestral mining was carried out in the province (without consequences for the environment due to the use of nets and pans), but in recent years it has been practiced in the form of small- and medium-scale mining (with machinery and mercury in the open air), which is harmful to the ecosystem, which is why the precautionary measure was established—in order to minimize the contamination of rivers and the inhabitants of communities that consume the affected water [45,46]. A similar situation occurred in the city of Zaruma, where the Government of the Republic declared a state of emergency for sixty days on 14 September 2017 due to the fact that excavations were being carried out in soil declared to be the cultural heritage of the country. According to investigations carried out, there were areas exploited for gold mining that represented a risk for the inhabitants, which is why the State took the decision to intervene by taking preventive measures [14].

3.1.2. Socioeconomic Sector

At the end of the 1990s, Ecuador was going through a financial crisis. To remedy the situation, the government (during the period 2000–2006) developed processes to offer the country's mining resources to foreign investors interested in carrying out this activity. Thus, with the intention of achieving the country's development, and together with pro-extractivist reforms, the Ecuadorian territory moved towards an authoritarian political system known as "political domination". As a consequence, a social resentment of the people towards the political institutions originated, which has been increasing over the years due to several factors [30].

On several occasions, there have been social, political, economic and environmental confrontations in communities due to mining projects. In some cases, the inhabitants of the mining areas say that they are not convinced by the land negotiation agreements and that they are often pressured to sell. In addition, there have been reports of destruction of the inhabitants' buildings by the companies in charge of the megaprojects [31]. Likewise, there have been evictions of families without any relocation agreement to make way for the mining companies [47]. In addition, indigenous movements have been unleashed against mining companies, causing clashes with the military that have resulted in injuries and even the death of leaders [35,36].

Esmeraldas is one of the most conflictive areas, and is also considered one of the most excluded and poorest regions in Ecuador, due to the lack of opportunities in the area for 22 years. At that time, there was a shortage of labor, so gold mining emerged as a job option for people because timber and palm oil activities did not require more labor. Social conflicts caused by gold mining include lack of access to water, distrust between families, and broken social relations. Armed conflicts have also occurred in the province over the last 50 years, mainly due to the illegal appropriation of gold material, resulting in the death of people due to confrontations [37]. Economic consequences include chronic poverty, in addition to illicit activities such as tax evasion, smuggling of mining inputs, capital flight and money laundering, import of unregistered machinery, etc. [45].

3.1.3. Environmental Impacts

In general, the most significant impact caused by gold mining is the contamination of soil and water resources. This can occur due to improper dumping or infiltration of toxic substances used in mining [28]. Sometimes, no remediation measures are applied after the mineral extraction process, leaving the pools of contaminated water exposed to the open air (Figure 2a). As a result, there is evidence of enormous extensions of brick-colored soil, fish and macroinvertebrates affected by the accumulation of heavy metals such as mercury, cyanide and arsenic [46,47]. These contaminants can also cause respiratory, skin and carcinogenic diseases that affect people's wellbeing. The debris generated by mining waste (Figure 2b) is also a problem, because it can cause the death of aquatic species and even humans [29]. Another impact related to gold mining is the change in the geographic landscape, as it implies changes in land use such as degradation and erosion, modifications to the relief, and large-scale deforestation to make way for the machinery used in mining activities [28].

Figure 2. Gold mining impacts. (**a**) Pools of contaminated water exposed in the open pit; (**b**) debris generated by the extraction of material for gold.

The Puyango-Tumbes watershed serves as a border between northern Peru and the provinces of Loja and El Oro in southeastern Ecuador, covering a total area of 5494.57 km² [48]. According to studies carried out since the beginning of the 20th century in the upper part of the Puyango River, high concentrations of contaminants from mining have been found in Portovelo-Zaruma, which has had an impact on the Peruvian territory. In 2001, a study was carried out in the area; the results showed that gold mining activities were affecting the ecosystem because of the tailings in the rivers, which subsequently released cyanide, mercury and other metals that exceeded the established quality criteria. The effects consisted of: reduced aquatic diversity, elevated levels of metals in larvae, large amounts of metals bound to suspended sediments at ambient pH conditions, and the levels of metals in carnivorous fish were low, with the exception of mercury [33]. In

2010, another was carried out, which showed the presence of about 1.5 tons of mercury being released per year in the city, of which 70% evaporated and 30% was released with tailings [49]. Later, in 2017, another investigation was conducted to analyze heavy metal contamination; the results indicated that the concentrations of Pb and As exceeded the permissible limits for metals in soils. In addition, biological samples (blood, urine and hair) were collected and, the results determined that inhabitants residing near the Puyango River contained small amounts of Hg and Pb inside their organism [32]. As a consequence of all the contamination that the river receives, no aquatic life forms develop; in addition, water consumption, and the use of water for agriculture and livestock were disabled. Due to all of these effects, in 2018, the Federation of Farmers of Tumbes filed a petition for a lawsuit against the government of Ecuador for the alleged contamination of the Tumbes River.

In 2010, an analysis was conducted in water bodies in Esmeraldas, in the Mario Unión and Sabaleta estuaries, and in the Santiago and Sabaleta rivers. Abnormalities were identified in the fish, such as deformities, as well as the presence of heavy metals inside the fish due to illegal gold mining. Later, in 2013, the Catholic University of Ecuador conducted a study that showed that the Maria estuary contained exaggerated amounts of aluminum, which exceeded the limits established in the regulations by 580 times, as well as exceeding the standard of copper by 2.4 times, the standard of iron by 33 times, and the standard of manganese by 1.3 times. It was determined that all these substances were the result of illegal mining carried out in these zones [45].

3.2. Current Situation

There is currently a total of 264 legally registered gold mines in the Litoral region, including artisanal mining, small-scale mining, medium-scale mining, large-scale mining and general regime (Table 1). The province with the highest number of concessions is El Oro, with a total of 213 gold mines, followed by Guayas, with 24 mines, Esmeraldas, with 16 mines, Los Ríos, with 5 mines, Santo Domingo de los Tsáchilas, with 4 mines, and Manabí, with 2 mines. Unlike the other provinces mentioned, Santa Elena does not have any registered gold mines, so there are no problems related to gold mining in the area.

3.2.1. Mining Policy

In recent months, the Ministry of Energy and the National Police have registered an alarming increase in illegal mining throughout Ecuadorian territory due to the lack of regularization of the extractive activity of gold mining during the COVID-19 pandemic. Among the provinces with clear evidence of illegal mining is El Oro, specifically in the San Lorenzo canton; this indicates that measures established in the canton in 2011, which prohibit any type of extraction in the area, are being disregarded. As a consequence of the activities carried out in the sector, on 18 November 2020 there was a landslide in an illegal mine that resulted in the death of several people [46].

With respect to political issues, the Organization of American States (OAS) in 2020 admitted the petition filed by the Federation of Farmers of Tumbes (Peru) against the government of the Republic of Ecuador for the alleged contamination of the Tumbes River from mining activity in the Portovelo-Zaruma district. Likewise, farmers and environmentalists from the neighboring country carried out a peaceful protest on 25 January 2021 in order to demand a pronouncement from the Inter-American Court of Human Rights, since the presence of heavy metals in the river continues despite the lawsuit filed, and there is a risk to the flora, fauna and health of approximately 200,000 people [50].

3.2.2. Socio-Economic Sector

In Ecuador, cantons such as Zaruma, Portovelo, and Camilo Ponce Enríquez extract 86% of the gold exported in the country through small-scale mining; however, only 0.38% of the territory's population resides in these areas. Despite the large exports generated by the cantons, they have average poverty rates for unsatisfied basic needs, 62% in Zaruma and

58% in Portovelo, according to data from the 2010 Population and Housing Census [51]. On the other hand, 84.6% of the inhabitants of San Lorenzo canton live in poverty, a percentage that exceeds 60% nationally and 51% in the province of Esmeraldas; only 23% of households have basic services in the canton [38,52].

In the interviews conducted, some miners stated that, due to the Mining Agreement decreed in 2009, many people were displaced to make way for large mining concessions. For this reason, many opted to engage in illegal mining (inside and outside the canton), and with all this came violence and poverty in the exclusion zones. Faced with the various problems and conflicts caused by illegal activity, they preferred to sell their land and move away from the concentration zones in search of a better way of life, leaving behind the place that had been their home for a long time. On the other hand, interviewees also stated that there are people constantly joining the mining areas from other regions of Ecuador and Peru who are willing to work for lower wages due to the lack of job opportunities and offers.

In the country, informal miners are known as "sableros". They are generally trapped between poverty and violence, which is why they risk their lives performing dangerous activities to obtain gold and thus earn an income to sustain themselves. In Zaruma and Portovelo, the sableros often enter subway mines that do not have oxygen, so they use long hoses to breathe while searching for gold veins. Some say that they only know how to "sabre" because they have worked in the mines since they were teenagers and that they have often broken into legal and illegal concessions to steal gold material. They also comment that sometimes they apply for jobs in formal companies, and once inside, they ingest small pieces of gold to later process them in their bodies and keep the material to make a profit by selling it.

Regarding the cost of living, some residents living near the mining concessions said that they have benefited from the job opportunities in the gold mines, but that salaries are minimal. While residents who own commercial establishments (stores) stated that mining activities do not influence their economy, because sales have not increased in their establishments and, in addition, government agencies are very strict in the regulation of prices implemented in sales.

One of the main reasons why people resort to informal and illegal gold mining activities is related to poverty and lack of job opportunities. According to a report by the National Institute of Statistics and Census of Ecuador (INEC), the province of Esmeraldas is one of the poorest and most excluded provinces in the country. Thus, three of its cantons have the highest poverty rate, Rio Verde (63%), Eloy Alfaro (64%) and Muisne (65%) of the coastal region. Meanwhile, in the province of Manabí, two cantons, Pichincha and Olmedo have a 63 and 61 percent poverty rate. It is important to note that these two provinces have the highest poverty rates in terms of multidimensional aspects, unsatisfied basic needs and access to basic services. These are incentives for the population of these cantons to migrate and go to develop gold mining activities in a precarious and unsafe way in areas with a high gold content, such as Alto Tambo in the province of Esmeraldas, and in the neighboring province of Imbabura in the area of Buenos Aires. Thus, after the province of El Oro in terms of mining history, the province of Esmeraldas in the last 10 years has shown an exponential increase in gold mining activities in the coastal region.

3.2.3. Environmental Impacts

In Ecuador, the use of mercury for gold amalgamation is prohibited; however, its use is still very frequent in the country, such that the substance is spread by rivers and ravines [17]. In this regard, activist Nathalia Bonilla stated in 2020 that due to the increase in illegal mining activities during the pandemic, the water sources of San Lorenzo and Eloy Alfaro have become contaminated. She also added that 80% of the rivers are polluted, but people do not have good water, so they have to continue using water resources with toxic substances [53]. On the other hand, 19.45% of the water in the province of El Oro is contaminated due to the activities carried out in Zaruma and Portovelo [14]. All these data are of great concern, since they show that most of the populations settled on the banks of the rivers in the gold ore mining areas do not have adequate drinking water systems, which represents a serious risk to the health of the people.

Exposure of people to water contaminated with toxic substances such as mercury can cause illnesses, as reported by gold mining workers in the province of Esmeraldas. Some of the diseases they reported contracting included: skin diseases, intestinal disorders, infections, vaginal problems (in the case of women), stomach diseases, and respiratory problems, among others. In addition, they mentioned knowing people who acquired long-term diseases such as Alzheimer's and cancer.

In 2020, a study was conducted in the Portovelo-Zaruma area, specifically in the Puyango-Tumbes river. The results showed that there was cyanide contamination at concentrations 9088 times higher than the established CCME standard of 5 g/L, and 1136 times higher than the 24-h LC 50 concentration of 40 g/L free of CN- for some species. The levels of affectation arose due to mining malpractice and environmental management in 87 gold processing centers, which used mercury tailings to obtain the ore in the river [39]. Another effect that could be seen in the city of Zaruma was large deformations due to land subsidence generated mainly by illegal subway gold mines [40]. The impact of these mining activities was so great that on 2 July 2021, a sinkhole was created (Figure 3a), which reached a diameter of 20 m and a depth of 30 m, putting at risk 50 homes located near the affected area (Figure 3b).

Figure 3. Sinkhole in the province of El Oro, city of Zaruma. (**a**) Sinkhole formation; (**b**) houses at risk due to sinkhole formation.

In the Los Ajos sector of the San Lorenzo canton, illegal mining activities were registered in January 2021, despite the fact that the measures imposed prohibit it. As a result, the Ecuadorian Chocó (one of the areas with the highest endemism of species in the country) is being affected by indiscriminate deforestation to make way for roads for illegal mining machinery [54–56]. Inhabitants of the area state that not even the COVID-19 situation has stopped the "sableros", because there are 52 active mining camps in the area that are causing great damage to the area, while also creating insecurity, so they demand that the authorities control the permits granted for these activities. The Alto Tambo sector was also affected, because after inspections carried out by the authorities in May, illegal miners invaded the area and destroyed a large part of the zone, putting the Awá Nationality that resides in the surrounding area at high risk (Figure 4a,b).

(a) (b)

Figure 4. Invasion by illegal miners. (**a**) Machinery used for gold extraction; (**b**) deforestation of the affected area.

3.3. Challenges

The findings suggest that Ecuador faces many challenges, both in the political sphere and in socioeconomic and environmental respects. On the one hand, it is evident that the country lacks policies that direct the mining sector towards more responsible and environmentally friendly activities. There is also a need for greater control of illegal mining due to its exponential increase as a result of the COVID-19 pandemic and the lack of job offers, causing various forms of damage to the ecosystem. The provinces belonging to the Litoral region with the most conflict are Esmeraldas and El Oro, which, according to a systematization developed by the Ombudsman's Office and the Colectivo de Geografía Crítica, occupy the first place with respect to the violation of human rights and nature at the national level [38]. Considering the problems that have arisen in the two provinces, it should be considered a priority to establish laws in order to improve the situation in the two zones and avoid conflicts with neighboring countries due to the contamination generated by the zones.

Since the exploration and extraction of gold ore in the country began, there have been socioeconomic conflicts that have harmed workers and residents in the gold mining areas. The Ecuadorian State should give priority to the people involved in gold mining activities in the country, implementing controls on the remuneration of mining workers, because, as has been stated, they do not receive a fair salary despite the risk they run when carrying out extraction activities. Strict plans should also be implemented to prevent mining concessions from using mercury to extract gold, as the contamination of water sources means that the local population has no water for drinking or other activities. It would also be advisable to offer jobs and invest in education, given the lack of opportunities in conflict zones such as Esmeraldas and El Oro.

Mining sector regulators should socialize with all the social actors involved in the mining processes about the lack of knowledge of the mining formalization and regulation processes. This will help to ensure that these processes are carried out in accordance with the law and comply with environmental regularization processes. In addition, the Ecuadorian government should promote specialized technical assistance programs on issues related to environmental management, mining safety and professional training for ASM; to achieve this, academia should provide support in the education and training of miners, making them aware of the risk of using prohibited substances such as Hg, as well as providing instruction on the implementation of more environmentally friendly extraction strategies. On the other hand, it is essential to establish incentives focused on environmental protection and entrepreneurship generation. These alternatives could be an option that would allow the adoption of better environmental practices in the gold extraction process and improve the local economy of the miners.

Perhaps the most serious challenge faced in gold mining concentration zones is the lack of awareness and knowledge regarding exposure to heavy metals such as mercury, as people are often unaware of the effects they can cause to the environment. This process, together with the lack of control and monitoring in mining areas, as well as the lack of management of mining waste generated in the extraction areas and the lack of legal regulations for the management of mercury, have a significant influence on the environmental aspect. An effective measure for minimizing environmental impacts could be to carry out government supervision to ensure that gold mining concessions mitigate the damage caused by their extractions, thus ensuring that they remediate (or at least minimize) the damage caused to the ecosystem.

4. Discussion

In recent decades, Ecuador has made great efforts to implement legislation to regulate the activities developed by the mining sector. However, it is evident that in recent years, there has been a loss of interest in the political aspect of mining, since the same law has been in force in the country since 2009, without taking into account the effects that have been generated over the years. It is also clear that there is a lack of rigorous control in the application of the laws; this was evidenced by informal talks with the gold mine workers (outside the interview questions), since on several occasions they mentioned that the visits by the authorities were scarce, which indicates the lack of monitoring to verify whether the mines comply or not with the established regulations and, most importantly, to verify that they do not use mercury to extract the mineral. According to [54], Ecuadorian legislation has its own dynamics in situ, and it is naïve to believe that the laws of the country can control the sector [54]. Other authors, such as [55], mention that legislation often involves two factors that generate negotiations in the legal framework of the mining sector: national laws and socio-normative agreements, which in this case are those that are applied in the country, although not in a recognized manner [56].

At this point, it is important to note that legislation should require the implementation of mitigation measures. However, this could mean a loss, since there is a great controversy in controlling and regulating gold mining practices, because if mitigation measures are applied, the potential to placate marginalized groups will be reduced, while if restoration measures are not used, environmental contamination impacts will also be present, as mentioned in [57]. To maintain greater control over gold mining in Ecuador, mining activities should be formalized. However, ref. [58] state that attempts to formalize the mining sector are generally unsuccessful, and as a consequence, pollution levels increase. According to the authors, in order to carry out an effective formalization process, education must first be implemented to prevent the continuation of incorrect extraction methods, and thus reduce environmental damage [58].

The perspective has not changed much in the historical and current situation regarding environmental impacts in gold mining areas in the coastal region of Ecuador, since the same aspects are present: contamination of rivers, water scarcity for people, reduction or lack of aquatic diversity and people with the presence of mercury in their bodies. The most worrisome aspect is that, according to the analyses carried out, all of these impacts have been occurring in the Puyango River since 2000, which is indicative of the lack of control in the Portovelo-Zaruma areas and the lack of awareness on the part of the illegal miners. However, ref. [15] mentions that studies conducted on ASM show that mining has not caused water contamination. Furthermore, he adds that in cities such as Portovelo, Zaruma, Nabija, Ponce Enriquez and Santa Rosa, metallic substances and metalloids have been detected at well below the limits established in the regulations [15]. Although the studies carried out show low levels of mercury, it must be taken into account that in general, the laboratories responsible for the analyses only have the capacity to measure 0.005 ppm, while the maximum permissible parameters in the country are established at 0.0002 ppm. There is a great difference between these two levels, so the probability of detecting high levels of mercury is very low, not because it is not present, but because of the lack of specialized equipment able to provide concrete results.

Although Ecuador has major shortcomings in the mining sector due to poor extraction practices or illegality, it is not the only country of which this is the case. In other countries around the world, these situations are also present. Such is the case in Colombia, where there have also been cases of mercury contamination of rivers due to the amalgamation processes used to extract gold, the introduction of solid loads, metals and waste used in mining, and the use of mercury in the mining industry [58]. On the other hand, in Ghana, there are cases of soil contamination with averages of 0.024 mg kg^{-1}, which is a very low concentration, well below the established criteria for human health [59–62].

5. Conclusions

Ecuador has had the same law regulating the mining sector since 2009, which indicates that new legislation is needed to adapt to the current situation with COVID-19. There is a major conflict regarding the preventive norms applied to the San Lorenzo canton in the province of El Oro, since, despite the fact that gold mining (legal or illegal) is prohibited in the area, there are still cases of environmental impacts caused by the influence of mining activities. Therefore, the authorities should carry out more frequent inspections to detect illegal gold mines in order to dismantle them and reduce pollution, especially of water resources, especially in the provinces of El Oro and Esmeraldas, which have proven to be the areas with the most conflict in the Litoral region.

Due to the decrease in employment in the provinces of El Oro and Esmeraldas, illegal mining activities have emerged as an option for survival and daily sustenance. However, these activities cause major conflicts that lead to the insecurity of the people living in the extraction zones. In the first instance, there is the violation of people's rights due to their eviction from their homes in order to make way for the megaprojects; there are also reports of armed conflicts, labor exploitation associated with minimum wages, lack of access to water, and high risk of diseases related to direct contact with heavy metals.

The analysis of the environmental aspect suggests that mercury is present in some areas of the Litoral region, specifically in El Oro and Esmeraldas. These contamination problems are not a recent issue, as illegal mining has been evident in these areas for several years. As a consequence, water resources, aquatic and terrestrial ecosystems are affected due to gold mining activities which, although they are small-scale in most cases, the contamination is due to the malpractice of the activities at the time of processing the mineral through the tailings in which mercury is used.

Author Contributions: Conceptualization, C.M.-R., G.D. and S.S.; methodology, C.M.-R., G.D. and S.S.; software, C.M.-R.; formal analysis, C.M.-R., S.P.-M., S.S., C.L.-P., M.J.-G., G.H.-M. and G.D.; investigation, C.M.-R., S.P.-M.; writing—original draft preparation, C.M.-R. and S.P.-M.; writing—review and editing, C.M.-R., S.P.-M., S.S. and G.D.; supervision, S.S., and G.D.; project administration, C.M.-R., S.S. and G.D.; Resources, C.M.-R., G.D., S.S., C.L.-P., M.J.-G., and G.H.-M. All authors have read and agreed to the published version of the manuscript.

Funding: This research received the financial support of the European Commission through the projects: H2020-MSCA-RISE REMIND "Renewable Energies for Water Treatment and Reuse in Mining Industries" (Grant agreement ID: 823948).

Institutional Review Board Statement: Not applicable.

Informed Consent Statement: Not applicable.

Data Availability Statement: Not applicable.

Acknowledgments: The authors are grateful for the financial support of GREEN AMAZON ECUADOR and the Escuela Superior Politécnica de Chimborazo (ESPOCH) in the field work. As lead author, C.M.-R., I thank the Doctoral School of the University of Calabria for allowing me to conduct my doctoral research. Thanks to Santiago Logroño for his collaboration in the digitization of figures and software handling.

Conflicts of Interest: The authors declare no conflict of interest.

References

1. Boyle, R. *La Economía de la Minería de Oro y Oro*; Springer: Boston, MA, USA, 1987; ISBN 978-1-4613-1969-6.
2. Dougherty, M. La industria global de la minería de oro: Materialidad, búsqueda de rentas, empresas junior y ciudadanía corporativa canadiense. *Compet. Chang.* **2013**, *17*, 339–354. [CrossRef]
3. Organización Mundial de la Salud. *La Minería Aurífera Artesanal o de Pequeña Escala y la Salud*; WHO: Geneva, Switzerland, 2017; ISBN 978-92-4-351027-9.
4. Newman, P.; Meader, N.; Swarts, W.; Klapwijk, P.; Liang, J.; Chou, E.; Gao, Y.; Barot, H.; Furuno, A.; Rey, F.; et al. *Gold Focus 2020*; Valcambi Suisse: London, UK, 2020; ISBN 9781916252608.
5. Ministerio del Ambiente. *Línea de Base Nacional para la Minería Artesanal y en Pequeña Escala de Oro en Ecuador, Conforme la Convención de Minamata sobre Mercurio*; United Nations Industrial Development Organization: UNIDO: Vienna, Austria, 2020.
6. El Comercio Mingaservice. 2013. Available online: https://www.mingaservice.com/web/index.php/noticia/tag/noticias?page=12&ipp=20 (accessed on 5 November 2021).
7. Banco Central del Ecuador. *Reporte de Minería*; Banco Central del Ecuador: Quito, Ecuador, 2020; Available online: https://contenido.bce.fin.ec/documentos/Estadisticas/Hidrocarburos/ReporteMinero062020.pdf (accessed on 5 November 2021).
8. Toapanta, R.A.R. Política minera y sostenibilidad ambiental en Ecuador. *FIGEMPA Investig. Desarro.* **2017**, *1*, 41–52. [CrossRef]
9. Ministerio de Minería. *Plan Nacional de Desarrollo del Sector Minero*; Ministerio de Minería: Santiago, Chile, 2016; p. 120.
10. Foro Intergubernamental sobre Minería Minerales Metales y Desarrollo Sostenible. In *Evaluación del Marco de Políticas Mineras del IGF*; IGF: Ottawa, ON, Canada, 2019; p. 63.
11. Correa Guaicha, H.M.; Alvarado Correa, L.E. Impactos ambientales en la explotación minera aurífera y al ser humano. Caso de estudio. *Desarro. Local Sosten.* **2017**, *29*, 22.
12. Martín, A.; Arias, J.; López, J.; Santos, L.; Venegas, C.; Duarte, M.; Ortiz-Ardila, A.; de Parra, N.; Campos, C.; Celis Zambrano, C. Evaluation of the effect of gold mining on the water quality in Monterrey, Bolívar (Colombia). *Water* **2020**, *12*, 2523. [CrossRef]
13. Gafur, N.A.; Sakakibara, M.; Sano, S.; Sera, K. A case study of heavy metal pollution in water of bone river by artisanal small-scale gold mine. *Water* **2018**, *10*, 1507. [CrossRef]
14. Vilela-Pincay, W.; Espinosa-Encarnación, M.; Bravo-Gonzales, A. La contaminación ambiental ocasionada por la minería en la provincia de El Oro. *Estud. Gestión. Rev. Int. Adm.* **2020**, *8*, 215–233. [CrossRef]
15. Wingfield, S.; Martínez-Moscoso, A.; Quiroga, D.; Ochoa-Herrera, V. Challenges to water management in Ecuador: Legal authorization, quality parameters, and socio-political responses. *Water* **2021**, *13*, 1017. [CrossRef]
16. Vinueza, D.; Ochoa-Herrera, V.; Maurice, L.; Tamayo, E.; Mejía, L.; Tejera, E.; Machado, A. Determining the microbial and chemical contamination in Ecuador's main rivers. *Sci. Rep.* **2021**, *11*, 17640. [CrossRef] [PubMed]
17. Rivera-Parra, J.L.; Beate, B.; Diaz, X.; Ochoa, M.B. Artisanal and small gold mining and petroleum production as potential sources of heavy metal contamination in Ecuador: A call to action. *Int. J. Environ. Res. Public Health* **2021**, *18*, 2794. [CrossRef] [PubMed]
18. Jiménez-Oyola, S.; Chavez, E.; García-Martínez, M.-J.; Ortega, M.F.; Bolonio, D.; Guzmán-Martínez, F.; García-Garizabal, I.; Romero, P. Probabilistic multi-pathway human health risk assessment due to heavy metal(loid)s in a traditional gold mining area in Ecuador. *Ecotoxicol. Environ. Saf.* **2021**, *224*, 112629. [CrossRef]
19. Banco Central del Ecuador. *Reporte de Minería*; Banco Central del Ecuador: Quito, Ecuador, 2019; Available online: https://www.bce.fin.ec/ (accessed on 5 November 2021).
20. Empresa Nacional Minera Ecuador Minero. *ENAMI EP* **2016**. Available online: https://www.enamiep.gob.ec/doc/2016/enero/GPR2016.pdf (accessed on 5 November 2021).
21. Navarrete Bastidas, R. La Preservación de la Biodiversidad, el Medio Ambiente y la Utilización de los Recursos Naturales para Impulsar el Desarrollo Sustentable y la Seguridad. Master's Thesis, Quito, Ecuador, 2005; pp. 1–17. Available online: https://www.flacsoandes.edu.ec/buscador/Record/iaen-24000-342/Details (accessed on 5 November 2021).
22. Secretaría Nacional de Planificación y Desarrollo. *Toda una Vida Contigo. Plan Nacional de Desarrollo 2017–2021*; Secretaría Nacional de Planificación y Desarrollo: Quito, Ecuador, 2017.
23. Sevillano Guetierrez, E. Prácticas Constructivas Locales de Bajo Costo, Estrategias Locales de Respuesta a Desastres Naturales & Capacidad de Inversión en Hábitat de la Población Desfavorecida. 2016. Available online: https://www.sheltercluster.org/sites/default/files/docs/ecuador_costa_habitat_local_y_estrategias_de_respuesta_craterre310516_1.pdf (accessed on 5 November 2021).
24. Winckfll, A. Relieve y geomorfologia. *Geomorfología* **1982**, *17*, 19.
25. Rivera Grunauer, R.E. *Características Físicas, Ubicación Geográfica y Calidad del Suelo Agrícola de las Provincias de la Costa Ecuatoriana*; Universidad Técnica de Machala: Machala, Ecuador, 2019.
26. Rea, R.; Paspuel, V.; Tobar, L. Inversión extranjera y política minera en Ecuador. *Rev. Publicando* **2017**, *4*, 375–396.
27. Carrión, A. Procesos. *Rev. Ecuat. Hist.* **2017**, *53*, 95.
28. Calderón Robles, P. Estado Acual de la Minería del Oro en Ecuador: Gran Minería vs. Minería Artesanal. Master's Thesis, Universitat Politécnica de Catalunya, Barcelona, Spain, 2020.
29. Cao, Y.; Zhu, X.; Liu, B.; Nan, Y. A qualitative study of the critical conditions for the initiation of mine waste debris flows. *Water* **2020**, *12*, 1536. [CrossRef]
30. Alvarado Vélez, J.A. Impactos económicos y sociales de las políticas nacionales mineras en Ecuador (2000–2006). *Rev. Cienc. Soc.* **2018**, *23*, 53–64. [CrossRef]

31. Massa-Sánchez, P.; del Arcos, R.C.; Maldonado, D. Minería a gran escala y conflictos sociales: Un análisis para el sur de Ecuador. *Probl. Desarro.* **2018**, *49*, 119–141. [CrossRef]
32. Oviedo-Anchundia, R.; Moina-Quimí, E.; Naranjo-Morán, J.; Barcos-Arias, M. Contaminación por metales pesados en el sur del Ecuador asociada a la actividad minera. *Bionatura* **2017**, *2*, 437–441. [CrossRef]
33. Tarras-Wahlberg, N.; Flachier, A.; Lane, S.N.; Sangfors, O. Environmental impacts and metal exposure of aquatic ecosystems in rivers contaminated by small scale gold mining: The Puyango River basin, southern Ecuador. *Sci. Total Environ.* **2001**, *278*, 239–261. [CrossRef]
34. Sandoval, F. La pequeña minería en el Ecuador. *Min. Miner. Sustain. Dev.* **2001**, *75*, 30. [CrossRef]
35. Betancourt, M. *Minería, Violiencia y Criminalización en América Latina. Dinámicas y Tendencias*; Broederlijk Delen: Bogotá, Colombia, 2016; ISBN 978-958-58470-9-5.
36. Observatorio de Conflictos Mineros de América Latina. *Conflictos Mineros en América Latina: Extracción, Saqueo y Agresión*; Observatorio de Conflictos Mineros de América Latina: Santiago, Chile, 2017.
37. Lapierre Robles, M.; Macías Marín, A. *Extractivismo, (Neo) Colonialismo y Crimen Organizado en el Norte de Esmeraldas*; Ediciones Abya-Yala: Quito, Ecuador, 2018; ISBN 978-9942-09-584-8.
38. Moreno Parra, M. Racismo ambiental: Muerte lenta y despojo de territorio ancestral afroecuatoriano en Esmeraldas. *Íconos—Rev. Cienc. Soc.* **2019**, 89–109. [CrossRef]
39. Marshall, B.G.; Veiga, M.M.; da Silva, H.A.M.; Guimarães, J.R.D. Cyanide contamination of the puyango-tumbes river caused by artisanal gold mining in Portovelo-Zaruma, Ecuador. *Curr. Environ. Health Rep.* **2020**, *7*, 303–310. [CrossRef]
40. Cando Jácome, M.; Martinez-Graña, A.M.; Valdés, V. Detection of terrain deformations using InSAR techniques in relation to results on terrain subsidence (Ciudad de Zaruma, Ecuador). *Remote Sens.* **2020**, *12*, 1598. [CrossRef]
41. Vásconez, I.A. *Historia de las Normas Mineras en Ecuador*; PBP Law: Athens, Greece; Available online: https://www.pbplaw.com/es/historia-de-las-normas-mineras-en-ecuador/ (accessed on 5 November 2021).
42. Paz, V. *Legislación Vigente en el Ecuador Sobre Minas, Inclusive el Código de Minería*; Imprenta y Libería Ecuatorianas: Guayaquil, Ecuador, 1886.
43. República del Ecuador. *Ley Reformatoria del Código de Minería*; Imprenta del Gobierno: Quito, Ecuador, 1896.
44. Asamblea Nacional de la República del Ecuador. *Ley Orgánica Reformatoria a la Ley de Minería, a la Ley Reformatoria para la Equidad Tributaría en el Ecuador y la Ley Orgánica de Régimen Tributario Interno*; Asamblea Nacional de la República del Ecuador: Quito, Ecuador, 2013; pp. 1–7.
45. Ponce, I. MONGABAY LATAM. May 2018. Available online: https://es.mongabay.com/ (accessed on 5 November 2021).
46. *La Actividad Minera Ilegal en el Norte de Esmeraldas se Realiza en 52 Frentes*; El Comercio: Lima, Peru, 2020.
47. Bayón, A.; Japhy, W. Tundayme: El Despojo Minero Avanza. Available online: https://www.planv.com.ec/historias/sociedad/tundayme-el-despojo-minero-avanza (accessed on 5 November 2021).
48. Humberto, C.; Cáceres, M.; Lorenz, S. *Ordenamiento Ambiental Integral en la Cuenca del Río Puyango*; Ministerio del Ambiente: Bogota, Colombia, 2013.
49. Velásquez, P.; Veiga, M.; López, K. Mercury balance in amalgamation in artisanal and small-scale gold mining: Identifying strategies for reducing environmental pollution in Portovelo-Zaruma, Ecuador. *J. Clean. Prod.* **2010**, *18*, 226–232. [CrossRef]
50. Torres, W. *Protesta en Perú por Contaminación de Río por Minería en Ecuador*; Primicias: Quito, Ecuador, 2021.
51. Machado, J. *Periodismo de Investigación de las Américas*; International Center for Journalists: Washington, DC, USA; Available online: https://www.icfj.org/our-work/investigative-journalism (accessed on 5 November 2021).
52. INEC. *Pobreza por Necesidades Básicas Insatisfechas*; INEC: Quito, Ecuador, 2010.
53. "Catástrofe Ambiental" por Actividad Minera en Esmeraldas. *La Hora* **2020**. Available online: https://lahora.com.ec/noticia/1102333875/catastrofe-ambiental-por-actividad-minera-en-esmeraldas (accessed on 5 November 2021).
54. Bogoni, J.A.; Peres, C.A.; Ferraz, K.M. Extent, intensity and drivers of mammal defaunation: A continental-scale analysis across the Neotropics. *Sci. Rep.* **2020**, *10*, 14750. [CrossRef] [PubMed]
55. Tubb, D. *Shifting Livelihoods: Gold Mining and Subsistence in the Chocó, Colombia*; University of Washington Press: Seattle, WA, USA, 2020; ISBN 0295747544.
56. Lara-Rodríguez, J.S. All that glitters is not gold or platinum: Institutions and the use of mercury in mining in Chocó, Colombia. *Extr. Ind. Soc.* **2018**, *5*, 308–318. [CrossRef]
57. Frækaland Vangsnes, G. The meanings of mining: A perspective on the regulation of artisanal and small-scale gold mining in southern Ecuador. *Extr. Ind. Soc.* **2018**, *5*, 317–326. [CrossRef]
58. De Theije, M.; Kolen, J.; Heemskerk, M.; Duijves, C.; Sarmiento, M.; Urán, A.; Lozada, I.; Ayala, H.; Perea, J.; Mathis, A.; et al. *Conflicts over Natural Resources in the Global South: Conceptual Approaches*; CRC Press: Boca Raton, FL, USA, 2014.
59. Sørhaug, T. Oro, Trabajo y Locura: Problemas de Dinero y Objetos. *Fagbokforlaget* **2016**. Available online: https://www.colibri.udelar.edu.uy/jspui/bitstream/20.500.12008/8576/1/Acosta%2C%20Matilde.pdf (accessed on 5 November 2021).
60. Marshall, B.; Veiga, M. Formalization of artisanal miners: Stop the train, we need to get off! *Extr. Ind. Soc.* **2017**, *4*, 300–303. [CrossRef]

61. Gallo Corredor, J.A.; Pérez, E.H.; Figueroa, R.; Figueroa Casas, A. Water quality of streams associated with artisanal gold mining; Suárez, Department of Cauca, Colombia. *Heliyon* **2021**, *7*, e07047. [CrossRef]
62. Yevugah, L.L.; Darko, G.; Bak, J. Does mercury emission from small-scale gold mining cause widespread soil pollution in Ghana? *Environ. Pollut.* **2021**, *284*, 116945. [CrossRef]

Article

Most Attractive Scenic Sites of the Bulgarian Black Sea Coast: Characterization and Sensitivity to Natural and Human Factors

Alexis Mooser [1,2], Giorgio Anfuso [2,*], Hristo Stanchev [3], Margarita Stancheva [3], Allan T. Williams [4] and Pietro P. C. Aucelli [1]

[1] Department of Science and Technology (DiST), Parthenope University, 80143 Naples, Italy; alex.moosr@gmail.com (A.M.); pietro.aucelli@uniparthenope.it (P.P.C.A.)
[2] Faculty of Marine and Environmental Sciences, University of Cádiz, Polígono Río San Pedro s/n, 11510 Puerto Real, Spain
[3] Center for Coastal and Marine Studies (CCMS), 9000 Varna, Bulgaria; stanchev@ccms.bg (H.S.); stancheva@ccms.bg (M.S.)
[4] Department of Architecture, Computing and Engineering, University of Wales: Trinity Saint David (Swansea), Mount Pleasant, Swansea SA1 6ED, UK; allanwilliams512@outlook.com
* Correspondence: giorgio.anfuso@uca.es

Citation: Mooser, A.; Anfuso, G.; Stanchev, H.; Stancheva, M.; Williams, A.T.; Aucelli, P.P.C. Most Attractive Scenic Sites of the Bulgarian Black Sea Coast: Characterization and Sensitivity to Natural and Human Factors. *Land* 2022, *11*, 70. https://doi.org/10.3390/land11010070

Academic Editor: Luís Carlos Loures

Received: 5 December 2021
Accepted: 30 December 2021
Published: 3 January 2022

Publisher's Note: MDPI stays neutral with regard to jurisdictional claims in published maps and institutional affiliations.

Copyright: © 2022 by the authors. Licensee MDPI, Basel, Switzerland. This article is an open access article distributed under the terms and conditions of the Creative Commons Attribution (CC BY) license (https://creativecommons.org/licenses/by/4.0/).

Abstract: Beach management is a complex process that demands a multidisciplinary approach, as beaches display a large variety of functions, e.g., protection, recreation and associated biodiversity conservation. Frequently, conflicts of interest arise, since management approaches are usually focused on recreation, preferring short-term benefits over sustainable development strategies; meanwhile, coastal areas have to adapt and face a changing environment under the effects of long-term climate change. Based on a "Sea, Sun and Sand (3S)" market, coastal tourism has become a major economic sector that depends completely on the coastal ecosystem quality, whilst strongly contributing to its deterioration by putting at risk its sustainability. Among beach users' preferences, five parameters stand out: safety, facilities, water quality, litter and scenery (the "Big Five"), and the latter is the focus of this paper. Bulgaria has impressive scenic diversity and uniqueness, presenting real challenges and opportunities as an emerging tourist destination in terms of sustainable development. However, most developing countries tend to ignore mistakes made previously by developed ones. In this paper, scenic beauty at 16 coastal sites was field-tested by using a well-known methodology, i.e., the Coastal Scenic Evaluation System (CSES), which enables the calculation of an Evaluation Index "D" based on 26 physical and human parameters, utilizing fuzzy logic matrices. An assessment was made of these high-quality sites located in Burgas (8), Varna (3) and Dobrich (4) provinces. Their sensitivity to natural processes (in a climate change context) and human pressure (considering tourist trends and population increases at the municipality scale) were quantified via the Coastal Scenic Sensitivity Indexes (CSSIs) method. The CSES and CSSI methods allowed us to conduct site classification within different scenic categories, reflecting their attractiveness (Classes I–V; CSES) and level of sensitivity (Groups I–III; CSSI). Their relationship made it possible to identify management priorities: the main scenic impacts and sensitivity issues were analyzed in detail and characterized, and judicious measures were proposed for the scenic preservation and enhancement of the investigated sites. Seven sites were classified as extremely attractive (Class I; CSES), but with slight management efforts; several Class II sites could be upgraded as top scenic sites, e.g., by cleaning and monitoring beach litter. This paper also reveals that investigated sectors were more sensitive to environmental impacts than human pressure; for example, eight were categorized as being very sensitive to natural processes (Group III; CSSI).

Keywords: landscape; beach; management; climate change; erosion; tourism pressure; sustainability; developing country

1. Introduction

Coastal areas host relevant aquatic and terrestrial ecosystems located at the interface between water and land [1] and have an intrinsic environmental value due to their great biodiversity that supports the provision of several ecosystem services and related functions essential for human subsistence [2,3]. Recreational and cultural activities are also relevant in coastal areas from several decades [4]. Such sensible and valuable environments are often threatened by natural processes and, in past decades, by an increasing level of population and human pressure [5–9].

Natural processes, such as coastal erosion and flooding, are often exacerbated by human-related activities [10] and linked to chronic erosion processes [11] and/or the impact of very energetic events, such as storms and hurricanes [12,13]. They constitute a rising issue enhanced by sea-level rise and other climatic-change-related processes, such as the increasing height of extreme waves, or a change in wave tracks, the intensity/frequency of storms and hurricanes [13–16]. Coastal sensitivity is the susceptibility of coastal environments to be affected by either inundation or erosion processes, and many studies have shown that over 70% of global shorelines are currently retreating because of climate change's processes [17]. This is a relevant problem that affects the majority of global coastal areas and is reflected in the reduction or complete loss of beach and dune systems and other relevant coastal environments, such as salt marshes and mangrove swamps [18,19]. Lincke and Hinkel [20] projected that, by 2100, there will be a global coastal land loss of around 60,000–415,000 km^2 and associated migration of 17–72 million people. The total or partial degradation of such environments would involve a loss of associated tourist, aesthetic and natural values [21–24]. Such a trend is emphasized when landward migration of coastal ecosystems is impeded because of the presence of seawalls or human settlements [25], a process known as "coastal squeeze" [26,27].

The population is expected to rise, and projections postulate an increase from 625 million (in 2000) up to 949 and 1388 million people in 2030 and 2060, respectively [28]. Infrastructure and activities related to human developments (tourist, fishing, industrial, etc.) therefore represent a worrying and increasing threat to coastal environments [28,29]. Land alteration is one of the most critical issues to manage, contributing to coastal erosion; land fragmentation; loss of habitats, biodiversity and ecosystem services; and landscape degradation. As an example, benefits derived from ecosystem services for the Mediterranean Sea, which represents 0.82% of the global ocean surface, are estimated to be over €26 billion a year [30]. Despite the coastal population carrying out a broad range of economic and environmental activities, in past decades, the increase of population and development has been greatly related to tourism, which is one of the fastest-growing industries worldwide: in 2019, it generated 10.3% of global Gross Domestic Product (GDP) and supported 330 million jobs [31], with coastal and marine tourism being the largest segment of this industry [32]. This is mainly due to the attraction of the "Sea, Sun and Sand" ("3S") tourism [33,34], which mostly characterized by peaks of visits limited to the summer season, due to a strong dependence on local weather conditions and partially to the coincidence of long breaks in schools, firms, etc. [8]. This can lead to overcrowded scenarios, a situation particularly affecting a site's scenic attractiveness and its associated prestige and positive image. Commonly, conflicts and adverse effects on the environment are caused by a lack of knowledge and weak management strategies, which frequently are non-existent. Coastal visitors are especially interested in beaches, a market that annually involves billions of US dollars [35]. This raised the following question: what are the visitors' preferences on beach choice? Research indicated five main parameters, the "Big Five", i.e., safety, no litter, water quality, facilities and scenery [33], and the latter, which is the main concern of this paper, is described as "an area, as perceived by people, whose character is the result of the action and interaction of natural and/or human factors" [36].

In this paper, the most attractive scenic sites of the Black Sea Bulgarian coast have been characterized and their sensitivity to climate-change-related processes and human pressure analyzed according to the methodology proposed by Mooser et al. [37,38]. The Coastal

Scenery Evaluation System (CSES) [39–41] was used to characterize the most attractive coastal scenic sites along the investigated area, together with the Coastal Scenic Sensitivity Index (CSSI) [37] to determine present and future coastal scenic sites' sensitivity to both natural processes and human interventions. Both are linked to the increasing coastal development associated with tourist demand in Bulgaria, where tourist arrivals grew in 2018 at a rate of 8% [42]. The results obtained in this paper constitute a useful tool for the preservation and enhancement of coastal beauty of the investigated areas and provide a basis for any sound foreseeable development plan devoted to landscape conservation.

2. Study Area

Bulgaria, which covers an area of 110,842 km^2, is situated on the Balkan Peninsula in Southeast Europe and has a 432 km in length microtidal coastline, which faces the western part of the Black Sea between Cape Rezovo to the south (at the border with Turkey) and Cape Sivriburun to the north (at the border with Romania) [43]. Most rivers flowing into the Black Sea along the Bulgarian coast are small, except for the Kamchia River. The coast comprises a variety of geomorphological features, showing a wide variety of coastal scenery: rock cliffs, sandy beaches, low-lying areas with estuaries and lagoons. The location of the investigated sites is presented in Figure 1.

Figure 1. Location map of the 16 investigated sites: (1) Silistar, (2) Lipite, (3) Lipite Pocket Beach (PB), (4) Veleka, (5) Koral, (6) Ropotamo, (7) Arkutino, (8) Irakli, (9) Vaya PB, (10) Karadere, (11) Kamchia, (12) Rakitnika, (13) Cape Shabla Shore Platform (SP), (14) Shabla–Ezerets Lakes, (15) Durankulak Lake and (16) Durankulak North.

Cliffs are the most common feature, covering 49.3%, or 213 km, of the whole shoreline. Sand beaches constitute 34.5% (149 km) of the coast and the armored/engineered coast occupies 16.2% (70 km) [43]. Beach erosion and cliff retreat, both natural and human-induced, are the main hazards affecting the coastline [44]. Such retreat is partially linked to sea-level rise that, along the Bulgarian Black Sea coast, varies from 1.5 to 3 mm/y [45].

Around 20% of the Bulgarian coast has been identified as vulnerable to inundation at given scenarios of sea-level rise (0–5 m) [46]. Coastal storms, which acquire great relevance in coastal erosion, are extreme meteorological events that mainly occur in winter and are associated with severe N and NE winds.

In the recent past, the Bulgarian coast was covered by large dune systems that, despite being protected environmental areas, have experienced relevant reductions because of human activities and development; today, they occupy only 10% of the entire country's coastline [47]. The existing diversity of coastline features/landscapes makes the Bulgarian coast a popular destination for both homes of local people and accommodation for domestic and foreign tourists. Further, the coast shows favorable natural conditions for the seaside tourism development, as a result of a temperate climate and wide beaches with fine-grained sand. In the southern part of the country, pocket beaches are quite common, while in the central and northern parts, large sand beaches are common.

There are 14 (out of 264) coastal municipalities occupying an area of 5770 km^2, corresponding to 5.2% of the entire country's territory, that accommodates 726,923 residents, i.e., 9.8% of the national population according to the 2011 Census data [48].

Heavily concentrated in the Black Sea coast, tourism is a central pillar of the Bulgarian economy: in 2018, it formed 10.4% of the GDP (€6.46 billion) and provided a total of 346,800 jobs. Coastal tourism is today the most significant subsector and the fastest-growing part of the local economy involving almost 2/3 of the tourist infrastructure and tourists who mostly relate to beach-based activities, as in the "3S" tourism [49]. It contributed 66% to the Blue economy jobs (48,300 persons employed) and 55% to the Gross Value Added (or €399 million) in 2017 [50]. The first large sea resorts were established during the 1950s and 1960s. The most significant influence from coastal tourism development began at the end of the 1990s and has been expanding steadily since 2005 [47]. In 2018, Bulgaria accommodated more than 12 million international visitors, which increased by over one million during the 2012–2018 period, with half originating from the EU. Tourism is essential to many local economies, but to preserve such economic benefits, it is mandatory to soundly manage destinations and to conserve the natural aspects in which tourists are interested: this is the challenge for coastal managers in the 21st century [33,49].

Finally, cultural heritage is intrinsically connected to Bulgaria as a Black Sea country. It is part of its history, daily life, culture and tourism. Eleven Bulgarian sites are included in the UNESCO list of tangible and intangible world cultural and natural heritage. The coast is also an archaeology important area, where numerous underwater and coastal archaeological sites from different periods have been discovered—Prehistory, Antiquity (ancient Greek, Hellenistic and Roman), Mediaeval (Early Byzantium and Bulgarian). This rich concentration of submerged sites provides a worldwide unique archive of data to investigate social and economic aspects of ancient civilizations and cultures, but today such remains are at risk because of intensive human activities, e.g., fishing, hydrocarbon exploration, dredging of ports, etc. [51,52].

3. Methods

Along the Black Sea Bulgarian coast, 16 of the most attractive coastal scenic sites were evaluated by using the Coastal Scenic Evaluation System (CSES) [39–41] and the Coastal Scenic Sensitivity Index (CSSI) [37,38], as sites can potentially be affected in future decades by climate-change-related processes and increased tourist pressure. The key aspects of both methods, as stated in Mooser et al. [37,38], are presented in Figure 2.

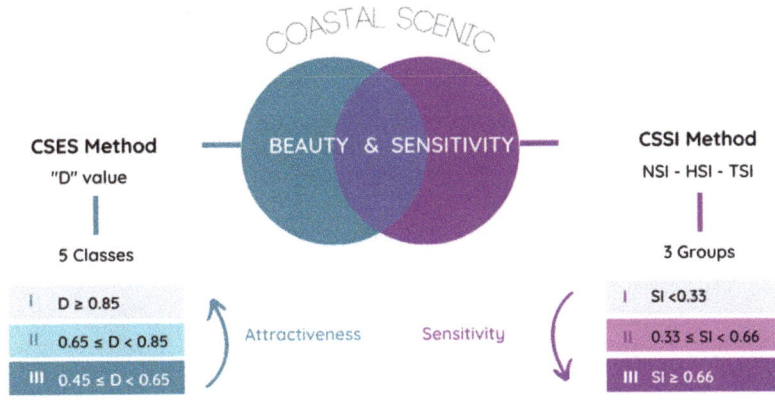

Figure 2. Summary of methods: indexes, scenic classes (Coastal Scenic Evaluation System; CSES) and sensitive groups (Coastal Scenic Sensitivity Indexes, CSSI).

Scenic beauty assessment. The focus of this paper is the research of Class I and II, sites and, secondary, Class III sites (Figure 2), that are accessible for beach users by a walk <1.5 h from the nearest car parking. The selection process was realized according to the same standard applied in the Balearic Islands and Andalusia, Spain [38,53], as detailed in the following lines:

(i) A first approximation on the location of the most attractive coastal areas was obtained via land-cover viewers and satellite images, e.g., Google Earth, Copernicus land viewer, etc. The images were used to eliminate urban and village areas, and preselect areas that appeared of great scenic values conforming to the 26 physical and human parameters (Table 1; CSES), e.g., a site which shoreline consists of cliff formations, extensive dunes system and/or shows high vegetation cover and, at the same time, records a low visual impact of human activities. If doubts arose relating the access difficulty or attractiveness, locations were automatically preselected.

(ii) Consequently, 26 sites respectively located in remote areas (17) and rural areas (9), were initially chosen after image viewing, irrespective as to whether they were located or not in protected areas.

(iii) After discussion with local coastal experts and detailed investigation of preselected areas, e.g., by consulting official webs of tourism and location of protected areas, review of published papers and grey literature, etc., a total of 21 locations were selected for field surveying. It should be noted that the criteria used to determine the distribution of preselected sites ranged according to the scenic variety of the shoreline investigated. Selected sites presented the greatest spatial density along heterogeneous scenic shorelines, such as Burgas province (points 1–4, Figure 1), whilst the opposite was true for homogenous scenic shorelines, e.g., Kamchia or Durankulak (points 11 and 15, Figure 1).

(iv) Field surveys were carried out in June 2021 between 10 a.m. and 6 p.m., during normal weather conditions, when stable conditions ruled (e.g., a storm affects water color, point 16 in Table 1) and over beach sectors 400–500 m in length; that is, when a long shoreline is assessed, it can be divided into different 400–500 m sectors. A few preselected sites such as Butamyata (Sinemorets) or Blatoto Alepu (also known as drivers' beach; Primorsko) were visited but, finally, not chosen, because of their low scenic quality. Cape Emine was finally not assessed because of the very regrettable condition of the access pathway, which required a walk estimated > 1.5 h. When constant alongshore scenic conditions were observed, adjacent sectors were joined together [54,55], giving finally a total of 16 coastal sites with different coastal lengths, from a 62 m in length pocket beach, i.e., Lipite PB (point 2, Figure 1), to a 6770 m long beach, i.e., Durankulak (point 15, Figure 1), and covering a total of c. 32 km, i.e., 8% of the total coastal length of Bulgaria.

Table 1. Coastal Scenic Evaluation System (CSES) parameters with their corresponding weight and attribute scale. * Cliff special features: indentation, banding, folding, screes and irregular profile. ** Coastal landscape features: Peninsulas, rock ridges, irregular headlands, arches, windows, caves, waterfalls, deltas, lagoons, islands, stacks, estuaries, reefs, fauna, embayment, tombola, etc. *** Utilities: power lines, pipelines, street lamps, groins, seawalls, revetments, restaurants, etc.

No.	Physical Parameters		Weight	Rating				
				1	2	3	4	5
1	CLIFF	Height (m)	0.02	Absent	$5 \leq H < 30$	$30 \leq H < 60$	$60 \leq H < 90$	$H \geq 90$
2		Slope	0.02	<45°	45–60°	60–75°	75–85°	circa vertical
3		Features *	0.03	Absent	1	2	3	Many (>3)
4	BEACH FACE	Type	0.03	Absent	Mud	Cobble/Boulder	Pebble/Gravel	Sand
5		Width (m)	0.03	Absent	W < 5 or W > 100	$5 \leq W < 25$	$25 \leq W < 50$	$50 \leq W \leq 100$
6		Color	0.02	Absent	Dark	Dark tan	Light tan/bleached	White/gold
7	ROCKY SHORE	Slope	0.01	Absent	<5°	5–10°	10–20°	20–45°
8		Extent	0.01	Absent	<5 m	5–10 m	10–20 m	>20 m
9		Roughness	0.02	Absent	Distinctly jagged	Deeply pitted and/or irregular	Shallow pitted	Smooth
10	DUNES		0.04	Absent	Remnants	Fore-dune	Secondary ridge	Several
11	VALLEY		0.08	Absent	Dry valley	(<1 m) Stream	(1–4 m) Stream	River/limestone gorge
12	SKYLINE LANDFORM		0.08	Not visible	Flat	Undulating	Highly undulating	Mountainous
13	TIDES		0.04	Macro (>4 m)		Meso (2–4 m)		Micro (<2 m)
14	COASTAL LANDSCAPE FEATURES **		0.12	None	1	2	3	>3
15	VISTAS		0.09	Open on one side	Open on two sides		Open on three sides	Open on four sides
16	WATER COLOR and CLARITY		0.14	Muddy brown/grey	Milky blue/green	Green/grey/blue	Clear/dark blue	Very clear turquoise
17	NATURAL VEGETATION COVER		0.12	Bare (<10% vegetation)	Scrub/garigue (marram, gorse)	Wetlands/meadow	Coppices, maquis (±mature trees)	Variety of mature trees
18	VEGETATION DEBRIS		0.09	Continuous (>50 cm high)	Full strand line	Single accumulation	Few scattered items	None
	Human Parameters							
19	NOISE DISTURBANCE		0.14	Intolerable	Tolerable		Little	None
20	LITTER		0.15	Continuous accumulations	Full strand line	Single accumulation	Few scattered items	Virtually absent
21	SEWAGE DISCHARGE EVIDENCE		0.15	Sewage evidence		Same evidence (1–3 items)		No evidence of sewage
22	NON-BUILT ENVIRONMENT		0.06	None		Hedgerow/terracing/monoculture		mixed cultivation ± trees/natural
23	BUILT ENVIRONMENT		0.14	Heavy Industry	Heavy tourism and/or urban	Light tourism and/or urban	Sensitive tourism and/or urban	Historic and/or none
24	ACCESS TYPE		0.09	No buffer zone/heavy traffic	No buffer zone/light traffic		Parking lot visible from coastal area	Parking lot not visible from coastal area
25	SKYLINE		0.14	Very unattractive		Sensitively designed high/low	Very sensitively designed	Natural/historic features
26	UTILITIES ***		0.14	>3	3	2	1	None

As previously stated, the Coastal Scenic Evaluation System (CSES) [39–41] is a method based on the evaluation of 26 parameters, 18 of which are physical and eight are human (Table 1). Such parameters were selected according to the results of numerous interviews of beach visitors in Turkey, Malta, Croatia, Portugal and the UK [39–41], and after discussion between coastal experts, rated on a five-point attribute scale, with 1 indicating "absence" or "poor" quality and 5 "excellent/outstanding" quality (Table 1). Each parameter had a different weight (Table 1); that is, not all parameters are the same, but the weighting of all physical components is equal to that of the human parameters. The method is based on fuzzy logic mathematics and parameter weighting matrices, which allow one "to overcome subjectivity and quantify uncertainties" [39]. Fuzzy Logic Assessment (FLA) [56] is a scientific approach used to limit any mistake that the scenic value assessor makes. The assessor must tick one box for each parameter on the checklist (Table 1) and could tick the wrong attribute box (see corrections coefficients in Table 1). After fieldwork, results were presented as follows:

(i) Histograms, which provided a visual summary of both physical and human parameters obtained from Table 1 scores;
(ii) A weighted average of attributes, which delineated relative comparison of physical and human parameters;
(iii) Membership degree of attributes, which represented overall scenic assessment over the attributes.

All the above allow for the calculation of a scenic evaluation value "D" for each site that, according to the "D" value obtained, is categorized into five distinct classes (Figure 2), from Class I, i.e., extremely attractive natural sites with very high landscape values ($D \geq 0.85$), to Class V, i.e., very unattractive urban sites with intensive development ($D \leq 0.0$; see Anfuso et al. [57] for a detailed description and > 1000 study cases around the world).

Scenic sensitivity assessment. Present and future coastal scenic sites' sensitivity to natural processes and human interventions were obtained by using the Coastal Scenic Sensitivity Index (CSSI) [37,38] (Figures 2 and 3). The method allows for the determination of the intrinsic sensitivity of coastal scenic parameters as follows:

(i) Erosion/flooding processes in a climate change context,
(ii) Unsustainable coastal population and level of development—very often linked to the tourism industry and the lack of management.

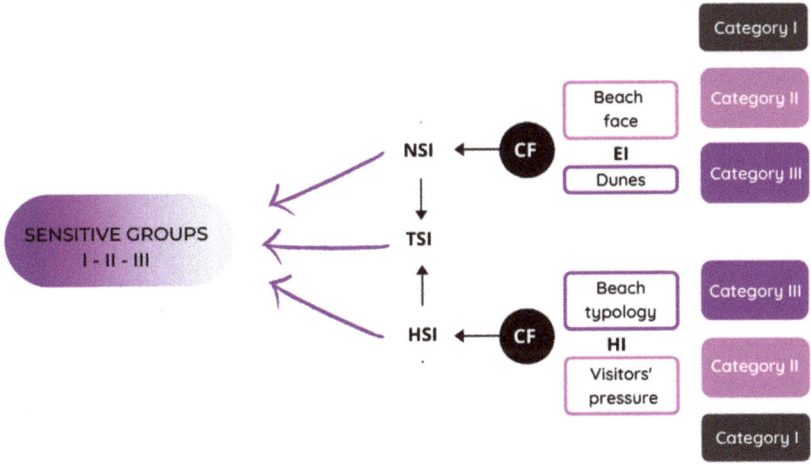

Figure 3. CSSI method: an overview of steps and parameters used for Natural Sensitivity Index (NSI), Human Sensitivity Index (HSI) and Total Sensitivity Index (TSI) assessment.

Concerning physical parameter sensitivity, Mooser et al. [37] considered that "Beach face" and "Dune" parameters (Table 2) were the most sensitive natural features to erosion/flooding processes that favor beach width reduction, i.e., low score at point 5 in

Table 1, sediment coarsening and darkening (points 4 and 6, Table 1) and dune erosion/disappearance (point 10, Table 1). Thereafter, the investigated areas were assessed according to the following standard (Figure 3):

(i) During the first phase, sites were classified within three categories according to the presence/absence of the abovementioned parameters (Figure 3);
(ii) In a second phase, the level of sensitivity of each site was determined by utilizing the Erodibility Index (EI) [37], which considers beach face and dune characteristics on a 1–5 scale (Table 2);
(iii) In a third phase, forcing variables and predicted changes of sea-level rise and storm surge were used to calculate a Correction Factor (CF, Table 2), since the effects of forcing factors on coastal environments are affected by future variations of those two variables;
(iv) In a fourth phase, the final sensitivity of natural parameters, i.e., the Natural Sensitivity Index (NSI), was obtained by considering all the above in a 0–1 range of values (Table 2 and Figure 3), allowing for the categorization of the sites into three sensitive groups (Table 2).

Table 2. Erodibility Index (EI) parameters and Correction Factors (CFs) used for NSI assessment. * Only for Category III sites. ** Imminent Collapse Zone. *** Estimation expected by the end of the 21st century.

Indexes and CF		Parameter	Null/Very Low (1)	Low (2)	Medium (3)	High (4)	Very High (5)
Natural Sensitivity Index	Erodibility Index — Beach face	Dry beach as a multiple of the ICZ **	Accretion/ >5 times ICZ	4 times ICZ	3 times ICZ	2 times ICZ	≤ ICZ
		Sediment grain size	Gravel/pebbles		Medium/ coarse sand or mixed		Fine sand
		Rocky shore — Width	>80	80–60	60–40	40–20	<20
		Rocky shore — Location	Nearshore		Foreshore		Absent
	Dunes *	Dune height (m)	≥6	≥3	≥2	≥1	<1 or absent
		Dune width (m)	>100	>75	>50	>25	<25
		Vegetation cover	Complete with fixed dune (forest)	Complete with fixed dune (shrub)	Semi-complete (without fixed dune)	Semi-completed (without embryo dune)	Incomplete or absent
		Washovers (%)	0	≤5	≤25	≤50	≥50
Correction Factor	Forcing	Significant wave height (m)	<0.75		0.75–1.5		>1.5
		Angle of approach	10°–45° (Oblique)		0°–10° (Sub-parallel)		0° (Parallel)
		Tidal range	Macro tidal		Meso tidal		Micro tidal
	Trends	Sea level rise (cm) ***	<0		0–40		>40
		Storm surge (m) ***	<1.5		1.5–3		>3

Concerning the sensitivity of human parameters, Mooser et al. [37] suggested that "Noise disturbance", "Litter" and "Sewage discharge evidence" (points 19–21, Table 1) were essentially linked to the influence of beach visitors; meanwhile, parameters such as "Non-built environment", "Built environment", "Access type" and "Utilities" (points 22–24 and 26, Table 1) were principally related to the site protection feature (if any); and "Skyline" (point 25, Table 1) was used to the urbanization level of the surrounding areas. All the above essentially depend on land use and beach typology [37]. Therefore, we employed the following:

(i) In a first phase, according to the level and typology of human pressure, each site was classified within one of the three pre-established categories (Figure 3);
(ii) In a second phase, "Visitors pressure" and "Beach typology" (Table 3) were determined on a 1–5 scale, and the Human Impact Index (HI, Table 3 and Figure 3) was calculated;

(iii) In a third phase, a Correction Factor (CF, Table 3) for human pressure was established considering trends of tourists and locals at municipality scale;
(iv) In a fourth phase, the Human Sensitivity Index (HSI) (Figure 3) was determined and places investigated were categorized into three sensitive groups, on a 0–1 range of values, according to the same standard previously established for the sensitivity to natural processes (Figure 2).

Table 3. Human Impact Index (HI) parameters and Correction Factors used for HSI assessment. * Values used in Mooser et al. [38]. ** Values used for this parameter were slightly modified from the original method [37]. *** New parameter considered for this study.

Indexes and CF			Parameter	Null/Very Low (1)	Low (2)	Medium (3)	High (4)	Very High (5)
Human Sensitivity Index	Human Impact Index	Visitor pressure	Access difficulty (min)	>45 or only accessible by sea	25–45	10–25	5–10	<5
			Protected Area Management Category	Ia & Ib	II & III	IV, V & VI	Only local designation	No
			Tourism Intensity Rate and Population density * — TIR: tourist beds per 1000 inhabitants **	<150	150–300	300–600	600–1000	>1000
			Tourism Intensity Rate and Population density * — PD: persons per km²	<70	70–150	150–300	300–700	>700
	Correction Factor		Beach typology **	Remote		Rural		Village or Resort
			Evolution of the number of beds in tourist establishments (%) **	Decrease	Minor increase	Increase 15–50%	50—100%	>100%
			Evolution of the number of inhabitant (%) ***	Major decrease >25%	Decrease 5–25%	Stable ±5%	Increase 5–25%	Major increase >25%

Finally, the combination of scores previously calculated for Natural and Human Sensitivity Indexes (NSI and HSI) allowed a Total Sensitivity Index (TSI) to be obtained enabling sites to be classified within corresponding sensitive groups (Figures 2 and 3). Equations employed for the assessment of the indexes, namely EI, NSI, HI, HSI, and TSI and Correction Factors (natural and human) are presented in Appendix A Table A1, and a detailed description of concepts and parameters used can be found in Mooser et al. [37].

4. Results and Discussion

4.1. Coastal Scenic Beauty (CSES Method)

In total, 16 sites respectively located in Burgas (9), Varna (3) and Dobrich (4) provinces were field-tested during June (2021). The Evaluation Index scores (D) and site scenic characteristics (relating physical and human parameters) are presented in Figure 4 and Table 4. Most sites showed very high values of "D": seven belonged to Class I, corresponding to extremely attractive scenic sites ($D \geq 0.85$); eight to Class II ($0.65 \leq D < 0.85$); and a single one to Class III ($0.40 \leq D < 0.65$), i.e., Vaya PB (Figures 2 and 4). It is noteworthy to mention that most Class II sites could be upgraded to Class I just by applying a few judicious measures (further detailed). Regarding the Vaya PB site, it could easily be upgraded to Class II by reducing beach litter amounts.

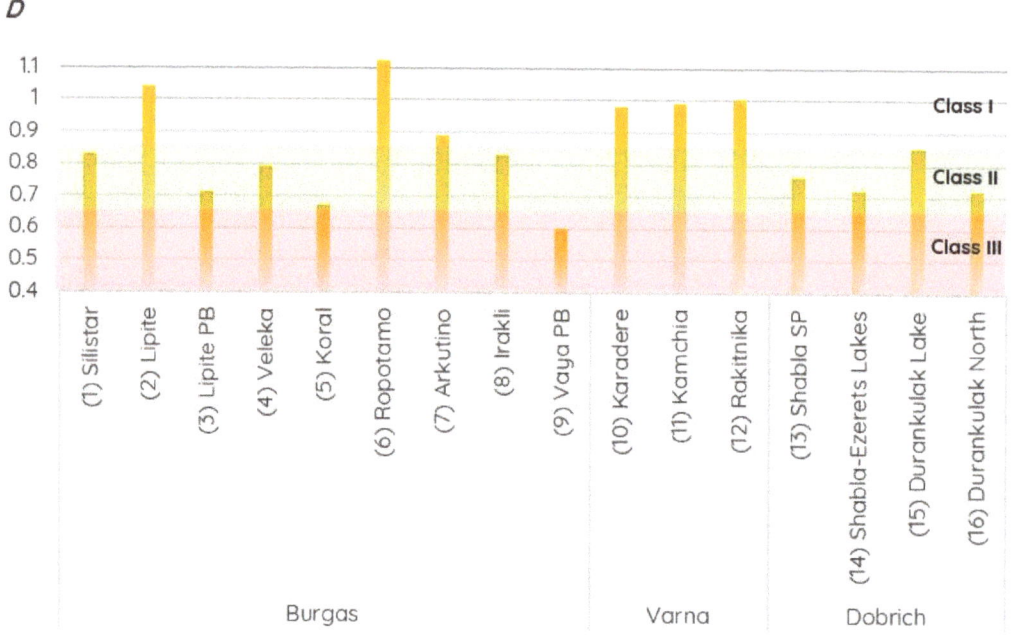

Figure 4. Site scores obtained for the Coastal Scenic Beauty Index ("D"), along with the corresponding location map number.

Bulgaria exhibits a splendid variety of scenery from vast plains with many coastal lakes (point 14, Tables 1 and 4; e.g., at Durankulak, Shabla), extensive sand coastlines surrounded by remarkable developed dune systems (point 10; e.g., Ropotamo and Arkutino), shore platform (points 7–9; e.g., Cape Shabla) and high cliff systems (points 1–3; e.g., Rakitnika), to impressive oak pristine forests located in the southern coast of Burgas (point 17; e.g., Silistar, Lipite). Such diversity makes Bulgaria an ideal place to assess coastal scenic beauty. Some sites clearly stand out from the rest with very high scenic values such as Ropotamo (D: 1.12), Lipite (1.04), Rakitnika (1.00), Kamchia (0.99) or Karadere (0.98) (Figure 4). Others showed medium or poor scores for human aspects and were consequently ranged in Class II or III, e.g., Koral (D: 0.67; Class II). General physical and human characteristics were analyzed on the following lines.

4.1.1. Physical Parameters

Excellent scenic values are often linked to the geomorphological setting, e.g., beach type/color and presence of developed dune systems, and the varied coastal physiography. Three basic physiographic systems run from east to west, splitting the whole country into three different regions, including the Danubian plain in the north, undulating and mountain plateaus in the southern part with a transitional area between them. Thus, good scores for "Skyline Landform" (point 12; Tables 1 and 4) were obtained for the Southern Burgas province, i.e., Silistar, Veleka, Arkutino and Ropotamo, whilst the Dobrich province (in the northern part of the coast) was characterized by low values related to flat landforms. Coastal relief along the southern coast of Varna also favored high ratings for "Cliff" parameters at Karadere, Vaya PB and Rakitnika (Figure 5A).

Table 4. Site scores obtained from CSES parameters: physical (1–18) and human aspects (19–26).

Parameter		1. Silistar (0.83)	2. Lipite (1.04)	3. Lipite PB (0.71)	4. Veleka (0.79)	5. Koral (0.67)	6. Ropotamo (1.12)	7. Arkutino (0.89)	8. Irakli (0.83)	9. Vaya PB (0.60)	10. Karadere (0.98)	11. Kamchia (0.99)	12. Rakitnika (1.00)	13. Cap Shabla SP (0.79)	14. Shabla-Ezerets Lakes (0.75)	15. Durankulak Lake (0.85)	16. Durankulak North (0.72)
1–3 Cliff	Height	1	1	2	1	1	1	1	1	3	2	1	3	1	1	1	2
	Slope	1	1	4	1	1	1	1	1	4	3	1	4	1	1	1	3
	Features	1	1	4	1	1	1	1	1	3	3	1	3	1	1	1	3
4–6 Beach face	Type	5	5	4	5	5	5	5	5	5	5	5	5	1	5	5	4
	Width	5	4	2	2	2	3	3	4	2	4	4	3	1	3	3	3
	Color	4	4	4	4	4	3	4	4	4	5	5	5	1	4	4	4
7–9 Rocky shore	Slope	1	1	3	1	1	1	1	1	1	1	1	1	5	1	1	1
	Extent	1	1	5	1	1	1	1	1	1	1	1	1	5	1	1	1
	Rough	1	1	1	1	1	1	1	1	1	1	1	1	2	1	1	1
10. Dunes		3	4	1	3	5	5	5	1	1	1	5	3	1	4	4	3
11. Valley		5	1	1	5	3	5	1	1	1	3	1	1	1	3	1	1
12. Skyline landform		3	1	1	3	3	4	4	3	1	3	3	1	1	2	2	1
13. Tides		5	5	5	5	5	5	5	5	5	5	5	5	5	5	5	5
14. Landscape features		3	4	3	4	3	3	4	1	3	2	1	3	3	3	3	2
15. Vistas		3	3	2	4	4	4	4	4	3	4	5	4	4	5	5	4
16. Water color		4	4	4	4	4	4	4	5	4	5	5	4	4	5	5	5
17. Vegetation cover		5	5	4	4	5	4	3	5	3	4	5	5	1	3	3	5
18. Vegetation debris		3	3	3	3	2	3	3	1	2	4	4	4	5	3	3	3
19. Noise disturbance		5	5	5	5	5	5	5	5	5	5	5	5	5	5	5	5
20. Litter		4	4	3	4	3	4	4	4	3	4	4	4	5	3	4	3
21. Sewage evidence		5	5	5	5	5	5	5	5	5	5	5	5	5	5	5	5
22. NB environment		5	5	5	5	5	5	5	5	5	5	5	5	5	5	5	5
23. Built environment		5	5	5	5	4	5	5	5	5	5	5	5	5	5	5	5
24. Access type		5	5	5	4	5	5	5	5	5	4	4	4	4	5	4	4
25. Skyline		5	5	5	3	3	4	3	4	5	5	4	4	4	4	4	5
26. Utilities		1	5	5	2	4	5	5	4	5	5	5	5	5	5	5	5

Figure 5. Scenic diversity of the Bulgarian coastline: clay cliffs of Karadere, Varna (**A**); river mouth at Silistar, Burgas (**B**); nearshore platform of Lipite PB, Burgas (**C**); Kamchia, the longest Bulgarian beach, Varna (**D**); dune system of Arkutino bordering the famous Ropotamo river, Burgas (**E**); Oak trees reaching the backshore in Lipite, Burgas (**F**); crystal water (**G**) and brackish lake of Sabla–Ezerets, Dobrich (**H**).

Top grades for "Vegetation cover" were particularly observed along the Burgas coastline. The Strandzha Nature Park offers a unique opportunity to see an extremely extensive oak forest very close to the beach that has been in existence since the end of the Tertiary period (2 million years ago) and is the only example of its kind in Europe. Considered as a "Tertiary living museum", Strandzha is one of the most relevant protected areas in the whole continent in terms of biodiversity (in all biological groups) [58]. The mouths of the Silistar and Veleka rivers are also considered as the most picturesque geotopes on the Bulgarian Black Sea coast [58], and this is reflected by top attribute rates for "Valley" (Figure 5B). The impressive sand spit (around 500 m in length) formed at the Veleka River's mouth was classified under "Special features". At places, rock sectors and headlands give rise to pocket beaches, i.e., at Lipite PB (Figure 5C).

Bulgaria also manifests a wide variety of dune systems. Their distribution commonly depends on the existence of strong onshore winds, coastline orientation, mineralogy and sediment grain size composition. The northern and southern dune systems are situated within the coastal sector of Kamchia and composed of gray dunes with wet dune slacks and forested dunes [58] (Figure 5D). Large dune systems are visible in the northern part of the country, e.g., at Durankulak and Shabla, but the most numerous systems are mainly located along the southern coast, e.g., at Ropotamo, Lipite or Arkutino (Figure 5E). The latter system reaches a maximum height of 50 m, as a result of abundant sediment supplies moving landward under the prevailing NE winds. In Lipite, the striking oaks reach the backshore (Figure 5F). North to Cape Shabla, fore-dunes, dune ridges and fixed stable dunes can be observed and reach 4 m in height at Shabla–Ezerets Lakes. A very detailed description of the Bulgarian dune systems can be found in Stancheva [59] and Stancheva et al. [60].

Most sites showed good scores for "Beach", since the coastline is predominantly composed of fine/medium-grained sand. In the northernmost area occur sand beaches consisting of organic medium-sized sands with high (93%) carbonate contents [61] because of the large mussel fields found in the nearshore. Beaches here have a low heavy mineral content, reflected by a top rating for "Beach color", e.g., Rakitnika or Karadere (white/gold color; Tables 1 and 4). Within the central coast, sand beach sediment input is basically from landslides and small rivers. Generally, beach sands are coarse to medium grain-sized, with low carbonate content consisting predominantly of quartz. At the southernmost section, beaches are composed of medium and fine grain sized magnetite–titanite sands, with a high content of heavy minerals (up to 75%) due to volcanic rocks [62], e.g., Ropotamo (dark tan color; rated 3, Table 4). Just one place had a large rock shore platform, i.e., Cape Shabla, and clear or crystal water was observed for "Water color and clarity" (point 16, Table 1) at almost all investigated sites (Figure 5G).

Finally, close to the border with Romania, coastal lakes were observed at Durankulak and Shabla–Ezerets sites (Dobrich) (Figure 5H). These brackish lakes are surrounded by fields, shrubs and separated from the Black Sea by narrow sand bars. In the case of Shabla, a connection between two lakes was made with a thin artificial canal. Situated on the Via Pontica, the second largest bird migratory route in Europe, they constitute an essential stopover or wintering refuge for many bird species and host a large variety of endemic plants, being both Ramsar sites. From a scenic perspective, these elements were reflected by good scores at "Coastal landscape features" (point 14, Table 1).

4.1.2. Anthropogenic Parameters

Scores obtained for human parameters (CSES method) [39–41] are presented in Figure 6. Located in remote (13) and rural areas (3), sites frequently showed top ratings for "Noise", "Sewage", "Built" and "Non-Built environment". However, some significant variances were observed for "Litter", "Skyline", "Utilities" and, to a lesser extent, "Access type" (Figure 6). The discussion is focused on these parameter scores and on the proposal of judicious interventions for their improvement.

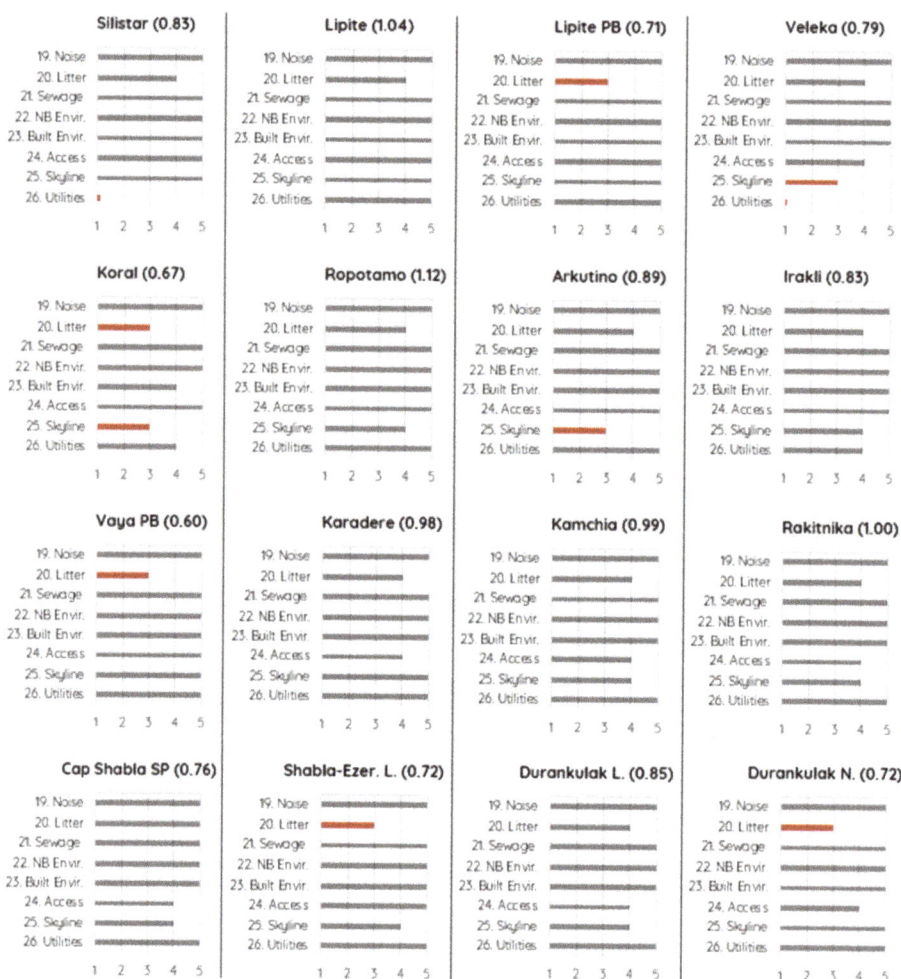

Figure 6. Scores obtained for human parameters by the CSES method (values ≤3 are in red).

4.1.3. Analysis and Suggestion Measures

(a) General analysis of Classes I and II

All investigated sites (apart from Koral) are located in protected areas belonging to different and complementary designations types, at regional, national (e.g., Nature Parks), European (e.g., SCI and SPA; Natura 2000) and/or international levels (e.g., Ramsar). The Strandzha Nature Park (Burgas) located along the southern coast, was the area most represented in this study with three sites, i.e., Silistar, Lipite and Lipite PB. Burgas was also the province that contained most sites (nine; Figure 1).

As stated previously, seven sites (out of 16) corresponded to Class I, eight to Class II and only one to Class III. Below, two distinctive examples of Class I and Class II, respectively Ropotamo and Veleka, were selected to characterize both classes. Their ratings, membership degree curves and weighted averages can give an immediate visual state of the scores obtained by relating the physical and anthropogenic parameters (Figures 7 and 8). Examples of Class III are not presented, because only one site was found, i.e., Vaya PB.

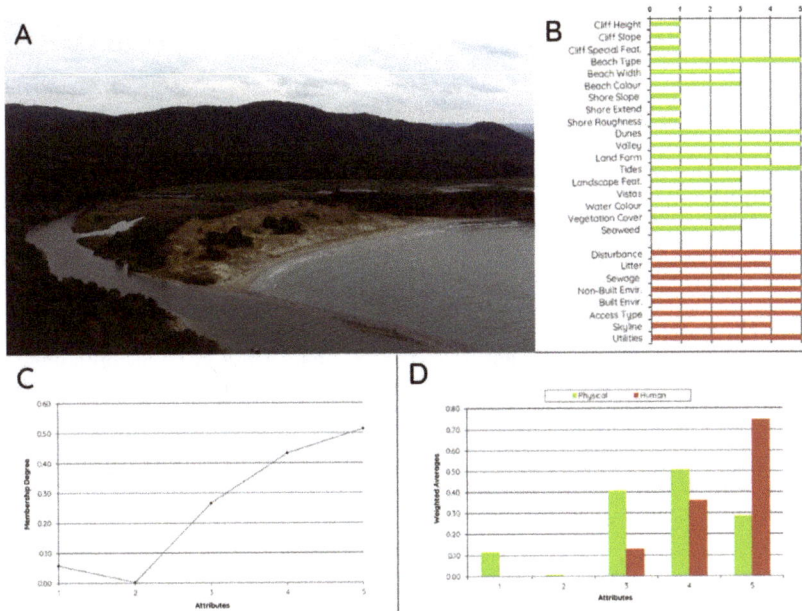

Figure 7. Ropotamo beach (**A**) and corresponding CSES ratings (**B**); membership degree vs. attributes (**C**) and weighted averages vs. attributes (**D**).

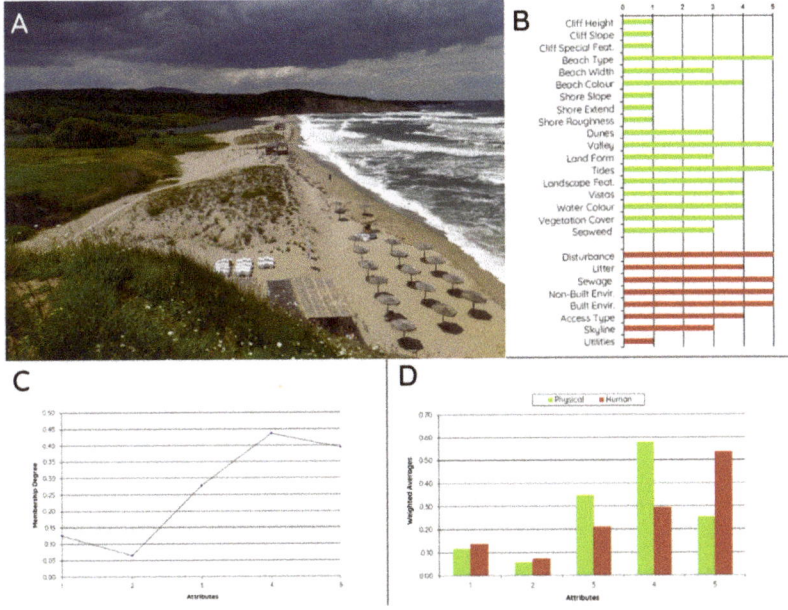

Figure 8. Veleka beach (**A**) and corresponding CSES ratings (**B**); membership degree vs. attributes (**C**) and weighted averages vs. attributes (**D**).

Class I sites are extremely attractive with outstanding features represented by physical and anthropogenic parameters (D ≥ 0.85). Located in a Strict Nature Reserve, Ropotamo

(D: 1.12) is the perfect illustration of a wild and remote area with restricted access, where human impact is almost non-existent (Figure 7A–C). Testing this place involved climbing the Arkutino' dune system, the crossing of a dense riparian forest of oak, ash, elm, etc., and the skirting of the eponymous river (entailing a 75-min walk). All human scores showed top grades, except for "Litter" and "Skyline", both rated 4. With regard to physical aspects, very good scores were observed for "Beach type" (sand), "Dunes", "Valley" (the river mouth is around 30 m in width), "Landform", "Vistas", "Water color" and "Vegetation cover" (Figure 7A,B). General physical and human ratings (Figure 7C,D) led to a very high "D" value (1.12). A few sites belonged to State Game Husbandries, e.g., Arkutino, a very common protection feature in Bulgaria that offers less protection than the previously cited features.

Class II refers to attractive natural sites with a low intrusion of human impact with "D" values $0.85 > D \geq 0.65$. These sites frequently rated lower than Class I due to a lower scoring of the physical parameters, e.g., Irakli (low scores for "Special Features", "Dunes", "Vegetation debris"), or because of the influence of human activities, e.g., Veleka. Chosen as an example in Figure 8, the latter showed top ratings for "Beach" (type and color), "Valley", "Coastal landscape features" (among them an impressive sand spit), "Vistas" and "Vegetation cover". However, low scores linked to the skyline quality and to very intrusive "Utilities" (Figure 8A–D) significantly downgraded the environmental richness. The same happened at Silistar, Koral, Shabla–Ezerets Lakes or Durankulak North, where human related activities critically lowered their natural attractiveness, downgrading them to Class II.

Finally, it is interesting to highlight which human impacts are observed in Class I and Class II (Figure 9A,B). Results clearly reflect how litter adversely affects the scenic beauty of investigated sites. Likewise, critical attention should be paid to any improvement in "Utilities", since low scores observed of this parameter enables a potential upgrade of sites with high physical values (Figure 9C) to Class I, e.g., Veleka or Silistar.

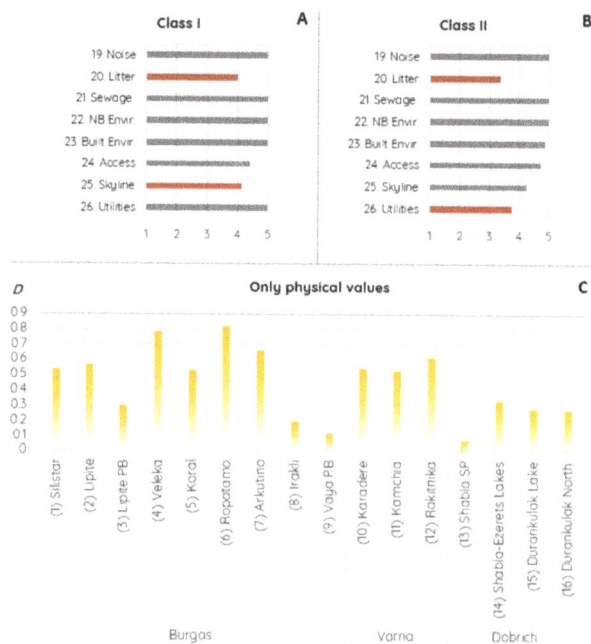

Figure 9. Human scenic impact observed in Class I and Class II (lowest scores in red) (**A,B**); sites' scenic beauty without considering human parameters (**C**).

Parameter values and suggestions for scenic enhancement of physical components, e.g., dune strengthening, rehabilitation and restoration, or beach nourishment. This is why emphasis should be placed on coastal managers to work out ways of upgrading anthropogenic parameter scores. Turning the clock back is not feasible, and certain scenic impacts, such as skyline quality, are virtually irreversible. However, several small and judicious interventions with high effectiveness may be achieved to upgrade any sites' attractiveness within determined areas, e.g., by reducing litter or utilities. Scores of human parameters were analyzed, and suggestions to managers are presented.

(1) **Noise disturbance** was non-existent at all investigated sites during field-work observations. However, it should be noted that scores obtained at Veleka (D: 0.79) and Silistar (0.83) could substantially vary during the peak tourist season. Indeed, both sites tend to considerably increase their number of visitors, as they are easily accessible by a <10-min walk and allow for the presence of several beach bars, together with a large number of sunbeds. Many tourists could decrease both sites' attractiveness to 0.73 and 0.79, respectively.

(2) **Litter,** linked to discharged items proceeding from different sources, was mainly characterized by single accumulations (rated 3; Table 1 and Figures 6 and 9) and a few scattered items (rated 4). Litter items observed along the study area were mainly stranded by sea currents and rivers, and they were usually composed of plastic items (bottles, bags and cups), glass drinks bottles, fast-food packaging, cans, foamed polystyrene and ship waste (e.g., shipping rope) (Figure 10A–D). Their presence critically lowered a site's rating. As observed in Andalusia or the Balearic Islands, the absence of periodic cleaning operations is probably due to the difficult access for cleanup machines [38,53]. However, the fact that most items have been lying on the beach for several years constitutes stark evidence of the low interest of competent authorities and managers. Litter has a very large impact on coastal tourism and recreation [63]. As a way of illustrating the relationship between beach litter and its scenic impact, if the current litter rating (3) observed at Lipite PB (D:0.71), Koral (0.67), Vaya PB (0.60), Shabla–Ezerets Lakes (0.75) and Durankulak North (0.72) is upgraded +1, the "D" value of these sites will respectively jump to 0.86, 0.80, 0.74, 0.89 and 0.88. These interventions would upgrade Lipite PB and the two latter to Class I, as well as Vaya PB to Class II. Only at Cap Shabla SP was litter virtually absent.

(3) **Sewage** was not evident at the sites investigated. Its presence is frequently visible in urban or village beach typologies [57], but hardly ever in remote areas.

(4) **Non-Built Environment** is the environment as perceived minus its buildings. In the case of Dobrich district, a very agricultural region located in the Northern Bulgaria, fields were relatively close but not visible from the beach, e.g., Durankulak North. All sites gave top scores (5) (Figure 6).

(5) **Built Environment** refers to surrounding anthropogenic structures, buildings, etc. Sites obtained top values (5), as they were located in the natural environment, except at Koral, where several bungalows are found around the beach (in its northern sector); this was characterized as "sensitive tourism" (rated 4).

(6) **Access type** usually showed good scores (≥4). Site lower scoring was principally due to four-wheel-drive vehicles that illegally crossed dune systems to get as close as possible to the beach to carry out recreational or fishing activities, among others, i.e., Kamchia, Rakitnika, Durankulak Lakes and Durankulak North. Beyond the scenic impact, this bad practice raises concerns about dunes and beach users' protection; this critical point is further discussed (cf. sensitivity section). Another curious case is Karadere, where the presence of motor homes and caravans were noticed in the backbeach, due to a lack of camping restrictions (Figure 10E). At Veleka, "Access type" was ticked attribute 4, since an unpaved road was visible from the beach.

(7) **Skyline** alludes to buildings' silhouettes not in harmony with the environment. Top grades are frequently related to sites having restricted views, e.g., Vaya and Lipite pocket beaches. Large coastal sectors, as Durankulak Lake, Shabla–Ezerets Lakes

or Irakli, located close to sensitively designed human settlements (without high buildings), obtained good scores (4). Kamchia rating (4) was linked to the presence of a 200 m pier emplaced in Shkorpilovtsi village. The worst scores (3) were noticed at Veleka, Arkutino and Koral (Figure 10F,G). The first two are located near the borders of protected areas, while the latter is out of any protected area. In the case of Veleka (D: 0.79; Class II), if the municipality would have not allowed the construction of a few elevated buildings (4 or 5 floors) near the beach rather than traditional houses (with low heights), the "D" value would reach 0.93 (Class I). At Arkutino (D: 0.89), a polemic unfinished resort complex whose construction was abandoned in the late 1980s remains relatively close to the beach (northern sector); without it, Arkutino would be one of the top Bulgarian scenic sites (D: 1.03; without skyline impact).

(8) **Utilities** is the parameter that covers a large variety of human items, e.g., power lines, lighting, pipelines, seawalls, revetments or temporary leisure facilities, amongst others. Most sites had good scores (4 or 5), except for Silistar and Veleka (rated 1; Figure 6). In both cases, their scenic impact was associated with intrusive structures devoted to seasonal use, i.e., several beach bars, beach umbrellas, first-aid stations and hundreds of sun beds (Figure 10G). This is the perfect illustration of one of the major issues that coastal managers must resolve in "3S" destinations where conflicts arise between scenic preservation and short-term benefits. Such a dilemma was also observed in Andalusia or Balearic Islands, among many other destinations [38,53]. If the administration of Strandzha Nature Park (both sites are in the Strandzha Nature Park) was not so permissive in relation to leisure facilities (but allowing first aid stations), the attractiveness of Silistar (0.83) and Veleka (0.79) could respectively jump to 0.97 and 0.92, upgrading both sites to Class I. At numerous places, lifeguard stations are certainly indispensable because of rip currents, but beach bars and other utilities (sun beds, beach umbrellas, beach kiosk, etc.) should be reduced and/or moved away from the beach, preserving the essence of natural sceneries. Beaches have to be managed according to their typologies and not as a whole. From a management approach, it is not rational that some remotes sites provide the same services as carried out on resort or urban beaches. Finally, at Koral, the presence of two old fallen lifeguard towers gave an attribute rating of 4 (Figure 10H); at Rakitnika, gas pipelines (linked to the offshore Galata Platform) were not considered, as they were not visible from the beach (covered by sand).

Figure 10. Human impact on scenery and examples of "Litter", "Access", "Skyline" and "Utilities": glass bottle (from Turkish producer) stranded by sea currents (**A**); cans, plastic bottles and foam polystyrene dropped off at the shoreline Shabla–Ezerets Lakes, Dobrich (**B**); discarded ship rope and can of adhesive paint probably dropped from a Turkish ship, Dobrich (**C**); inspiring good practices observed at Arkutino, Burgas (**D**); cars and motorhomes setting up camp near the beach of Karadere, Varna (**E**); detailed view of skyline impact caused by unfinished constructions at Arkutino (**F**); seasonal utilities at Silistar, Burgas (**G**); abandoned lifeguard tower and building construction in the background at Koral, Burgas (**H**).

4.2. Coastal Scenic Sensitivity (CSSI Method)

As mentioned above, scenic sensitivity to natural processes and human pressure was quantified by using the CSSI method [37,38]. General site characteristics and scores obtained for the Natural Sensitivity Index (NSI), Human Sensitivity Index (HSI) and Total Sensitivity Index (TSI) are presented in Table 5. The results can be useful to prevent and limit future environmental degradation linked to natural processes, in a climate-change context, and human activities in coastal areas of great scenic values, as well as to suggest measures to improve their resilience.

Table 5. Main site characteristics: provinces (Pr.), typologies, beach length, protected areas types (National, Natura 2000 and Ramsar) with corresponding IUCN categories, Scenic Sensitivity Indexes (CSSIs) and "D" values (CSES). * Acronyms: pocket beach (PB); shore platform (SP). ** Acronyms: Nature Park (NP); Site of Community Importance (SCI); Special Protection Area for Birds (SPA).

Site *	Pr.	Typology	Length (m)	Protected Areas and IUCN Category **	NSI	HSI	TSI	D
1. Silistar		Remote	462	Strandzha NP (V) SPA and SCI	0.84	0.59	0.72	0.83
2. Lipite		Remote	380	Strandzha NP (V) Silistar Protected Area (VI) SPA and SCI	0.77	0.47	0.62	1.04
3. Lipite PB		Remote	62		0.59	0.53	0.56	0.71
4. Veleka		Rural	838	Strandzha NP (V) Veleka Protected Area (VI) SPA and SCI	0.81	0.58	0.70	0.79
5. Koral	Burgas	Rural	765	None terrestrial (only SCI marine)	0.58	0.67	0.63	0.67
6. Ropotamo		Remote	547	Ropotamo Strict Nature Reserve (Ia) SCI and SPA Ramsar Estuary of the Ropotamo River Ropotamo State Game Husbandries	0.64	0	0.32	1.12
7. Arkutino		Remote	1005	Ropotamo State Game Husbandries SCI and SPA	0.66	0.63	0.65	0.89
8. Irakli		Remote	3829	Irakli Protected Site (VI; southern sector) Nessebar State Game Husbandries SCI and SPA	0.66	0.69	0.68	0.83
9. Vaya PB		Remote	328	Nessebar State Game Husbandries SCI and SPA	0.89	0.63	0.76	0.60
10. Karadere		Remote	3771	SCI and SPA Natural Monuments (III, southern sector)	0.92	0.63	0.78	0.98
11. Kamchia	Varna	Remote	6140	SPA and SCI Strict Nature Reserve (Ia, Kamchia River outlet) Protected Site (VI, Kamchia River outlet)	0.58	0.56	0.57	0.99
12. Rakitnika		Remote	1349	Rakitnika Protected Site (VI; northern sector) Liman Protected Site (VI; southern sector) SCI (northern sector) SPA	0.66	0.56	0.61	1.00
13. Cape Shabla SP		Rural	1738	Balchik State Game Husbandries SPA	0	0.59	0.30	0.76
14. Shabla–Ezerets Lakes	Dobrich	Remote	2820	Shablensko Ezero Protected Site (VI) SCI and SPA Ramsar Lake Shabla Balchik State Game Husbandries	0.50	0.38	0.44	0.72
15. Durankulak Lake		Remote	6770	Ezero Durankulak Protected Site (VI) SCI and SPA Ramsar Lake Durankulak Balchik State Game Husbandries	0.53	0.38	0.46	0.85
16. Durankulak North		Remote	1915	Balchik State Game Husbandries SCI and SPA	0.59	0.50	0.55	0.72

4.2.1. Sensitivity to Natural Processes

This index aims to determine the intrinsic scenic sensitivity of most attractive coastal sectors to erosion and/or flooding processes by considering their scenic characteristics, the level of potential stress caused by forcing variables and, finally, the predictions of Relative Sea-Level Rise (RSLR) and Storm Surge (SS) by 2100. The location, length and values obtained for NSI are presented in Figure 11. The sectors considered are characterized by homogeneous scenic values relating the physical and human aspects. For example, in the Kamchia study case, the most extensive Bulgarian beach (12.4 km), "only" 6.1 km of length was considered in this study (from the eponymous river mouth to the beginning of Shkorpilovski village), since its northern and southern sectors are surrounded by resorts and settlements linked to nearby villages, e.g., houses and a pier impacting on scenic beauty.

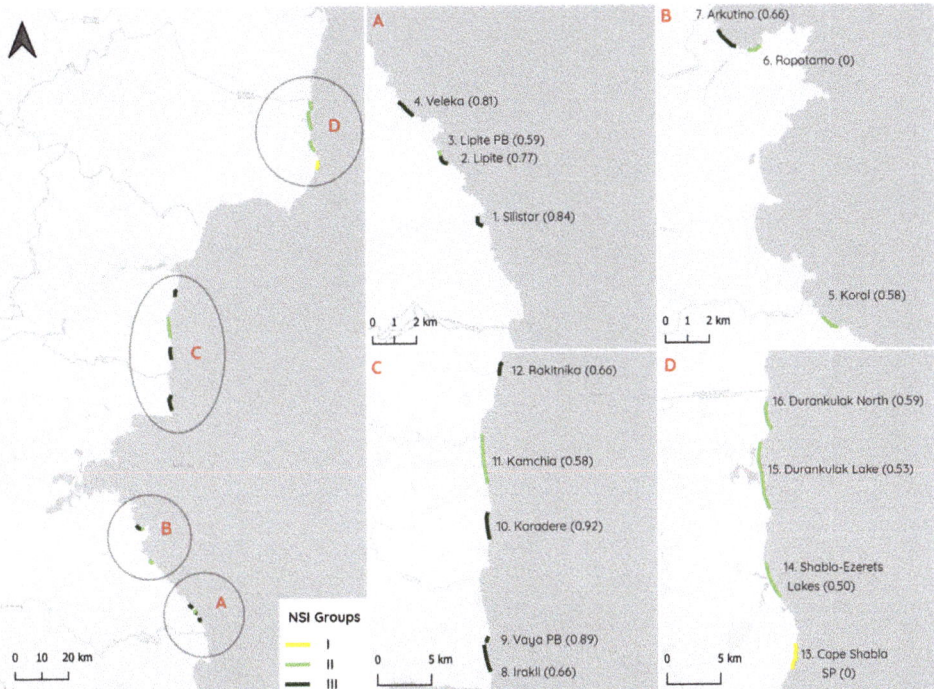

Figure 11. Investigated sectors, global view on the left map and zooms of (**A–D**) zones, with NSI values and the corresponding sensitive group.

First, the sites were divided into three categories according to their physical characteristics and scores previously obtained via the CSES method (Figure 3). Only Cape Shabla SP, characterized by a large shore platform, was considered as not sensitive and therefore included in Category I and not further investigated (Figure 11). A few sites that showed "Beach face" but no "Dunes" (≤2; CSES) were ranked in Category II (3), i.e., Lipite PB, Vaya PB and Karadere, whilst most of them belonged to Category III (12), as they presented good scores for both parameters (Tables 2 and 6). Next, the following parameters related to "Beach face" and "Dunes" were assessed, enabling an Erodibility Index (EI) for locations classified in Categories II and III (Table 6). All parameters were rated on a five-point attribute scale, with 1 indicating a great contribution of a specific key variable to site resilience and 5 indicating a low contribution/a high sensitivity.

Table 6. Site scores for NSI parameters. * Category I site: No further investigation is required.

Site	Province	Category	Dry Beach	Sediment	RS Width	RS Location	Dunes Height	Dunes Width	Vegetation Cover	Washovers	EI	Hs	Angle of Approach	Tidal Range	Storm Surge	Sea-Level Rise	NSI	Group
1. Silistar	Burgas	III	5	5	5	5	4	4	3	4	0.90	3	5	5	1	5	0.84	III
2. Lipite		III	5	5	5	5	3	4	2	1	0.79	3	5	5	1	5	0.77	III
3. Lipite PB		II	5	3	3	1					0.58	3	3	5	1	5	0.59	II
4. Veleka		III	5	5	5	5	3	4	3	4	0.88	3	3	5	1	5	0.81	III
5. Koral		III	2	5	5	5	2	2	1	2	0.56	3	3	5	1	5	0.58	II
6. Ropotamo		III	5	5	5	5	1	1	1	1	0.67	3	3	5	1	5	0.64	II
7. Arkutino		III	5	5	5	5	1	1	1	1	0.67	3	1	5	1	5	0.66	III
8. Irakli		III	1	5	5	5	4	4	3	3	0.65	3	3	5	1	5	0.66	III
9. Vaya PB		II	5	5	5	5					1.00	3	1	5	1	5	0.89	III
10. Karadere	Varna	II	5	5	5	5					1.00	3	5	5	1	5	0.92	III
11. Kamchia		III	1	5	5	5	2	2	2	3	0.55	3	5	5	1	5	0.58	II
12. Rakitnika		III	1	5	5	5	4	4	3	4	0.67	3	3	5	1	5	0.66	III
13. Shabla-Ezerets SP *	Dobrich	I									0.00						0.00	I
14. Shabla-Ezerets Lakes		III	1	3	5	5	2	4	2	2	0.46	3	3	5	1	5	0.50	II
15. Durankulak Lake		III	1	3	5	5	3	4	3	2	0.50	3	3	5	1	5	0.53	II
16. Durankulak North		III	2	3	5	5	4	4	3	2	0.58	3	3	5	1	5	0.59	II

(1) **Dry beach width as a multiple of the ICZ** was calculated comparing shorelines for the period 1972–2011, using topographic maps (1:5000) and orthophotos (images from 2019 were only available for Dobrich province). Half of the sites were rated 5, since high recorded erosion rates indicated significant beach width loss ("Dry beach", Table 6). Sectors such Silistar, Karadere or Vaya PB lost respectively 8.47 m (i.e., a beach width of 40 m), 8.36 m (21.5 m) and 14.48 m (4 m) during the investigated period. Only five, mainly located in the central–northern part of the country, manifested stability or accretion rates (rated 1), i.e., Irakli, Kamchia, Rakitnika, Shabla–Ezerets Lakes and Durankulak Lake (Table 6). The last two abovementioned sites increased their width by 8.62 m and 6.72 m respectively. Two sites, Koral and Durankulak North, gave a rating of 2, as they presented slight erosion rates coupled with values of "Dry beach" >4 times the ICZ (Table 6).

(2) **Sediment grain size** showed high ratings, since most sites (11) were composed of fine-grained sand (rated 5, Table 6). Four mixed beaches, mainly consisting of sand and, to a lesser extent, pebbles, gravel and/or broken shells, obtained intermediate scores (3) (Table 6). Curious cases were noticed in the northern sectors of Shabla and Durankulak where very impressive accumulations of black shell mussels remained on the beach shoreline (Figure 12A). At these places, reefs constitute the main source of beach material, providing over 90% of sediments [61].

Figure 12. Fragmented shells at Shabla–Ezerets Lakes (**A**), mixed-beach and nearshore platform dissipating wave energy at Lipite PB (**B**) and illustration of adverse effects linked to vehicle circulation on the dune system at Kamchia (**C**).

(3) **Rocky shore width and location**—only Lipite PB exhibited a large and emerged nearshore platform dissipating wave energy that provided a natural defense to erosion processes (Figure 12B). This was reflected by good scores for both parameters (Table 6).

(4) **Dune parameters**, including dune height, width, vegetation cover and washovers, were considered for sites belonging to Category III. Very strong and healthy dune systems that are highly resilient to potential stressing events were recorded at Arkutino and Ropotamo, giving the lowest values (1) for each parameter (Table 6). Low grades were also observed at Kamchia and Koral. However, in the case of Kamchia (and the rest of the northern and central part of the country), the illegal use of vehicles and their

associated adverse effects, e.g., fragmentation, loss of vegetation and biodiversity, displacement, compaction, etc., is a very serious issue that beach managers must resolve. For example, at Kamchia and Rakitnika, washover fans, whose formation was favored by this bad practice, broke the dune ridge continuity forming sensitive hot spots to coastal erosion (Figure 12C). At many sites, dune width was also considerably reduced or fragmented by trails parallel to the coast that are mainly used by off-road vehicles, i.e., Kamchia, Shabla–Ezerets Lakes, Durankulak Lake and Durankulak North. To give an instance of effective dune management, a place such as Rakitnika (NSI: 0.66) could improve its general NSI to 0.61 only by reducing washovers <25% and increasing dune width up to 50 m. In the southern part of the country, Veleka's (NSI: 0.81) and Silistar's (NSI: 0.84) high scorings were partially associated with the high level of recreational activities and related impacts forming critical gaps in the dune ridge, leading to a loss of vegetation cover and dune width. By mitigating the cumulative effects of pedestrians and beach bars presence, both sites could respectively decrease their sensitivity to 0.75 and 0.77. The real effectiveness of these measures is relative (and probably undervalued), as it is hard to predict how they could influence/reduce the rates of shoreline erosion in future decades. The lowering of the current scores recorded in the first parameter, "Dry Beach as multiple of the ICZ" (rated 5 for both sites, Table 6), would greatly increase the resilience of such coastal features.

Finally, a Correction Factor (CF) was estimated by taking into account forcing variables and regional predictions of RSLR and SS by the end of the century [37].

(1) **Forcing variables** include "Wave characteristics" and "Tidal range" parameters. The second was characterized by the highest grade (5) (Table 6), since microtidal coasts are most exposed to potential storm events, as they are always near high tide—a large amount of the literature supports this viewpoint [64–66]. Results for "Significant wave height" (H_s) and "Angle of wave approach" were extracted from the Bulgarian National Oceanographic Data Centre [67]. Only three virtual buoys, respectively located in Burgas, Shkorpilovski and Varna, were analyzed during the winter period from October 2020 to March 2021—due to the lack of long time series and scarcity of virtual buoys along the study area—leading to the following values: Varna (H_s: 1.01 m; 90–95°), Shkorpilovski (H_s: 1.14 m; 80–90°) and Burgas (H_s: 0.97 m; 75–85°). Given this context, all sites obtained a rating of 3 for H_s (0.75–1.5 m) (Table 6), whilst the "Angle of wave approach" was judged relating to each site location, varying from 1 to 5, e.g., Ropotamo (1; oblique 40°), Silistar (5; parallel 0°) (Table 6). It should be noted that the Shkorpilovski buoy recorded the highest energy event, with waves reaching around 5.70 m in height in March 2021.

(2) **Ongoing changes in RSLR and SS** are due to anthropogenic climate change and other factors, and they present a global challenge to coastal managers. It is acknowledged that the Black Sea and its coastal zones are one of the most sensitive areas in Europe at risk for coastal erosion and saltwater intrusion [68]. For European countries, Mean Sea Level (MSL) is expected to reach around 53 cm and 77 cm, under the Representative Concentration Pathways 4.5 and 8.5 (RCP), while projections for the Black Sea give around 59 cm and 80 cm by 2100 [69]. According to Volkov and Landerer [70], the forcing of sea level in the Black Sea is dominated by the basin freshwater budget and water exchange through the Bosporus Strait, as well as depth-integrated changes in seawater density. Many studies have reported that MSL reaches the highest levels during the May–June period [68,70]. RSLR predictions obtained from the LISCOAT database [71] gave a rating of 5 for all investigated sites (Table 6), with values varying from 0.44 (RCP 4.5) to 0.71 (RCP 8.5). Data gaps in tide gauge stations did not allow for estimates of a reliable trend for potential local subsidence effect [68,72]. The storm surge level, defined as the difference between the pure tide and the total water-level simulations, was estimated by using the Copernicus dataset of "Sea level indicators for the European coast from 1977 to 2100" [73]. Based upon past observational data and future climate projections at any regional scale, predictions around 35 cm (for 2100)

were recorded for the entire Bulgarian coast; this was reflected by a low scoring (1) (Table 6).

Combining the EI and the CF values as specified by Mooser et al. [37], we obtained the following NSI values for the 16 investigated sites (Figures 12 and 13 and Table 6), enabling their classification into one of the three sensitive groups (Figure 3).

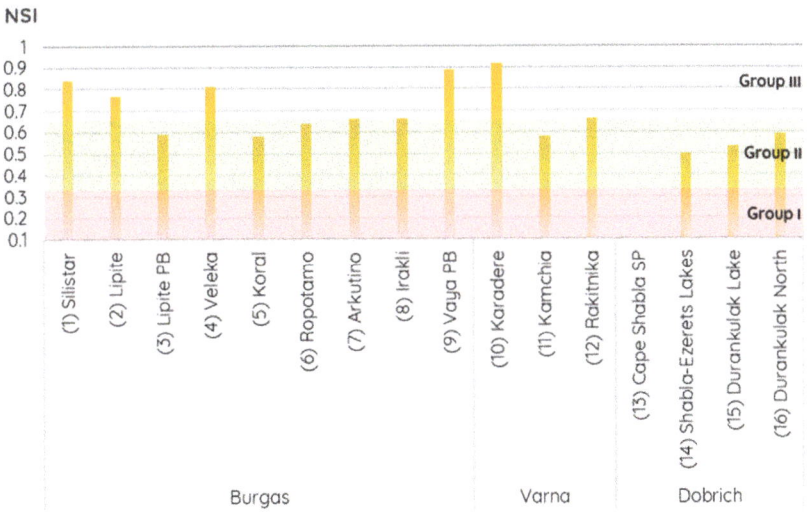

Figure 13. NSI values recorded at sites investigated, along with corresponding sensitive group.

4.2.2. Sensitivity to Human Pressure

In a global scenario of coastal unsustainable growth, this section aims to determine a sites scenic sensitivity to visitor pressure and their perception to scenery and human settlements [74], considering, as Correction Factors, local trends at the municipality scale of tourists and locals. The above enabled the calculation of a Human Sensitivity Index (HSI) presented in Figure 14 with their corresponding location and sensitive group.

Sites were firstly included in one of the three scenic categories detailed by Mooser et al. [37], in agreement with their location and ratings previously obtained for human parameters by the CSES method. Only Ropotamo belonged to Category I (Table 7 and Figure 14), as it is the only place located in a very remote area (a walk of around 1-h walking) and under a strong protection category, i.e., Strict Natural Reserve—its sensitivity to human factors was not investigated. Almost all locations were considered as Category II (12), showing low scoring for human impact, mainly associated with "Litter" and "Utilities" (Table 7 and Figure 14). Only three fell into Category III, due to a major scenic impact at "Skyline", "Built Environment" or "Access type", principally linked to their typology, i.e., Veleka, Koral and Cape Shabla SP (Table 7 and Figure 14). The following parameters were evaluated for all investigated sectors (except Ropotamo), allowing for the calculation of a Human Impact Index (HI) (Table 7).

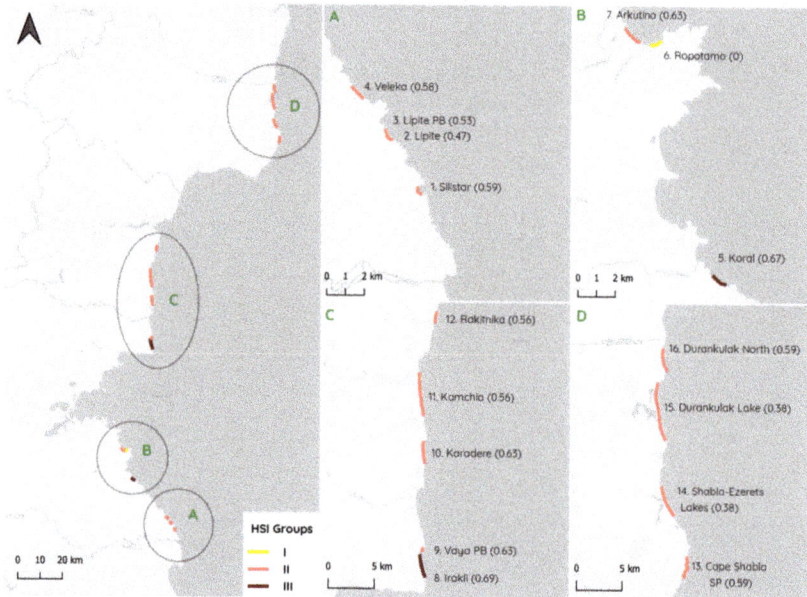

Figure 14. HSI values along with corresponding sensitive group. Global view on the left map and zooms of (**A–D**) zones.

Table 7. Site scores for HSI parameters. * Site belonging to a Strict Nature Reserve (IUCN; Ia) and accessible by a walk >45 min (Category I site; no further investigation is required). ** For sites located in two municipalities, the highest value of HSI was considered.

Site	Province	Municipality	Category	Access	PAMC	TIR	PD	Beach Typology	HI	Beds	Population	HSI	Group
1. Silistar		Tsarevo	II	4	3	5	1		0.58	4	3	0.59	II
2. Lipite		Tsarevo	II	2	3	5	1		0.42	4	3	0.47	II
3. Lipite PB		Tsarevo	II	3	3	5	1		0.50	4	3	0.53	II
4. Veleka		Tsarevo	III	4	3	5	1	3	0.58	4	3	0.58	II
5. Koral	Burgas	Tsarevo	III	3	5	5	1	3	0.67	4	3	0.67	III
6. Ropotamo *		Primorsko	I										I
7. Arkutino		Primorsko	II	3	4	5	1		0.58	5	3	0.63	II
8. Irakli		Nesebar	II	3	4	5	2		0.63	4	5	0.69	III
9. Vaya PB		Nesebar	II	3	4	5	2		0.63	4	5	0.63	II
10. Karadere		Byala	II	3	4	5	1		0.58	5	3	0.63	II
11. Kamchia **		Dolni Chiflik	II	3	4	2	3		0.54	3	2	0.50	II
	Varna	Avren	II			3	1		0.50	4	4	0.56	II
12. Rakitnika **		Avren	II	4	3	3	1		0.50	4	4	0.56	II
		Varna	II			2	5		0.63	1	4	0.56	II
13. Shabla–Ezerets SP		Shabla	III	5	4	3	1	3	0.67	5	1	0.59	II
14. Shabla–Ezerets Lakes	Dobrich	Shabla	II	2	3	3	1		0.33	5	1	0.38	II
15. Durankulak Lake		Shabla	II	2	3	3	1		0.33	5	1	0.38	II
16. Durankulak North		Shabla	II	3	4	3	1		0.50	5	1	0.50	II

(1) **Access difficulty** is an essential component of management approaches to regulate and protect sites from too many tourists. Among the 16 investigated sectors, only three sites were easily accessible by a <10-min walk from the nearest car park, i.e., Cape Shabla SP (rated 5, Table 7), Rakitnika and Veleka (both rated 4), and eight required a 10–25 min promenade (rated 3). Lower scoring was noticed for Lipite, Irakli, Shabla Ezerets Lakes and Durankulak Lake, which demanded at least 25 min of walking (rated 2, Table 7).

(2) **Protected Area Management Category** was assessed accordingly to the standard methodology provided by the International Union for Conservation of Nature (IUCN) [75], ranging from protected areas very strictly managed, e.g., Ropotamo Strict Nature Reserve (Ia), to ones managed in a relatively permissive way, e.g., Silistar Protected Area (VI). As shown in Table 5, sectors were partially or completely covered by several national and international designations, e.g., Nature Parks, Natura 2000, apart from Koral beach (rated 5, Table 7). All sites belonged to the Natura 2000 network characterized by 26 Marine Protected Areas (MPAs), with most including a coastal land area with only a narrow strip protruding into the sea, 11 SPAs (under the Birds Directive), 13 SCIs (under Habitats Directives) and two SCI–SPAs under both directives [51]. However, the practical application of Natura 2000 still poses major problems, since its process of implementation is coordinated and managed by the Ministry of Environment and Water, while CDDA (Nationally designated areas) is managed by different Institutions. Today, there is still a lack of approved and operational management plans for coastal protected areas and MPAs [51]. Because of a lower grade of protection, sites located within State Game Husbandries combined with MPAs obtained a rating of 4, i.e., Vaya PB, Arkutino or Cape Shabla (Table 7), whilst sites within a Nature Park (Category V, IUCN), i.e., Strandzha, Protected Area/Site (VI), e.g., Silistar, Veleka and Shablensko Ezero, gave intermediated values (3) (Table 7). At Kamchia (rated 4), the eponymous Protected Site (VI) and Strict Nature Reserve (Ia) (Table 5), were not considered, as they only related to the river outlet and not the beach. Irakli is a similar case, as a Natural Monument area (III) situated along the southern sector was not reflected in its rating (3), since it only represents a minor part of the total beach length.

(3) **Tourism Intensity Rate (TIR) and Population Density (PD)** were evaluated by using the dataset provided by the Ministry of Tourism [76] and the National Statistical Institute [77] (2021), both at municipality scale (Nomenclature of Territorial Units for Statistics, NUTS 5), given that provincial averages bear the risk of misleading disparities. Top grades for TIR (5) were registered at several municipalities (Table 7), suggesting that tourist capacity is superior to that of the permanent population. Highest values were noticed at Primorsko, i.e., 4346 tourist beds per 1000 inhabitants (2020), Nessebar (3216 beds per 1000 inhabitants), Tsarevo (1757 beds) and Byala (1088 beds), whilst Avren and Varna showed the lowest ratings (<30% beds per inhabitants; rated 2, Table 7). However, an opposite trend was recorded for PD. Bulgaria is experiencing a decline in population, which began at the beginning of the 1990s, and currently is losing roughly around 50,000 citizens per year [78], this being one of the major issues/challenges that the governing authority has to deal with. With regard to coastal municipalities, low values were commonly observed ($\leq$2), except at Varna (1438 inhabitants/km^2; rated 5, Table 7) and, to a lesser extent, at Dolni Chiflik, with 152 inhabitants per km^2 (rated 3, Table 7). Lastly, in the case of Kamchia and Rakitnika, since both sectors belonged to two different municipalities, the highest values obtained for the TIR and PD were chosen for the HI assessment.

A Correction Factor value (CF) was then calculated for each site by considering the following variables, once again, at NUTS 5.

(1) **Evolution of tourist beds,** obtained from the Ministry of Tourism [76], was generally characterized by high values ($\geq$4) for most sites (Table 7). Only two municipalities presented lower scores, respectively, Dolni Chiflik (rated 3, Table 7), with an increase of 44% during the period 2006–2021, and Varna, which stands out from the rest with a

10% decrease (rated 1, Table 7). Opposite results were noticed at Shabla and Byala, both rated 5, with, respectively, an increase of 588% (538 in 2006, and 3702 in 2021) and 434% (from 292 to 1561).

(2) **Evolution of the resident population** was also considered to complement the latter variable, since a decrease/increase of the resident population can also have a significant impact on coastal areas. In this case, and considering the current Bulgarian situation, a stable evolution was reflected by intermediate scores (3), whereas an increase >25% obtained the top rating (5), and vice versa, for a decrease >25% (1) (Table 7). The municipality of Nessebar (rated 5) recorded a 42% rise in inhabitants from 2005 (20,938) to 2020 (29,814) [77]. A slight increase was also registered at Varna (rated 4, Table 7). The lowest rate corresponded to Shabla, which showed a 27% decrease, with 5959 inhabitants in 2005 and 4337 in 2020. Finally, Tsarevo, Primorsko and Byala municipalities maintained a stable population in the last 15 years (rated 3, Table 7).

As stated for natural systems, a Human Sensitivity Index (HSI) was obtained by linking values obtained for HI and CF established by Mooser et al. [37] (Table 7 and Figure 15).

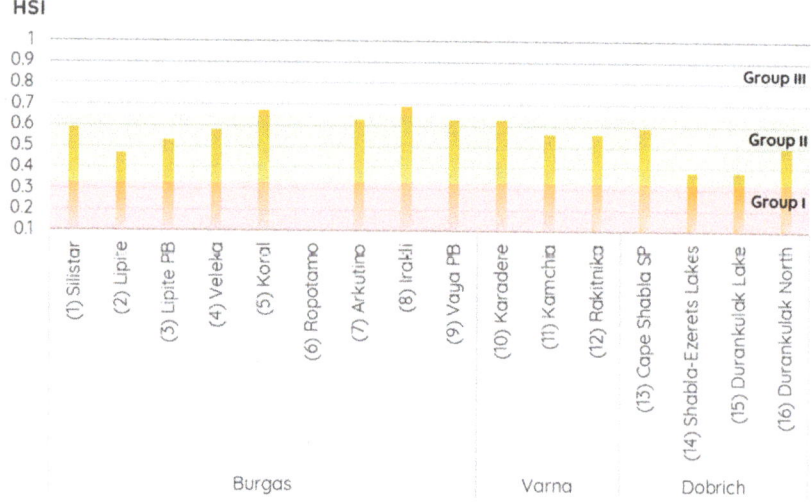

Figure 15. Site scores obtained for the Human Sensitivity Index (HSI).

Finally, a Total Sensitivity Index (TSI) was estimated by associating the values formerly achieved for NSI and HSI. Scores are presented in Table 5 and Figure 16. High values of TSI enabled us to identify/highlight sites that are very sensitive to both natural processes and human impacts. Only two sectors were included in the Group I, i.e., Ropotamo and Cap Shabla SP (both 0.32). The sites mostly belonged to Group II (9), but five fell within Group III, being broadly exposed to both factors: Silistar (0.72), Veleka (0.70), Irakli (0.68), Vaya PB (0.76) and Karadere (0.78).

4.3. Beauty versus Sensitivity: Priorities in Terms of Management

The above lead to the following question: which scenarios are the most sensitive to human and/or natural processes among the top scenic sites? To answer this, the relationship between NSI/HSI (CSSI) and "D" value (CSES) was analyzed below, with the aim of making data easier to read and interpret in order to identify priorities in terms of management (Figure 16). Our comparison of NSI and HSI clearly shows that investigated sectors were substantially much more exposed to natural processes than human pressure (Figure 16A). This is because investigated beach typologies were predominantly remote, and their overall assessment was considerably lowered by the "Population Density" parameter. Limits of Class I (D: 0.85; CSES) and Group III (SI: 0.66) were signaled in Figure 16B–D to identify the most attractive and sensitive sites.

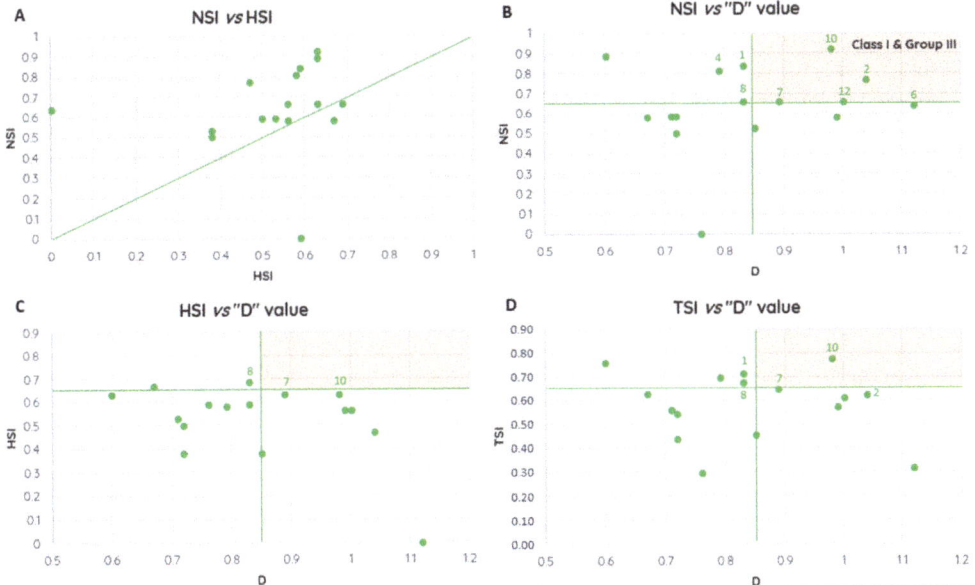

Figure 16. CSES versus CSSI: Natural Sensitivity Index (NSI) versus Human Sensitivity Index (HSI) (**A**), NSI versus "D" value (CSES) (**B**), HSI versus "D" value (**C**) and Total Sensitivity Index (TSI) versus "D" value (**D**), with corresponding location map number for highly sensitive sites of great scenic value.

4.3.1. NSI versus "D" Value

Only one site belonged to Group I (Cape Shabla SP; Category I) and seven to Group II, whilst half (8) were considered as very sensitive to natural processes (NSI $\geq$ 0.66; Group III). The most sensitive sceneries were Karadere (NSI: 0.92) and Vaya PB (NSI: 0.89), as both recorded the highest values for EI parameters: dry beach < ICZ (5), fine sand (5) and absence of shore platform (5) (Table 6). For either, very few sustainable measures could be carried out, since both are predominantly surrounded by cliff formations.

Four sites belonged to Class I and Group III and, thus, require specific attention from managers (Figure 16B): Arkutino (point 7, Figure 16B), Rakitnika (12), Karadere (10) and Lipite (2). With regard to Rakinika (12), one of the most attractive places (D: 1.00), it would be essential to control illegal access of vehicles, mainly four-wheel-drive vehicles, to preserve the dune system, together with beach users' security. From a scenic approach, the beach of Arkutino was divided into two different sectors during the field surveys: a recreational one (not investigated because of its low scenic value) and a "natural" one, 1 km in length (Table 5). The natural sector, despite showing a strong and resilient dune complex (rated 1; Table 6), recorded very high erosion rates ($\leq$ICZ; rated 5) associated with the presence of a breakwater in its northern limit. Regarding Lipite, apart from artificial beach nourishment, very few effective measures could be carried out to reduce its sensitivity, as Lipite and Karadere are not able to migrate landward because they are backed by cliffs and bluffs.

However, a few places included in Group III and close to the limits of Class I could upgrade/strengthen their attractiveness and resilience by limiting or avoiding human trampling on dunes and recreational activities on the beach (e.g., bars, etc.), i.e., Silistar (1) and

Veleka (4). With these interventions, both sites could be upgraded to Class I, accomplishing, in this way, two objectives with one action.

4.3.2. HSI versus "D" Value

Most investigated sectors belonged to Group II (13), and no sites apart from Ropotamo, initially included in Category I, were considered as "not sensitive" (Group I) (Figures 14 and 15). Within Group II, the lowest values were found at Durankulak and Shabla–Ezerets Lakes (both 0.38; Table 7 and Figures 14 and 15), because of their restrictive access (2), intermediate values for PAMC (3), very weak density of population (1) and a decline of this latter counterbalancing a significant increase of tourist beds (5) (Table 7). In a similar case, Lipite showed low scores for HI parameters but higher values related to CF, slightly raising the final HSI value to 0.47. Regarding the most sensitive sceneries, two places stood out from the rest: Koral and Irakli (Group III), both located in Burgas province (Figures 14 and 15). At Koral, high grades observed for PAMC (rated 5) and TIR, corresponding to the municipality of Tsarevo (1757 beds/1000 inhabitants; rated 5, Table 7), lead the site to Group III, with a HSI value of 0.67. The Koral HSI value could be reduced to 0.58, thus downgrading it to Group II, by securing this sector within an adequate PAMC.

No site belonged to Class I and Group III, but three sites seemed to need more management involvement: Arkutino (point 7, Figure 16C), Irakli (8) and Karadere (10). For the first two abovementioned sites, management interventions should be required to improve their Protected Area Management Category, as they only belong to State Game Husbandries and Natura 2000 SCI and SPA. Furthermore, it would be interesting to extend the current Irakli Protected Area (VI) toward the northern coastal sector, thereby including most of the beach. Nessebar municipality (where belongs Irakli) also recorded the highest increase of locals and registered an increase in tourist beds of 57% (period 2005–2020), which corresponded to a capacity three times above the permanent population. Concerning Karadere, high scores were also recorded for CF parameters, and particularly at tourist beds, i.e., an increase of 434% in the last 15 years corresponding to the municipality of Byala (rated 5; Table 7). Karadere is also the only place with a TSI and "D" value falling into both categories (Figure 16D).

Likewise, it would be convenient to determine the tourism carrying capacity of several destinations, i.e., Tsarevo, Nessebar, Primorsko and Byala (Table 7), which present a tourism flux widely superior to the number of its resident population. Local management systems are generally not prepared to manage the associated environmental stress, e.g., the insufficiency of wastewater treatment plants, etc., resulting in even greater damages for nearby ecosystems [49]. Local municipalities need to understand the carrying capacity of each site and ways to spread tourists over a greater area, e.g., by introducing marketing strategies to promote "Local jewels" to reach a large public and attract new types of tourists. This can be performed by using coastal scenic beauty as a qualitative resource, i.e., Class I sites (CSES), within the aims of the ECOTOUR-NET project framework (Development of the ecotourism network in the Black Sea region) [79] funded by the "Joint Operational Program Black Sea region 2014–2020". Some countries have already proved how important it is to use labels as a marketing brand that indicates quality, e.g., in the UK, with the eco-label "Areas of Outstanding Natural Beauty" (AONB) [80].

5. Conclusions

This paper is a contribution focused on the preservation/enhancement of the natural scenic beauty of investigated coastal areas by providing the following:

(i) The characterization of most attractive coastal scenic sites and associated weakness and sensitivity to natural and human induced factors,
(ii) The promotion of their potential development under ecotourism principles.

Bulgaria offers an impressive scenic diversity along a limited coastline length (432 km), having unique places, such as Strandzha, with a remarkable oak forest dating from the Tertiary. Seven sites were classified as "extremely attractive with outstanding features"

(Class I), but with slight management efforts, several Class II sites could efficiently improve their attractiveness and be upgraded to Class I. The results showed how litter generally downgraded the scenic value of sites. Emphasis should be also devoted to reducing intrusive "Utilities", essentially linked to recreational activities, as observed at Silistar and Veleka. Further, it is fundamental to manage each beach according to its typology, bearing in mind that scenic quality has to prevail over recreational services in remote areas.

This paper also reveals that investigated sites were generally more sensitive to environmental impacts than human pressure, and half were categorized as very sensitive to natural processes (NSI; Group III). In this regard, special attention should be paid to protect dune systems by reducing the illegal circulation of vehicles that leads to the loss of biodiversity, dune erosion and fragmentation at places linked to the enhancement of existing washover fans. By mitigating bad human practices and uses, lots of coastal areas would jointly increase their resilience to erosion in a climate-change context with critical predictions of RSLR in order to ensure the safety of beach users and improve their scenic attractiveness. Therefore, it is of the utmost importance to strengthen areas with low (or no) grades of protection by extending existing protected areas, e.g., Irakli, Karadere, Kamchia, or by creating new ones, i.e., Koral. Another concern is the inability to implant management plans to ensure the sustainable conservation of Natura 2000 sites. The growing tourist capacities of Byala, Primorsko and Nessebar should be also controlled in order to avoid future scenarios of overcrowding and related adverse effects, e.g., landscape and environmental degradation, uncontrolled mass tourism, loss of quality in services provided to visitors and loss of its positive image as a pleasant tourist destination.

Finally, the results obtained in this paper could be used as a baseline for the establishment of a novelty "coastal scenic award" to (i) promote extremely attractive sites along the Bulgarian coast under the umbrella of sustainable tourism, e.g., Class I sites; and (ii) increase the interest of local managers in landscape preservation, within the scope of the ECOTOUR-NET project.

Author Contributions: A.M. and G.A. designed the study and participated in all phases. H.S. and M.S. participated in the development of results/discussion and provided specific information related to the study area, i.e., erosion rates, categories of protection and tourism trends at municipality scale. A.T.W. provided a global structural discussion and made English corrections. P.P.C.A. made contributions regarding the conceptual approach (ideas, formulation of research goals) and provision of resources (materials, literature review). A.M., H.S. and M.S. carried out the field work observations. All authors have read and agreed to the published version of the manuscript.

Funding: This research was partially funded by Università degli Studi di Napoli Parthenope (D.R. 890/19) as the first author was supported by a PhD scholarship under the program "Environmental Phenomena and Risk", cycle 35th.

Acknowledgments: Thanks to the Center for Coastal Marine Studies (CCMS, Varna, Bulgaria) for the strong support given during the research stay (2021). This is a contribution to the PROPLAYAS Network and the Andalusia PAI Research Group RNM-328 (Spain).

Conflicts of Interest: The authors declare no conflict of interest.

Appendix A

Table A1. Equations used for the assessment of EI, NSI, HI, HSI, TSI and Correction Factors (natural and human; CSSI method).

Indexes and Categories	Equations	Parameters
Erodibility Index (1) for Category II sites (EI_{C2})	$EI_{C2} = E_{BF} = \dfrac{\dfrac{Pn_1 + Pn_2 + \dfrac{Pn_{3a} + Pn_{3b}}{2}}{n_{Pn}} - 1}{A - 1}$	E_{BF}: erodibility of beach face parameters Pn: natural parameter Pn_1: dry beach evolution Pn_2: sediment grain size Pn_{3a}: rocky shore width Pn_{3b}: rocky shore location n_{Pn}: number of natural parameters (3) A: maximum attribute value (5)
Erodibility Index (2) for Category III sites (EI_{C3})	$EI_{C3} = E_{BF} \times \dfrac{2}{3} + E_{DS} \times \dfrac{1}{3}$	E_{DS}: erodibility of dune system parameters
Erodibility of dune system (3) (E_{DS})	$E_{DS} = \dfrac{\dfrac{Pn_4 + Pn_5 + Pn_6 + Pn_7}{n_{Pn}} - 1}{A - 1}$	Pn_4: dune height Pn_5: dune width Pn_6: vegetation cover Pn_7: washovers
Natural Correction Factor (4) (CF_N)	$CF_N = \dfrac{\dfrac{\dfrac{c_{1a} + c_{1b}}{2} + c_2 + c_3 + c_4}{n_c} - 1}{A - 1}$	c_{1a}: significant wave height c_{1b}: angle of wave approach c_2: tidal range c_3: sea-level rise c_4: storm surge
Natural Sensitivity Index (5) (NSI)	$NSI = EI \times \dfrac{3}{4} + CF_N \times \dfrac{1}{4}$	
Human Impact Index (6) for Category II sites (HI_{C2})	$HI_{C2} = \dfrac{\dfrac{Ph_1 + Ph_2 + \dfrac{Ph_{3a} + Ph_{3b}}{2}}{n_{ph}} - 1}{A - 1}$	Ph: human parameter Ph_1: access difficulty Ph_2: protected area management category Ph_{3a}: tourism intensity rate Ph_{3b}: population density n_{Ph}: number of human parameters A: maximum attribute value (5)
Human Impact Index (7) for Category III sites (HI_{C3})	$HI_{C3} = \dfrac{\dfrac{Ph_1 + Ph_2 + \dfrac{Ph_{3a} + Ph_{3b}}{2} + Ph_4}{n_{ph}} - 1}{A - 1}$	Ph_4: beach typology
Human Correction Factor (8) (CF_H)	$CF_H = \dfrac{\dfrac{c_1 + c_2}{n_c} - 1}{A - 1}$	c_1: tourism trend c_2: population trend
Human Sensitivity Index (9) (HSI)	$HSI = HI \times \dfrac{3}{4} + CF_H \times \dfrac{1}{4}$	
Total Sensitivity Index (10) (TSI)	$TSI = \dfrac{NSI + HSI}{2}$	

References

1. Carter, R.W.G. *Coastal Environments*; Academic Press: Cambridge, MA, USA, 1988; p. 617.
2. Reid, W.V.; Mooney, H.A.; Cropper, A.; Capistrano, D.; Carpenter, S.R.; Chopra, K.; Dasgupta, P.; Dietz, T.; Duraiappah, A.K.; Hassan, R.; et al. *Ecosystems and Human Well–Being—Synthesis: A Report of the Millennium Ecosystem Assessment*; Island Press: Washington, DC, USA, 2005.
3. Maes, J.; Teller, A.; Erhard, M.; Liquete, C.; Braat, L.; Berry, P.; Egoh, B.; Puydarrieux, P.; Fiorina, C.; Santos, F.; et al. Mapping and Assessment of Ecosystems and Their Services. In *An Analytical Framework for Ecosystem Assessments under Action 5 of the EU Biodiversity Strategy to 2020*; Publications Office of the European Union: Luxembourg, 2013; pp. 1–58.
4. Fabbri, P. (Ed.) *Recreational Uses of Coastal Areas: A Research Project of the Commission on the Coastal Environment, International Geographical Union*; The GeoJournal Library; Springer: Dordrecht, The Netherlands, 1990; Volume 12, p. 287. [CrossRef]
5. Hughes, Z.; Duchain, H. Tourism and climate impact on the North American Eastern sea-board. In *Disappearing Destinations: Climate Change and Future Challenges for Coastal Tourism*; Jones, A., Phillips, M.R., Eds.; CABI: Oxford, UK, 2011; pp. 161–176.

6. Pranzini, E.; Wetzel, L.; Williams, A.T. Conclusions. In *Coastal Erosion and Protection in Europe*; Pranzini, E., Williams, A.T., Eds.; Routledge/Earthscan: London, UK, 2013; pp. 427–445.
7. Pilkey, O.H.; Cooper, J.A. *The Last Beach*; Duke University Press: Durham, UK, 2014.
8. EC (European Commission). *Study on Specific Challenges for a Sustainable Development of Coastal and Maritime Tourism in Europe*; Final Report; EC (European Commission): Luxembourg, 2016.
9. Williams, A.T.; Rangel-Buitrago, N.; Pranzini, E.; Anfuso, G. The management of coastal erosion. *Ocean Coast. Manag.* **2018**, *156*, 4–20. [CrossRef]
10. Molina, R.; Anfuso, G.; Manno, G.; Gracia, F.J. The Mediterranean Coast of Andalusia (Spain): Medium-Term Evolution and Impacts of Coastal Structures. *Sustainability* **2019**, *11*, 3539. [CrossRef]
11. Anthony, E.; Sabatier, F. Coastal Stabilization Practice in France. In *Pitfalls of Shoreline Stabilization*; Coastal Research Library; Cooper, J., Pilkey, O., Eds.; Springer: Dordrecht, The Netherlands, 2012; Volume 3, pp. 303–321.
12. Santos, V.M.; Haigh, I.D.; Wahl, T. Spatial and Temporal Clustering Analysis of Extreme Wave Events around the UK Coastline. *J. Mar. Sci. Eng.* **2017**, *5*, 28. [CrossRef]
13. Anfuso, G.; Loureiro, C.; Taaouati, M.; Smyth, T.; Jackson, D. Spatial Variability of Beach Impact from Post-Tropical Cyclone Katia (2011) on Northern Ireland's North Coast. *Water* **2020**, *12*, 1380. [CrossRef]
14. Lozano, I.; Devoy, R.; May, W.; Andersen, U. Storminess and vulnerability along the Atlantic coastlines of Europe: Analysis of storm records and of a greenhouse gases induced climate scenario. *Mar. Geol.* **2004**, *210*, 205–225. [CrossRef]
15. Komar, P.D.; Allan, J.C. Increasing hurricane-generated wave heights along the US East Coast and their climate controls. *J. Coast. Res.* **2008**, *24*, 479–488. [CrossRef]
16. Beudin, A.; Ganju, N.K.; Defne, Z.; Aretxabaleta, A. Physical response of a back-barrier estuary to a post-tropical cyclone. *J. Geophys. Res. Ocean.* **2017**, *122*, 5888–5904. [CrossRef]
17. Bird, E. *Coastal Geomorphology: An Introduction*; John Wiley & Sons: Hoboken, NJ, USA, 2011.
18. Nicholls, R.J.; Hoozemans, F.M.; Marchand, M. Increasing flood risk and wetland losses due to global sea-level rise: Regional and global analyses. *Glob. Environ. Chang.* **1999**, *9*, S69–S87. [CrossRef]
19. Kulp, S.A.; Strauss, B.H. New elevation data triple estimates of global vulnerability to sea-level rise and coastal flooding. *Nat. Commun.* **2019**, *10*, 1–12.
20. Lincke, D.; Hinkel, J. Coastal Migration due to 21st Century Sea-Level Rise. *Earth's Future* **2021**, *9*, e2020EF001965. [CrossRef]
21. Mir Gual, M.; Pons, G.X.; Martín Prieto, J.A.; Rodríguez Perea, A. A critical view of the blue flag beaches in Spain using environmental variables. *Ocean Coast. Manag.* **2015**, *105*, 106–115. [CrossRef]
22. Semeoshenkova, V.; Newton, A. Overview of erosion and beach quality issues in three southern European countries: Portugal, Spain and Italy. *Ocean Coast. Manag.* **2016**, *118*, 12–21. [CrossRef]
23. Anfuso, G.; Williams, A.T.; Martínez, G.C.; Botero, C.; Hernández, J.C.; Pranzini, E. Evaluation of the scenic value of 100 beaches in Cuba: Implications for coastal tourism management. *Ocean Coast. Manag.* **2017**, *142*, 173–185. [CrossRef]
24. European Environement Agency. The Millennium Ecosystem Assessment. Available online: https://www.eea.europa.eu/policy-documents/the-millennium-ecosystem-assessment (accessed on 22 July 2020).
25. Cinelli, I.; Anfuso, G.; Privitera, S.; Pranzini, E. An Overview on Railway Impacts on Coastal Environment and Beach Tourism in Sicily (Italy). *Sustainability* **2021**, *13*, 7068. [CrossRef]
26. Doody, J.P. "Coastal squeeze" an historical perspective. *J. Coast. Conserv.* **2004**, *10*, 129–138. [CrossRef]
27. Martínez, M.L.; Mendoza-Gonzalez, G.; Silva, R.; Mendoza, E. Land use changes and sea level rise may induce a "coastal squeeze" on the coasts of Veracruz, Mexico. *Glob. Environ. Chang.* **2014**, *29*, 180–188. [CrossRef]
28. Neumann, B.; Vafeidis, A.T.; Zimmermann, J.; Nicholls, R.J. Future coastal population growth and exposure to sea-level rise and coastal flooding-a global assessment. *PLoS ONE* **2015**, *10*, e0118571. [CrossRef] [PubMed]
29. European Environmental Agency. *The Changing Faces of Europe's Coastal Areas*; Office for Official Publications of the European Communities: Bruxelles, Belgium, 2006.
30. Claudet, J.; Fraschetti, S. Human-driven impacts on marine habitats: A regional meta-analysis in the Mediterranean Sea. *Biol. Conserv.* **2010**, *143*, 2195–2206. [CrossRef]
31. UNWTO (United Nations World Tourism Organization). UNWTO World Tourism Barometer and Statistical Annex, Jan 2020. *Barom* **2020**, *18*, 1–6.
32. Honey, M.; Krantz, D. *Global Trends in Coastal Tourism*; Center on Ecotourism and Sustainable Development: Washington, DC, USA, 2007.
33. Williams, A.T.; Micallef, A. *Beach Management: Principles and Practices*; Earthscan: London, UK, 2009; p. 480.
34. Williams, A.T. Definitions and typologies of coastal tourism beach destinations. In *Disappearing Destinations: Climate Change and Future Challenges for Coastal Tourism*; Jones, A., Phillips, M., Eds.; CABI: Wallingford, UK, 2011; p. 296.
35. Houston, J.R. The Economic Value of Beaches a 2013 Update. Available online: https://www.researchgate.net/publication/284772036_The_economic_value_of_beaches_a_2013_update (accessed on 1 December 2021).
36. Council of Europe. *European Landscape Convention*; Council of Europe: Florence, Italy, 2000.
37. Mooser, A.; Anfuso, G.; Williams, A.T.; Molina, R.; Aucelli, P.P.C. An Innovative Approach to Determine Coastal Scenic Beauty and Sensitivity in a Scenario of Increasing Human Pressure and Natural Impacts due to Climate Change. *Water* **2021**, *13*, 49. [CrossRef]

38. Mooser, A.; Anfuso, G.; Gómez-Pujol, L.; Rizzo, A.; Williams, A.T.; Aucelli, P.P.C. Coastal Scenic Beauty and Sensitivity at the Balearic Islands, Spain: Implication of Natural and Human Factors. *Land* **2021**, *10*, 456. [CrossRef]
39. Ergin, A.; Karaesmen, E.; Micallef, A.; Williams, A.T. A new methodology for evaluating coastal scenery: Fuzzy logic systems. *Area* **2004**, *36*, 367–386. [CrossRef]
40. Ergin, A.; Williams, A.T.; Micallef, A. Coastal scenery: Appreciation and evaluation. *J. Coast. Res.* **2006**, *22*, 958–964. [CrossRef]
41. Ergin, A. Coastal Scenery Assessment by Means of a Fuzzy Logic Approach. In *Coastal Scenery: Evaluation and Management*; Rangel-Buitrago, N., Ed.; Springer: Dordrecht, The Netherlands, 2019; pp. 67–106.
42. UNWTO (United Nations World Tourism Organization). *Tourism Highlights*; UNWTO: Madrid, Spain, 2019.
43. Stanchev, H.; Young, R.; Stancheva, M. Integrating GIS and high-resolution orthophoto images for the development of a geomorphic shoreline classification and risk assessment—A case study of cliff/bluff erosion along the Bulgarian coast. *J. Coast. Conserv.* **2013**, *17*, 719–728. [CrossRef]
44. Stancheva, M. Bulgaria. In *Coastal Erosion and Protection in Europe*; Pranzini, E., Williams, A., Eds.; Routledge: Abingdon, UK, 2013; pp. 378–395.
45. Pashova, L.; Yovev, I. Geodetic studies of the influence of climate change on the Black Sea level trend. *J. Environ. Prot. Ecol.* **2010**, *11*, 791–801.
46. Palazov, A.; Stanchev, H. Risk for the population along the Bulgarian Back Sea Coast from Flooding caused by Extreme Rise of Sea Level. *Inf. Secur. Int. J.* **2009**, *24*, 65–75.
47. Stancheva, M.; Ratas, U.; Orviku, K.; Palazov, A.; Rivis, R.; Kont, A.; Peychev, V.; Tonisson, H.; Stanchev, H. Sand dune destruction due to increased human impacts along the Bulgarian Black Sea coasts. *J. Coast. Res.* **2011**, *64*, 324–328.
48. National Statistical Institute (NSI). Regions, Districts and Municipalities in the Republic of Bulgaria 2011 (in Bulgarian). Available online: www.nsi.bg (accessed on 16 November 2021).
49. Stanchev, H.; Stancheva, M.; Young, R. Implications of population and tourism development growth for Bulgarian coastal zone. *J. Coast. Conserv.* **2015**, *19*, 59–72. [CrossRef]
50. EC (European Commission). *The EU Blue Economy Report 2019*; Publications Office of the European Union: Luxembourg, 2019.
51. Stancheva, M.; Stanchev, H.; Peev, P.; Anfuso, G.; Williams, A.T. Coastal protected areas and historical sites in North—Challenges, mismanagement and future perspectives. *Ocean Coast. Manag.* **2016**, *130*, 340–354. [CrossRef]
52. Peev, P.; Farr, R.H.; Slavchev, V.; Grant, M.J.; Adams, J.; Bailey, G. Bulgaria: Sea-Level Change and Submerged Settlements on the Black Sea. In *The Archaeology of Europe's Drowned Landscapes*; Bailey, G., Galanidou, N., Peeters, H., Jöns, H., Mennenga, M., Eds.; Coastal Research Library, Springer: Cham, Switzerland, 2020; Volume 35. [CrossRef]
53. Mooser, A.; Anfuso, G.; Mestanza, C.; Williams, A.T. Management Implications for the Most Attractive Scenic Sites along the Andalusia Coast (SW Spain). *Sustainability* **2018**, *10*, 1328. [CrossRef]
54. Cristiano, S.; Rockett, G.; Portz, L.; Anfuso, G.; Gruber, N.; Williams, A.T. Evaluation of coastal scenery in urban beaches: Torres, Rio Grande do Sul, Brazil. *JICZM* **2016**, *16*, 71–78.
55. Alcérreca-Huerta, J.C.; Montiel-Hernández, J.R.; Callejas-Jiménez, M.E.; Hernández-Avilés, D.A.; Anfuso, G.; Silva, R. Vulnerability of Subaerial and Submarine Landscapes: The Sand Falls in Cabo San Lucas, Mexico. *Land* **2021**, *10*, 27. [CrossRef]
56. Patel, A. Analytical structures and analysis of fuzzy PD controllers with multifuzzy sets having variable cross-point level. *Fuzzy Sets Syst.* **2002**, *129*, 311–334. [CrossRef]
57. Anfuso, G.; Williams, A.T.; Rangel-Buitrago, N. Examples of Class Divisions and Country Synopsis for Coastal Scenic Evaluations. In *Coastal Scenery*; Springer: Berlin/Heidelberg, Germany, 2019; pp. 143–210.
58. Ministry of Environment and Waters. Strandja Nature Park. Available online: https://www.strandja.bg/en/info/40-organizacionna-struktura-i-administracia-en (accessed on 24 November 2021).
59. Stancheva, M. Sand dunes along the Bulgarian Black Sea coast. *Compt. Rend. Acad. Bulg. Sci.* **2010**, *63*, 1037–1048.
60. Stancheva, M.; Stanchev, H.; Palazov, A.; Young, R. Coastal dune changes under natural/human hazards. In Proceedings of the 12th International Conference on the Mediterranean Coastal Environment, MEDCOAST 15, Varna, Bulgaria, 6–10 October 2015.
61. Dachev, V.Z.; Trifonova, E.V.; Stancheva, M. Monitoring of the Bulgarian Black Sea Beaches. In *Maritime Transportation and Exploitation of Ocean and Coastal Resources*; Soares, C.G., Garbatov, Y., Fonseca, N., Eds.; Taylor & Francis Group/Balkema: Leiden, The Netherlands, 2005; pp. 1411–1416.
62. Sotirov, A. Division of the Bulgarian Black Sea coast according the type of the beach sands and their supplying provinces. *Rev. Bulg. Geol. Soc.* **2003**, *64*, 39–43.
63. Botero, C.M.; Anfuso, G.; Milanes, C.; Cabrera, A.; Casas, G.; Pranzini, E.; Williams, A.T. Litter assessment on 99 Cuban beaches: A baseline to identify sources of pollution and impacts for tourism and recreation. *Mar. Pollut. Bull.* **2017**, *122*, 47–64. [CrossRef] [PubMed]
64. Thieler, E.R.; Hammar-Klose, E.S. National Assessment of Coastal Vulnerability to Future Sea-Level Rise: Preliminary Results for the U.S. Pacific Coast. USGS 2000; Fact Sheet 076-00. Available online: https://pubs.usgs.gov/dds/dds68/reports/westrep.pdf (accessed on 28 December 2020).
65. Pendleton, E.A.; Thieler, E.R.; Williams, S.J.; Beavers, R.S. Coastal Vulnerability Assessment of Padre Island National Seashore (PAIS) to Sea-Level Rise. USGS Rep. 2004; Open File Report 2004–1090. Available online: https://pubs.usgs.gov/of/2004/1090/ (accessed on 15 November 2020). [CrossRef]

66. McLaughlin, S.; Cooper, J. A multi-scale coastal vulnerability index: A tool for coastal managers? *Environ. Hazards* **2010**, *9*, 233–248. [CrossRef]
67. Bulgarian National Oceanographic Data Centre. Available online: http://bgodc.io-bas.bg/ (accessed on 24 November 2021).
68. Avşar, N.B.; Kutoğlu, Ş.H. Recent Sea Level Change in the Black Sea from Satellite Altimetry and Tide Gauge Observations. *ISPRS Int. J. Geo-Inf.* **2020**, *9*, 185. [CrossRef]
69. Vousdoukas, M.; Voukouvalas, E.; Annunziato, A.; Giardino, A.; Feyen, L. Projections of extreme storm surge levels along Europe. *Clim. Dyn.* **2016**, *47*, 3171–3190. [CrossRef]
70. Volkov, D.L.; Landerer, F.W. Internal and external forcing of sea level variability in the Black Sea. *Clim. Dyn.* **2015**, *45*, 2633–2646. [CrossRef]
71. EC (European Commission). Large Scale Integrated Sea-level and Coastal Assessment Tool, JRC Data Catalogue. Available online: http://data.jrc.ec.europa.eu/collection/LISCOAST (accessed on 17 October 2021).
72. Pashova, L. Assessment of the sea level change on different timescales from varna and burgas tide gauge data. *CR Acad. Bulg. Sci.* **2012**, *65*, 193–202.
73. CDS—Copernicus Data Store. Available online: https://cds.climate.copernicus.eu/portfolio/dataset/sis-water-level-change-indicators (accessed on 10 October 2021).
74. Botero, C.; Anfuso, G.; Duarte, D.; Palacios, A.; Williams, A.T. Perception of coastal scenery along the Caribbean littoral of Colombia. *J. Coast. Res.* **2013**, *65*, 1733–1738. [CrossRef]
75. Dudley, N. (Ed.) *Guidelines for Applying Protected Area Management Categories*; IUCN: Gland, Switzerland, 2008. Available online: https://portals.iucn.org/library/sites/library/files/documents/PAG-021.pdf (accessed on 5 October 2021). [CrossRef]
76. Ministry of Tourism. National Tourist Register. Available online: https://ntr.tourism.government.bg/CategoryzationAll.nsf/mn.xsp (accessed on 23 November 2021).
77. National Statistical Institute (NSI). Population by Districts, Municipalities, Place of Residence and Sex. Available online: https://www.nsi.bg (accessed on 23 November 2021).
78. United Nations. Department of Economic and Social Affairs, World Population Prospects 2019. Available online: https://population.un.org/wpp/ (accessed on 18 November 2021).
79. Ecotour-net. Ecotourism in Black Sea. Available online: https://ecotournet.net/ (accessed on 4 December 2021).
80. North Devon Coast AONB. *Area of Outstanding Natural Beauty Management Plan 2019–2024*; The North Devon Coast AONB: Devon, UK, 2019. Available online: https://www.northdevon-aonb.org.uk/about/management-plan-2019-2024 (accessed on 4 December 2021).

Article

Using Spatial Planning Tools to Identify Potential Areas for the Harnessing of Ocean Currents in the Mexican Caribbean

Isabel Bello-Ontiveros [1], Gabriela Mendoza-González [2,3,*], Lizbeth Márquez-Pérez [1] and Rodolfo Silva [4]

[1] Facultad de Ciencias, UMDI-Sisal, Universidad Nacional Autónoma de México, Carretera Sierra Papacal-Chuburná Puerto Km 5, Sierra Papacal 97302, Yucatán, Mexico; isabo2911@gmail.com (I.B.-O.); lizbmarq@gmail.com (L.M.-P.)

[2] Conacyt–Facultad de Ciencias, UMDI-Sisal, Universidad Nacional Autónoma de México, Carretera Sierra Papacal-Chuburná Puerto Km 5, Sierra Papacal 97302, Yucatán, Mexico

[3] Laboratorio Nacional de Ciencias de la Sostenibilidad (LANCIS), Instituto de Ecología, ENES-Mérida, Unam, Mexico City 04510, Yucatan, Mexico

[4] Instituto de Ingeniería, Universidad Nacional Autónoma de México, Edificio 17, Ciudad Universitaria, Mexico City 04510, Yucatan, Mexico; rsilvac@iingen.unam.mx

* Correspondence: gabriela.mendoza@ciencias.unam.mx; Tel.: +521-55-2517-2255

Abstract: A spatial analysis was carried out to evaluate the compatibility of human activities and biophysical characteristics in the Mexican Caribbean Sea, in order to identify the most viable areas for energy generation from ocean currents and the areas where the population would most benefit from such energy projects. Of the study area, 82% have some form of protection legislation. Tourism is the main economic activity in the area and this is reflected in a wide range of activities and services that often overlap within the same spatial area. In the case study, the use of renewable ocean energies is seen as an important innovation to reduce fossil fuel dependency. These energies have the potential to meet the demands of the region. However, it is vital to seek for potential areas for this type of energy harvesting where the social, economic and environmental impacts would be minimal. The lack of marine policies and land-use planning processes in Mexico is a major obstacle in avoiding land use conflicts.

Keywords: ocean energy harvesting; marine spatial planning; environmental impact; mitigation strategies

1. Introduction

In Mexico, oil and natural gas reserves are decreasing, from peak production in 2004; the oil era is in its final stage [1]. Figures from the Ministry of Energy [2] show that at the beginning of 2019 the proven reserves, of 6.66 billion barrels of oil and 9.7 trillion cubic feet (tcf) of gas, would last for approximately 9 and 5 years, respectively if the current rate of extraction continued (oil: 1.833 mb/day; gas: 4.847 bcf/day in 2018) [1]. An energy transition is thus needed in Mexico, from fossil fuels to sustainable energy. This would contribute significantly to achieving the climate goals set out in the General Law on Climate Change, reduce the looming energy poverty and facilitate access to energy in the region studied [3]. Renewable sources of energy from the ocean are an innovative source of great importance, thanks to their magnitude and the fact that they are found in all latitudes, that that would allow us to reduce fossil fuel consumption and meet increasing energy demands [4–6]. According to estimates, ocean currents and tidal energy have an annual global potential of 800 TWh and 300 TWh, respectively [4]. However, only 1TWh of energy is currently generated from the ocean globally [7]. Marine renewable energies are generally considered to have a low environmental impact, thanks to their low or zero greenhouse gas emissions [8]. However, some fundamental questions arise, including the formulation of standards for this industry [9], economic aspects, and environmental consequences of any mass deployment of energy generation from these sources.

Due to their inherent reliability, predictability and sustainability, ocean currents are an attractive option within marine renewables. In locations where flow acceleration is exacerbated as a consequence of the geomorphology and topography of the seabed, such as straits and channels, there is greater potential for exploitation [4,10]. The marine currents that flow through the Cozumel Channel in the Mexican Caribbean Sea have been the subject of a study by the Mexican Centre for Ocean Energy Innovation (CEMIE-Océano), defining it as a key site and pilot area for the installation of an energy harvesting device of this type. [11] examined areas, where it is possible to harvest energy for low-current hydrokinetics for approximately 50% of the time, finding that near-permanent energy extraction of ~32–215 W/m^2 would be possible in the Mexican Caribbean Sea. Fossil fuels are the main energy source in this area (59%), mostly coming from natural gas, and mainly used in the transport sector, and have a very negative impact on the environment [12,13].

Our oceans are spaces where there is great diversity of economic activities, such as tourism, fishing and transport. The lack of adequate regulations and the absence of marine policies in land-use planning in Mexico may generate local conflicts, with the result that changes or adaptations are needed in some of these activities so that they can continue to thrive [14].

Marine Spatial Planning (MSP) is a tool that is widely used to carry out a diagnosis of an area to define and analyse existing and future conditions [14]. For the deployment and operation of current energy projects in the Mexican Caribbean Sea, MSP can be very useful in identifying the areas most feasible for renewable energy conversion [14]. As they take into account the integrity of marine-coastal ecosystems as well as human activities, MSP help to avoid many antagonistic conflicts.

As part of the strategy for energy transition, this study describes a geospatial analysis, using MSP, carried out in the Mexican Caribbean Sea, in order to foresee possible environmental, social and economic impacts in areas where harnessing of energy from ocean currents is most feasible. With this information, more harmonious, sustainable and integrated decision-making is possible.

2. Materials and Methods

The study area (Figure 1) is in the northwest of Quintana Roo state, Mexico, encompassing parts of six municipalities which vary considerably in size, economic development and social characteristics. As any energy generation devices must be connected to the mainland to transfer the electricity to the national grid, a 10 km strip, along the coast was chosen, from the town of Holbox, in the north, to the town of Tulum, in the south, 895 km in all [15]. The marine area was delimited by the Cozumel Channel, and the boundary of the Caribbean Sea Ecoregion, as established by the Commission for Environmental Cooperation (CEC), which is roughly 55 km off the coast (http://www.cec.org/ accessed on 1 May 2020).

2.1. Biophysical Characteristics for Marine Energy Extraction

The Mexican Caribbean is a region of great environmental interest, with a range of sensitive coastal and marine ecosystems that are vulnerable to changes in the environment. These are of importance, both economically and socially, due to their biological productivity and the human activities that take place here [16]. These ecosystems include coral reefs, seagrass meadows, coastal beaches and dunes, coastal lagoons and mangroves. They are interconnected and act as habitats for a wide range of marine and terrestrial flora and fauna. Some of these species are at risk, according to NOM-059-SEMARNAT-2010, including the four species of sea turtles that nest in the area (*Eretmochelys imbricata, Caretta caretta, Chelonia mydas* and *Dermochelys coriacea*), and four of the six mangrove species found here (*Rhizophora mangle, Laguncularia racemosa, Avicennia germinans* and *Conocarpus erectus*) [15,17,18].

The study area harbors significant biodiversity, particularly on the coral reefs that belong to the Mesoamerican Reef System (MAR), and stretch 300 km, from Cabo Catoche in the north to Xcalak in the south. Mangroves, seagrass beds and deep-sea flora and fauna communities are also plentiful [19–21]. Several legal instruments currently exist which aim to protect areas of Mexico, like this, that are rich in significant ecosystems and biodiversity characteristics.

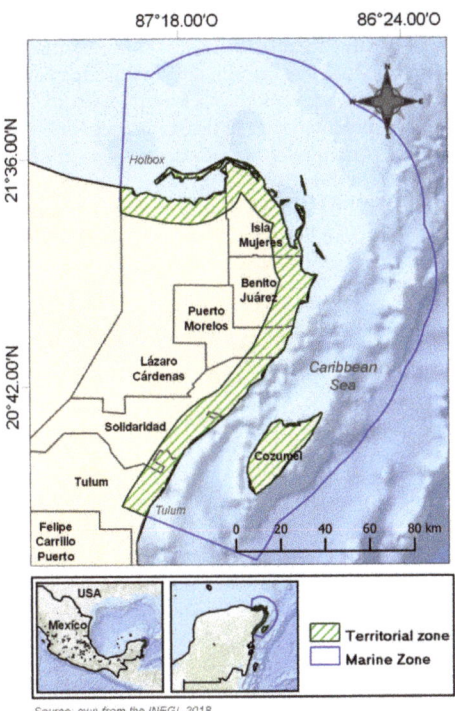

Figure 1. Study Area.

The terrestrial part of the study area is low lying, flat land, and includes three islands: Cozumel, Mujeres and Contoy. There are diverse geographical features including bays, dune systems, coves, cays, coastal reef lagoons, islands and sea cliffs [16].

In the study area, land use consists mainly of human settlements, natural ecosystems and secondary vegetation (Figure 2). Land use is important in relation to the need to connect the marine energy produced to the electricity grid. The characteristics of the area near to the plant must be considered when the installation of the electrical infrastructure required is being planned. It is also important to identify the areas of urban development that require electricity. In areas that already have an environmental impact, it is easier to plan the infrastructure needed to connect the plant to the existing electricity grid, thus avoiding more environmentally sensitive areas that can be preserved in better condition, and areas of potential socio-environmental conflict.

The bathymetry of the marine area helps identify currents and their power [4]. In the marine part of the study area the continental shelf covers approximately 6% of the ecoregion; 20 km wide near Cancun and less than 3 km in the Sian Ka'an region. The continental slope has depths of up to 3000 m (36%); and an abyssal plain over 3000 m deep [21].

Within the Caribbean Sea ecoregion there are two important channels: the Cozumel channel, approximately 50 km long and 18 km wide (Alcérreca-Huerta et al., 2019), with a depth of ~400 m; and the channel to the east of Cozumel Island, ~1000 m deep (Figure 2). The Yucatan Current flows through both channels, with an oceanic transport of 23 Sv and an average velocity of 1.5 m s^{-1} [22]. A part of this current flows eastward from the island of Cozumel, while ~5 Sv and 20% of the mean transport of the current flows through the Cozumel Channel [23]. The channels merge eastward at a depth of 2040 m, to form the Yucatan Channel (196 km wide), where velocities increase to 2.5 m s^{-1} [4,24]. This zone is the connection between the Caribbean Sea and the Gulf of Mexico, with a water flux of 23.8 ± 1 Sv [25] and depths of over 3500 m [26]. The widths of the channels are 50 and 100 km, respectively, and the currents recorded in both are semi-permanent and intense, always greater than 0.6 m s^{-1}, with maximum speeds near the surface. The speed and variability of these currents mean this area has potential for energy generation and has therefore been identified as a suitable area for energy harvesting [22,26]. Silva et al. [27] identified potential sites in the study area for the harvesting of marine renewable energies from five potential energy sources: thermal gradients (Ocean Thermal Energy Conversion, OTEC), winds (Nearshore Wind Power Converters, NSWPC), waves (Wave Energy Converters, WEC), salinity gradient (Pressure Retarded Osmosis, PRO), and marine currents (Marine Current Energy Converters, MCEC) (Figure 3). This last was calculated for 2 to 7 km off the coastline, at 50 m depth.

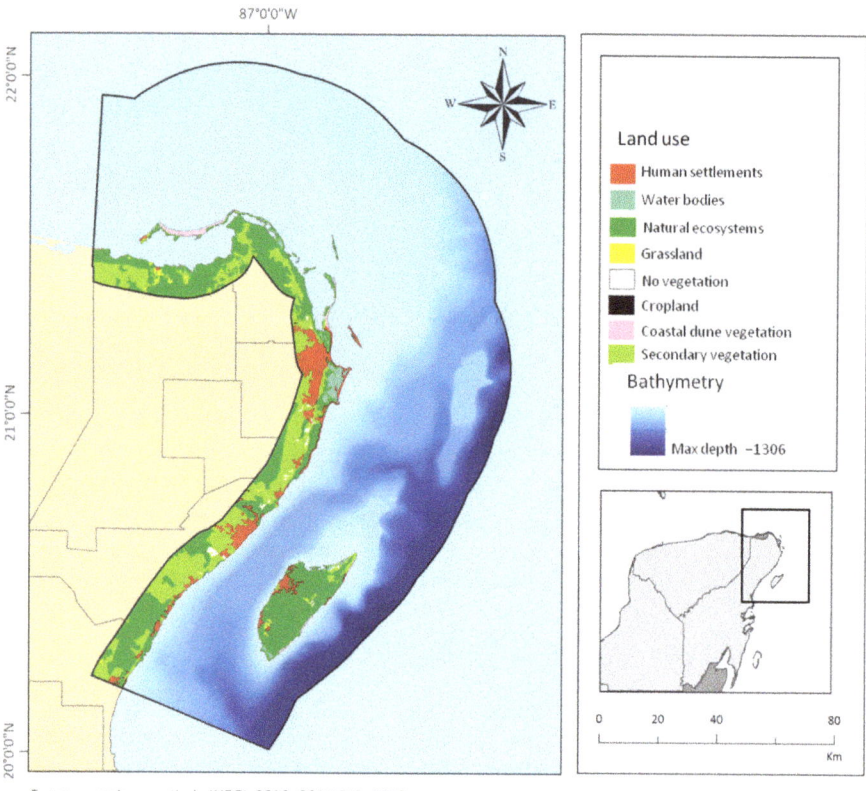

Fuente: propia a partir de INEGI, 2018; CONABIO, 2018.

Figure 2. Biophysical characteristics.

The study area has a warm humid climate, with an average annual temperature of 26 °C, with summer rainfall, occasionally accompanied by extreme weather events such as tropical storms and hurricanes (June to September) [28,29]. The risk level for hurricane formation in the region is classified as very high, making this a potential threat to the operation of some ocean energy technologies [30]. The average annual surface water temperature is 27 °C and 7.7 °C at a depth of 700 m [30].

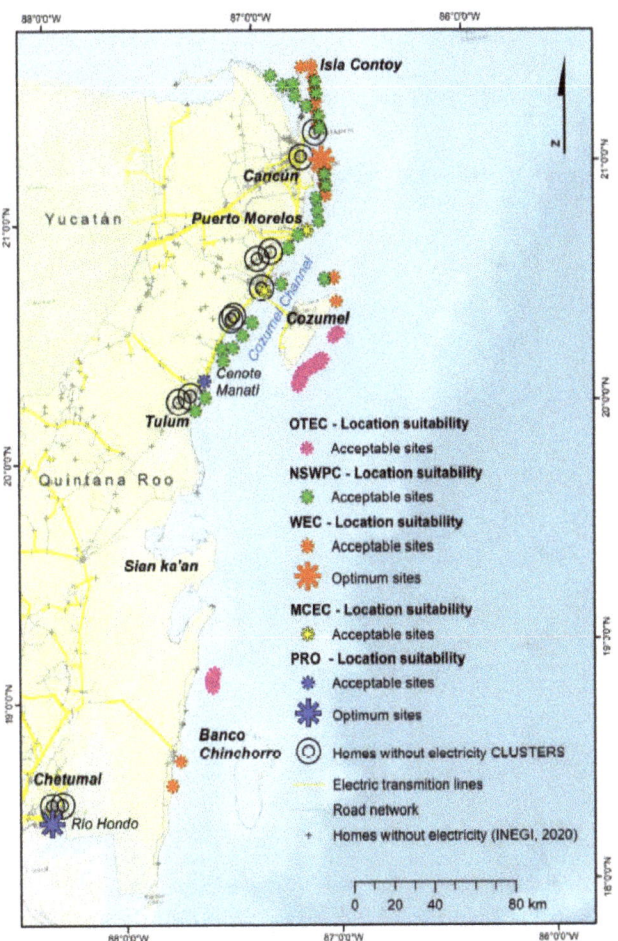

Figure 3. Suitable locations for energy in the study area for five potential energy sources from [27]. Sites for Marine Current Energy Converters (MCEC) are shown by yellow asterisks.

2.2. Marine Spatial Planning

The methodology of this work was based on the UNESCO [31] guide for Marine Spatial Planning (MSP), in which a spatial diagnosis is developed to plan integrated management in a given marine area. This consists of a definition and analysis of existing conditions in the Mexican Caribbean Sea and a compatibility analysis between existing characteristics/activities and energy needs.

2.2.1. Analysis of Existing Conditions in the Mexican Caribbean Sea

The various activities that take place in the region were defined, as well as the physical or social characteristics that could be related to the potential ocean energy harvesting. Scientific literature, Governmental databases (CONAPESCA, CONANP, INEGI, CONABIO), and websites of regional, local hotels, companies and NGOs were consulted to compile information in three key categories, as shown in Table 1:

Table 1. Characteristics/activities assessed for compatibility in the Caribbean Sea.

Category	ID	Characteristics/Activities	Objectives
(1) Protected areas and areas of environmental importance	1	Federal NPAs	Conservation of ecosystems and biodiversity
	2	State NPAs	
	3	Voluntary NPAs	
	4	Fisheries refuge zones	
	5	Wildlife refuge areas	
	6	Protected beach (turtle nesting)	
	7	RAMSAR Sites	
	8	IBAs	
	9	Priority Marine Sites	
(2) Human activities	10	Selected Tourist Destinations	Economic use/recreation
	11	Tourist beaches	
	12	Certified tourist beaches	Economic use/recreation/conservation
	13	Dive sites	Economic use/recreation
	14	Sport Fishing	
	15	Archaeological sites	Economic use/preservation of cultural heritage/recreation
	16	Permitted Fishing	Economic use
	17	Lobster concessions	
	18	Mining	
(3) Port and urban infrastructure	19	Decks	Berthing/transport
	20	Marinas	
	21	Port infrastructure	

NPA: Natural protected area; IBA: Important Bird Area.

Category (1) includes the protected areas and areas of environmental importance, significant for their biodiversity, richness, abundance, endemism, etc. (Table 2). In Mexico, Natural Protected Areas (NPA) are used as a conservation tool to protect marine and terrestrial ecosystems that shelter wild flora and fauna, natural landscapes, ecological processes, recreation opportunities, etc. as goods and ecosystem services that provide benefits for local inhabitants, for the region, and for the country. In the Mexican Caribbean, all these protected areas are important support instruments for the integrated management of coastal zones, and are designed to stimulate good practices in fisheries management, tourism, governance, etc. A considerable part of these areas has some kind of protection policy. They may be legally protected or categorised as environmentally important (although they have significant environmental characteristics, they do not have official regulations governing them). The specific objectives of each protected area depend on the goals proposed for each. Their administration is the responsibility of three governmental agencies: The National Commission on Natural Protected Areas (CONANP), Ministry of Ecology and the Environment of Quintana Roo (SEMAQRoo) and The National Commission of Aquaculture and Fisheries (Conapesca), each of which has its own respective legislation, as well as of civil organisations, whose practices are aligned to the legal and regulatory framework (Table 2).

Category (2) concerning human activities, economic activities, activities that concern the preservation of cultural heritage, and recreational activities are shown. Tourism and fishing are the most important activities, because of the economic and social benefits they generate. In 2018, Quintana Roo received 16,675 million visitors, (top ranking, nationally), generating almost 9 billion USD [32]. About half of the employed population in Quintana Roo work in tourism [33]. With respect to fishing, the state ranks 21st nationally, producing 3571 tonnes in 2018, generating over $181 million Mexican pesos. Its share of national production was 0.17% in 2018, with octopus, grouper and lobster being the main commercial species [34].

Table 2. Types of Protected Areas.

Environmental Protection Instrument	Number of Sites in the Study Area	Area (km²)	Administrative Body	Legislation Applicable
Federal NPAs	12	14,317.09	National Commission of Natural Protected Areas	General Law on Ecological Balance and Environmental Protection. Regulation of the General Law of Ecological Balance and Environmental Protection of Natural Protected Areas.
State NPAs	7	234.91	SEMAQRoo	General Law of Ecological Balance and Environmental Protection of Quintana Roo. Regulation of the General Law of Ecological Equilibrium and Environmental Protection of Natural Protected Areas of Quintana Roo.
Voluntary NPAs	3	3.42	Owners	General Law on Ecological Balance and Environmental Protection. Regulation of the General Law of Ecological Balance and Environmental Protection of Natural Protected Areas.
Fisheries Refuge Zones (FRZ)	2	9.96	National Fisheries Commission (CONAPESCA)	Law on Sustainable Fisheries and Aquaculture. NOM-049-SAG/PESC-2014.
Refuge areas (RA)	2	5728.31	National Commission of Natural Protected Areas	General Wildlife Law.
Protected Beaches	51	3.31	National Commission of Natural Protected Areas Sea Turtle Protection and Conservation Centres (MTSPCs) Committee for the Protection of Sea Turtles in Quintana Roo	General Wildlife Law. NOM-059-SEMARNAT-2010.

Category (3) contains port and urban infrastructure, important in both the installation of marine energy harvesting devices, which requires the transport of technical personnel and supplies, as well as the existing port and urban infrastructure, such as access roads, dock facilities or the feasibility of building a dock, and proximity to services.

The information available was downloaded in vector and raster format and integrated with processed information to generate maps for geospatial analysis to calculate delimited areas and generate assessment categories through ArcMap 10.4 software. The information without a spatial format was processed and edited in spreadsheets for subsequent conversion to vector data in the Mexico_ITRF2008_UTM_Zone_16N coordinate system. The location software Google Earth was used to rectify the maps or to obtain specific coordinates for specific locations.

2.2.2. Analysis of Existing Characteristics/Activities in the Mexican Caribbean Sea

From the overlaid vector layers, new maps were generated to show the degree of compatibility between the 21 characteristics/activities described (Table 1). This information was organised and analysed by means of three matrices: compatibility between objectives, spatial intersection and objectives compatibility special intersection, in three consecutive steps:

1. The compatibility between the objectives of the characteristics/activities was analysed by first constructing a compatibility matrix (Table 3). Three degrees of compatibility were considered: (a) compatible objectives, when the two characteristics/activities evaluated can be developed in the same time and space, represented by green cells; (b) poorly compatible objectives, when there may be conflicts when developing both characteristics/activities in the same time and space, yellow cells, and; (c) incompatible objectives, when the characteristics/activities cannot, or should not, be developed in the same time and space, red cells. The frequency, or number of cells, with which each degree of compatibility occurred was also calculated.
2. The spatial information was analysed to construct a second intersection matrix, in which the number of spatial intersections between the types of characteristics/activities with the same and other types, was calculated to determine the number of coincidences (Figure 4, letter "a"). For this, the "Selection by location" tool of ArcGis 10.4 was used. With this information, the percentage of spatial intersections between each of the features/activities presented in the area was counted.
3. Finally, a third matrix was obtained calculating the frequency of cells with 0% and 100% intersection percentages, and then the frequency of cells for each percentage of the compatibility degree matrix (Compatible, Poorly compatible and Incompatible) was calculated. Subsequently, the percentages of intersection 0, 100 and >0 to 100 were calculated, for all the characteristics/activities found in the area, for each degree of compatibility of objectives (compatible, poorly compatible and not compatible) (Figure 4, letters "b" and "c"). These results were then summarised in a table of compatibility degree percentages identifying two types: the percentage of intersection (>0–100%) and the percentage of non-intersection (0%) of characteristics/activities.

Table 3. Compatibility matrix between the objectives of the characteristics/activities.

Objectives	Conservation	Economic Use	Recreation	Preservation of Cultural Heritage	Berth/Transport
Conservation	Compatible	Poorly or Incompatible	Compatible or poorly compatible	Compatible	Poorly or Incompatible
Economic use	Poorly or Incompatible	Compatible	Compatible or poorly compatible	Compatible or poorly compatible	Compatible or poorly compatible
Recreation	Compatible or poorly compatible	Compatible or poorly compatible	Compatible	Compatible or poorly compatible	Compatible
Preservation of cultural heritage	Compatible	Compatible or poorly compatible	Compatible or poorly compatible	Compatible	Incompatible
Berth/transport	Poorly or Incompatible	Compatible or poorly compatible	Compatible	Incompatible	Compatible

In order to identify areas with potential for energy harvesting from ocean currents with respect to all the activities in the study area, a map was generated to spatialize the information obtained in the compatibility matrix. The same colour code (green, potential areas; yellow, little potential; and red, no potential) was used to indicate the respective areas. Then the "merge" polygon tool was used to construct a single layer for each degree of compatibility obtained in the matrix. Since it is not possible to merge points and polygons with the tool used, the characteristics/activities were assigned the specific colour according to the compatibility matrix.

2.2.3. Energy Supply Needs of Communities

For this analysis, government databases for the study area [35–37] and a civil organization [38] were consulted. Information on access to electricity, the degree of social marginalization, and features of the electricity infrastructure (power plants, electricity substations, electricity grid) were compiled. Using the INEGI Catalogue of localities [39] settlements were identified that are less than 10 km from the coast, at less than 100 m above sea level and with more than 100 inhabitants. These socio-economic criteria have previously been used for the deployment of marine energy harvesting devices [30,40].

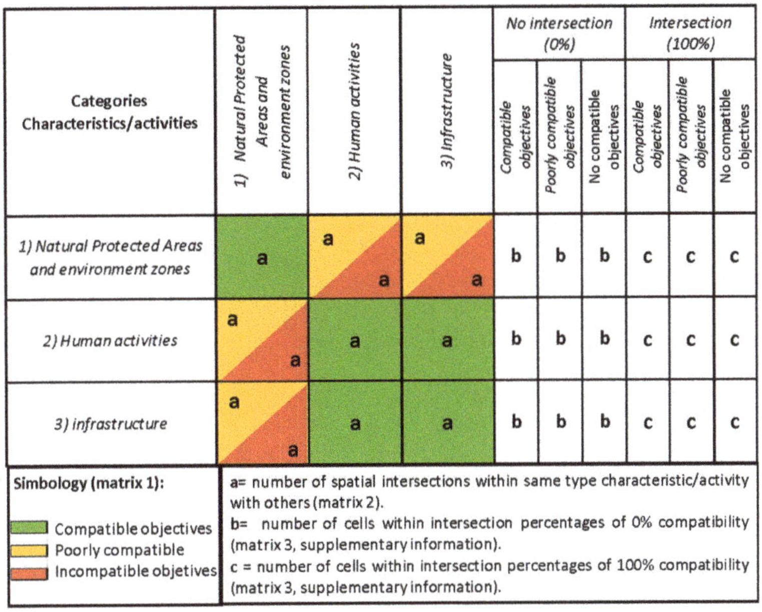

Categories Characteristics/activities	1) Natural Protected Areas and environment zones	2) Human activities	3) Infrastructure	No intersection (0%)			Intersection (100%)		
				Compatible objectives	Poorly compatible objectives	No compatible objectives	Compatible objectives	Poorly compatible objectives	No compatible objectives
1) Natural Protected Areas and environment zones	a	a	a	b	b	b	c	c	c
2) Human activities	a	a	a	b	b	b	c	c	c
3) infrastructure	a	a	a	b	b	b	c	c	c

Simbology (matrix 1):

■ Compatible objectives
■ Poorly compatible
■ Incompatible objetives

a= number of spatial intersections within same type characteristic/activity with others (matrix 2).
b= number of cells within intersection percentages of 0% compatibility (matrix 3, supplementary information).
c = number of cells within intersection percentages of 100% compatibility (matrix 3, supplementary information).

Figure 4. Summary of matrices that show the degree of compatibility between the characteristics/activities objectives in the Mexican Caribbean represented by colors: green for compatible, orange for poorly compatible and red for incompatible objectives (matrix 1). Number of spatial intersections (matrix 2), and number of cells with intersection percentages of 0 and 100% (matrix 3 in Supplementary materials).

The information was organized into databases in.cvs format for specialization in a GIS, by category and by municipality, using ArcMap 10.4 software. Subsequently, a digital map of the study area was produced showing the spatial distribution of electricity needs (% of dwellings without electricity) and the degree of social marginalization of the municipalities. Regarding settlements, 34 were found to be less than 10 km from the coast, with more than 100 inhabitants, where the implementation of ocean energy technologies would be beneficial.

3. Results

3.1. Analysis of Existing Conditions in the Mexican Caribbean

3.1.1. Protected Areas and Areas of Environmental Importance

(a) Protected Areas

82% of the study area is under legal protection, 77% is marine territory and 5% terrestrial. There are six types of protected areas: federal Natural Protected Areas (NPAs), state NPAs, Voluntary NPAs, fisheries refuge zones (FRZs), refuge areas for the protection of aquatic species (RAs) and protected beaches (Figure 5). In total, 77 protected areas were recorded in the area (Table 2). Some of these overlap, so that in the same location more than one legal regulation is applicable, as is the case of refuge areas that are in federal NPAs or FRZs.

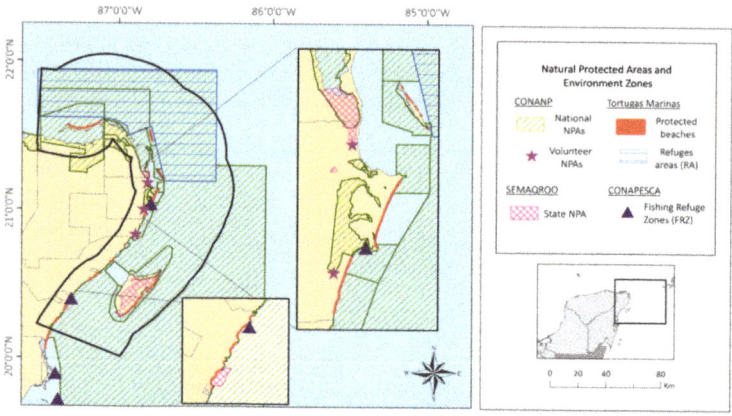

Figure 5. Protected Areas.

81% of the study area is part of a federal NPA, protecting 92% of the marine territory; the rest of the territory is administered by the Law of National Waters and other laws, such as the General Law of Ecological Balance and Environmental Protection (Ley General de Equilibrio Ecologico y Protección al Ambiente, LGEEPA). Refuge Areas (RA) cover 32% of the area, with their objectives focused on one or more species. There are two marine RAs in the study area: one for the protection of the Whale Shark (north) and one for the protection of Akumal Bay marine species (south). Both are located within a Federal NPA, and so their protection programme is governed by the corresponding NPA Management Plan; Article 68 of the General Wildlife Law. The rest of the areas (state and voluntary NPAs; PRZs and protected beaches) make up less than 2% of the study area, for example, state NPAs cover 1.32%, with the largest covering the forests and wetlands of Cozumel.

The objectives of the Fisheries Refuge Zones (FRZ) are to conserve and promote the reproduction, growth and recruitment of fishery resources [41]. In the north these are: (1) the FRZ Canal Nizuc, in the municipality of Benito Juárez and (2) the FRZ Akumal, off the coast of Tulum. In the first, no commercial, didactic, promotional, sport-recreational or fishing for self-consumption is allowed of any species or aquatic flora and fauna. In the second, fishing is allowed periodically, with specific fishing gear for commercial, sport-recreational fishing, or for self-consumption.

Finally, voluntary NPAs and protected beaches together represent only 0.04% of the study area. NPAs of this type are private properties, where the owners are interested in conservation. In accordance with the law established by SEMARNAT, they are granted a certificate. In the study area, these NPAs are terrestrial and owned by community landowners, or "*ejidatarios*". Protected beaches, on the other hand, are nesting sites for sea turtles and are coordinated by a turtle camp, generally managed by CONANP or a civil organization, under NOM-059-SEMARNAT-2010. They are guarded and monitored during the nesting season (if they are not within an NPA).

(b) Areas of Environmental Importance

These are defined following studies involving national and international agencies, with the aim of conserving and maintaining the connectivity of sites considered a priority for hosting ecosystems and wildlife under threat. In the study site, three types of environmentally important areas were identified: Sites of Marine Priority (SMP), Important Bird Areas (IBAs) and RAMSAR sites (Figure 6). These cover 28.5% of the study area, mainly marine territory (Figure 5).

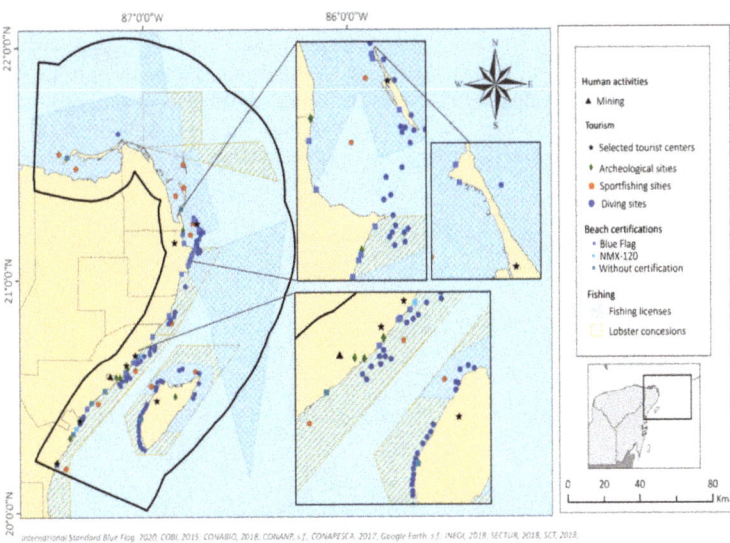

Figure 6. Areas of environmental importance.

The SMPs were defined in 2005, based on the Priority Marine Regions (PMRs) [42,43]. They are the result of collaborative work between various governmental organisations and civil associations (CONABIO, CONANP, Pronatura and TNC), and aim to conserve sites of coastal and marine biodiversity in Mexico. There are 11 SMPs in the study area, of which two are entirely marine and nine are coastal, covering 24% of the study area. The "Coastal Wetlands and the Continental Shelf of Cabo Catoche", SMP 68, is the largest in the area located in the municipalities of Lázaro Cárdenas and Isla Mujeres (continental zone).

IBAs are part of an internationally initiated project that aims to create a regional network of important areas for bird conservation [44]. Some are the result of collaboration between governmental bodies and civil organisations (CIPAMEX, CONABIO and SEO Birdlife). Seven IBAs are located in the study area, and it is worth noting that nearly half of the bird species recorded in Mexico (483) have been recorded in Quintana Roo [45]. The seven IBAs cover 9.7% of the study area, mostly terrestrial (60%). The entire islands of Cozumel and Contoy are IBAs (Figure 6).

The RAMSAR sites are recognised internationally by the "RAMSAR Convention" as being of international importance, and their objective is the conservation and wise use of wetlands. All RAMSAR sites in the country have been designated as federal NPAs, which is why they were considered as sites of environmental importance (since there is no exclusive legislation for wetlands in the country). Eight RAMSAR sites are located in the study area (Figure 6).

As with the protected areas, some of the sites of environmental importance overlap spatially, which is why these areas could be considered to be more important, such as the islands of Cozumel and Contoy, declared SMPs, IBAs and RAMSAR sites. 82% of the surface area of environmentally important sites is a protected area.

3.1.2. Human Activities

(a) Tourist activities

Tourism is the activity that generates most employment and income in the state. It is the core of the Mexican Caribbean economy and demand for energy here is high. The region is known as the "Riviera Maya", and has fourteen established tourist destinations on the coast of Quintana Roo, of which six are located in the study area. These include Cancun,

Mexico's main "sun and beach" destination, and are known as Selected Tourist Centres (STCs). They each have one, or more, of the following conditions: (1) over 2000 hotel rooms, (2) permanent or periodically significant tourist inflows, (3) participation in the "Mundo Maya" or "Centros de Playa" development programmes, and (4) are part of a tourist complex planned by the Fondo Nacional de Fomento al Turismo (Fonatur). In 2018, the six STCs in the study area received over 12 million tourists, 70% to Cancun [46] (Figure 7).

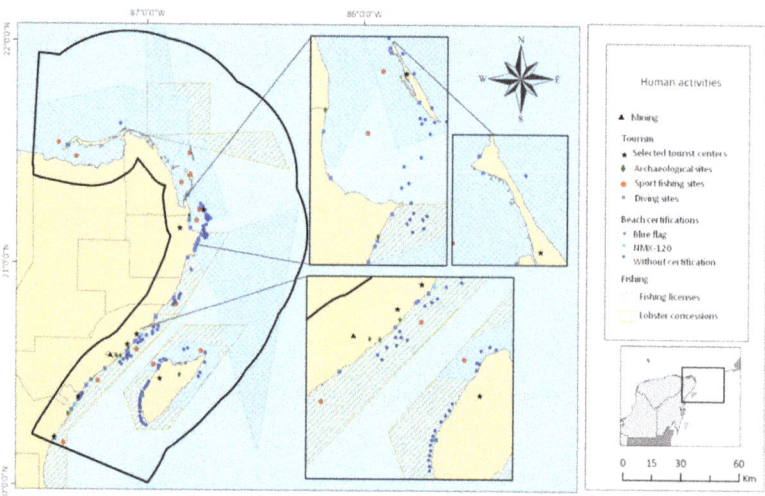

Figure 7. Human Activities.

The STCs are also classified as "Beach Centres", as the "sun and beach binomial" is the fundamental feature of recreation and leisure in the study area [47]. The coastline of the study area varies from solitary beaches to modern architectural complexes, including hotels and marinas, where a range of leisure and recreational activities take place [48].

Beaches with and without certification were considered. In the study area, there are 23 certified beaches. Of these there are two types of certification: Blue Flag, an international distinction awarded to beaches for their environmental education and information, water quality, environmental management, safety and services; and secondly, the "Playa Limpia Sustentable" certification, of the Mexican government, NMX-120, which defines the environmental quality, health, safety and services for beach sustainability.

There are also 10 beaches without certification that are registered as "sites of tourist interest" by the Mexican Institute of Transport in its National Road Network. (http://189.254.204.50:83/ accessed on 1 November 2020). These are mostly in the south of the study area, between Akumal and Playa del Carmen (Figure 7). Of the certified beaches, 43% are located in Cancun, and 87% coincide with one of the six STS within the study area.

Diving in the study area is a high value activity that is growing in popularity. It is closely related to the Mesoamerican Barrier Reef System. One of the dive magazines most internationally circulated, "Scuba Diving Mag", describes the Mayan Riviera as a world-class diving destination, with Isla Mujeres and Cozumel defined as "Best Diving Destinations" in the categories l arge animals and advanced diving, respectively [48–50]. Most dive sites are found off the coasts of Cozumel (34%), Solidaridad (19%) (mainly off Playa del Carmen), Akumal (13%), Benito Juárez (13%), Isla Mujeres (11%) and Puerto Morelos (7%) (Figure 7). Since there is no official repository of dive sites, the maps provided by diving companies in the study area were used to identify 90 dive sites. All of these are located within an ANP, so although the dive service is private, it is regulated by CONANP.

Sport fishing is a very important activity in the region and is regulated by CONAPESCA and SEMARNAT. In 2017, 3373 permits were granted in Quintana Roo, representing just over half a million Mexican pesos as revenue for the state. The state was ranked nationally in seco nd and fif th place, respectively, for number of permits and amount of money collected [34]. Annually, Quintana Roo hosts at least 25 tournaments, 21 of which are of international stature. According to the 2020 calendar published by CONAPESCA, 26 tournaments were planned in the state, of which 58% would be within the study area [32,51]. The study site has 13 of the 15 sport fishing locations reported in the state.

Finally, there are nine archaeological zones (A.Z.) in the study area, of which Tulum is by far the most visited; in fact it is the third most important A.Z. in Mexico. the National Institute of Anthropology and History (INAH) records the number of visitors to archaeological sites, and figures for seven of the nine sites in 2018 are seen in Table 4.

Table 4. Number of visitors to archaeological sites 2018.

Rank	Archaeological Zone	Municipality	Number of Visitors
1	Tulum	Tulum	2,189,536
2	San Gervasio	Cozumel	203,042
3	Mayan Museum of Cancun and Z. A	Benito Juarez	72,302
4	El Rey	Benito Juarez	20,975
5	El Meco	Benito Juarez	15,074
6	Xelhá	Tulum	2949
7	Xcaret	Solidarity	134 *
8	Calica		0 *
9	Playa del Carmen		0 *

* Lack of registration of tourists at the site (Pers. Comm., 2021).

(b) Fishing

In the study area, fishing is both coastal and oceanic. In 2017, in Quintana Roo, 3800 tones were fished, with grouper being the species with the highest volume reported, 530 tones, followed by lobster, 490 tones. The total value of the production that year was almost 200 million Mexican pesos. There are 2910 fishermen, 889 coastal vessels, 29 larger vessels, and 10 fishing plants registered in the state [34].

In Quintana Roo there are nine areas where fishing is permitted and all of them converge in the authorized environmental buffer zones of a natural protected area, so they must comply with the provisions of the LGEEPA and the Management Plan of the respective NPAs. There are 24 permitted fishing zones in force in the study area, 92% are based in Puerto Juárez, 4% at the dock in Cozumel, and 4% at the dock in Isla Mujeres. 67% of these permits are for fishing. 67% of permitted fishing is for deep-sea shrimp, 13% for Caribbean lobster and sea scales, 4% for lobster and 4% for octopus.

CONAPESCA has extended the concessions for fishing cooperatives to carry out commercial lobster fishing, which has become a growing market in Quintana Roo.

3.1.3. Infrastructure

(a) Port infrastructure

In the study area there are 20 ports for tourist, industrial, fishing, military and ferry terminals, as well as 39 piers and 11 marinas. In 2018, Quintana Roo had eight registered tourist marinas, with facilities for pleasure boats or yachts, both public and private (SCT 2014) and twelve berths for tourist cruise ships and ferries [46]. Six ports have cruise ship docking infrastructure, three of which are on the island of Cozumel, the most important tourist port in Mexico in terms of the number of ships that dock. In 2019 1366 cruisers arrived, carrying 4,569,853 tourists [52]. Puerto Morelos is deemed a tourist destination with port infrastructure, but it is exclusively for cargo, although there are plans to expand it, to receive cruise ships.

Connections between the islands and the mainland are needed for the transport of people, both tourists and workers. 10 ferry terminals exist, among which the most important routes are: Chiquilá-Holbox; Cancún-Isla Mujeres; Isla Mujeres-Isla Contoy; and Playa del Carmen–Cozumel (Figure 8). According to estimates from the port administration agency [53], approximately 11 million passengers, tourists and local inhabitants [52], used these ferries (Table 5). Given the uneven economic activity associated with COVID-19 mobility restrictions, the data for 2020 is poorly representative of normal maritime mobility in the study area.

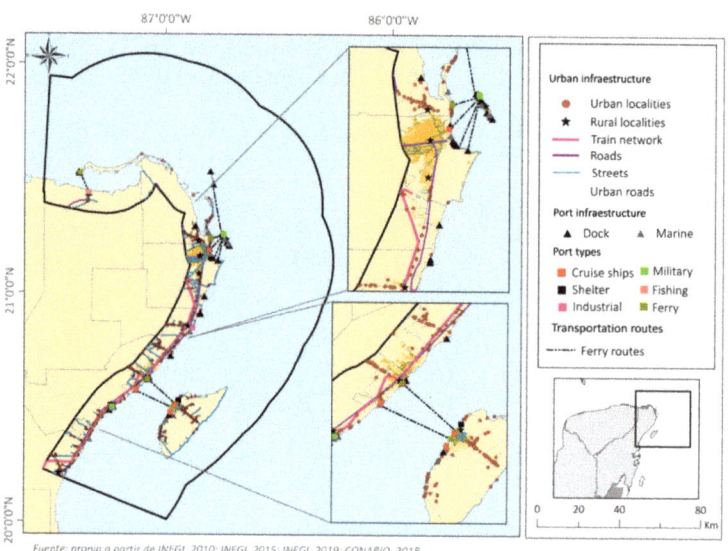

Figure 8. Urban infrastructure.

Table 5. Maritime passenger movement.

	2015	2016	2017	2018	2019	2020	2021
Cruises							
Cozumel							
Passengers	3,403,414	3,645,576	4,106,849	4,299,871	4,578,142	1,132,101	652,007
Arrivals	1079	1116	1243	1298	1366	371	354
Playa del Carmen							
Passengers	1471	NA	NA	NA	NA	NA	NA
Arrivals	1	NA	NA	NA	NA	NA	NA
Ferry							
Cozumel	1,755,906	2,041,156	2,472,927	2,510,352	2,510,461	164,888	177,019
Isla Mujeres	2,020,530	2,296,185	2,496,192	2,831,118	2,935,381	85,181	147,950
Playa del Carmen	1,580,271	1,822,305	2,249,636	2,241,087	2,185,722	NA	NA
Puerto Juarez	1,912,346	2,184,002	2,319,399	2,707,187	2,757,898	NA	NA
Punta Sam	111,770	123,651	166,582	133,431	239,030	71,434	135,287
Punta Venado	128,514	138,072	153,380	160,612	202,440	NA	NA

Source: CruisSource: Cruise and Ferry Data from 2015 to 2019 own elaboration with information obtained from [52,54,55]. Cruise and Ferry data for 2020 to 2021 were obtained from [53,55]. NOTE: worldwide, cruise activity was paralysed from January to May 2021, due to the Covid-19 pandemic [55].

(b) Urban Infrastructure

The study area contains 917 localities, with 1.4 million inhabitants, only nine of which are urban: Cancun, Playa del Carmen, Cozumel, Tulum, Puerto Aventuras, Alfredo V.

Bonfil, Puerto Morelos, Isla Mujeres and Zona Urbana Ejido Isla Mujeres. However, 98.7% of the state population live in these centres. Similarly, of the total number of inhabited private dwellings, 90% are in these areas (Table 6) [39].

Table 6. Rural and urban localities within the study area.

	No. of Locations	Total Population	Homes
Rural	908	20,533	5948
Urban	9	1,393,346	443,546
Total	917	1,413,379	449,494

Source: own elaboration with information obtained from [39].

A motorway runs along the coastline in the study area, with occasional intersecting roads giving access to the sea and inland. The main roads connect Cancun, Playa del Carmen and Cozumel, where the major tourist developments are found. From the end of the 19th century to the middle of the 20th century, railway lines were built mainly for commercial transport, but they are all abandoned nowadays. Currently the "Tren Maya" a railway line linking Cancun and Chetumal, and these two cities with neighbouring states, is under construction, due to be completed in 2023.

3.2. Compatibility Analysis of Existing Activities in the Caribbean Sea

The results of the analysis showing compatibility and the number of spatial intersections between the objectives of the 21 activities which currently occur in the study area are shown in Table 7. Activities related to protected areas/areas of environmental importance have objectives which are compatible with those in archaeological sites and certified beaches. These are shown in green.

Poorly compatible objectives (yellow) are seen between the activities of protected areas/areas of environmental importance with those of socio-economic activities, such as tourism and fishing (Figure 9).

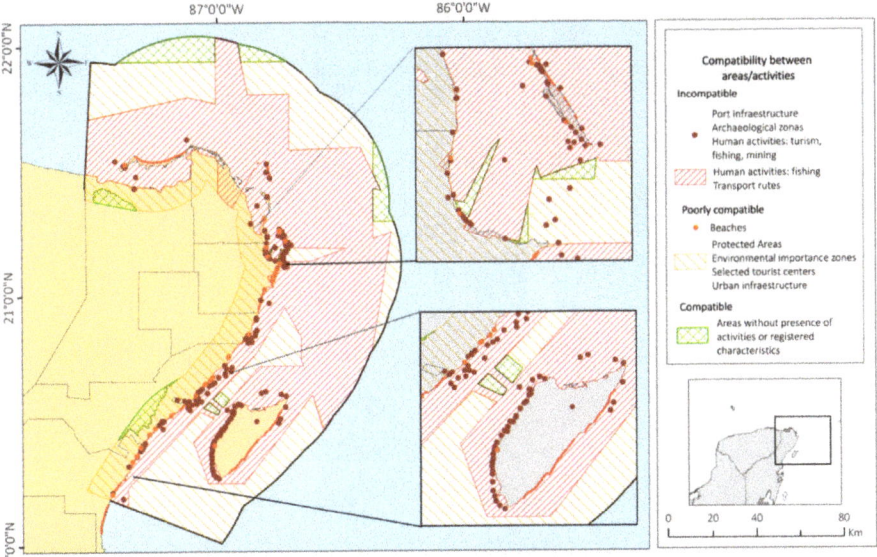

Figure 9. Compatibility Map.

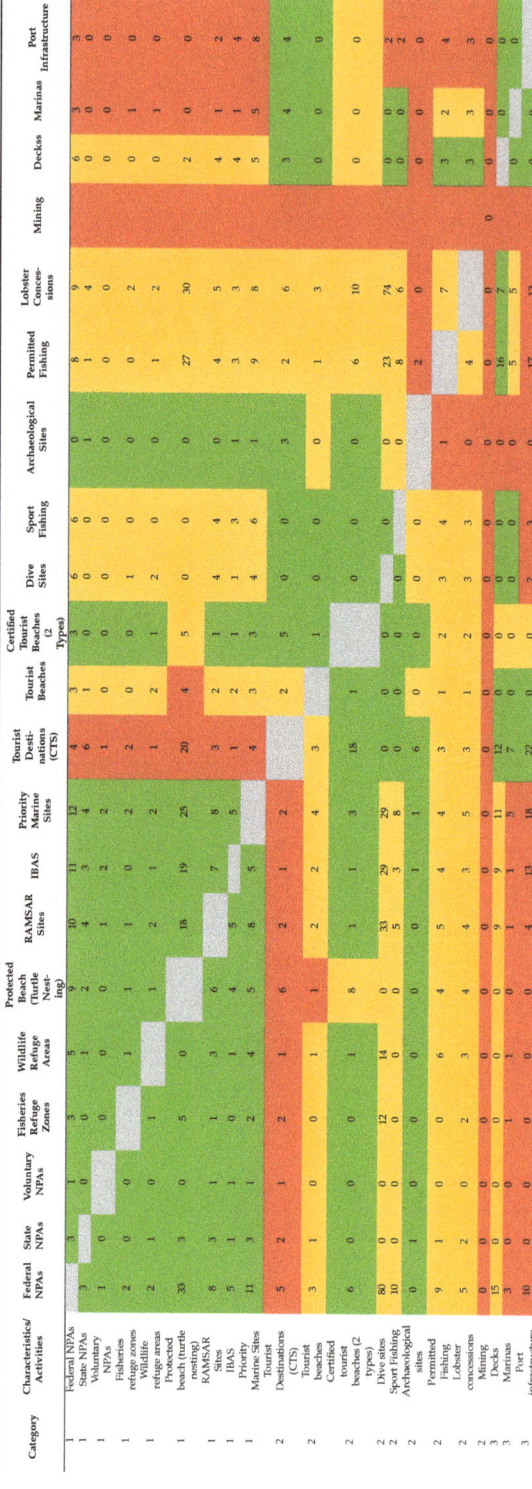

Table 7. Interaction between compatibility (colours) and the number of spatial intersection (represented by "a" in Figure 4) between characteristics/activities. The colours represent the degree of compatibility between the objectives of the characteristics/activities. Compatible objectives (green), Poorly compatible objectives (yellow) and Incompatible objectives (red). Categories: 1: Natural Protected Areas and Environmental zones, 2, Human activities, 3. Infra structure.

Objectives that are non-compatible (red) are related to protected areas/areas of environmental importance with activities of tourist destinations, tourist beaches, marinas and other port infrastructure (Figure 9). Incompatibility of objectives was also found between tourist beaches and protected beaches (sea turtle nesting), as well as between port infrastructure, even piers and marinas, with other marine activities, such as diving and both commercial and sport fishing.

With geospatial analysis, intersections between the types of activity and land uses in a given area can be identified (Table 7, Figure 9). Table 8 summarises the percentage of spatial intersections between characteristics and activities into two types: no intersection (=0% intersections) or to some degree (>0% intersections), according to the degree of compatibility of the land use objectives.

Table 8. Summary of the results of the interaction between compatibility and spatial intersection (Table 7 and Supplementary Materials).

Type of Intersection	Degree of Compatibility		
	Compatible Objectives	Poorly Compatible Objectives	Non-Compatible Objectives
No intersection	42.4%	33.1%	49.5%
Intersection (>0 to 100)	57.6%	66.9%	50.5%
Total	100%	100%	100%

There are two wildlife refuge areas in the study area that intersect completely with the federal NPAs, the SMPs, the tourist destinations and the areas with lobster concessions. The objectives and activities of the first two are compatible, but of the third and fourth the objectives and activities are contradictory, and poorly compatible, respectively.

Regarding the port and urban infrastructure, there are 19 decks and 20 marinas for the berthing of maritime transport in the study area. For decks, 15 coincide with federal NAPs, 11 are in SMPs, 12 in tourist destinations, and 16 in areas with fishing permits. 66.9% (see Figure S1, Table 8) of the characteristics/activities of poor compatibility intersect in one or more occasions. Tourist destinations and fishing zones are highly compatible with port and urban infrastructure.

All Fishing Refuge Zones and Wildlife Refuge Areas intersect with Lobster Concession sites, whose management objectives are poorly compatible (2 and 3 times respectively, see Figure S1). The same is true for Permitted Fishing and Lobster Concessions (seven times, Figure S1), which intersect in their entirety with Federal NPAs, which they have poorly compatible management objectives.

50.5% of activities with non-compatible management objectives intersect on one or more occasions. For example, the total number of Fisheries Refuge Zones (2) were found to intersect with Selected Tourism Destinations (2), with which they have non-compatible management objectives (Figure S1).

It was also found that some characteristics/activities which are seen in large land and marine areas, such as federal NPAs, RAMSAR Sites and Marine Protected Sites have up to 100% intersections with other characteristics/activities in the categories: (1) protected areas and areas of environmental importance and (2) human activities (Figure S1). However, these intersections occurred with characteristics/activities with compatible conservation objectives.

3.3. Energy Needs of the Population

The municipality of Isla Mujeres is that with the highest percentage of households without electricity (6.05%) in the study area, followed by the Lázaro Cárdenas (3.58%) and Tulum (2.35%) municipalities [36]. The spatial relationship of energy needs with social marginalization shows that there is a high degree of marginalization in Lázaro Cárdenas, while the other municipalities have Low, or Very Low, degrees of marginalization [35] (Figure 10, Table 9).

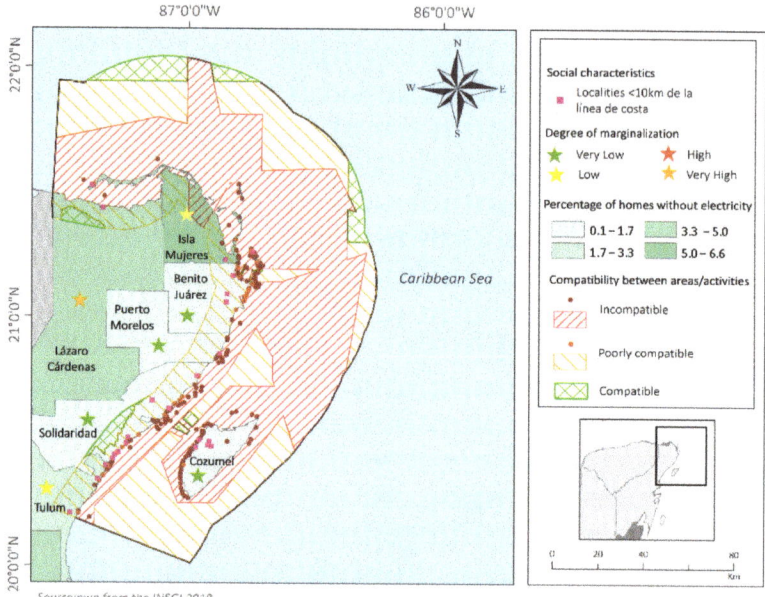

Figure 10. Percentage of households without electricity.

Table 9. Municipalities in Quintana Roo with the highest percentage of inhabited private dwellings without electricity and their degrees of marginalisation. Own elaboration with information obtained from [35,36,39].

Municipalities	Inhabited Private Dwellings	Homes without Electricity (%)	Degree of Marginalisation	Number of Localities <10 km from the Coast with >100 Inhabitants
Isla Mujeres	5889	6.05	Low	6
Lázaro Cárdenas	6991	3.58	High	2
Tulum	9385	2.35	Low	5
Solidarity	68,501	0.78	Very Low	11
Benito Juarez	221,950	0.78	Very Low	3
Cozumel	24,146	0.32	Very Low	7

The highest concentration of transmission grids, substations and power plants was found in the municipality of Benito Juárez, followed by Solidaridad and Cozumel. Isla Mujeres and Lázaro Cárdenas are the municipalities with fewest of these features.

Analysis of the criteria of priority (greatest social benefit) and proximity (close to the ocean) shows that in the three municipalities with greatest energy needs, Isla Mujeres, Lázaro Cárdenas and Tulum, there are 13 coastal localities with over 100 inhabitants that are less than 10 km from the coastline (Figure 10, Table 9).

4. Discussion

There is ample evidence of a future crisis in energy based on fossil fuels [1,56]. Added to this, greenhouse gas emissions, global warming and environmental contamination, increase the need to develop alternative technologies that allow us to harness energy from the ocean [3,56]. Energy from ocean currents is a promising option, given the widespread availability of the resource, worldwide [5,6]. In Mexico today, large coastal areas are still without electric energy supply, causing great socioeconomic inequalities [4,57].

There are many types of devices being developed to harness ocean energy [56], and the exploration of potential areas for energy harvesting must also be prioritised [27]. The ocean currents in the Mexican Caribbean were analysed to find the areas where conditions are most favourable for energy generation [4,57,58]. A subsequent geospatial analysis shows the great diversity in terms of human development and biodiversity in the study area. In 82% of the study area, 77 NPAs and areas of environmental importance exist, overlapping with each other. Legal regulatory instruments exist within these areas that establish the activities that drive economic development and the generation of services. In some cases, the objectives of the various activities that take place there are compatible, but in others this is not so and can mean that environmental degradation is likely in protected areas.

In regard to human activities, tourist activities in the region are very diverse, from beach visits to sport fishing and visits to archaeological sites. Limestone extraction, in one specific area, near Playa del Carmen, in the municipality of Solidaridad is a special case. Since 1986 the renovation of local, state and federal authorization has meant the mining activities were expanded every 15 years. However, since 2016 these permits have not been renewed, as the activity is incompatible with local regulations and with the 2009 ecological zoning policy of the municipality. In 2017 the environmental authorities partially closed down a coastal rock mining following inspections of the site after an application for a renovation of the permit was turned down. The main environmental damage caused by the mining is the complete removal of vegetation.

Regarding existing infrastructure, it is important that any infrastructure related to a marine energy generation project does not affect existing transport routes, nor put at risk the activities currently taking place in the area.

Section 5 of the "Tren Maya", the new rail line presently under construction, runs through the study area, connecting Tulum, Playa del Carmen, Puerto Morelos and Cancún. The infrastructure associated with the construction of the railway is severely modifying subsurface water flows that are of utmost importance for the preservation of coastal wetlands.

Socio-environmental conflicts are defined as "mobilizations of local communities, social movements, which may include support for national or international networks against particular economic activities, infrastructure construction or waste disposal/pollution, whereby environmental impact is a key element of their claims" [59]. In the study area there are several hotspots where there is already dissatisfaction with public policies associated with national level projects. Tourism and fishing depend on healthy ecosystems, as scenic beauty and biodiversity are the main factors in the development of these activities. On the other hand, these activities can produce negative environmental impacts if they are not properly regulated. As an example, the Atlas of Environmental Justice [59] describes the "recreational tourism" on Holbox Island, related to the "La Ensenada" project as a cause for concern. A luxury tourism development was planned which would entail 70 community landowners, or "*ejidatarios*" being dispossessed of their land with a payment of less than 5% of the value of the land. Although the project has been stopped and the conflict is reported to have ended in 2016, this area is vulnerable to this type of conflict due to its status as a tourist attraction [60].

The second area of conflict is related to the "extraction of minerals and construction materials", classified as medium intensity, described earlier. The extraction of limestone near Playa del Carmen from previously undeveloped areas of jungle has exceeding the limits authorised and affected the communities living in the region. The main impacts are pollution, loss of biodiversity, landscape degradation, water contamination [61].

As marine currents influence the distribution of commercially important species and impact on the marine and coastal environment, a current energy project could only be successful in an area where there are no repercussions for fishing, commercial and recreational. Bárcenas et al. [58] analysed the areas where marine energy harnessing would be feasible in part of our study area, considering environmental conditions and a range of floating and fixed devices. Their work, in front of Cozumel island suggests that the areas with greatest potential are at 30–50 metres depth. They also examined the NPAs, land use policies of the area (one regional, for the Gulf of Mexico and the Caribbean, and the other at local scale), shipping routes, infrastructure, tourist areas, military facilities, academic and research institutions, concluding that restrictions are needed in these areas and that it is important to take into consideration information from maps and internet pages for decision making regarding possible marine energy projects.

In Mexico the environmental impact assessment of marine energy power generation projects is still in process [62]. This work should be carried out prior to the installation of any marine energy project to minimize anthropogenic impacts. The areas with greatest need of electricity in the study area are in the north, Isla Mujeres, followed by locations in the interior and centre of Lázaro Cárdenas, and finally in Tulum, in the south (Figure 10, Table 9). Any project focused on improving socioeconomic change needs the participation of local actors for its acceptance, development and success. If it does not have this, socio-environmental conflicts may arise which limit the possible benefits of the project.

5. Conclusions

With geo-visualization tools, a diagnosis was made of areas in which marine energy generation is feasible as a way of ensuring environmental friendliness and economic viability. The extraction of marine energy through currents, would require a stable, safe physical space. Power generation projects must be able to be properly integrated into existing socio-environmental processes, for example, in the case of diving and sport fishing activities where safety must be guaranteed.

The development of human activities in the study area should be achieved by seeking a positive impact on the populations with the greatest need for electricity. The generation of electricity through ocean energy will have to coexist with economic activities both in the marine and terrestrial space.

The results of the present work provide a tool for spatial and marine planning that would enable the development and installation of such projects, that considers the needs of all, without forgetting that a potential energy generation project can be seen as an opportunity for socioeconomic development. This can serve as a basis for similar studies in other parts of the world, and to enable decision-makers and stakeholders in Mexico to make better use of the Mexican Caribbean's biological resources for a fairer society and a less polluted world.

Supplementary Materials: The following supporting information can be downloaded at: https://www.mdpi.com/article/10.3390/land11050665/s1. Figure S1: Matrix 3 for the calculation of intersection percentages of 0% and 100%, related to the frequency of intersections between the characteristics/activities objectives.

Author Contributions: Conceptualization, I.B.-O. and G.M.-G.; methodology, I.B.-O.; software, I.B.-O.; validation, R.S., formal analysis, I.B.-O., G.M.-G. and L.M.-P.; investigation, I.B.-O., G.M.-G., L.M.-P. and R.S.; resources, G.M.-G. and R.S.; data curation, I.B.-O. and L.M.-P.; writing—original draft preparation, I.B.-O. and G.M.-G.; writing—review and editing, I.B.-O., G.M.-G., L.M.-P. and R.S.; visualization, I.B.-O. and G.M.-G.; supervision, G.M.-G.; project administration, G.M.-G. and R.S.; funding acquisition, G.M.-G. and R.S. All authors have read and agreed to the published version of the manuscript.

Funding: This research was funded by the Mexican Centre for Innovation in Ocean Energy (CEMIE-Océano, CONACYT project 249795).

Institutional Review Board Statement: Not applicable.

Informed Consent Statement: Not applicable.

Data Availability Statement: Not applicable.

Acknowledgments: This study was supported by CEMIE-Océano. We also thank Maribel Badillo Alemán and Alfredo Gallardo Torres for the technical support in the Biological Conservation Laboratory.

Conflicts of Interest: The authors declare no conflict of interest.

References

1. Kühne, K.; Sanchez, L.; Roth, J.; Tornel, C.; Ivetta, G. *Beyond Fossil Fuels: Fiscal Transition in Mexico*; International Institute for Sustainable Development (IISD): Winnipeg, MB, Canada, 2019; p. 38.
2. Sener. *Reporte de Avance De Energías Limpias Primer Semestre 2018*; Secretaría de Energía del Gobierno de México: Ciudad de México, México, 2018; p. 21. Available online: https://www.gob.mx/cms/uploads/attachment/file/418391/RAEL_Primer_Semestre_2018.pdf (accessed on 18 March 2021).
3. Congreso de la Unión. Ley General de Cambio Climático (LGCC). In *Diario Oficial de la Federación-06-2012, última reforma DOF-06-11-2020*; Congreso de la Unión: Ciudad de México, México. Available online: https://www.diputados.gob.mx/LeyesBiblio/pdf/LGCC_061120.pdf (accessed on 19 March 2021).
4. Alcérreca-Huerta, J.C.; Encarnacion, J.I.; Ordoñez-Sánchez, S.; Callejas-Jiménez, M.; Barroso, G.G.D.; Allmark, M.; Mariño-Tapia, I.; Casarín, R.S.; O'Doherty, T.; Johnstone, C.; et al. Energy yield assessment from ocean currents in the insular shelf of Cozumel Island. *J. Mar. Sci. Eng.* **2019**, *7*, 147. [CrossRef]
5. Calero, R.; Viteri, D. Energía Undimotriz, alternativa para la producción de Energía Eléctrica en la Provincia de Santa Elena. *Rev. Cient. Tecnol. UPSE* **2013**, 1–12. [CrossRef]
6. Hucherby, J.; Jeffrey, H.; de Andres, A.; Finlay, L. *An International Vision for Ocean Energy, Version III: February 2017*; Ocean Energy Systems Technology Collaboration Programme: Paris, France, 2011; Volume 7, p. 28. Available online: www.ocean-energy-systems.org (accessed on 20 March 2021).
7. International Renewable Energy Agency. Renewable Energy Highlights—July 2020. Available online: https://www.irena.org/-/media/Files/IRENA/Agency/Publication/2021/Apr/IRENA_-RE_Capacity_Highlights_2021 (accessed on 21 May 2021).
8. Pérez, H.B.; Ramírez, J.C.C.; Andrade, M.Á.G.; Pulido, E.P.O. Evaluacion de una política de sustitución de energías fósiles para reducir las emisiones de carbono. *Trimest. Econ.* **2017**, *84*, 137–164. [CrossRef]
9. Noble, D.R.; O'shea, M.; Judge, F.; Robles, E.; Martinez, R.; Khalid, F.; Thies, P.R.; Johanning, L.; Corlay, Y.; Gabl, R.; et al. Standardising marine renewable energy testing: Gap analysis and recommendations for development of standards. *J. Mar. Sci. Eng.* **2021**, *9*, 971. [CrossRef]
10. Yang, X.; Haas, K.A.; Fritz, H.M. Theoretical assessment of ocean current energy potential for the Gulf Stream system. *Mar. Technol. Soc. J.* **2013**, *47*, 101–112. [CrossRef]
11. Hernández-Fontes, J.V.; Felix, A.; Mendoza, E.; Cueto, Y.R.; Silva, R. On the marine energy resources of Mexico. *J. Mar. Sci. Eng.* **2019**, *7*, 191. [CrossRef]
12. Hernández-Rodríguez, J.; Acosta-Olea, R.; Barbosa-Pool, G.R.; Aguilar-Aguilar, J.O.; Chargoy-Rosas, M.A.; Quinto-Diez, P. Indicadores de Desarrollo Energético Sustentable. Caso: Quintana Roo, México. *Quivera Univ. Autónoma Estado México* **2016**, *18*, 111–129.
13. Lozano, L.; 3 desafíos del sector eléctrico de la Península de Yucatán y cómo superarlos. El Financiero Península Home Page (Mérida, Yucatán, México). 28 de enero de 2019. Available online: https://www.elfinanciero.com.mx/peninsula/3-desafios-del-sector-electrico-de-la-peninsula-de-yucatan-y-como-superarlos/ (accessed on 15 May 2019).
14. Aldana, O.; Hernández, A. La Planificación Espacial Marina: Marco Operativo Para Conservar la Diversidad Biológica Marina y Promover el Uso Sostenible del Potencial Económico de los Recursos Marinos en el Caribe. In *Adaptación Basada en Ecosistemas: Alternativa para la Gestión Sostenible de los Recursos Marinos y Costeros del Caribe*; Instituto de Oceanología: La Habana, Cuba, 2016; pp. 1–15. ISBN 978-959-298-036-5.
15. Comisión Nacional para el Conocimiento y Uso de la Biodiversidad (Conabio). Geoportal del Sistema Nacional de Información sobre Biodiversidad Home Page. In *Subdirección de Sistemas de Información Geográfica.*. Available online: http://www.conabio.gob.mx/informacion/gis/ (accessed on 15 May 2021).
16. Rioja-Nieto, R.; Garza-Pérez, R.; Álvarez-Filip, L.; Mariño-Tapia, I.; Enríquez, C. The Mexican Caribbean: From Xcalak to Holbox. In *World Seas: An Environmental Evaluation*, 2nd ed.; Elsevier: Warwick, UK, 2018; Volume I: Europe, the Americas and West Africa, pp. 637–653. [CrossRef]
17. Comisión Nacional de Áreas Naturales Protegida (Conanp). *Programa de Manejo de la Reserva de la Biosfera Caribe Mexicano*; Comisión Nacional de Áreas Naturales Protegidas, Secretaría de Medio Ambiente y Recursos Naturales del Gobierno de México: Ciudad de México, México, 2018; p. 375. Available online: https://simec.conanp.gob.mx/pdf_libro_pm/191_libro_pm.pdf (accessed on 20 March 2022).
18. Sandoval Herazo, E.J.; Lizardi Jiménez, M.A. Hydrocarbons: Pollution at the Mexican Caribbean. *Rev. Digit. Univ.* **2019**, *20*, 1–10. [CrossRef]
19. Ardisson, P.L.; May-Kú, M.A.; Herrera-Dorantes, M.T.; Arellano-Guillermo, A. El Sistema Arrecifal Mesoamericano-México: Consideraciones para su designación como Zona Marítima Especialmente Sensible. *Hidrobiologica* **2011**, *21*, 261–280.
20. Comisión Nacional para el Conocimiento y Uso de la Biodiversidad (Conabio). Mares Mexicanos. Available online: https://www.gob.mx/semarnat/articulos/mares-mexicanos; http://www.biodiversidad.gob.mx/pais/mares/Nuestros; (accessed on 19 March 2021).
21. Wilkinson, T.; Wiken, E.; Bezaury Creel, J.; Hourigan, T.F.; Agardy, T.; Herrmmann, H.; Janishevsji, L.; Madden, C.; Morgan, L.; Moreno, P. *Ecorregiones Marinas de América del Norte*; Comisión para la Cooperación Ambiental: Montreal, QC, Canada, 2009; p. 200. ISBN 978-2-923358-72-7.

22. Badan, A.; Candela, J.; Sheinbaum, J.; Ochoa, J. Upper-layer circulation in the approaches to Yucatan channel. In *Circulation in the Gulf of Mexico: Observations and Models*; American Geophysical Union: Washington, DC, USA, 2005; Volume 161, pp. 57–69. Available online: https://ui.adsabs.harvard.edu/abs/2005GMS...161...57B/abstract (accessed on 20 March 2022). [CrossRef]
23. Chávez, G.; Candela, J.; Ochoa, J. Subinertial flows and transports in Cozumel Channel. *J. Geophys. Res. Ocean.* **2003**, *108*, 1–11. [CrossRef]
24. Abascal, A.J.; Sheinbaum, J.; Candela, J.; Ochoa, J.; Badan, A. Analysis of flow variability in the Yucatan Channel. *J. Geophys. Res. Ocean.* **2003**, *108*, 11. [CrossRef]
25. Sheinbaum, J.; Candela, J.; Badan, A.; Ochoa, J. Flow structure and transport in the Yucatan Channel. *Geophys. Res. Lett.* **2002**, *29*, 10-11–10-14. [CrossRef]
26. Athié, G.; Candela, J.; Sheinbaum, J.; Badan, A.; Ochoa, J. Yucatan Current variability through the Cozumel and Yucatan channels. *Cienc. Mar.* **2011**, *37*, 471–492. [CrossRef]
27. Silva, R.; Zúñiga, A.; Guimarais, M.; Barcenas, J.F.; Chávez, V.; Martínez, M.L.; Wojtarowski, A. Marine energy in the Mexican Caribbean: Needs and resources. In Proceedings of the SEEP2021, Boku, Vienna, Austria, 13–16 September 2021; pp. 600–605.
28. Hernández, M.L. Evaluación Del Riesgo y Vulnerabilidad Ante la Amenaza de Huracanes en Zonas Costeras del Caribe Mexicano: Chetumal y Mahahual. Ph.D. Thesis, Universidad de Quintana Roo, Chetumal, México, 2014; p. 398. Available online: http://repobiblio.cuc.uqroo.mx/handle/20.500.12249/98 (accessed on 20 March 2022).
29. Instituto Nacional de Estadística y Geografía (Inegi). Inegi Home Page. *Climatología; Mapas Climatológicos*. Available online: https://www.inegi.org.mx/temas/climatologia/ (accessed on 20 May 2021).
30. Garduño-Ruiz, E.P.; Silva, R.; Rodríguez-Cueto, Y.; García-Huante, A.; Olmedo-González, J.; Martínez, M.L.; Wojtarowski, A.; Martell-Dubois, R.; Cerdeira-Estrada, S. Criteria for optimal site selection for ocean thermal energy conversion (Otec) plants in Mexico. *Energies* **2021**, *14*, 2121. [CrossRef]
31. Intergovernmental Oceanographic Commission; Directorate General for Fisheries and Maritime, Affairs. MSPglobal: International Guide on Marine/Maritime Spatial Planning. 2021. Available online: https://unesdoc.unesco.org/ark:/48223/pf0000379196 (accessed on 4 April 2021).
32. Secretaría de Desarrollo Agropecuario Rural y Peasa. PESCA DEPORTIVA. 2018. Available online: https://qroo.gob.mx/sedarpe/pesca-deportiva/ (accessed on 1 December 2020).
33. Inegi. Censos Económicos 2014. 2014. Available online: https://www.inegi.org.mx/programas/ce/2014/ (accessed on 20 October 2020).
34. Conapesca. Anuario Estadístico de Acuacultura y Pesca. 2018. Available online: https://nube.conapesca.gob.mx/sites/cona/dgppe/2018/ANUARIO_2018.pdf (accessed on 23 November 2020).
35. Conapo. Índice de Marginación por Entidad Federativa y municipio 2015 | Consejo Nacional de Población | Gobierno | gob.mx. 2015. Available online: https://www.gob.mx/conapo/articulos/indice-de-marginacion-por-entidad-federativa-y-municipio-2020-271404?idiom=es (accessed on 9 November 2020).
36. Instituto Nacional de Estadística y Geografía. Encuesta Intercensal 2015. Available online: https://www.inegi.org.mx/programas/intercensal/2015/ (accessed on 20 March 2022).
37. Sener. *Programa Nacional para el Aprovechamiento Sustentable de la Energía 2014–2018. Avances y Resultados* **2018**, *42*.
38. Geocomunes, C. Geovisualizador-Alumbrar las Contradicciones del Sistema Eléctrico Nacional y de la Transición Energética | Geocomunes. Available online: http://geocomunes.org/Visualizadores/SistemaElectricoMexico/ (accessed on 6 November 2020).
39. Inegi. Catálogo Único de Claves de Áreas Geoestadísticas Estatales, Municipales y Localidades. 2020, pp. 1–8. Available online: https://www.inegi.org.mx/app/ageeml/ (accessed on 6 February 2021).
40. Hernández-Fontes, J.V.; Martínez, M.L.; Wojtarowski, A.; González-Mendoza, J.L.; Landgrave, R.; Silva, R. Is ocean energy an alternative in developing regions? A case study in Michoacan, Mexico. *J. Clean. Prod.* **2020**, *266*, 121984. [CrossRef]
41. Conapesca. Zonas de Refugio Pesquero Vigentes en México al 11 de Diciembre de 2019. 2019; pp. 1–5. Available online: https://www.gob.mx/cms/uploads/attachment/file/516926/ZRP_VIGENTES_191211__2_.pdf (accessed on 17 January 2021).
42. Conabio, The Nature Conservancy-Programa México, Pronatura. Sitios Prioritarios Marinos Para la Conservación de la Biodiversidad. 2007. Available online: http://geoportal.conabio.gob.mx/metadatos/doc/html/spm1mgw.html (accessed on 20 November 2020).
43. Semarnat. Regiones Marinas Prioritarias. 1998. Available online: http://dgeiawf.semarnat.gob.mx:8080/ibi_apps/WFServlet?IBIF_ex=D3_BIODIV01_14&IBIC_user=dgeia_mce&IBIC_pass=dgeia_mce (accessed on 20 June 2021).
44. Conabio. Regionalización. Áreas de Importancia para la Conservación de las Aves (AICAS). 2004. Available online: http://conabioweb.conabio.gob.mx/aicas/doctos/aicas.html (accessed on 20 April 2021).
45. Pozo, C.; Armijo, N.; Calmé, S. *Mexico. Comisión Nacional para el Conocimiento y Uso de la Biodiversidad. Riqueza Biológica de Quintana Roo: Un Análisis Para su Conservación*; Comisión Nacional Para el Conocimiento y Uso de la Biodiversidad (Conabio): Mexico City, Mexico, 2011; p. 2.
46. Sectur. Compendio Estadistico del Turismo en México. 2018. Available online: https://www.datatur.sectur.gob.mx/SitePages/CompendioEstadistico.aspx (accessed on 23 December 2020).
47. Sectur. Centros de Playa. 2014. Available online: https://www.sectur.gob.mx/programas/programas-regionales/centros-de-playa/ (accessed on 23 April 2021).

48. Uqroo. Tourism Competitiveness Study of the Riviera Maya destination. 2013; p. 548. Available online: https://www.sectur.gob.mx/wp-content/uploads/2015/02/PDF-Riviera-Maya.pdf (accessed on 15 June 2021).
49. Gobierno de Quintana, R. Quintana Roo, Mejor Destino de Buceo del Mundo. 2019. Available online: qroo.gob.mx (accessed on 17 November 2020).
50. Santander, L.C.; Propín, E. Impacto ambiental del turismo de buceo en arrecifes de coral. *Cuadernos Turismo* **2009**, *24*, 207–227.
51. Conapesca. Calendar of the Mexican Caribbean Sport Fishing Tournaments 2020. 2020. Available online: https://www.pescandoenelcaribe.com/torneos.html (accessed on 5 November 2020).
52. Sct. Movimiento Marítimo de Pasajeros por Tipo de Embarcación, Litoral y Puerto, Serie Anual de 2015 a 2019. 2019, p. 240. Available online: www.sct.gob.mx/fileadmin/DireccionesGrales/DGP/PDF/DEC-PDF/Anuario_2019.pdf (accessed on 2 March 2021).
53. Apiqroo. Postal de estadísticas. 2017. Available online: http://servicios.apiqroo.com.mx/estadistica/index.php (accessed on 2 April 2021).
54. Sectur. Movimiento de Cruceros en los Principales Puertos del país: Anual. 2021. Available online: https://www.datatur.sectur.gob.mx/SitePages/CompendioEstadistico.aspx (accessed on 14 April 2021).
55. Semar. Actividades en Crucero 2021–2022. 2022. Available online: https://www.datatur.sectur.gob.mx/SitePages/Actividades%20En%20Crucero.aspx (accessed on 15 April 2021).
56. Ibrahim, W.I.; Mohamed, M.R.; Ismail, R.M.T.R.; Leung, P.K.; Xing, W.W.; Shah, A.A. Hydrokinetic energy harnessing technologies: A review. *Energy Reports.* **2021**, *7*, 2021–2042. [CrossRef]
57. Garcia-Reyes, L.A.; Beltrán-Telles, A.; Bañuelos-Ruedas, F.; Reta-Hernández, M.; Ramírez-Arredondo, J.M.; Silva-Casarín, R. Level-Shift PWM Control of a Single-Phase Full H-Bridge Inverter for Grid Interconnection, Applied to Ocean Current Power Generation. *Energies* **2022**, *15*, 1644. [CrossRef]
58. Bárcenas Graniel, J.F.; Fontes, J.V.H.; Gomez Garcia, H.F.; Silva, R. Assessing hydrokinetic energy in the mexican caribbean: A case study in the cozumel channel. *Energies* **2021**, *14*, 4411. [CrossRef]
59. Ejatlas. Atlas de Justicia Ambiental. 2021. Available online: https://ejatlas.org/?translate=es (accessed on 24 March 2021).
60. Ejatlas. Proyecto la Ensenada en Holbox, México | EJAtlas. 2018. Available online: https://ejatlas.org/conflict/proyecto-la-ensenada-en-holbox-mexico?translate=es (accessed on 24 March 2021).
61. Ejatlas. MINA CALICA DE VULCAN Materiales Empresa EN PAPAA DEL CARMEN, MÉXICO | EJAtlas. 2020. Available online: https://ejatlas.org/conflict/devastacion-mina-calica-de-vulcan-materials-company-en-playa-del-carmen?translate=es (accessed on 24 March 2021).
62. IMP. Gaceta IMP. Available online: https://backend.aprende.sep.gob.mx/media/uploads/proedit/resources/gaceta_instituto_mex_7bdb5f2f.pdf (accessed on 18 March 2021).

Article

Implications of Spatio-Temporal Land Use/Cover Changes for Ecosystem Services Supply in the Coastal Landscapes of Southwestern Ghana, West Africa

Stephen Kankam [1,2,*], Adams Osman [3], Justice Nana Inkoom [1] and Christine Fürst [1,4]

1 Department of Sustainable Landscape Development, Institute for Geosciences and Geography, Martin Luther University Halle-Wittenberg, Von-Seckendorff-Platz 4, 06120 Halle, Germany
2 Hen Mpoano (Our Coast), 38. J Cross Cole Street, Windy Ridge Extension, Takoradi P.O. Box AX 296, Ghana
3 Department of Geography Education, University of Education, Winneba P.O. Box 25, Ghana
4 German Centre for Integrative Biodiversity Research (iDiv) Halle-Jena-Leipzig, Puschstraße 4, 04103 Leipzig, Germany
* Correspondence: stephen.kankam@student.uni-halle.de

Abstract: Land use/land cover change (LULCC) is an important driver of ecosystem changes in coastal areas. Despite being pervasive in coastal Ghana, LULCC has not been investigated to understand its effects on the potential for coastal landscapes to supply ecosystem services (ES). In this study, the impacts of LULCC on the potential supply of ES by coastal landscapes in Southwestern Ghana was assessed for the years 2008 and 2018 by using remote sensing and benefit transfer approaches. Based on available data, relevant provisioning and regulating ES were selected for the assessment while indicators to aid the quantification of the ES were obtained from literature. Supervised classification methods and maximum likelihood algorithms were used to prepare land use/land cover (LULC) maps and the derived LULC categories were assigned according to the descriptions of the Land Cover Classification System (LCCS). Potential supply of provisioning (food, fuelwood) and regulating (carbon storage) services was quantified and the spatial and temporal distributions of these ES illustrated using maps. The results show variations in food and fuelwood supply and carbon storage potentials over the study period and across different locations on the landscape. Potentials for fuelwood supply and carbon storage in mangrove forests indicated declining trends between 2008 and 2018. On the other hand, food-crop supply and carbon storage potential in rubber plantations depicted increasing patterns over the same period. Population, slope and elevation exhibited strong effects on LULC conversions to food crop and rubber plantations whereas these factors were less important determinants of mangrove forest conversions. The findings of the study have implications for identifying and addressing tradeoffs between land uses for agriculture, industrial development and conservation of critical coastal ES within the context of rapid land system transformations in the study region.

Keywords: ecosystem services; land use/land cover change; benefit transfer; coastal landscapes; quantification; spatio-temporal; West Africa; Ghana

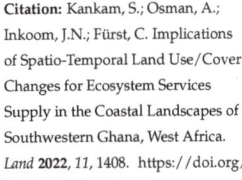

Citation: Kankam, S.; Osman, A.; Inkoom, J.N.; Fürst, C. Implications of Spatio-Temporal Land Use/Cover Changes for Ecosystem Services Supply in the Coastal Landscapes of Southwestern Ghana, West Africa. *Land* 2022, *11*, 1408. https://doi.org/10.3390/land11091408

Academic Editors: Pietro Aucelli, Angela Rizzo, Rodolfo Silva Casarín and Giorgio Anfuso

Received: 3 August 2022
Accepted: 22 August 2022
Published: 27 August 2022

Publisher's Note: MDPI stays neutral with regard to jurisdictional claims in published maps and institutional affiliations.

Copyright: © 2022 by the authors. Licensee MDPI, Basel, Switzerland. This article is an open access article distributed under the terms and conditions of the Creative Commons Attribution (CC BY) license (https:// creativecommons.org/licenses/by/ 4.0/).

1. Introduction

During the past half century, coastal zones have witnessed unprecedented transformation, due in part to the increasing impacts of human activities in these regions [1–3]. Urbanization patterns, natural resources exploitation, infrastructure development, industrial and commercial activities are concentrated on a narrow strip of land in the coastal zone [3]. In West Africa, rapid land use/land cover changes (LULCC) have gained prominence in the coastal zone as associated demographic, socioeconomic, technological and political drivers of change interact within the coastal socioecological system [4–6]. The accelerating pace of land use change coupled with the resulting impacts on coastal ecosystems has heightened concerns among planners, policy-makers and scientists about the sustainability of coastlines and associated ecosystem services (ES). Research on land use/land cover dynamics and related impacts on coastal ES is therefore gaining traction in scientific discourse (e.g., [6–9]). Global-scale assessments estimate annual ES losses due to land use changes at approximately $20.2 trillion [10,11]. Increasingly, individuals and societies respond to opportunities created by globalization processes, including market conditions by altering land uses [12,13]. These changes trigger degradation and conversion of high-value ecosystems such as forests, cropland, water and grasslands to low-value land uses [10,13]. Over the last two decades, ecosystem degradation have heightened due to an exponential increase in population and doubling of economic activities with attendant increase in the demand for ecosystem goods and services [2,4]. In many developing countries, weak or absent land use regulatory institutions are critical among the conditions giving rise to rapid modifications of ecosystems and landscapes [12]. While focusing assessments on Africa, the Intergovernmental Science-Policy Platform on Ecosystem Services (IPBES) identifies significant risks of ES and biodiversity losses in the face of poorly regulated and unregulated LULCC in the continent [14].

Nevertheless, there is also universal recognition that impacts of LULCC on ES are not always negative as society–nature interactions are integral to the processes of ES coproduction [15,16]. Indeed, landscapes' potential to supply food, fiber and fodder, and perform other functions such as climate regulation are inexorably linked to LULCC [17]. Thus, availability of land use/land cover data and related ES information underpin landscape planning and land use decision-making to sustain ES [18]. Decisions to protect ecosystems are taken expeditiously and aided by the availability of low-cost information [19]. However, application of field-based measurements for biophysical data acquisition and information gathering on ES is expensive and time consuming. Additionally, while field measurements are conducted at the local scale, land use decisions to protect ES are made at relatively larger scales [19]. In West Africa, challenges associated with lack of data at the appropriate planning scale and resolution hinder the conduct of ES studies and also their integration into land use decisions [20–22]. Where there is a paucity of data on specific ES, proxies have been utilized for mapping broad-scale trends in ES supply [23,24]. Applications of such proxy-based techniques for quantifying and mapping ES involve benefits transfer [25,26]. In benefits transfer, biophysical measures or economic estimates from a previously estimated site are extrapolated to another site with similar conditions [27–29]. A key element of benefit transfer applications is homogeneity in land cover characteristics between the site from where data is transferred and the site to which the transfer data is applied [18,19]. Thus, the existence of homogenous land cover types between sites facilitates the transfer of data to aid ES quantification [30]. Yet, such proxy-based data transfers result in error propagation from the transfer site [24,31]. Relatedly, benefit transfer applications enable the use of single point estimates or average values as a basis to transfer empirical data from one site in order to estimate ES values for another site [18,20]. Provisioning ES is amenable to quantification using benefits transfer as they represent long-standing economic sectors and traditional research areas such as agriculture, fisheries and forestry, for which a large body of datasets are available [19]. Similarly, "carbon sequestration", as a regulating ES, provides opportunities for quantification using biophysical units.

Quantification of ES is a useful process for raising awareness and providing insights about critical ecosystems in land use and spatial planning systems. However, in coastal areas, land use and spatial planning systems are challenged by complex and interrelated drivers of ecosystem changes. Particularly in the coastal landscapes of Southwestern Ghana, LULCC are consequences of an evolving oil and gas industry in the marine and coastal zones, population growth, urbanization and plantation agriculture development [32,33]. Increasingly, land losses from oil palm, cropland and shrubland favor gains in rubber plantation [34]. Nonetheless, rubber plantations are fragmented over the landscape as their establishment on few acres of land are determined by individual land owners in the context of an outgrower scheme [34]. Over the past decade, the region has been the focus of spatial planning and ecosystem-based management initiatives due to its importance for conservation and maintenance of a healthy small-scale fishery. Furthermore, the area is being explored for its potential contribution toward national climate change mitigation strategies such as REDD+ and other voluntary carbon offset programs [35]. However, with the advent of offshore oil and gas discovery in 2007, competition among land uses in Southwestern Ghana has intensified [33,36].

Recent impact assessments of LULCC in this region have focused analysis on the capacity for landscapes to supply cultural ES using participatory land use scenarios (e.g., [37]). Similar studies in the region also explored the provision of fisheries-related ES in support of decisions to establish marine protected areas (e.g., [38]). Nevertheless, the potential supply of provisioning and regulating ES by the coastal landscapes remains poorly understood. Meanwhile, such understandings are necessary to improve land use actors' awareness of potential critical losses of ES and of opportunities to sustain them or even increase the landscapes' potential to supply ES without producing tradeoffs. Potential ES supply is the maximum biophysically possible supply of a given ES in the absence of societal demand for, or benefits derived from, such services [39,40]. Coastal landscape boundaries are defined as the areas between 50 m below mean sea level and 50 m above the high tide level, or extending landward to a distance 100 km from shore [2].

To fill the aforementioned knowledge gap, this study investigates how LULCC in the coastal landscapes influence the quantities of provisioning and regulating ES supply and their spatial and temporal distribution over the landscape. It also explores how social and environmental drivers of LULCC affect ES supply potentials of the landscape. The implications of changes in ES supply potentials for land use planning in the region are discussed.

2. Materials and Methods

2.1. Study Area

This study was carried out in the Greater Amanzule Landscape located in Southwestern Ghana. This landscape falls within Ghana's Wet Evergreen Forest zone, which lies in the Upper Guinean Forest Ecosystem of West Africa. Covering approximately 60,000 ha, the landscape extends from the Ankobra River estuary, stretching to the Tano basin on Ghana's southwestern boundary with Cote d'Ivoire (Figure 1) [41]. The area is characterized by a bimodal rainfall regime, with peak rainfall occurring in May to June and October to November each year. Mean annual rainfall is 1600 mm with a relative humidity of 87.5% [42]. It encompasses a relatively pristine and vast expanse of coastal ecosystems comprising swamp forests, freshwater lagoons, rivers, mangrove forests, terrestrial forests, agricultural lands and grasslands. It is associated with a relatively high diversity of flora and fauna (237 species of plants, 27 species of mammals and 26 species of demersal fish) and known to be inhabited by most of Ghana's forest primate species [42,43]. The landscape traverses three district boundaries. It is a community-protected area and awaits official government designation as a conservation area. Farming is largely subsistence and a source of nutrition for the growing population. Increasingly, plantation agriculture, notably rubber and oil palm, are becoming attractive land use options for land owners and the agro-based private sector. Culturally, mangrove wood is the preferred fuelwood for smoking fish in traditional ovens [42]. The discovery of oil and gas in commercial quantities off the continental shelf in Southwestern Ghana ushered the region into a new wave of competition between industrial, residential and agricultural land uses [36]. This is manifested by the losses of farmland and forests in favor of built-up areas in the region's urban core and peripheries [36,44]. It is noteworthy that onshore oil and gas infrastructure is expanding into ecologically sensitive areas of this landscape, thereby causing further habitat fragmentation and threatening wildlife [45]. The population has doubled over the last decade and is increasing above the national average due to the region being a focal point for in-migration [44]. Historically, economic development in this area was driven by a vibrant fishing industry, but more recently the fisheries sector has suffered decline [46]. Similar to other coastal regions in

Ghana, the well-being of the local population is inexorably linked to natural resources which underpin their contentment with ES [47].

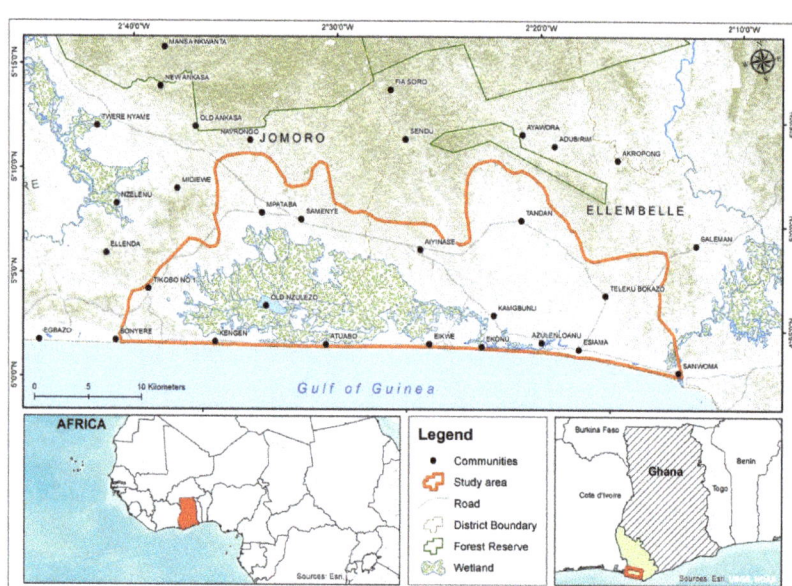

Figure 1. Map of the coastal landscapes of Southwestern Ghana showing the study area.

2.2. Methodological Framework

A stepwise and iterative process was utilized, as illustrated in the methodological framework to assess land use impacts on coastal ES (Figure 2). In the first step, we identified and selected relevant provisioning and regulating ES on the basis of available data and relevance to spatial planning in Southwestern Ghana. In the second step, we conducted land cover classification for the study landscape through application of remote sensing techniques. Thirdly, land use/land cover types were matched with ES, and finally, using benefit transfer approaches, the landscape's potential to supply provisioning and regulating ES was quantified. Results of the ES supply potential for the study landscape were spatially represented using 2000 and 2018 as temporal reference points.

2.3. Data Types and Sources

To enable assessment of land use impacts on ES supply potentials, a representative landscape of 362 km^2 was delineated on the basis of the following criteria: (a) representativeness of regional ecological (critical watersheds) and sociocultural characteristics (different land use intensities), and (b) availability of cloud-free satellite images for the assessment timeframe. We used two temporal reference points to depict important milestones in regional land uses in the study area, which in turn provided the basis for comparing land use impacts on the landscape potential to supply ES over time.

The study utilized two main remote sensing datasets (Landsat Thematic Mapper (TM) and Landsat Operational Land Imager (OLI)), which were acquired from the United States Geological Survey (USGS) web data repository. Landsat OLI was acquired for December 2016 and January 2018 and combined into a single image. The Landsat TM was acquired for February 2000 and January 2002. All images from Landsat sources had 30 m spatial resolution. Orthorectified images (at 5 m spatial resolution) were acquired for 2005 from the Ghana Geological Survey. Google Earth images were used for data verification. The study also relied on nonspatial data collected from published sources, gray literature and agricultural statistics. Assessment of the landscapes' potential to supply food-crop provisioning services was based on agricultural data collated by the Ministry of Food and Agriculture (MOFA) at the regional level. For estimating the landscapes' potential to supply fuelwood and sequester carbon, data from ecological surveys conducted in the study region and comparable ecosystems along Ghana's coast were utilized.

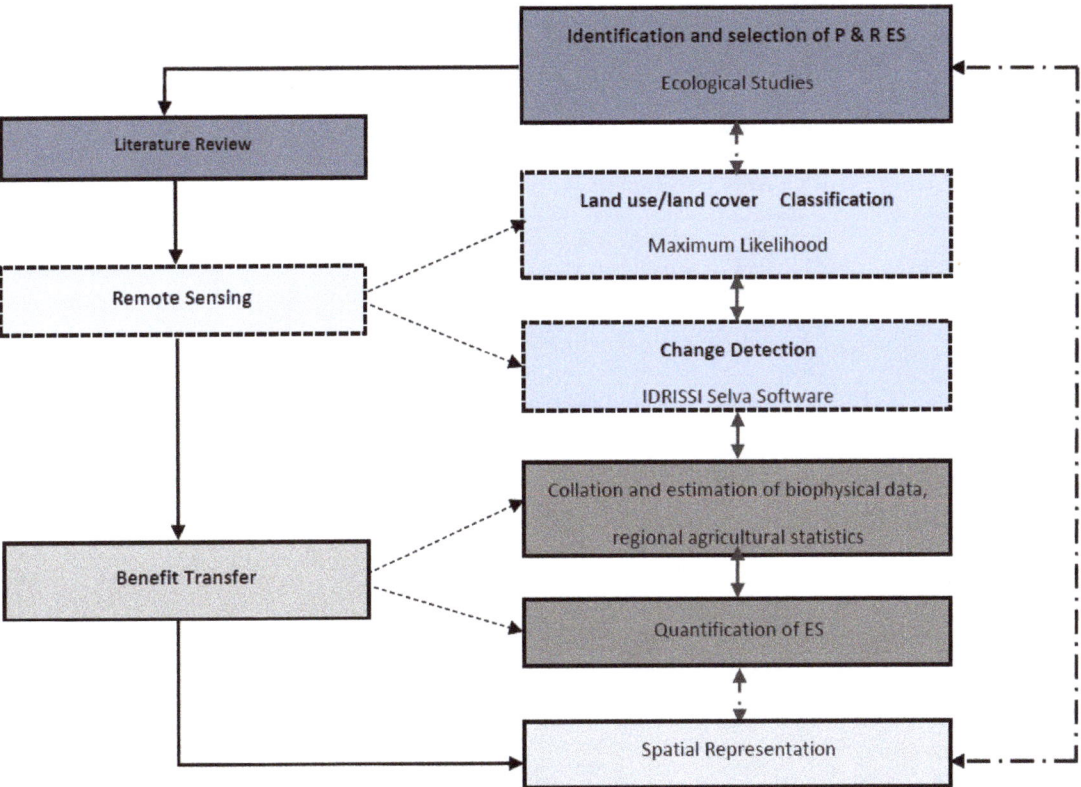

Figure 2. Methodological framework for assessing land use impacts on coastal ES in Southwestern Ghana.

2.3.1. Remote Sensing Data Processing and Analysis

Landsat 5 TM (Thematic Mapper) data from 2000/2002 and Landsat 8 OLI (Operational Land Imager) data from 2016/2018 provided by the United States Geological Survey (USGS) Earth Explorer database system were used for generating land use/land cover maps. All of the raw images were taken in the same season and nearly free of clouds. For each of these periods, tiles from 2 dates were selected and combined to provide a single image for the study area. All the processing and post-classification steps were completed using the software packages Erdas Imagine 2015 and ArcGIS 17.1. Prior to interpretation, image pre-processing including geometric and radiometric corrections was performed for each of the images. All of the data were geometrically corrected and projected to Universal Transverse Mercator (UTM) zone 30 N. After image pre-processing, supervised classification methods and maximum likelihood algorithms were used for preparing land use/land cover maps for two temporal reference points. The land use/land cover categories were assigned according to the descriptions of the Land Cover Classification System (LCCS), which is a hierarchical a priori classification scheme providing a flexible framework for identifying land use classes in highly heterogeneous landscapes such as those found in the study region [48]. Ten land cover classes were derived to match available data for ES quantification. Change analysis was conducted using the Land Change Modeler embedded in the IDRISI TerrSet software.

2.3.2. Selection of Provisioning and Regulating ES

Ecosystem services were selected from the list of land-cover-based proxies compiled in the literature and for mapping ES [17,49]. The types and sources of data and corresponding proxy indicators for ES quantification are presented in Table 1. The derived land cover classes (see Section 3.1) were the basis for representing ES supply with land cover types occurring in the study region [17]. Our proxy measures for carbon storage were based on data from primary ecological studies, which estimated aboveground carbon in multiple mangrove stands along Ghana's coast and also in rubber plantations [50–52]. Similarly, primary ecological studies that estimated aboveground tree biomass across mangrove stands were utilized as proxies for fuelwood supply. Food supply was based on land cover data combined with official agricultural statistics. Mangrove fuelwood was selected as it is a dominant source of fuelwood utilized in the local fishing industry, while plantain and cassava are the major staple food supply from cropland in the region. Additionally, mangroves store large quantities of carbon, and are increasingly receiving attention currently in Ghana's climate change mitigation strategies. In addition to latex production, rubber plantation development is arguably presented as a significant opportunity requiring inclusion in Ghana's climate change mitigation programs [53].

Table 1. Types and sources of data utilized for assessing ES in the coastal landscapes of Southwestern Ghana. LULC = land use/land cover types; USGS = United States Geological Survey; TM = Thematic Mapper; OLI = Operational Land Imager; MoFA = Ministry of Food and Agriculture; ES = ecosystem services; P = provisioning services; R = regulating services; Mg C_{org} = quantities of organic carbon stored in vegetation; - = not applicable.

Type of Data	Period	Sources of Data	Relevant LULC Types/ES	Proxy Indicator	Unit	References
Remote sensing	February 2000/2002	USGS Landsat TM/Landsat OLI	Mangrove, Rubber, Cropland	-	-	https://earthexplorer.usgs.gov/ (accessed on 12 August 2022)
	December 2016; January 2018					
Annual cassava and plantain yield	2000–2016	MoFA regional agricultural statistics	Food—P	Total crop yield	Tons	-
Mangrove forest stand biomass	2015, 2016	Ecological survey	Fuelwood—P	Total growing stock	Tons	[50,51,54]
		Ecological survey	Carbon storage—R	Carbon stored in aboveground vegetation	Mg C_{org}	[50,51,54]
Aboveground carbon in rubber tree stands	2017	Ecological survey	Carbon storage—R	Carbon stored in aboveground vegetation	Mg C_{org}	[52]

2.4. Benefit Transfer

Benefit transfer involves extrapolation of either biophysical measures or economic estimates from a previously estimated site to a study area of interest [27,28]. This is based on the assumption that spatial units are homogenous; hence, estimates from one area is transferable to the other [31]. We compiled biophysical values from ecological studies conducted in mangrove forests and rubber plantations along with crop yield estimates from regional agricultural statistics (Supplementary Table S1). Extrapolated mean values from the ecological studies and agricultural statistics were assigned to the corresponding land use/land cover types in GIS. This ensured that generalization errors were minimized and better correspondence was achieved in the biophysical characteristics between the previously estimated sites and the study landscape [55]. Using the image resampling tool in ArcGIS Pro, we resampled the 30 m × 30 m land cover data to hectares. In estimating the landscapes' potential to supply ES, we multiplied the mean values computed from the ecological studies and agricultural statistics by the resampled land cover data.

2.4.1. Quantification of Provisioning Services Supply Potentials

Mangrove Fuelwood

Mangrove forests are adapted to tropical and subtropical coastal environments [38,39]. There is widespread harvesting and utilization of mangroves as sources of fuelwood for fish smoking in the coastal areas of Ghana. Mangrove fuelwood is hereby defined as wood harvested from live trees and standing dead wood. The dominant mangrove species found in the study region are *Rhizophora mangle, Avicennia germinans* and *Laguncularia racemosa* [50,51]. Adotey [50] and Nortey et al. [51] utilized allometric equations derived from diameter at breast height (DBH) and height (H) to estimate aboveground (standing dead wood and live trees) biomass of mangroves found along four river estuaries (*Whin, Amanzule, Kakum* and *Nyan*) located on Ghana's western and central coasts. In this study, we estimated the potential of the landscape to supply mangrove fuelwood by calculating the mean aboveground biomass across all the four sites sampled by Adotey [50] and Nortey et al. [51] according to the formula:

$$M_{ABG} = B_{S1} + B_{S2} + B_{S3} + B_{S4}/N_s, \quad (1)$$

where M_{ABG} = mean aboveground biomass, B_{S1} = biomass at site 1, B_{S2} = biomass at site 2, biomass at site 3, B_{S3} = biomass at site 4 and N_s = number of sites. The mangrove fuelwood supply potential of the landscape was mapped in GIS and the results compared over the two temporal reference points.

Food Production

Staple food crops in Southwestern Ghana comprise cassava, yam, cocoyam, rice, maize and plantain. At the regional level, the Ministry of Food and Agriculture (MOFA) maintains a database of crop production, crop yield and area cultivated. Cassava and plantain comprise over 80% of total food-crop production in the region. In assessing the food-crop supply potential of the landscape, we extracted the staple crop yield statistics of the three districts—Ellembelle, Nzema East and Jomoro—that span the study landscape. The mean yield of cassava and plantain was estimated for the period 2000 to 2018 according to the formula:

$$MC_{yield} = [Y_1 + \ldots\ldots Y_n]/N_y prod, \quad (2)$$

$$MPl_{yield} = [Y_1 + \ldots\ldots Y_n]/N_y prod, \quad (3)$$

where MC_{yield} = mean cassava crop yield in tons^{-ha}, Y_1 = yield in the first year of production in tons^{-ha}, Y_n = yield in last year of production in tons^{-ha}, MPl_{yield} = mean plantain crop yield in tons^{-ha} and N_y = number of years of staple food-crop production.

The estimated mean yield of the major staple crops was multiplied by the area of rainfed cropland in the land use/land cover map to estimate the potential food supply in tons, as per the formula:

$$P_{fs} = M_{yield} \times A_{cropland} \quad (4)$$

where P_{fs} = potential food-crop supply and $A_{cropland}$ = area of cropland. The food-crop production potential of the landscape was mapped using GIS and the results compared over the two temporal reference points.

2.4.2. Quantification of Regulating Services Supply Potentials

Mangrove Carbon Storage

Mangrove ecosystems are globally recognized for their significant contribution to carbon cycling and sequestration [56–58]. Mangrove ecosystem carbon pools are stored in aboveground biomass, belowground biomass, litter and soil organic matter components [43,45]. While protection of mangroves contributes to attainment of climate change mitigation objectives, their conversion to other land cover types is a significant source of carbon emissions into the atmosphere. Carbon quantity stored in mangroves is estimated using allometric equations that relate biomass with parameters such as diameter at breast height, height and density of mangrove trees [59]. Adotey [50] and Nortey et al. [51] utilized species and site-specific allometric equations to quantify mangrove carbon stocks across sample plots established in mangrove ecosystems found in the Amanzule, Kakum, Nyan and Whin river estuaries located along the coast of Ghana. In this study, we derived proxy data from Adotey [50] and Nortey et al. [51] to quantify aboveground carbon stored in mangrove ecosystems found in the study landscape. We narrowed and focused on aboveground carbon pools as these are better reflected in the mangrove vegetation captured using satellite data. Per hectare aboveground mangrove biomass estimates from the abovementioned sample sites were summed and multiplied by a conversion factor [50] and the average determined according to the formula:

$$M_{AGC} = [B_{S1} + B_{S2} + B_{S3} + B_{S4}] \times 0.46/N_s, \qquad (5)$$

where M_{AGC} = Mean aboveground carbon; B_{S1} = biomass at site 1; B_{S2} = biomass at site 2; B_{S3} = biomass at site 3; B_{S4} = biomass at site 4; N_s = number of sites; 0.46 = conversion factor for tropical mangroves. The mangrove carbon storage potential of the landscape was quantified by multiplying the estimated mean aboveground carbon stored in mangroves (M_{AGC}) by the mangrove extent in the land use/land cover map using the formula;

$$P_{cs} = M_{AGC} \times A_{mangrove} \qquad (6)$$

where P_{cs} = landscape potential to store carbon and $A_{mangrove}$ = area of mangrove. The mangrove carbon storage potential of the landscape was mapped using GIS and the results compared over the two temporal reference points.

Rubber Carbon Storage

Growing market demand for natural rubber on the international market is partly driving expansion of rubber plantations in the tropics [60]. Because of land use competition between rubber plantation and tropical forestry, the potential role of rubber plantations in ecosystem services provisioning is gaining scholarly attention (e.g., [46,47]). Carbon in a rubber plantation is stored in aboveground and belowground biomass and in latex and soil [52]. The quantity of carbon stored in rubber varies with the age of trees in a plantation. Site specific allometric equations have been developed for estimating carbon sequestered in rubber plantations [53]. Using data collected from 25 sample plots in rubber plantations located in Ghana's western region, Tawiah et al. [52] integrated data on the age of rubber plantations, field-measured diameter at breast height and latex production in allometric equations to estimate aboveground, belowground and latex carbon. In this study, we utilized proxy values from Tawiah et al. [52] to estimate aboveground carbon for the study landscape according to the formula:

$$M_{AGC} = M_F + M_S + M_L \qquad (7)$$

where M_{AGC} = mean aboveground carbon, M_F = mean foliage carbon, M_S = mean stem carbon and M_L = mean latex carbon. Potential carbon storage in rubber plantation was estimated by multiplying the estimated mean aboveground carbon by the extent of rubber in the land use/land cover map using the formula;

$$P_{cs} = M_{AGC} \times A_{rubber} \qquad (8)$$

where P_{cs} = landscape potential to store rubber carbon and A_{rubber} = area of rubber.

2.5. Social and Environmental Drivers of LULCC

We utilized the Exploratory Regression tool in Arc GIS Pro to explore the relationships between pre-selected independent variables (elevation, slope, rainfall, fishing population, farming population, community population, distance from gas pipeline, distance from oil processing plant, distance from rubber processing facility, distance from river and distance from sea) and the dependent variables (land use/land cover transitions) in the study region. Pre-selection of independent variables was informed by literature on driving forces of LULCC in tropical regions [61]. The dependent variables were conversions to cropland, rubber plantations and mangroves since they constitute land use/land cover classes for the supply of relevant ES. Rapid landscape transformations emanating from industrial activities provided an additional basis for exploring the effects of the independent variables on conversions to artificial/bare areas [34].

Subsequently, we ran geographically weighted regression (GWR) to predict the effect of changes in the high-performing independent variables on land use/land cover outcomes. The model parameters for transitions to artificial/bare areas were elevation, slope, distance from road, fishing population, farming population and total resident population. Parameters for transitions to rubber plantation were elevation, slope, distance from road, total resident population and farming population. Parameters for transitions to cropland were elevation, slope and distance from road. The aim of the GWR was to explain the spatial variations in the relationships between the independent variables and the dependent land use/land cover transitions.

3. Results

3.1. Land Use/Land Cover Changes in 2000 and 2018

The land use/land cover maps shown in Figure 3 and the extent of changes in the land use/land cover classes depicted in Table 2 were generated as inputs for assessment of the selected ES. Next to wetlands, cropland dominated the land use/land cover situation in the study area, representing approximately 24 and 35% coverage in 2000 and 2018, respectively. Within the cropland category are cassava and plantain, which are the major staple crops in the region and "others" subclasses. Cassava dominated the cropland category, increasing from 10 to 15% of total land use/land cover in 2000 and 2018, respectively. This was followed by the "others" subclass, which occupied 9% of the area in 2000 and increased to 12% in 2018. Plantain occupied 5% and increased to 8% of total land use/land cover in 2000 and 2018, respectively. Grassland decreased markedly, from 11 to 4%, and shrubland/sparse vegetation reduced from 19 to 8% in 2000 and 2018, respectively. Artificial/bare areas and rubber plantation showed sharp increases from 7 to 11% and from 1 to 5%, respectively, in 2000 and 2018. However, mangrove cover remained relatively stable at approximately 1% in 2000 and 2018.

3.2. Quantities, Spatial and Temporal Distribution of ES Supply Potentials

3.2.1. Mangrove Fuelwood Supply

The landscape's potential to supply mangrove fuelwood ranged from a minimum of 0.01 tons to a maximum of 87.19 tons, as shown in Figure 4. For the two temporal reference points, marked differences in the spatial distribution of mangrove fuelwood supply potential were also observed across the landscape (Figure 4). The potential mangrove fuelwood supply was concentrated along the intertidal areas extending eastward from Sanzule and Essiama. Similarly, between 2000 and 2018, mangrove fuelwood supply potential was relatively high on the coastlines stretching eastward from Sanzule, while the supply remained relatively stable over these two temporal horizons for the same location, as depicted in Figure 4A. Comparatively, mangrove fuelwood supply potential sharply decreased along the coastal stretches westward and eastward from Essiama, as shown in Figure 4C. At these locations, the decrease in mangrove fuelwood supply potential was even more pronounced between 2000 and 2018, as illustrated by Figure 4B,C.

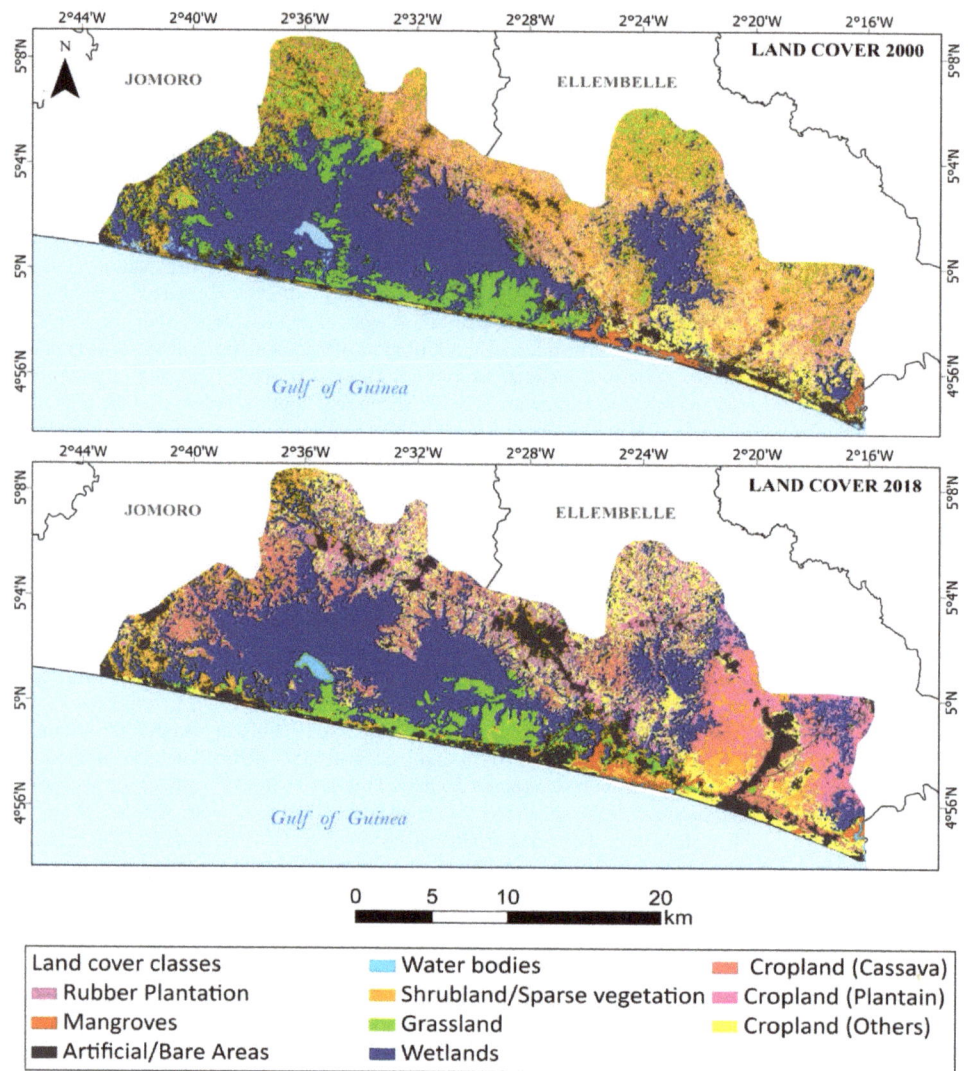

Figure 3. Land use/land cover types in the study area.

Table 2. Main land use/land cover types, their extent and description.

LULC Types	Extent (%)		Description
	2000	2018	
Mangroves	1.18	1.02	Coastal forests of stilted shrubs or trees bordering the ocean or coastal estuaries, composed of one or several mangrove species.
Wetlands	34.92	34.15	Herbaceous or aquatic vegetation in permanent or semipermanent swamps.
Rubber plantation	1.49	5.38	Regular stands of trees planted for the purpose of producing materials for industry.

Table 2. *Cont.*

LULC Types	Extent (%)		Description
Artificial areas/bare areas	7.02	11.27	Cover resulting from human activities such as urban development, extraction or deposition of materials. It comprises areas that are not covered by vegetation, such as rocky or sandy areas.
Grassland	10.71	4.04	Mixed mapping unit that consists of 50–70% grassland.
Shrubland/sparse vegetation	19.38	8.34	A class representing a mapping unit which contains 20–10% to 1% vegetative cover.
Water bodies	1.39	0.46	Areas covered by natural water bodies such as ocean, lakes, ponds, rivers or streams.
Cropland_Cassava	9.55	15.44	Mix crops and nonforest vegetation with cassava representing more than 90% of the cover.
Cropland_Plantain	4.94	7.62	Mix crops and nonforest vegetation with plantain representing more than 90% of the cover.
Cropland_Others	9.40	12.27	Mix crops and nonforest vegetation with croplands representing more than 50% of the cover.

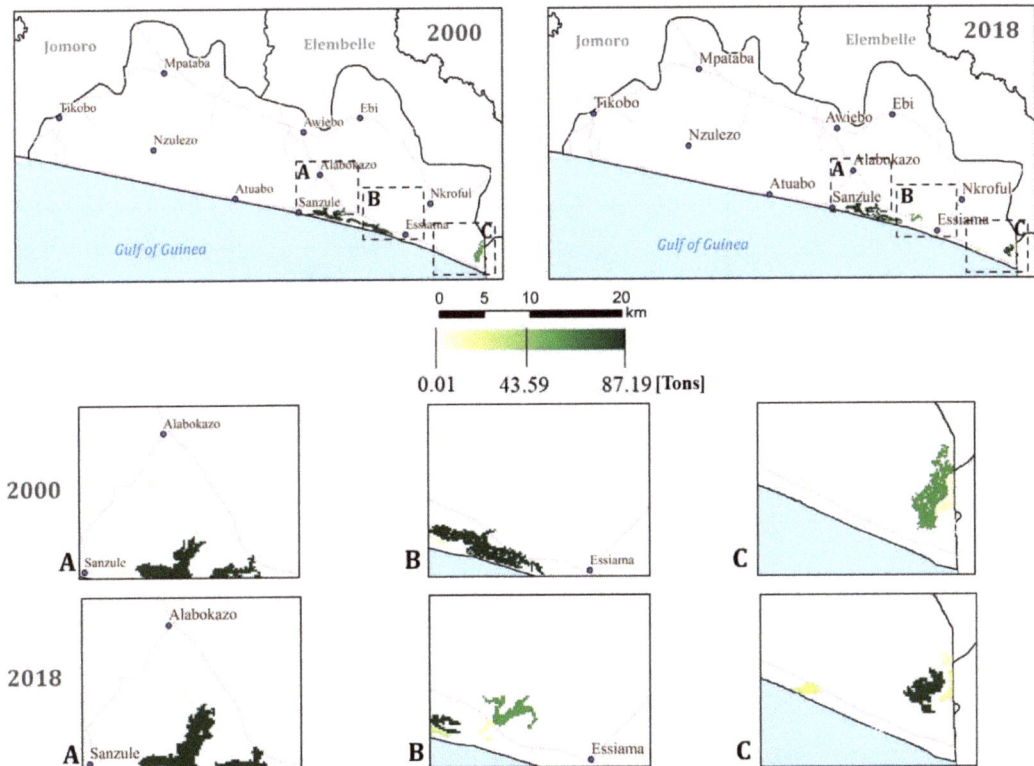

Figure 4. Quantities of mangrove fuelwood supply and distribution across the landscape in 2000 and 2018; (**A–C**) compares the spatial and temporal distribution of mangrove fuelwood supply potential in segments of the landscape.

3.2.2. Food-Crop Supply

Cassava Supply

The minimum and maximum potentials of the landscape to supply cassava food crop ranged from 1.7 to 23.4 Gt, as shown in Figure 5. Generally, the spatial distribution patterns of potential cassava food-crop supply showed skewness toward the southeastern, northeastern and northwestern portions of the landscape. Comparison of cassava food-crop supply potential between the two temporal reference points also showed higher potential supply in 2018 than 2000, as depicted in Figure 5A–D. Furthermore, during 2018, cassava supply potential was more spatially concentrated within the northern sections of the landscape and in areas within close proximity to road networks and major towns such as Tikobo, Nkroful, Alabokazo and Essiama.

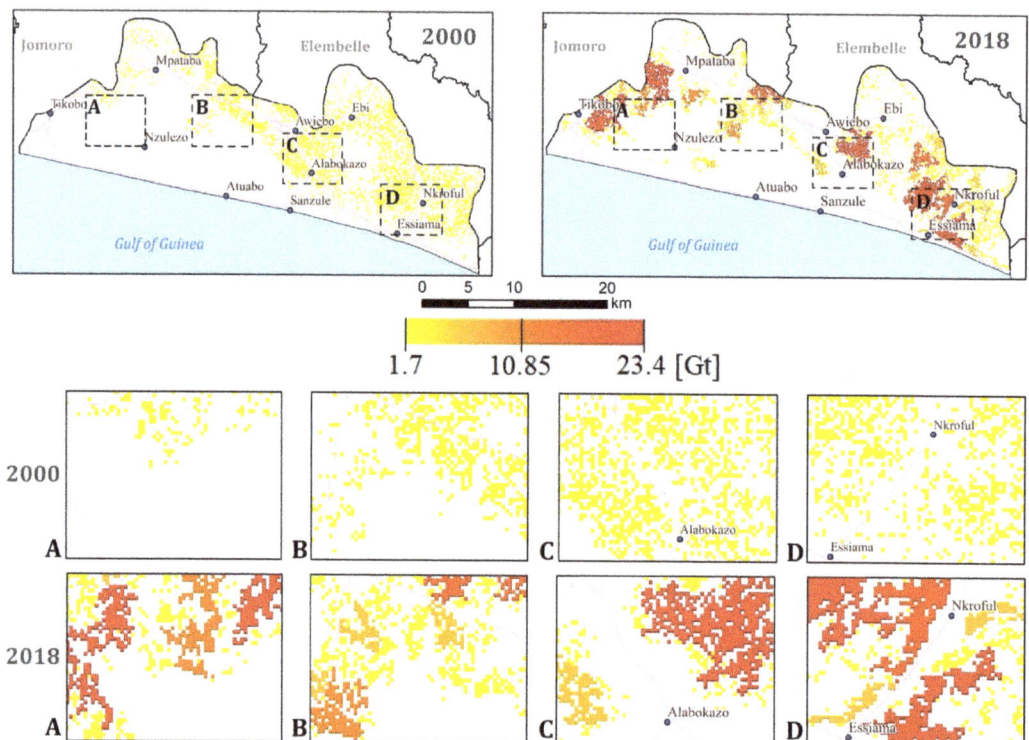

Figure 5. Quantities and distribution of food supply from cassava across the landscape in 2000 and 2018; (**A–D**) compares the spatial and temporal distribution of cassava food-crop supply potential for segments of the landscape.

Plantain Supply

The potential of the landscape to supply plantain food crop is depicted in Figure 6. While the minimum and maximum potential supply ranged from 0.001 to 3.3 Gt for the two temporal reference points, potential supply was found to be higher during 2018 than 2000. However, as shown in Figure 6D, the distribution patterns of potential plantain supply in Nkroful and Essiama was less in 2018 compared to 2000.

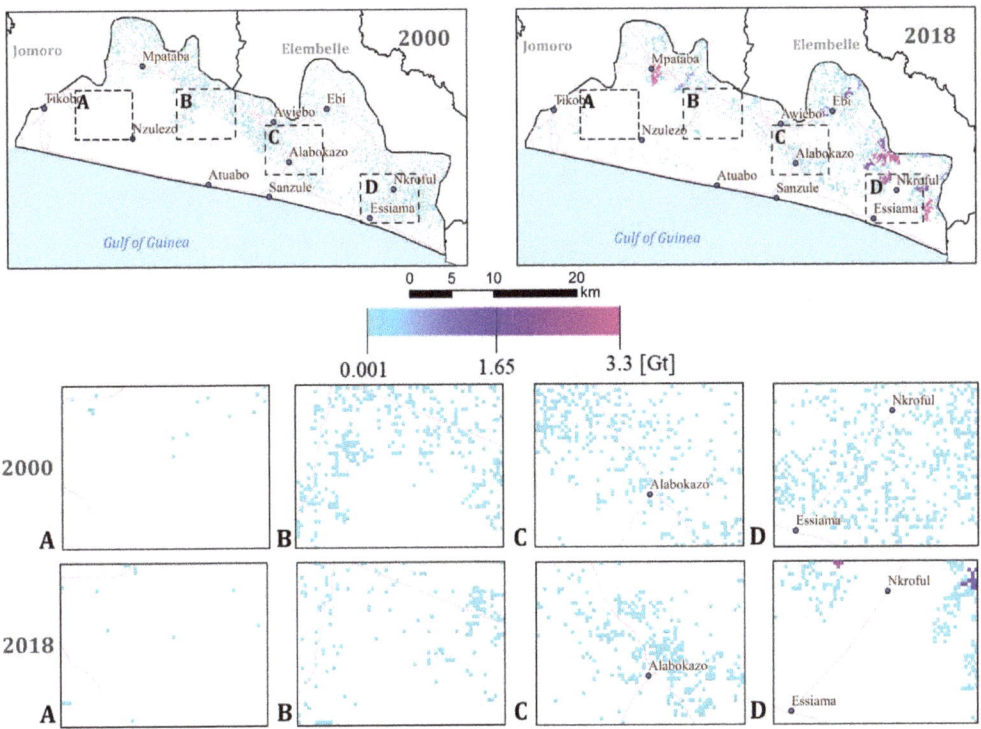

Figure 6. Quantities and distribution of food supply from plantain across the landscape in 2000 and 2018; (**A–D**) compares the spatial and temporal distribution of plantain food-crop supply potential for segments of the landscape.

3.2.3. Carbon Storage
Mangrove Carbon

The aboveground carbon storage potential of mangrove forests in the study landscape ranged from a minimum of 6.6 Mg C to a maximum of 87,196 Mg C, as shown in Figure 7. Spatial and temporal patterns of mangrove carbon storage potential remained relatively unchanged across the landscapes. However, the spatial distribution pattern of aboveground mangrove carbon storage potential showed variations along the coast. Potential for mangrove carbon storage was concentrated in the southeastern portions of the landscape: along the coastal stretch between Sanzule and Essiama, eastward from Essiama and along the lower Ankobra riparian areas (Figure 7A–C), whereas the landscape showed no potential for mangrove carbon storage along the southwestern side of the coast.

Rubber Carbon

The aboveground rubber carbon storage potential of the landscape is shown in Figure 8. This ranged from a minimum of 0.05 Mg C to a maximum of 429 Mg C. Nonetheless, as indicated in Figure 8, rubber carbon storage potential skewed toward the lower limit, between 0.05 to 257.40 Mg C. The distribution pattern of rubber carbon storage potential also showed clustering on the central and northern portions of the landscape. Additionally, this potential increased markedly in 2018 compared to the situation in 2000 (Figure 8A–D).

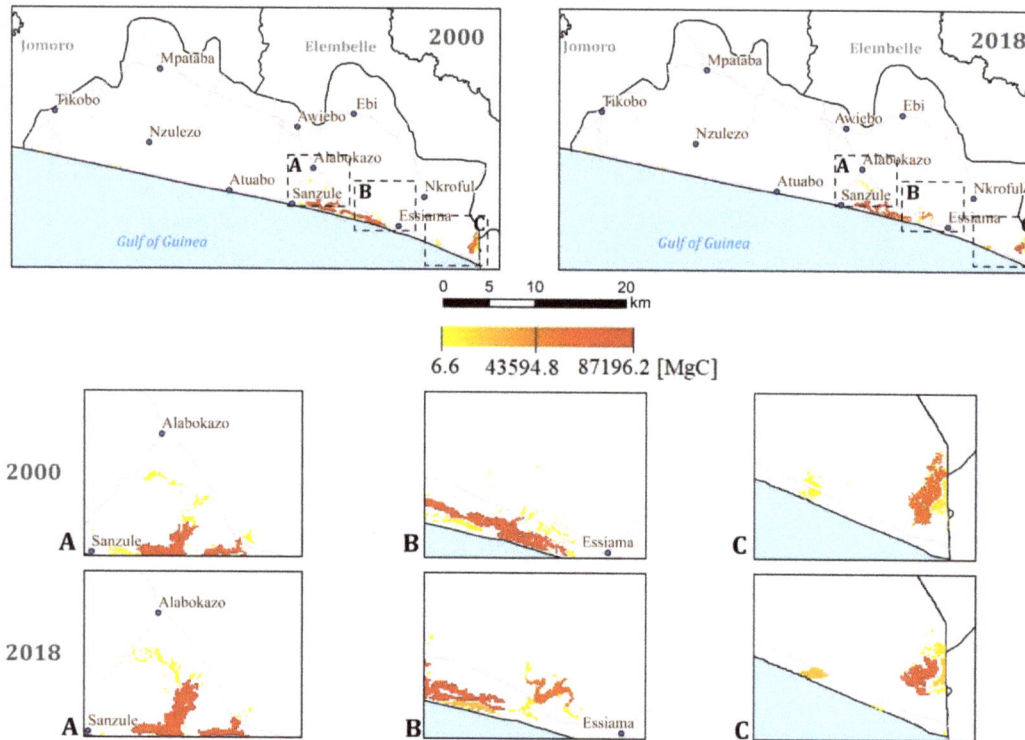

Figure 7. Quantities and distribution of aboveground carbon stored in mangrove forest across the landscape in 2000 and 2018; (**A–C**) compares the spatial and temporal distribution of aboveground carbon storage potential of sections of the landscape.

3.3. Predictors of LULCC

Geographically weighted regression models showing predictors of transitions to artificial/bare areas, rubber plantation and cropland are depicted in Figures 9–11, respectively. The predictors—farming population, total resident population, elevation and slope—exhibited spatial variations in their relationships with transitions to artificial/bare areas. Within the major towns such as Atuabo, Beyin and Krisan, there were strong spatial effects between farming population (β = 3.09–16.31, R^2 = 0.6–0.9) and transitions to artificial/bare areas. Similarly, as shown in Figure 9, resident population and slope positively affected transitions to artificial/bare areas. These effects were strong in Awiebo and surrounding towns. On the other hand, elevation, total resident population and farming population showed strong effects with land use/land cover transitions to rubber plantation. These effects were exhibited at the northwestern and northeastern portions of the landscape (Figure 10). Elevation exhibited strong effects with transitions to cropland at Tikobo No. 1, Mpataba and Awiebo. Similarly, slope showed strong spatial effects with transitions to cropland within the foregoing locations and around Nkroful, Esiama and Asanta (Figure 11).

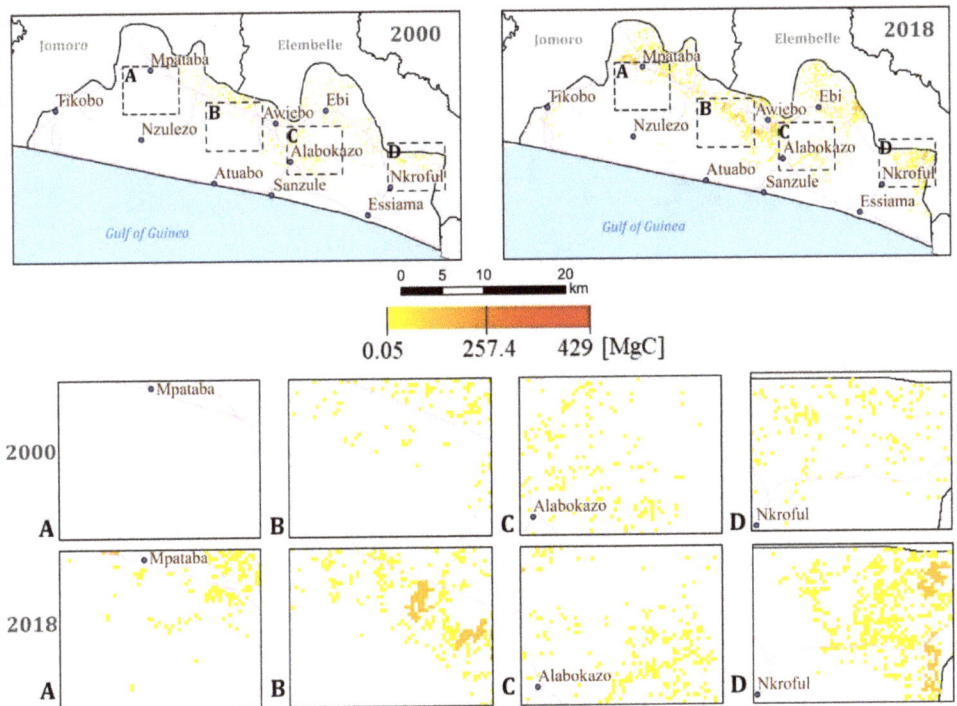

Figure 8. Quantities and distribution of aboveground carbon stored in rubber plantation across the landscape in 2000 and 2018; (**A–D**) compares the spatial and temporal distribution of aboveground rubber carbon storage potential of segments of the landscape.

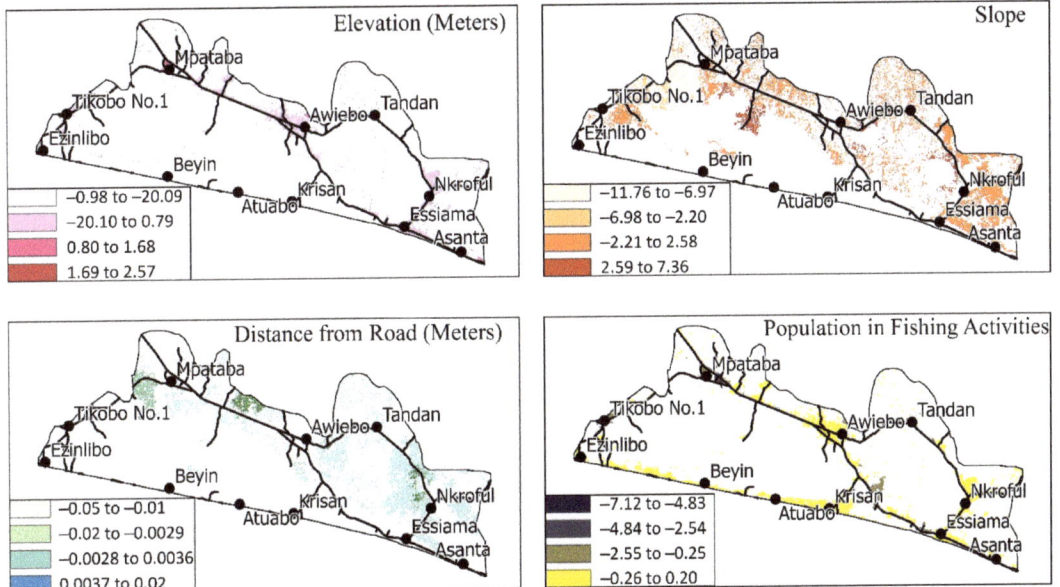

Figure 9. Cont.

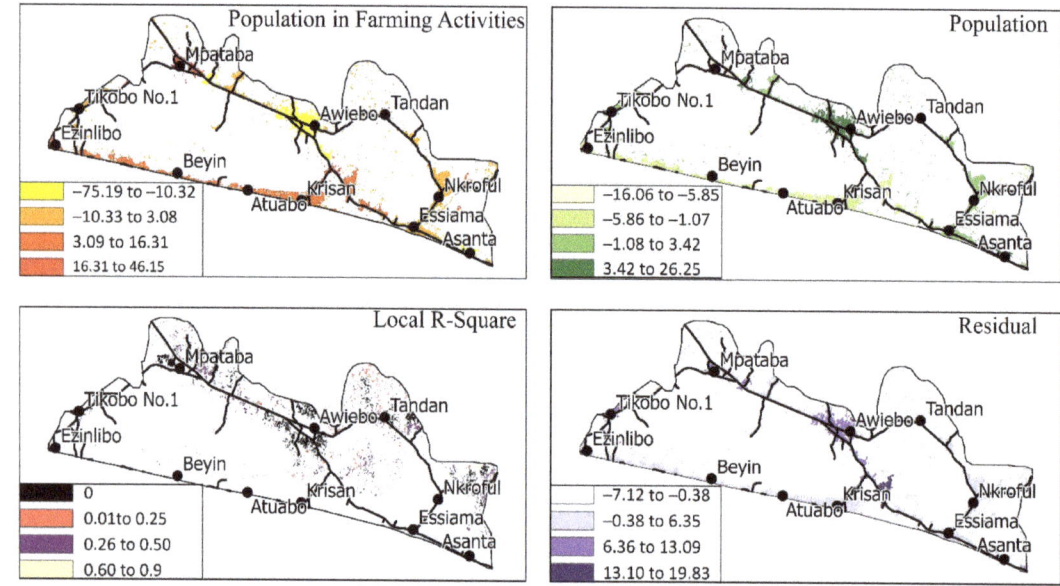

Figure 9. Results of geographically weighted regression showing spatial variations in the relationships between elevation, slope, distance from road, fishing population, farming population, total resident population and land use/land cover transitions to artificial/bare areas. Positive coefficients indicate the magnitude of spatial variation. Negative coefficients indicate no spatial correlations between the factor and transitions to artificial/bare areas.

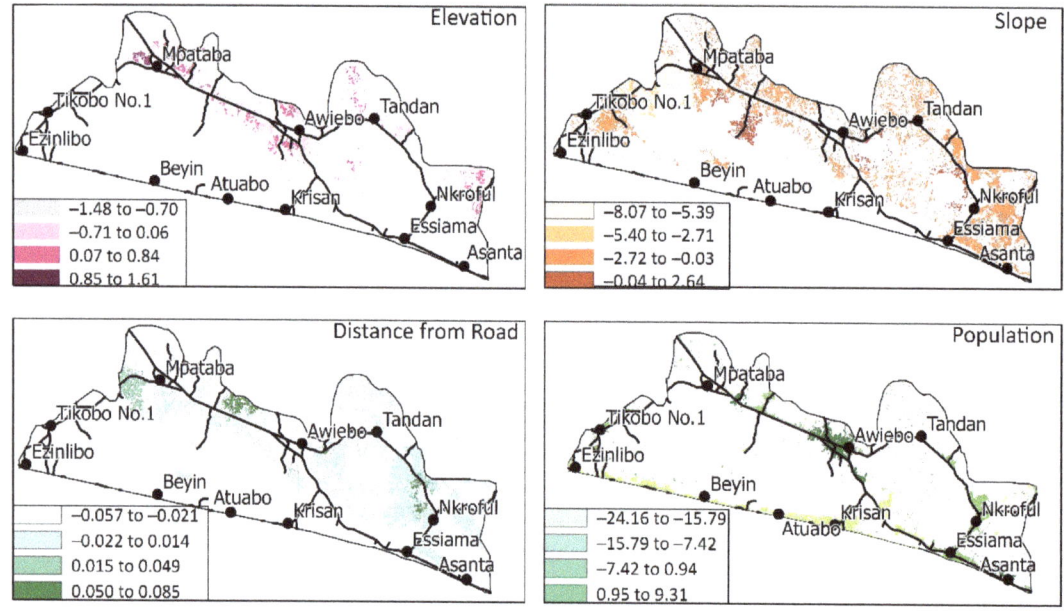

Figure 10. *Cont.*

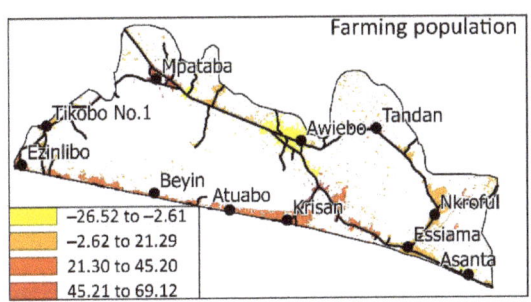

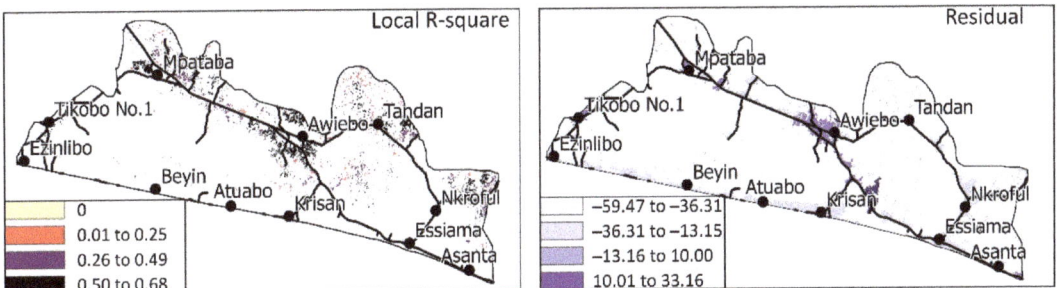

Figure 10. Results of geographically weighted regression showing spatial variations in the relationships between elevation, slope, distance from road, farming population, total resident population and land use/land cover transitions to rubber plantation. Positive coefficients indicate the magnitude of spatial variation. Negative coefficients indicate no spatial correlations between the factor and transitions to rubber plantation.

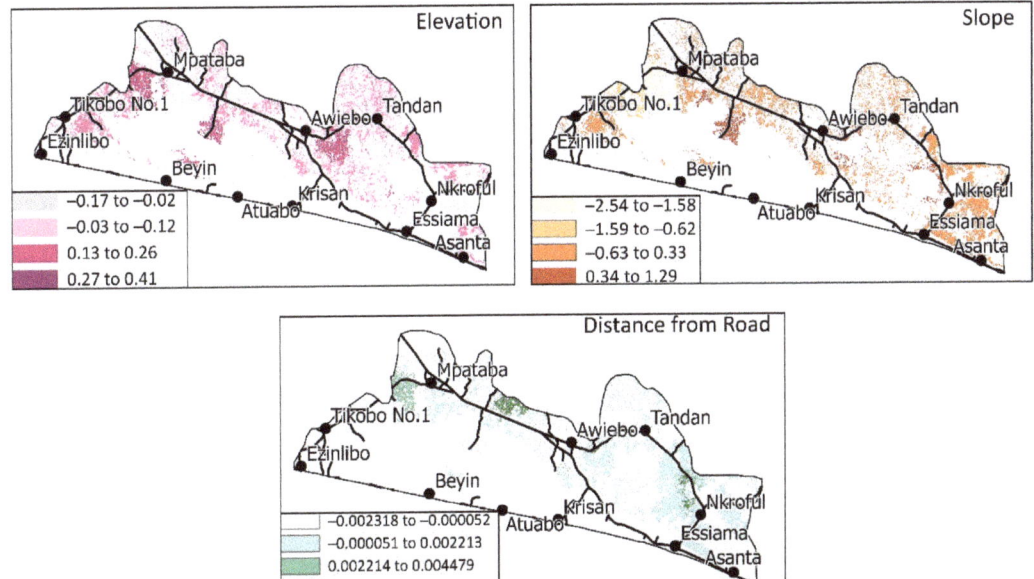

Figure 11. *Cont.*

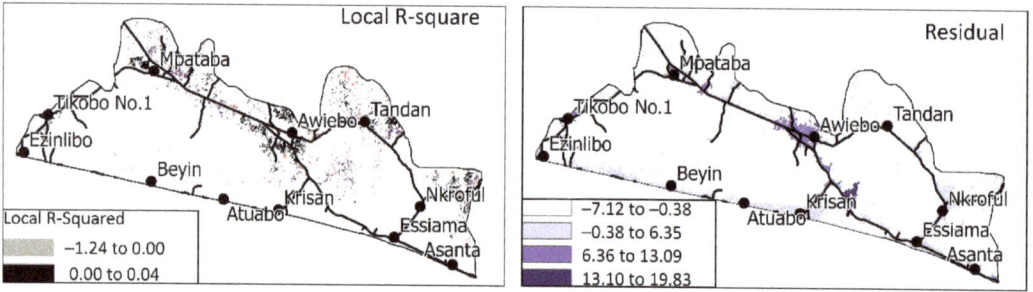

Figure 11. Results of geographically weighted regression showing spatial variations in the relationships between elevation, slope, distance from road and land use/land cover transitions to cropland. Positive coefficients indicate the magnitude of spatial variation. Negative coefficients indicate no spatial correlations between the factor and transitions to cropland.

4. Discussion

4.1. Land Use/Land Cover Changes and ES Supply Potentials

Coastal areas are rich in biodiversity and provide a variety of ES that are the basis for human well-being and sustainable development in most West African countries. The potential of coastal landscapes to sustainably supply provisioning services such as food and fuelwood are fundamental to livelihood improvement for the majority of coastal dwellers in the subregion. However, the growing conflicts between uses and among users of coastal resources require a better understanding of the impacts of human modification of coastal environments on the supply of ES [62]. In the coastal landscapes of Southwestern Ghana, land systems transformation is rapid and characterized by degradation of mangrove forests, rapid expansion of rubber plantation and infrastructure projects related to oil and gas activities [63,64]. Despite the potential negative impacts of such land conversions on ES, our findings point to increasing landscape potential to supply food. Increase in food production is inevitable, considering the food requirements of the growing population in the study region. Expansion of the area under cultivation and adoption of sustainable agricultural intensification practices while reducing land degradation are probable pathways for enhancing food production. Cropland constitutes the largest share of LULC type in the region, and this share has increased during the study period. Expansion in the area utilized for crop cultivation is a major contributing factor to the increase in food supply. This finding is congruent with similar studies that reported consistent increases in the value of ecosystem services for food production due to agricultural land expansion [13,65]. We narrowed and focused assessments to determine the relative potential to supply cassava and plantain, since they constitute the major staple crops in the study region. Our findings showed that food supply potential of the landscape was higher for cassava compared to plantain. Maximum cassava supply potential was seven times greater than the maximum plantain supply. The higher landscape potential to supply cassava is likely the result of expansion in the area of land under cassava cultivation between the different time periods. In the study region, land use for the purposes of agriculture is unregulated as the present land use system fails to allocate areas for specific crop types. Land allocation for arable crop production needs to consider biophysical constraints such as slope, elevation and soil suitability [66,67]. The absence of such land use guidelines increases the probability for conversion of fertile agricultural lands for food production in favor of industrial development, thereby risking food insecurity. Findings of a study conducted in adjacent coastal landscapes of Southwestern Ghana suggest that rapid spatial transformations characterized by rubber plantation and settlement expansion into traditional food-crop lands threatens local food production [34].

Mangroves provide wood and nonwood products, and are also critical for conservation of biological diversity while providing habitats and spawning grounds for a variety of fish and shellfish [68]. Similar to many coastal areas in West Africa, coastal dwellers in the study region rely heavily on mangrove fuelwood for fish smoking [42,69]. However, we found that the landscapes' potential to supply mangrove fuelwood in the region is decreasing through time. Threats to mangrove forests vary across locations on the study landscape. For mangrove forest stands within proximity to small-scale fish markets, a supply chain for mangrove fuelwood is enabled by localized exploitation for fish processing. This supply chain is characterized by harvesting from the Ankobra riparian system, over relatively short distances by river and road transport to small scale fish processing destinations. Such mangrove fuelwood supply chains offer short-term economic returns to harvesters, thereby incentivizing further exploitation. Studies conducted in the eastern coast of Ghana also found that well-developed, local fuelwood markets motivates mangrove harvesters to prioritize short-term economic gains over long-term ecological benefits from mangrove protection [70]. Similarly, evidence from the Cameroon estuary indicates that growing wood markets in nearby cities is a major cause of mangrove forest overexploitation, as the sale of mangrove wood is a major source of household incomes [69]. In the study region, urbanization was evidenced by population changes between Essiama and Atuabo and related land use/land cover conversions to artificial/bare areas. Such changes were also influenced by elevation. The aforementioned locations are the centers for oil and gas infrastructure development. Urbanization trends in this area also account for mangrove forest losses.

Mangrove ecosystems store large quantities of carbon in below- and aboveground components, hence, if disturbed, will result in the release of high levels of greenhouse gas emissions [57]. In Ghana, mangroves are of interest for inclusion in climate mitigation strategies. The mean aboveground carbon storage potential of the coastal landscapes in Southwestern Ghana was estimated at 114.66 Mg C ha^{-1} (Supplementary Table S2). This lies within the reported range from 5.2 to 312 Mg C ha^{-1} aboveground carbon pools for mangroves in West-Central Africa [71]. Considering the total extent of the landscape, the aboveground mangrove carbon storage potential ranged from 6.6 to 87,196.2 Mg C. This potential showed a decreasing trend over the study period (Supplementary Table S2). Additionally, there were spatial variations in the distribution patterns of potential carbon storage in the study landscape. Generally, areas along the estuaries of the Ankobra and Amanzule rivers recorded relatively higher potential carbon storage compared to areas farther away. Global studies reporting on sites in West Africa found direct correlations between spatial variations in mangrove carbon storage potential and the geomorphic positions of mangrove stands [59]. Other environmental factors such as soil properties, salinity and precipitation also influence mangrove carbon storage [71].

Contrary to mangroves, aboveground rubber carbon storage potential showed increasing temporal trends and ranged from 0.05 to 429 Mg C. Despite the increasing carbon storage potential and the relatively large extent of rubber plantations, mangrove carbon storage potential was two hundred-fold greater than rubber carbon storage during the study period. This finding reinforces the evidence for investments in mangrove blue carbon as a viable option for achieving climate change mitigation targets in tropical coastal landscapes [72].

4.2. Implications for Land Use Planning

Within the context of Ghana's three-tiered land use planning approach, spatial development frameworks are prepared at the strategic level, which in turn guide the preparation of structure and local plans at lower tiers [73,74]. This approach to land use planning hinges on setting goals and objectives and developing future scenarios as a basis for expression and implementation of spatial social, environmental and economic policies. The spatial development framework for the coastal subregion of Ghana's western region identifies the need to reconcile industrial development and conservation of sensitive coastal habitats. Yet this spatial plan falls short of identifying areas within the landscape for the supply of critical ES as a basis for conservation decisions. Tradeoffs between land use for agriculture, industrial development and conservation of coastal ecosystems will likely arise with the current pace of land transformation dominated by market-oriented drivers such as increasing monoculture of rubber plantations, mining activities and oil and gas infrastructure development [63,64]. Moreover, tradeoffs in the supply of relevant ES will also emerge and therefore require prioritization of ES supply. In the study region, tradeoffs exist between mangrove conservation as a regulating service and mangrove fuelwood supply as provisioning services. Similarly, understanding the carbon storage potential of rubber plantation versus mangrove forests will support prioritization of land use investments for implementation of climate mitigation strategies. Understanding and addressing these tradeoffs in a rapidly transforming land system requires integration of ES perspectives into the planning process [21]. Comparative analysis of the quantities of ES supply will support pragmatic objectives and decisions regarding areas where critical ES supply require maintenance and also areas where tradeoffs can be minimized. Prime areas for agriculture can be maintained considering biophysical attributes supporting food production. This will also strengthen biophysical justifications for land use scenarios presented during the planning process while proactively supporting decisions to protect ecosystems.

Additionally, development of zoning regulations as part of structure plan preparation processes will benefit from landscape scale estimations of ES. Areas of high biodiversity and ES will be apparent, leading to their protection through the use of appropriate regulatory instruments. Areas where there are risks due to tensions between conservation and coastal development can be identified and mitigated.

4.3. Study Limitations

Despite the similarities between the ecosystems from which data were transferred and that of the study area, the application of benefit transfer in this study risked introduction of errors due to landscape heterogeneity. This implies that transfer sites were not truly representative of the corresponding sites in the study landscape. This is because within each broad land cover class there are subclasses with different attributes. Thus, implementation of benefit transfer in this study assumed that all land cover units within a broad land cover class are the same. Moreover, key variations in ES supply were not evident due to the coarse temporal resolution of ten-year intervals of the remote sensing datasets. Finally, results of the study were not validated to confirm or not confirm, the estimates of ES values in the study region. Despite these challenges, the applied methodology provides a quick first step toward quantifying, mapping and including information on ES into coastal land use planning in the study region.

4.4. Conclusions and Future Outlook

Mapping and assessing ES require data at the appropriate scale and spatial and temporal resolutions [19]. In data-poor regions, such as Southwestern Ghana, the absence of such datasets hinders ES mapping and valuation. However, availability of data collected using primary studies and agricultural production statistics compiled by government institutions presents opportunities for mapping, especially, relevant regulating and provisioning ES. This study demonstrated the use of remote sensing data, cassava and plantain yield statistics and estimates of mangrove forest stand biomass and aboveground carbon in rubber plantation to quantify the landscapes' potential to supply ES. Relevant ES supply potentials quantified were food supply, fuelwood supply and carbon storage, and their spatial distribution patterns. The mean aboveground carbon storage potential of the coastal landscapes in Southwestern Ghana was estimated at 114.66 Mg C ha^{-1}. However, considering the total extent of the landscape, the aboveground mangrove carbon storage potential ranged from 6.6 to 87,196.2 Mg C with spatial and temporal variations in the distribution of mangrove carbon storage potential. Relatedly, mangrove fuelwood supply potential ranged from 0.01 to 87.19 tons and varied over the study period and across different locations on the landscape. Overall, potential to supply mangrove fuelwood and also aboveground carbon storage in mangrove ecosystems depicted decreasing trends. The potential for food supply from cropland and carbon storage in rubber plantation increased during the study period. Cassava supply ranged from 1.7 to 23.4 gigatons and plantain supply ranged from 0.001 to 3.3 gigatons. Rubber carbon storage potential ranged from 0.05 to 429 Mg C. Population, slope and elevation exhibited strong effects on LULC conversions to food crop and rubber plantations, whereas these factors were less important in determining mangrove forest conversions.

Rapid transformation of the land system in the study region is a major risk to sustainable supply of ES and to the minimization of tradeoffs in land use decision-making. Integration of ES perspectives will strengthen the biophysical basis of land use planning and decision-making in the region. Future ES mapping should take into account estimation of regional balance in food supply, as this will be necessary for optimal allocation of land for food production and conservation of critical coastal ecosystems.

Supplementary Materials: The following supporting information can be downloaded at: https://www.mdpi.com/article/10.3390/land11091408/s1, Table S1: Regional food-crop production statistics; Table S2: Mangrove carbon estimates from selected forest stands; Table S3: Mangrove biomass estimates from selected forest stands; Table S4: Rubber carbon estimates from selected plots.

Author Contributions: Conceptualization, S.K.; methodology, S.K.; software, A.O. and S.K.; formal analysis, S.K., A.O. and J.N.I.; writing—original draft preparation, S.K.; writing—review and editing, C.F. and J.N.I.; supervision, C.F.; funding acquisition, C.F. All authors have read and agreed to the published version of the manuscript.

Funding: This research was funded by the German Federal Ministry of Environment and Research (BMBF) through a long term EU-Africa research and innovation Partnership on food and nutrition security and sustainable Agriculture (LEAP-Agri) under the grant number at Martin Luther University Halle-Wittenberg [01DG18020].

Institutional Review Board Statement: Not applicable.

Informed Consent Statement: Not applicable.

Data Availability Statement: Not applicable.

Acknowledgments: We acknowledge the support of Hen Mpoano (Our Coast) in providing access to spatial datasets for this study. We express our appreciation to the Regional Director of the Ministry of Food and Agriculture (MoFA), Western Region, for sharing regional agricultural statistics. Messrs. Daniel Nii Doku Nortey, Justice Mensah and Joshua Adotey are duly acknowledged for their support in providing data for this study. We also thank the four anonymous reviewers for their helpful comments and suggestions on a draft version of this manuscript.

Conflicts of Interest: The authors declare no conflict of interest.

References

1. He, Q.; Silliman, B.R. Climate Change, Human Impacts, and Coastal Ecosystems in the Anthropocene. *Curr. Biol.* **2019**, *29*, R1021–R1035. [CrossRef] [PubMed]
2. Millennium Ecosystem Assessment. In *Ecosystems and Human Well-Being*; Island Press: Washington, DC, USA, 2005; Volume 5, ISBN 1559634022.
3. Ramesh, R.; Chen, Z.; Cummins, V.; Day, J.; D'Elia, C.; Dennison, B.; Forbes, D.L.; Glaeser, B.; Glaser, M.; Glavovic, B.; et al. Land-Ocean Interactions in the Coastal Zone: Past, Present & Future. *Anthropocene* **2015**, *12*, 85–98. [CrossRef]
4. Herrmann, S.M.; Brandt, M.; Rasmussen, K.; Fensholt, R. Accelerating Land Cover Change in West Africa over Four Decades as Population Pressure Increased. *Commun. Earth Environ.* **2020**, *1*, 53. [CrossRef]
5. Dada, O.A.; Agbaje, A.O.; Adesina, R.B.; Asiwaju-Bello, Y.A. Effect of Coastal Land Use Change on Coastline Dynamics along the Nigerian Transgressive Mahin Mud Coast. *Ocean Coast. Manag.* **2019**, *168*, 251–264. [CrossRef]
6. Zhang, Y.; Zhao, L.; Liu, J.; Liu, Y.; Li, C. The Impact of Land Cover Change on Ecosystem Service Values in Urban Agglomerations along the Coast of the Bohai Rim, China. *Sustainability* **2015**, *7*, 10365–10387. [CrossRef]
7. Yirsaw, E.; Wu, W.; Shi, X.; Temesgen, H.; Bekele, B. Land Use/Land Cover Change Modeling and the Prediction of Subsequent Changes in Ecosystem Service Values in a Coastal Area of China, the Su-Xi-Chang Region. *Sustainability* **2017**, *9*, 1204. [CrossRef]
8. Brown, G.; Helene, V. Ocean & Coastal Management an Empirical Analysis of Cultural Ecosystem Values in Coastal Landscapes. *Ocean Coast. Manag.* **2017**, *142*, 49–60. [CrossRef]

9. Zhao, B.; Kreuter, U.; Li, B.; Ma, Z.; Chen, J.; Nakagoshi, N. An Ecosystem Service Value Assessment of Land-Use Change on Chongming Island, China. *Land Use Policy* **2004**, *21*, 139–148. [CrossRef]
10. Aziz, T. Changes in Land Use and Ecosystem Services Values in Pakistan, 1950–2050. *Environ. Dev.* **2021**, *37*, 100576. [CrossRef]
11. Costanza, R.; de Groot, R.; Sutton, P.; van der Ploeg, S.; Anderson, S.J.; Kubiszewski, I.; Farber, S.; Turner, R.K. Changes in the Global Value of Ecosystem Services. *Glob. Environ. Change* **2014**, *26*, 152–158. [CrossRef]
12. Lambin Eric, G. *Land Use Cover Change Local Processes Global Challenges*; Springer: Berlin/Heidelberg, Germany, 2006; Volume 53, ISBN 9788578110796.
13. Lambin, E.F.; Turner, B.L.; Geist, H.J.; Agbola, S.B.; Angelsen, A.; Folke, C.; Bruce, J.W.; Coomes, O.T.; Dirzo, R.; George, P.S.; et al. The Causes of Land-Use and Land-Cover Change: Moving beyond the Myths. *Glob. Environ. Change* **2001**, *11*, 261–269. [CrossRef]
14. IPBES. *The Regional Assessment Report on Biodiversity and Ecosystem Services for Africa of the Intergovernmental Science-Policy Platform on Biodiversity and Ecosystem Services*; IPBES: Bonn, Germany, 2018; ISBN 978-3-947851-05-8/978-3-947851-02-7.
15. Osman, A.; Mariwah, S.; Oscar, D.; Kankam, S. Broadening the Narratives of Ecosystem Services: Assessing the Perceived Services from Nature and Services to Nature. *J. Nat. Conserv.* **2022**, *68*, 126188. [CrossRef]
16. Comberti, C.; Thornton, T.F.; Wylliede Echeverria, V.; Patterson, T. Ecosystem Services or Services to Ecosystems? Valuing Cultivation and Reciprocal Relationships between Humans and Ecosystems. *Glob. Environ. Change* **2015**, *34*, 247–262. [CrossRef]
17. Burkhard, B.; Kroll, F.; Müller, F.; Windhorst, W. Landscapes' Capacities to Provide Ecosystem Services—A Concept for Land-Cover Based Assessments. *Landsc. Online* **2009**, *15*, 1–22. [CrossRef]
18. Von Haaren, C.; Albert, C. Integrating Ecosystem Services and Environmental Planning: Limitations and Synergies. *Int. J. Biodivers. Sci. Ecosyst. Serv. Manag.* **2011**, *7*, 150–167. [CrossRef]
19. Syrbe, R.; Schroter, M.; Grunewald, K.; Waltz, U.; Burkard, B. *What to Map?* Pensoft Publishers: Sofia, Bulgaria, 2017; ISBN 9789546428295.
20. Seppelt, R.; Dormann, C.F.; Eppink, F.V.; Lautenbach, S.; Schmidt, S. A Quantitative Review of Ecosystem Service Studies: Approaches, Shortcomings and the Road Ahead. *J. Appl. Ecol.* **2011**, *48*, 630–636. [CrossRef]
21. Inkoom, J.N.; Frank, S.; Fürst, C. Challenges and Opportunities of Ecosystem Service Integration into Land Use Planning in West Africa—An Implementation Framework. *Int. J. Biodivers. Sci. Ecosyst. Serv. Manag.* **2017**, *13*, 67–81. [CrossRef]
22. Koo, H.; Kleemann, J.; Fürst, C. Impact Assessment of Land Use Changes Using Local Knowledge for the Provision of Ecosystem Services in Northern Ghana, West Africa. *Ecol. Indic.* **2019**, *103*, 156–172. [CrossRef]
23. Eigenbrod, F.; Armsworth, P.R.; Anderson, B.J.; Heinemeyer, A.; Gillings, S.; Roy, D.B.; Thomas, C.D.; Gaston, K.J. The Impact of Proxy-Based Methods on Mapping the Distribution of Ecosystem Services. *J. Appl. Ecol.* **2010**, *47*, 377–385. [CrossRef]
24. Nolander, C. *Spatial and Economic Values of Ecosystem Services*; Lulea University of Technology: Lulea, Sweden, 2018; ISBN 9789177900696.
25. Costanza, R.; Arge, R.; De Groot, R.; Farber, S.; Grasso, M.; Hannon, B.; Limburg, K.; Naeem, S.; Neill, R.V.O.; Paruelo, J.; et al. The Value of the World's Ecosystem Services and Natural Capital. *Nature* **1997**, *387*, 253–260. [CrossRef]
26. Andrew, M.E.; Wulder, M.A.; Nelson, T.A.; Coops, N.C. Spatial Data, Analysis Approaches, and Information Needs for Spatial Ecosystem Service Assessments: A Review. *GIScience Remote Sens.* **2015**, *52*, 344–373. [CrossRef]
27. Brown, G.; Pullar, D.; Hausner, V.H. An Empirical Evaluation of Spatial Value Transfer Methods for Identifying Cultural Ecosystem Services. *Ecol. Indic.* **2016**, *69*, 1–11. [CrossRef]
28. Rosenberger, R.S.; Loomis, J.B. *Benefit Transfer of Outdoor Recreation Use Values: A Technical Document Supporting the Forest Service Strategic Plan*; U.S. Department of Agriculture, Forest Service, Rocky Mountain Research Station: Fort Collins, CO, USA, 2001; p. 59. Available online: https://www.fs.usda.gov/rm/pubs/rmrs_gtr072.pdf (accessed on 24 July 2022).
29. Plummer, M.L. Assessing Benefit Transfer for the Valuation of Ecosystem Services. *Front. Ecol. Environ.* **2009**, *7*, 38–45. [CrossRef]
30. Koschke, L.; Fürst, C.; Frank, S.; Makeschin, F. A Multi-Criteria Approach for an Integrated Land-Cover-Based Assessment of Ecosystem Services Provision to Support Landscape Planning. *Ecol. Indic.* **2012**, *21*, 54–66. [CrossRef]
31. Eigenbrod, F.; Armsworth, P.R.; Anderson, B.J.; Heinemeyer, A.; Gillings, S.; Roy, D.B.; Thomas, C.D.; Gaston, K.J. Error Propagation Associated with Benefits Transfer-Based Mapping of Ecosystem Services. *Biol. Conserv.* **2010**, *143*, 2487–2493. [CrossRef]
32. Coastal Resources Center. *Building Capacity for Adapting to a Rapidly Changing Coastal Zone*; Graduate School of Oceanography, University of Rhode Island: Narragansett, RI, USA, 2010.
33. Asante-yeboah, E.; Ashiagbor, G.; Asubonteng, K.; Sieber, S.; Mensah, J.C.; Fürst, C. Analyzing Variations in Size and Intensities in Land Use Dynamics for Sustainable Land Use Management: A Case of the Coastal Landscapes of South-Western Ghana. *Land* **2022**, *11*, 815. [CrossRef]
34. Asante, W.; Jengre, N. *Carbon Stocks and Soil Nutrient Dynamics in the Peat Swamp Forest of the Amanzule Wetlands & Ankobra River Basin*; Graduate School of Oceanography, University of Rhode Island: Narragansett, RI, USA, 2012; p. 45.
35. Kleemann, J.; Inkoom, J.N.; Thiel, M.; Shankar, S.; Lautenbach, S.; Fürst, C. Peri-Urban Land Use Pattern and Its Relation to Land Use Planning in Ghana, West Africa. *Landsc. Urban Plan.* **2017**, *165*, 280–294. [CrossRef]
36. Kankam, S.; Inkoom, J.N.; Koo, H.; Fürst, C. Envisioning Alternative Futures of Cultural Ecosystem Services Supply in the Coastal Landscapes of Southwestern Ghana, West Africa. *Socio-Ecol. Pract. Res.* **2021**, *3*, 309–328. [CrossRef]
37. Daniels, T.; Chan, J.K.H.; Kankam, S.; Murphy, M.; Day, D.; Fürst, C.; Inkoom, J.N.; Koo, H. Four Shareworthy SEPR Scenario Ideas. *Socio-Ecol. Pract. Res.* **2021**, *3*, 9–15. [CrossRef]

38. Sagoe, A.A.; Aheto, D.W.; Okyere, I.; Adade, R.; Odoi, J. Community Participation in Assessment of Fisheries Related Ecosystem Services towards the Establishment of Marine Protected Area in the Greater Cape Three Points Area in Ghana. *Mar. Policy* **2021**, *124*, 104336. [CrossRef]
39. Maes, J.; Fabrega, N.; Zulian, G.; Barbosa, A.; Vizcaino, P.; Ivits, E.; Polce, C.; Vandecasteele, I.; Rivero, I.M.; Guerra, C.; et al. *Mapping and Assessment of Ecosystems and Their Services: Trends in Ecosystems and Ecosystem Services in the European Union between 2000 and 2010*; European Union: Rome, Italy, 2015; ISBN 978-92-79-46206-1.
40. Cord, A.F.; Brauman, K.A.; Chaplin-Kramer, R.; Huth, A.; Ziv, G.; Seppelt, R. Priorities to Advance Monitoring of Ecosystem Services Using Earth Observation. *Trends Ecol. Evol.* **2017**, *32*, 416–428. [CrossRef] [PubMed]
41. Amoakoh, A.O.; Aplin, P.; Awuah, K.T.; Delgado-fernandez, I.; Moses, C.; Peña, C. Testing the Contribution of Multi-Source Remote Sensing Features for Random Forest Classification of the Greater Amanzule Tropical Peatland. *Sensors* **2021**, *21*, 3399. [CrossRef] [PubMed]
42. Ajonina, G. *Rapid Assessment of Mangrove Status to Assess Potential for Payment for Ecosystem Services in Amanzule in the Western Region of Ghana*; Coastal Resources Center, Graduate School of Oceanography, University of Rhode Island: Narragansett, RI, USA, 2011; pp. 1–30.
43. Osei, D.; Horwich, R.H.; Pittman, J.M. First Sightings of the Roloway Monkey (Cercopithecus Diana Roloway) in Ghana in Ten Years and the Status of Other Endangered Primates in Southwestern Ghana. *Afr. Primates* **2015**, *10*, 25–40.
44. Adjei Mensah, C.; Kweku Eshun, J.; Asamoah, Y.; Ofori, E. Changing Land Use/Cover of Ghana's Oil City (Sekondi-Takoradi Metropolis): Implications for Sustainable Urban Development. *Int. J. Urban Sustain. Dev.* **2019**, *11*, 223–233. [CrossRef]
45. Coastal Resources Center; Friends of the Nation. *Assessment of Critical Coastal Habitats of the Western Region, Ghana*; Coastal Resources Center, Graduate School of Oceanography, University of Rhode Island: Narragansett, RI, USA, 2011; p. 132.
46. Finegold, C.; Gordon, A.; Mills, D.; Curtis, L.; Pulis, A.; Crawford, B. *Western Region Fisheries Sector Review*; Coastal Resources Center, Graduate School of Oceanography, University of Rhode Island: Narragansett, RI, USA, 2010; 82p.
47. Duku, E.; Agbeko, P.; Mattah, D.; Angnuureng, D.B.; Adotey, J. Understanding the Complexities of Human Well-Being in the Context of Ecosystem Services within Coastal Ghana. *Sustainability* **2022**, *14*, 10111. [CrossRef]
48. Di Gregorio, A.; Jansen, L.J.M. *Land Cover Classification System (LCCS): Classification Concepts and User Manual*; FAO: Rome, Italy, 2000; Volume 53, p. 179.
49. de Groot, R.S.; Alkemade, R.; Braat, L.; Hein, L.; Willemen, L. Challenges in Integrating the Concept of Ecosystem Services and Values in Landscape Planning, Management and Decision Making. *Ecol. Complex.* **2010**, *7*, 260–272. [CrossRef]
50. Adotey, J. Carbon Stock Assessment in the Kakum and Amanzule Estuary Mangrove Forests, Ghana. Master's Thesis, University of Cape Coast, Cape Coast, Ghana, 2015.
51. Nortey, D.D.N.; Aheto, D.W.; Blay, J.; Jonah, F.E.; Asare, N.K. Comparative Assessment of Mangrove Biomass and Fish Assemblages in an Urban and Rural Mangrove Wetlands in Ghana. *Wetlands* **2016**, *36*, 717–730. [CrossRef]
52. Tawiah, E.N. Assessing the Potential Contribution of Latex from Rubber (Hevea Brasiliensis) Plantations as a Carbon Sink. 2017. Available online: http://internationaljournalcorner.com/index.php/ijird_ojs/article/view/141495/99533 (accessed on 10 April 2022).
53. Wauters, J.B.; Coudert, S.; Grallien, E.; Jonard, M.; Ponette, Q. Carbon Stock in Rubber Tree Plantations in Western Ghana and Mato Grosso (Brazil). *For. Ecol. Manag.* **2008**, *255*, 2347–2361. [CrossRef]
54. Komiyama, A.; Poungparn, S.; Kato, S. Common Allometric Equations for Estimating the Tree Weight of Mangroves. *J. Trop. Ecol.* **2005**, *21*, 471–477. [CrossRef]
55. Troy, A.; Wilson, M.A. Mapping Ecosystem Services: Practical Challenges and Opportunities in Linking GIS and Value Transfer. *Ecol. Econ.* **2006**, *60*, 435–449. [CrossRef]
56. Kuenzer, C.; Bluemel, A.; Gebhardt, S.; Quoc, T.V.; Dech, S. Remote Sensing of Mangrove Ecosystems: A Review. *Remote Sens.* **2011**, *3*, 878–928. [CrossRef]
57. Donato, D.C.; Kauffman, J.B.; Murdiyarso, D.; Kurnianto, S.; Stidham, M.; Kanninen, M. Mangroves among the Most Carbon-Rich Forests in the Tropics. *Nat. Geosci.* **2011**, *4*, 293–297. [CrossRef]
58. Alongi, D.M. Carbon Cycling and Storage in Mangrove Forests. *Ann. Rev. Mar. Sci.* **2014**, *6*, 195–219. [CrossRef] [PubMed]
59. Kauffman, J.B.; Adame, M.F.; Arifanti, V.B.; Schile-Beers, L.M.; Bernardino, A.F.; Bhomia, R.K.; Donato, D.C.; Feller, I.C.; Ferreira, T.O.; Jesus Garcia, M.D.C.; et al. Total Ecosystem Carbon Stocks of Mangroves across Broad Global Environmental and Physical Gradients. *Ecol. Monogr.* **2020**, *90*, e01405. [CrossRef]
60. Wang, M.M.H.; Carrasco, L.R.; Edwards, D.P. Reconciling Rubber Expansion with Biodiversity Conservation. *Curr. Biol.* **2020**, *30*, 3825–3832.e4. [CrossRef]
61. Geist, H.J.; Lambin, E.F. Proximate Causes and Underlying Driving Forces of Tropical Deforestation. *Bioscience* **2002**, *52*, 143. [CrossRef]
62. Crain, C.M.; Halpern, B.S.; Beck, M.W.; Kappel, C.V. Understanding and Managing Human Threats to the Coastal Marine Environment. *Ann. N. Y. Acad. Sci.* **2009**, *1162*, 39–62. [CrossRef]
63. Otchere-Darko, W.; Ovadia, J.S. Incommensurable Languages of Value and Petro-Geographies: Land-Use, Decision-Making and Conflict in South-Western Ghana. *Geoforum* **2020**, *113*, 69–80. [CrossRef]
64. Asare-Donkor, N.K.; Adimado, A.A. Influence of Mining Related Activities on Levels of Mercury in Water, Sediment and Fish from the Ankobra and Tano River Basins in South Western Ghana. *Environ. Syst. Res.* **2016**, *5*, 5. [CrossRef]

65. Kindu, M.; Schneider, T.; Teketay, D.; Knoke, T. Changes of Ecosystem Service Values in Response to Land Use/Land Cover Dynamics in Munessa-Shashemene Landscape of the Ethiopian Highlands. *Sci. Total Environ.* **2016**, *547*, 137–147. [CrossRef] [PubMed]
66. Widiatmaka; Ambarwulan, W.; Setiawan, Y.; Walter, C. Assessing the Suitability and Availability of Land for Agriculture in Tuban Regency, East Java, Indonesia. *Appl. Environ. Soil Sci.* **2016**, *2016*, 7302148. [CrossRef]
67. Manyevere, A. An Integrated Approach for the Delineation of Arable Land and Its Cropping Suitability under Variable Soil and Climatic Conditions in the Nkonkobe Municipality, Eastern Cape, South Africa. Ph.D. Thesis, University of Fort Hare, Alice, South Africa, 2014.
68. Corcoran, E.; Ravilious, C.; Skuja, M. *Mangroves of Western and Central Africa*; UNEP World Conservation Monitoring Center: Cambridge, UK, 2007; ISBN 9789280727920.
69. Atheull, A.N.; Din, N.; Longonje, S.N.; Koedam, N.; Dahdouh-Guebas, F. Commercial Activities and Subsistence Utilization of Mangrove Forests around the Wouri Estuary and the Douala-Edea Reserve (Csameroon). *J. Ethnobiol. Ethnomed.* **2009**, *5*, 35. [CrossRef] [PubMed]
70. Aheto, D.W.; Kankam, S.; Okyere, I.; Mensah, E.; Osman, A.; Jonah, F.E.; Mensah, J.C. Community-Based Mangrove Forest Management: Implications for Local Livelihoods and Coastal Resource Conservation along the Volta Estuary Catchment Area of Ghana. *Ocean Coast. Manag.* **2016**, *127*, 43–54. [CrossRef]
71. Boone, J.K.; Bhomia, R.K. Ecosystem Carbon Stocks of Mangroves across Broad Environmental Gradients in West-Central Africa: Global and Regional Comparisons. *PLoS ONE* **2017**, *12*, e0187749. [CrossRef]
72. Alongi, D.M. Global Significance of Mangrove Blue Carbon in Climate Change Mitigation. *Sci* **2020**, *2*, 67. [CrossRef]
73. Acheampong, R.A.; Ibrahim, A. One Nation, Two Planning Systems? Spatial Planning and Multi-Level Policy Integration in Ghana: Mechanisms, Challenges and the Way Forward. *Urban Forum* **2016**, *27*, 1–18. [CrossRef]
74. Ministry of Environment, Science and Technology. Manual for the Preparation of Spatial Plans. 2011. Available online: https://www.luspa.gov.gh/media/document/PLANNING_MANUAL_final_DESIGN.pdf (accessed on 5 May 2022).

Review

Geo-Environmental Characterisation of High Contaminated Coastal Sites: The Analysis of Past Experiences in Taranto (Southern Italy) as a Key for Defining Operational Guidelines

Angela Rizzo [1,2], Francesco De Giosa [3], Antonella Di Leo [4], Stefania Lisco [1,2], Massimo Moretti [1,2], Giovanni Scardino [1,2,*], Giovanni Scicchitano [1,2] and Giuseppe Mastronuzzi [1,2]

1   Department of Earth and Geoenvironmental Sciences, Campus Universitario, University of Bari Aldo Moro, via E. Orabona, 4, 70125 Bari, Italy; angela.rizzo@uniba.it (A.R.); stefania.lisco@uniba.it (S.L.); massimo.moretti@uniba.it (M.M.); giovanni.scicchitano@uniba.it (G.S.); giuseppe.mastronuzzi@uniba.it (G.M.)
2   Interdepartmental Research Centre for Coastal Dynamics, Campus Universitario, University of Bari Aldo Moro, via E. Orabona, 4, 70125 Bari, Italy
3   Environmental Surveys s.r.l., via Renato Dario Lupo, 65, 74121 Taranto, Italy; francescodegiosa@ensu.it
4   Istituto di Ricerca sulle Acque (IRSA) Consiglio Nazionale delle Ricerche, via Roma, 3, 74123 Taranto, Italy; magda.dileo@irsa.cnr.it
*   Correspondence: giovanni.scardino@uniba.it

Citation: Rizzo, A.; De Giosa, F.; Di Leo, A.; Lisco, S.; Moretti, M.; Scardino, G.; Scicchitano, G.; Mastronuzzi, G. Geo-Environmental Characterisation of High Contaminated Coastal Sites: The Analysis of Past Experiences in Taranto (Southern Italy) as a Key for Defining Operational Guidelines. *Land* **2022**, *11*, 878. https://doi.org/10.3390/land11060878

Academic Editor: Andrea Belgrano

Received: 4 May 2022
Accepted: 6 June 2022
Published: 9 June 2022

Publisher's Note: MDPI stays neutral with regard to jurisdictional claims in published maps and institutional affiliations.

Copyright: © 2022 by the authors. Licensee MDPI, Basel, Switzerland. This article is an open access article distributed under the terms and conditions of the Creative Commons Attribution (CC BY) license (https://creativecommons.org/licenses/by/4.0/).

Abstract: Despite its remarkable geomorphological, ecological, and touristic value, the coastal sector of the Apulia region (Southern Italy) hosts three of the main contaminated Italian sites (Sites of National Interest, or SINs), for which urgent environmental remediation and reclamation actions are required. These sites are affected by intense coastal modification and diffuse environmental pollution due to the strong industrialisation and urbanisation processes that have been taking place since the second half of the XIX century. The Apulian coastal SINs, established by the National Law 426/1998 and delimited by the Ministerial Decree of 10 January 2000, include large coastal sectors and marine areas, which have been deeply investigated by the National Institution for the Environmental Research and Protection (ISPRA) and the Regional Agency for the Prevention and Protection of the Environment (ARPA) with the aim of obtaining a deep environmental characterisation of the marine matrices (sediments, water, and biota). More recently, high-resolution and multidisciplinary investigations focused on the geo-environmental characterisation of the coastal basins in the SIN Taranto site have been funded by the *"Special Commissioner for the urgent measures of reclamation, environmental improvements, and redevelopment of Taranto"*. In this review, we propose an overview of the investigations carried out in the Apulian SINs for the environmental characterisation of the marine matrices, with special reference to the sea bottom and sediments. Based on the experience gained in the previous characterisation activities, further research is aimed at defying a specific protocol of analysis for supporting the identification of priority actions for an effective and efficient geo-morphodynamic and environmental characterisation of the contaminated coastal areas, with special reference to geomorphological, sedimentological, and geo-dynamic features for which innovative and high-resolution investigations are required.

Keywords: coastal contaminated sites; geo-morphodynamic model; reclamation activities; Apulia region; Taranto

1. Introduction

The sustainable management of industrial and high-urbanised coasts is a significant issue globally. The U.S. Government Accounting Office [1] identified that approximately 60% of most contaminated sites are located along the coastal areas. Manzoor et al. [2] highlighted how the rapid economic development and industrialisation have caused an increase in metal concentrations in marine sediments in all the coastal regions of China.

High concentrations of pollutants were also found in the Don River estuarine region [3]. Currently, only less than 1% of the Mediterranean coasts remain relatively unaffected by human activities [4] and almost 200 petrochemical plants and energy systems are located along the Mediterranean coastal sectors [5]. A similar condition is experienced in the northern European countries, whose relatively long coastlines are negatively influenced by anthropogenic activities [6]. In England, more than 1200 landfills are located in coastal areas [7], while in Italy, 77,733 ha of marine and coastal areas are included in the perimeters of the Sites of National Interest (SINs) that, according to the national legislation (Legislative Decree 152/2006 and subsequent amendments and additions), represent *"a large portions of the national territory, which include all the different environmental matrices and entail a high health and ecological risk due to the density of the population or the extent of the site itself, as well as a significant socio-economic impact and a high risk to assets of historical and cultural interest."*

SINs management is entrusted to the Italian Ministry of Environment, Land and Sea (now Ministry for the Ecological Transition—MiTE), which uses the National Network System for Environmental Protection (SNPA, Rome, Italy) and the National Institute of Health (ISS, Rome, Italy), as well as other qualified public or private entities, for the technical investigation (Article 252-Legislative Decree 152/2006). The identification of the SINs and the definition of their boundaries started in 1998 in the frame of previous national regulations (MiTE, 2022—https://bonifichesiticontaminati.mite.gov.it/sin/istituzione-perimetrazione/, accessed on 3 May 2022 [8]). In 2012, 57 SINs were identified. With the entry into force of the Law 134/2012, which changed the criteria and parameters for the identification of SINs, the number of SINs decreased from 57 to 39. Then, a number of specific laws added further areas to the list. To date, the current number of SINs is 42 (Figure 1). With a total of 171,211 ha on land, SINs surface represents 0.57% of the Italian territory [9].

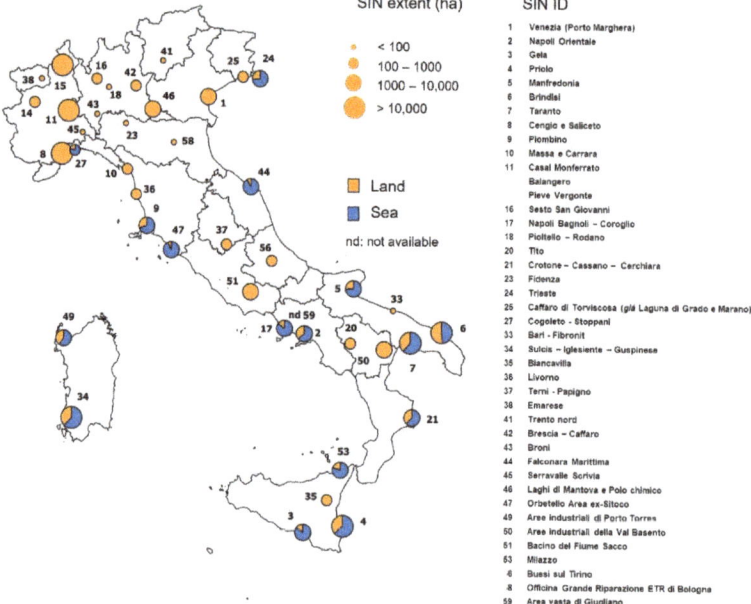

Figure 1. Italian Sites of National Interest (SINs). Last update: 2021 [9]. The numbers identify SIN_ID. The extent of the SINs coastal area is shown in the circles: the orange circles represent SINs whose perimeters do not include marine and coastal area.

As shown in Table 1, the environmental status of the marine and coastal areas in SINs is deeply affected by the direct and indirect impacts of different anthropogenic activities, including industrial plants (chemical, petrochemical, metallurgical, steel, mechanical, pharmaceutical, cement, thermal), uncontrolled landfills, military arsenals, shipyards, and harbour areas with high maritime traffic. To include all the potentially contaminated coastal matrices in the SINs, the boundaries of the marine areas were defined by extending the perimeters up to 3 km from the coastline, as seaward limit of the potential impact of anthropogenic activities [10,11].

Table 1. Main anthropogenic activities located in the Italian Sites of National Interest whose perimeters include marine areas [8].

Coastal SIN	Main Anthropogenic Activities
SIN_2 "Napoli"	Petrochemical, mechanical, and transport industries; manufacturing companies; mechanical office; disused thermoelectric power station and sewage treatment plant.
SIN_3 "Gela"	Industrial pole (petrochemical, hydrocarbons treatment and production).
SIN_4 "Priolo"	Refineries; petrochemical and cement industries; Landfills; Ex-Eternit plant.
SIN_5 "Manfredonia"	Fertiliser industry; urban and industrial waste landfill; agricultural land.
SIN_6 "Brindisi"	Petrochemical and electrical industries; agricultural land; areas belonging to the Harbour Authority.
SIN_7 "Taranto"	Iron and steel, petrochemical, and cement industries; Military shipyard; areas belonging to the Harbour Authority.
SIN_9 "Piombino"	Industrial pole; industrial waste landfill; areas with backfill material.
SIN_17 "Bagnoli"	Iron and steel industries; Ex-Eternit plant. All the activities are disused.
SIN_21 "Crotone"	Wide industrial pole, including disused plants to produce Zinc, phosphoric acid, complex fertilisers (nitrogen and phosphate), nitric acid, and sulphuric acid.
SIN_24 "Trieste"	Harbour zone; industrial pole; cast-iron industry.
SIN_27 "Cogoleto"	Industrial plant devoted to the production of sodium dichromate and other chromium derivatives. All the activities are disused.
SIN_34 "Sulcis-Iglesiente"	Mining and industrial activities related to the processing of extracted minerals; oil refining and petrochemical industries.
SIN_36 "Livorno"	Refinery and related facilities; thermoelectric power plant; areas belonging to the Harbour Authority.
SIN_44 "Falconara Marittima"	Refining and storage of petroleum products.
SIN_47 "Orbetello"	Mining; chemical plants to produce chemicals, glue, and fertilisers; dynamite and explosives factories; waste accumulation area. Mining and industrial activities are disused.
SIN_49 "Porto Torres"	Disused petrochemical plant; thermoelectric power station; active and disused chemical and mechanical industries; landfill.
SIN_53 "Milazzo"	Refining and storage of petroleum products; electricity power plant; asbestos processing (disused) industrial waste dumps.

In order to obtain a comprehensive environmental characterisation and to define tailored reclamation projects for the risk reduction, between the years 2004 and 2014, the Italian Institute for the Environmental Research and Protection (ISPRA, Rome, Italy), commissioned by the Italian Ministry of Environment (now MiTE) in the framework of the National Programme for Land Reclamation and Environmental Restoration (Ministerial Decree 468/2001), carried out large-scale investigation plans aiming to define the concentration, distribution, and potential pathways of organic and inorganic pollutants. During these activities, particular attention was paid to the characterisation of marine sediments, since they represent the final sink for a wide variety of chemicals [12]. At the same time, several natural factors, such as bioturbation and resuspension by waves, storms, and tidal currents, and different anthropic activities (e.g., dredging, trawling, and navigation activities) may cause contaminants to become mobilised and released from sediments, which therefore play a fundamental role as a secondary source of pollution for the aquatic environment and marine fauna [13–16]. In addition, sediments represent the most suitable matrix for the assessment and monitoring of the marine environmental quality because the concentration of contaminants is less variable in time and space than in seawater [17]. A detailed description of the methodological approach applied by ISPRA to characterise the environmental status of the marine areas included in the SINs is shown in Ausili et al. [10]. The strategy was defined accounting for the main European legislation in force in the early 2000s.

Based on the distribution and related concentrations of contaminants, Ausili et al. [10] highlighted similarities among SINs where the same type of anthropogenic activities was established. In fact, SINs characterised by the presence of large iron and steel plants (SIN_17, SIN_09, SIN_13, SIN_21, and SIN_07) were mainly contaminated by metals (Cd, Pb, Zn, As, Cu, and Hg), PAHs, TPHs, and TBTs, which show an almost homogeneous distribution and higher concentrations in the sediments sampled close to the plants. Metals and TPHs resulted to be the main contaminants in the SINs were both industrial and petrochemical activities were carried out (SIN_06, SIN_03, SIN_36, SIN_02, and SIN_04). According to the results of this characterisation, the SINs numbered SIN_27, SIN_05, SIN_47, and SIN_34 showed a single source of pollution, being characterised by the presence of factories related to the production of Cr compounds (SIN_27), nitrogenous fertilisers (SIN_5), and mining activities (SIN_47 and SIN_34). Finally, for the SINs numbered SIN_44, SIN_48, and SIN_10, the concentrations of chemical pollutants were lower than in the other sites (the last two ones are currently excluded from the national lists). Furthermore, with Ministerial Decree n. 222 of 22 November 2021, the MiTE identified, on the proposal of the regions, the list of "orphan sites" to be reclaimed and which can be redeveloped thanks to the investments provided for in the National Recovery and Resilience Plan. As indicated by Ministerial Decree n. 269 of 29 December 2020, the "orphan site" represents a potentially contaminated area for which the person responsible for the pollution is not identifiable. In fact, the sites are abandoned industrial or mining areas, illegal landfills, former incinerators or refineries. These areas are often covered with waste, polluted by various toxic substances, which pose a threat to human health as well as have a strong environmental impact, in particular on soil, water and air.

The Apulian coastal (southern Italy) sector extends along the Adriatic and Ionian Sea for approximately 900 km, which corresponds to 12% of the Italian littorals. It is characterised by a remarkable scenic and environmental value due to the high geomorphological and ecological diversity [18–20]. Its coastal environments host species and habitats of great importance and protected at the international level, for the protection and safeguarding of which, three marine protected areas and a number of coastal sites included in the Natura 2000 framework have been established. The presence of numerous scenic coastal sites has turned the Apulian region into a highly attractive tourist destination [21], leading to an increasing tourism demand that contributes significantly to the regional economy, in terms of job activities, facilities development, and foreign exchange. In contrast to the high natural, ecological, and tourist importance, the Apulian coasts host three sites included in the SINs list, being characterised by a high environmental risk [22–26] and requiring priority importance for their complex reclamation. The Apulian coastal SINs (SIN_05 "Manfredonia", SIN_06 "Brindisi", and SIN_07 "Taranto") were established by the National Law 426/1998 and delimited by the Ministerial Decree of 10 January 2000. They occupy a total land surface of 10,450 ha and 13,458 ha of sea surface. The marine-coastal and brackish areas included in the perimeters of the Apulian SINs were characterised following the standard procedure defined by ISPRA. The SIN in Taranto, due to the high environmental complexity that characterises its coastal sectors, has been object of further cognitive investigations promoted by the *Special Commissioner for the urgent measures of reclamation, environmental improvements, and redevelopment of Taranto"* for gathering new insight into the origin, distribution, and mobility of the contaminants within the Taranto coastal basins ("Mar Piccolo" (Little Sea), and "Mar Grande"(Big Sea) basins). In particular, in a first phase (2013–2014), the Regional Environmental Agency (ARPA Puglia, Bari, Italy), in collaboration with different scientific partners, developed a technical-scientific program of activities aimed at deepening the knowledge of environmental status of the Mar Piccolo basin [22,23,25–32]. In the second phase (2015–2017), new multidisciplinary surveys were funded by the Special Commissioner with the aim of collecting data for the definition of geological, sedimentological, mineralogical, chemical, and geotechnical features of both coastal basins included in the SIN area [33–39]. Collected data support the understanding of the interactions between contaminants and environmental matrices, which are of paramount importance in the definition of the conceptual model of the site [37,40].

Starting from the analysis of experience gained in the previous characterisation activities, ongoing research actions are aimed at developing a protocol of integrated investigations for the definition of priority analyses to support the effective geo-environmental characterisation of contaminated coastal areas, for which innovative and high-resolution investigation methodologies are required. The suggested investigations will allow the geological model of the investigated site to be obtained, as well as its stratigraphic, sedimentological, and geochemical features. This set of information, integrated with chemical and geochemical analysis, will represent the scientific basis for defining the most suitable remediation and protection strategies for reducing health and environmental risks. The characterisation surveys are also needed to define tailored actions for monitoring the long-term effects after remediation operations. In addition, harmonised guidelines for the management of contaminated sites would be beneficial for exchange of knowledge between national administrators and the international community [6].

2. Overview of Reports Concerning the Characterisation of the Contaminated Sites in Italy

At the national level, ISPRA has made available a number of manuals and reports to support the preliminary characterisation of potentially contaminated sites, which is defined as *"the set of activities that allow to reconstruct the contamination of environmental matrices, in order to obtain basic information on which to make decisions feasible and sustainable for the safety and/or remediation of the site itself"* (Annex 2 to Title V, Part Four of Legislative Decree 152/2006). The report 146/2017 [41] defines tools to guide the elaboration of characterisation plan (art. 239 paragraph 3 of Legislative Decree 152/2006) relating to remediation and management of areas characterised by diffuse pollution whose management is committed to regional authorities. This document also provides a summary of the technical documentation available to support the chemical, microbiological and ecotoxicological characterisation of contaminated sites.

Report 146/2017 updated a previous manual [42] that for the first time has addressed the issues related to contaminated sites, paying particular attention to the investigations needed for the characterisation of soil, subsoil, and groundwater. In 2018, SNAP published a document to collect the experiences developed by the regional environmental agencies with regard to the methodological aspects and procedures for the determination of background values for pollutants present in soils and groundwater [43]. This report complements the information of previous documents published by other national and regional authorities with regard to the determination of background values in different environmental matrices, such as agricultural lands included in contaminated sites, groundwater, and underground water bodies. With regard to the technical procedures for handling marine sediments, a specific manual was published by APAT/ICRAM in 2007 [44] to summarise actions to address issues related to the handling of sediments in the marine-coastal environment with particular reference to harbour dredging, beach nourishment, and immersion in the sea of excavated material. On the basis of the re-organisation of the Italian legislation regulating the handling of sediments in SINs (Ministerial Decree 172/2016) and the immersion in the sea of excavation materials (Ministerial Decree 173/2016), ISPRA has published a technical manual [45] (ISPRA, Manuals and Guidelines 169/2017) to support the use of mathematical models for the prediction and assessment of environmental effects related to the transport of sediments during the handling activities. However, the document does not address the aspects related to the analysis of the effects of the mobilisation of contaminants that may be present in the handled sediments. Finally, in 2021, ISPRA released a report in which reliable, homogeneous, and comprehensive data on the management of contaminated sites are provided [46]. The collection, systematisation, and analysis of a common dataset on the administrative procedures relating to the contaminated sites allowed both the management progress and the state of environmental contamination to be adequately described. The results of this analysis show that the total number of contaminated sites is 34.478 (updated to December 2019) and that, at the national level, there is a substantial balance between sites waiting for preliminary investigations (contamination not known; 35%), potentially contaminated sites (screening values exceeded; 33%) and contaminated sites (unacceptable risks; 29%). Nevertheless, the current version of this report does not include data related to the sites under the direct care of the MiTE (SINs).

3. Study Area

3.1. Physical and Environmental Setting of the Apulian Coastal SINs

The Apulian coastal SINs are located both on the Adriatic and the Ionian side of the Apulia region (Figure 2). From the geo-morphological point of view, the Apulia region presents varied characters that allow different districts characterised by specific physical features to be identified.

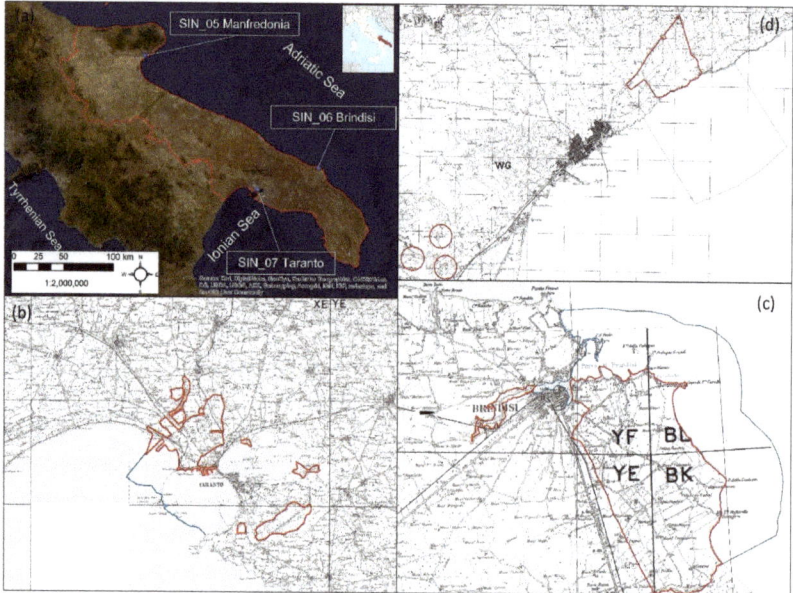

Figure 2. Study area location and Apulian coastal SINs perimeters. (**a**) Geographical location of the Apulia region, whose regional boundary is shown in red; (**b**) SIN "Taranto" (SIN_07); (**c**) SIN "Manfredonia" (SIN_05); (**d**) SIN "Brindisi" (SIN_06). The inland limit of each SIN is identified in red, while the blue lines show the perimeters of the marine areas included in the SIN. (**b–d**) were provided by the Italian Ministry of Ecological Transition (MiTE).

The SIN "Taranto" (SIN_07) is located on the northern Ionian coast of the Apulia region, between the south-western sector of the Apulian Foreland and the eastern Bradanic Trough, which represents the Pliocene-Pleistocene foredeep of the South Apennines orogenic system [47,48]. The Taranto coastal area is divided into two basins, called Mar Grande ("Big Sea") and Mar Piccolo ("Little Sea"). The latter is divided by the N–S Punta Penne promontory in two connected embayments: the First Bay and the Second Bay; see Figure 3a. The Mar Piccolo is connected with the Mar Grande by two channels: the shallow natural one ("Porta Napoli" channel, showed in Figure 3a) and the artificial one ("Navigabile" channel, Figure 3a), excavated during the XIX century in the Pleistocene calcarenite [49]. The stratigraphy of the area consists of Mesozoic limestone (Calcare di Altamura Fm.), and Upper Pliocene–Lower Pleistocene calcarenite (Calcarenite di Gravina Fm.) passing upwards and laterally to the interfingered argille subappennine informal unit. Marine, transitional, and continental terraced deposits occur in the surrounding foreland and foredeep sectors [50,51] and reference therein.

Figure 3. Taranto area. (**a**) Identification of the Mar Piccolo and Mar Grande with the channels who connecting the two basins (1, Porta Napoli channel; 2, Navigable channel). The main industrial sites are also indicated in red, while green and orange circles identify the areas pertaining to the Navy and ex-Tosi shipyards, respectively. (**b**) SCI IT9130004 "Mar Piccolo" perimeter. SCI area also includes the Regional Reserve "Palude la Vela". (**c**) Citri. (**d**) Cliff in the Mar Piccolo basin in which it was possible to recognise the argille subappennine unit and the overlying MIS 5 calcarenites.

The current landscape is dominated by a series of marine terraces, slightly dipping toward the sea whose deposits unconformably overlay the argille subappennine informal unit [48,50,52] (Figure 3b) forming quasi-flat surfaces consisting of marine terraces crossed by a fluvial network. The marine terraced deposits are located at different elevations (ranging from 2 m to 24 m above sea level (asl)) and show different lithostratigraphic features [49,53]. According to radiometric data [50], these marine terraces are connected to the sea-level highstand that occurred during the Last Interglacial (MIS 5) and they host a well-preserved marine record represented by a marly sandy unit with specimens of *Thetystrombus latus* (=*Strombus bubonius*, Gmelin, 1791) and other warm-water indicators such as *Cladocora caespitosa* (Linnaeus, 1767). Updated version of the geological and geomorphological map of the Taranto area have been proposed in Lisco et al. [54] at the 1:15,000 scale.

Due to its semi-enclosed features, the Taranto coastal area is characterised by a limited sea water circulation. Nevertheless, the presence of several submarine springs (Figure 3c), locally known as *"Citri"*, recharge the basins with freshwater. These submarine springs are characterised by a deep and steep inverted cone surface and by a high groundwater velocity determining an outflow visible on the seawater surface [55,56]. The outflows are characterised by high pressure and come from a karst aquifer developed into Mesozoic limestones. *Citri* are mainly located in the Mar Piccolo and are associated with subcircular depressions with variable depth up to 50 m in the Mar Grande and 30 m in the Mar Piccolo [57]. The peculiar hydrogeological characteristics of the Taranto coastal area have determined typical lagoonal features, which have favoured the distribution of transitional habitats and a very high biodiversity [34] for whose conservation and protection a Site of Community Importance (SCI-IT9130004 "Mar Piccolo") of about 14 km^2 has been established (Figure 3d).

The SIN "Brindisi" (SIN_06) is located on the eastern part of the Apulian region, on the Adriatic side. The marine coastal area included in the SIN_06 is constituted by a coastal stretch delimited to the North by Punta del Serrone and to the South by Cerano municipality; the area includes the Harbour of Brindisi and extends offshore up to 3 km. The Brindisi Harbour, one of the largest ports in the Mediterranean Sea, is divided into three main areas: Inner Harbour, Middle Harbour and Outer Harbour (Figure 4a). The Inner Harbour is formed by two long arms that encircle the city to the North and to the East, which are known, respectively, as "Seno di Ponente" and "Seno di Levante", with a surface of approximately 727,000 m^2. The seabed, with a depth varying between 2 and 7 m, is characterised by fine sediments. The Middle Harbour has an extension of 1,200,000 m^2 and overlooks the areas devoted to shipbuilding. Finally, the Outer Harbour occupies approximately 3,000,000 m^2. From a geological–structural point of view, the area represents a distensive tectonic depression, filled by the deposits of the "Bradanic Trough Cycle" and by subsequent "terraced marine deposits" [58]. In detail, starting from the most ancient unit, the following successions can be defined as follows: stratified Mesozoic limestone (Upper Cretaceous); sandy calcarenites (Upper Pliocene–Lower Pleistocene); blue-grey clays and sandy silty clays (Lower-Middle Pleistocene); yellowish sands with varying degrees of cementation or locally, organogenic calcarenites of a lithoid character (Upper Pleistocene); marine sands and sandy clays, marsh and lagoon silts (Holocene) [59]. In the area of Brindisi, calcarenitic facies belonging to the fourth lithostratigraphic unit outcrop. This unit is indicated in literature with the denomination of "terraced marine deposits". The morphological features of Brindisi area is determined by a deep *ria* located at the mouth of Pigonati river and by the swamps on the backdune areas [59]. The main reference related to the geological setting of the area is represented by the Geological Map "Foglio 204–Lecce" at the scale 1:100,000 and its related illustrative notes [60]. In the southernmost sector of the SIN "Brindisi", near the municipality of Cerano, a wide wetland area has been identified as a protected area (SCI/ZPS IT9140003 "Ponds and salt marshes of Punta Contessa"; see Figure 4a) due to its ecological value as a nesting and resting site of the migratory aquatic avifauna.

The SIN "Manfredonia" (SIN_05) is located in the Manfredonia Gulf (Adriatic Sea) on the southern part of the Gargano Promontory (Figure 2a,b). The marine coastal area in the SIN perimeter extends north of the Manfredoina industrial harbour and has an extension of about 4 km along the coast, for a total area of 860 ha (Figure 5, Table 2). Past geophysical surveys carried out by the Harbour Authority of Manfredonia allowed four lithological units to be identified, represented by: fine and homogeneous sediments (silts and clays) with a thickness up to 15 m; sands, silty sands, and gravelly sands (with limestone fragments) with a thickness varying from 10 to 15 m; intercalations of lenticular levels of clayey sands and gravels with compactness and resistance higher of the previous levels; limestones (Lower Cretaceous) covered by a cemented breccias of varying thickness. The depths at which the bedrock is reached increase towards the open sea. The main reference for the geological setting of the area is represented by the Geological Map "Foglio 164-Foggia" at the scale 1:100,000 and its related illustrative notes [61].

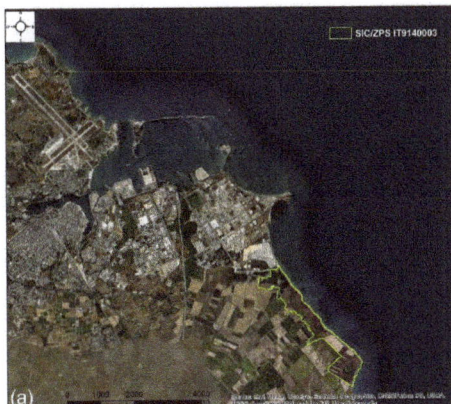

Figure 4. Brindisi area. (**a**) Localisation of the inner, middle and outer parts of the Brindisi Harbour. The perimeter of the SIC/ZPS IT9140003 "Ponds and salt marshes of Punta Contessa" is indicated in green. (**b**) Industrial plants included in the SIN "Brindisi" (Image credits: L'ora di Brindisi, available at: https://www.loradibrindisi.it/2020/09/on-gava-eliminati-33-mln-destinati-alle-bonifiche-del-sin-di-brindisi/, accessed on 3 May 2022).

Figure 5. Manfredonia area. Industrial plants included in the SIN "Manfredonia" (Image credits: Manfredonia news, available at: https://www.manfredonianews.it/2020/09/13/quali-le-sorti-dellarea-sin-ex-enichem/, accessed on 3 May 2022).

Table 2. Land and marine surface in the Apulian coastal SINs.

SIN	Land Surface (ha)	Marine Surface (ha)	Coastal Stretch (km)
Taranto (SIN_07)	4383	7005	approx. 40
Brindisi (SIN_06)	5851	5600	approx. 30
Manfredonia (SIN_05)	216	860	approx. 4

3.2. History of the Contamination and Status of Characterisation Procedures in the Apulian Coastal SINs

The area of Taranto has been affected by intensive environmental changes due to the strong industrialisation that has been taking place since the second half of the XIX century. The zone is characterised by a high concentration of industrial districts that have a high environmental impact, including the largest steelworks in Europe (Acciaierie d'Italia S.p.a.-ex-ILVA) inaugurated in 1965, the ENI refinery completed in 1964 and operative since 1967, the Taranto thermoelectric power plant (ex-Enipower S.p.A.), the Cemitaly plant (ex-Cementir), various landfills and numerous small and medium-sized manufacturing industries. Taranto also hosts one of the Italian Navy's most important and historic bases, with the related arsenal and shipyards (Figure 3a). In addition, the coastal basins of Taranto are significantly impacted by intensive mussel aquaculture activities [62,63]. The presence of numerous industrial plants has led to serious environmental problems, which affect all environmental matrices (soil, air, water and marine sediments) and, for this reason, Taranto is considered one of the most polluted cities in the western Europe [40,64].

The area included in the SIN_06 "Brindisi" is characterised by different industrial settlements mainly consisting of chemical and energy plants, including the thermoelectric power plant of Brindisi. The northern part of the SIN_06 also includes a craft-industrial agglomerate (Figure 4b), while, in the central and southern part, the site is occupied by agricultural areas, with intensive cultivations and vineyards. Finally, the ENEL thermoelectric plant of Cerano (built in the 1980s and at present the second largest thermoelectric plant in Italy) is located in the southernmost sector of the site and it is connected to the harbour for fuel supplies. The agricultural areas, located in the central part of the site, are characterised by intensive cultivation and vineyards and by a complex system of drainage channels that intersect the services connecting the thermoelectric power plant to the industrial area and representing, therefore, potential critical points of surface contamination, as well as discharge pathways for chemicals used in agricultural activities. The agricultural land has been included in the SIN perimeter since it is considered highly prone to be affected by pollutants produced by the surrounding industrial sites.

The area included in the SIN_05 "Manfredonia" covers 216 ha, of which about 96 ha are private areas consisting of the Polo Chimico (ex Agricoltura S.p.A. ex Enichem), currently owned by Eni Rewind S.p.A (ex Syndial). The industrial plants are devoted to the production of nitrogenous fertilisers, various chemical products, and by the petrochemical pole. Private areas located adjacently to Eni Rewind S.p.A plants are devoted to agricultural uses (Figure 5). In addition, the site includes public areas consisting of landfills built in calcarenite quarries used in the 1970s as storage sites for unauthorised solid urban waste. The main environmental problem in SIN_05 is due to the past activities aimed at the production of nitrogenous fertilisers for agricultural use, chemical products for artificial fibres, technopolymers and/or aromatic intermediates. In addition, in 1976, an explosion in the industrial plant resulted in a large arsenic leak. Since 1999, Agricoltura S.p.A. has suspended all production activities.

Regarding the state of the procedures for the reclamation of the land and groundwater in the SINs, the MiTE periodically publishes an updated document in which all the data available for the national sites are synthetised. In Figure 6, the most updated maps relating to the status of the reclamation of the Apulian coastal SINs are shown [8].

With reference to the SIN_07 "Taranto" (Figure 6a), the most updated data, which are refereed to May 2021, show that the characterisation plan was implemented for an area of 1997 ha; the soil reclamation project has been approved for an area of 329 ha while the reclamation of the water table has been approved for a surface of 341 ha. The data show that surfaces of 355 ha (0.18% of the characterised area) and 325 ha (0.16% of the characterised area) do not show contaminated soils water, respectively. Regarding the SIN_06 "Brindisi" (Figure 6b), the data, referring to May 2021, show that the characterisation plan has been concluded and approved for 5206 ha (0.89% of the total SIN land surface). The reclamation plan has been approved for a 689 ha of contaminated soil and 915 ha of contaminated ground water. According to these data, a surface of 386 ha (corresponding to the 0.07% of the characterised SIN land surface) contained uncontaminated soil and a surface of 486 ha (0.09%) contained uncontaminated groundwater. Finally, accounting for the SIN_05 "Manfredonia" (Figure 6c), whose data were updated in June 2020, the characterisation plan has been approved for the total surface included in the perimeter while the reclamation plan has been approved for 73 ha for the soil and 168 ha for the groundwater/aquifer. According to these data, a surface of 38 ha (corresponding to the 17.6% of the SIN land surface) presents uncontaminated soil while the whole groundwater surface resulted to be affected by pollution.

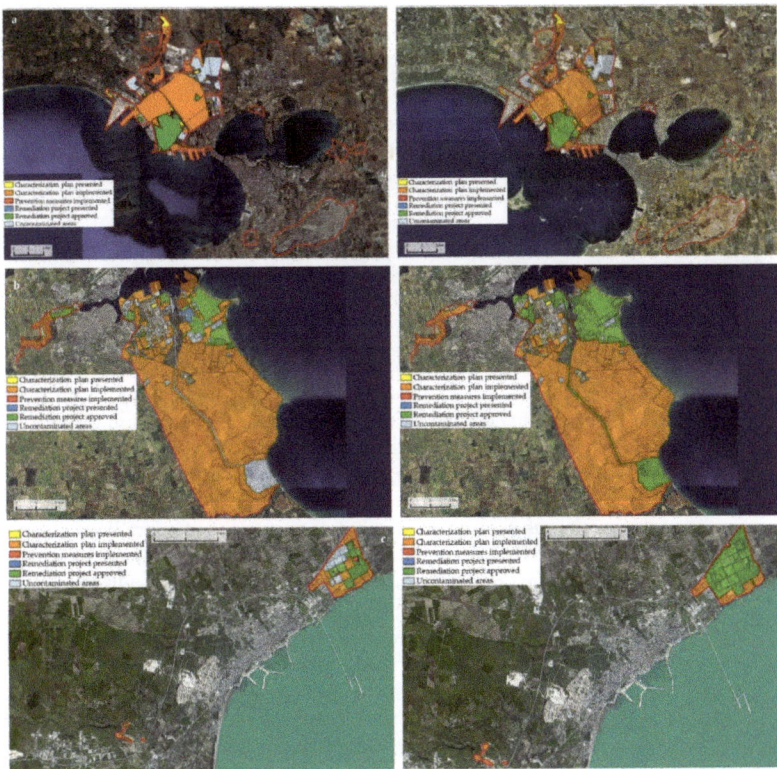

Figure 6. Areas in the Apulian coastal SINs for which characterisation plans and remediation projects have been approved. Pictures on the left panel are referred to soil while pictures on the right panel refers to the groundwater in (**a**) SIN "Taranto", (**b**) SIN "Brindisi", (**c**) SIN "Manfredonia". The red line identifies the SINs perimeters. Percentage data are provided in the main text. Images have been downloaded from MiTE website (https://bonifichesiticontaminati.mite.gov.it/sin/stato-delle-bonifiche/, last update: 5 June 2022).

4. Overview about the Results from the Characterisation Activities Carried out in the Apulian Coastal SINs

In 2001, the Italian Institute for the Marine Research (ICRAM, now ISPRA) was commissioned by the Ministry of Environment (now MiTE) to define a methodological approach for the integrated environmental characterisation of the marine and coastal areas included the SINs. ISPRA proposed a flexible and large-scale monitoring program that was applied in the period 2004–2014 for the analysis of almost all the coastal marine areas in the SINs [10]. In detail, the proposed investigation strategy envisaged a sampling scheme with coastal grids varying in size (from 50×50 m to 450×450 m) according to the type, extension, and complexity of the investigated area. In each grid, a sediment core was executed (with a depth from 2 to 5 m) and a number of superficial samples were planned along transects covering the areas not included in the grid. A physical, chemical, and ecotoxicological analysis was carried out in standardised levels defined for the sediment cores. In order to evaluate the contamination level, the concentrations of the investigated pollutants were compared with the site-specific reference values determined in 2004 by ICRAM for the SINs and with the "CSC—*Concentrazioni Soglia di Contaminazione*", which translates as "Contamination Threshold Concentrations" established by the national legislation (Legislative Decree 152/2006) for all the Italian industrial sites (Table 3).

Table 3. Threshold limit values identified at the national level (expressed in terms of CSC–cf. Legislative Decree 152/2006) and at the site-specific level for the SINs "Taranto" and "Brindisi". Concentration limits are in ppm (mg/kg dw). (*) refers to sediments with a silt distribution < 20% while (**) refers to sample with silt distribution > 20%. The site-specific threshold limits are not available for the SIN "Manfredonia".

Limit Values	OSn	PAHs	PCB	TPH	As	Cd	Cr	Hg	Ni	Pb	V	Cu	Zn
National thresholds (CSC)	-	100	5	750	50	15	800	5.0	500	1000	250	600	1500
Taranto site-specific action levels	0.07	4	0.19	-	20	1	70 */160 **	0.8	40 */100 **	50	-	45	110
Brindisi site-specific action levels	0.07	4	0.19	-	20	1	100	0.4	50	50	-	45	110

The characterisation plans proposed for the Apulian coastal SINs were tailored accordingly to the site-specific geo-morphological characteristics and to the potential extent of the contaminated areas. The characterisation plan proposed for the SIN_06 "Brindisi" was released by ISPRA in 2011 [65]. Due to its large extent, the marine and coastal area included in the perimeters was divided in "Harbour Area", which included the Inner, Middle, and Outer harbour, and "Coastal Area", which included marine and coastal areas external to the Harbour up to the offshore limit of the SIN. The characterisation plan envisages the following activities: (i) preliminary investigations aimed at the detection of weapons; (ii) geophysical analysis aimed at calibrating the sampling grid; (iii) sediments cores and extractions of samples for the chemical analysis. The characterisation plan of the Harbour Area consisted of a 150 × 150 m sampling grid corresponding to 252 cores with a length ranging from 2 m to 3 m. The characterisation of the coastal area external to the Brindisi Harbour up to 500 m from the coastline was based on a sampling grid of 150 × 150 m and 206 sampling stations. Furthermore, in the marine area over 500 m from the coastline, samples were collected along 700 m-long transects; along each transect, several sampling stations spaced 500 m apart were identified (49 cores with a depth of 2 m and 64 superficial samples). Finally, along the beaches, the characterisation plan consisted of 67 linear transects spaced 150 m apart, along which a single core was to be extracted with a depth of 2–3 m. In total, the proposed characterisation activity consisted of the collection of 322 cores and 64 superficial samples. In addition, a tailored characterisation plan was defined for the Sant'Apollinare area, located in the Outer Harbour, in the frame of a specific agreement signed between ISPRA and Brindisi Harbour Authority [66]. In this case, the characterisation plan consisted of 65 sediment sampling stations within a regular grid measuring 50 × 50 m.

Due to the lack of previous environmental characterisations for the coastal and marine area included in the SIN_05 "Manfredonia", no data for the identification of critical areas were available. For this reason, in 2004, ICRAM investigated the entire marine area included in the SIN [67]. The proposed characterisation plan consisted of a sampling grid of 150 × 150 m for the analysis of the marine area from the shoreline up to a distance of 600 m. In each cell, a sample core with a depth ranging from 2 m to 4 m was extracted. In the remaining marine area (from the distance of 600 m to the offshore SIN limit) eight sample transects perpendicular to the shoreline were defined with a spatial distance of 450 m. Along each transect, three or four sample stations (core and/or superficial sediments) were identified. In total, the characterisation activity consisted of the collection of 100 cores and 15 superficial samples.

The results of the characterisation plan proposed for the SIN_07 "Taranto" were released by ISPRA in 2011 [68] and was aimed at the characterisation of the entire marine and coastal area in the SIN perimeter which, due to its large area, was divided into four different sectors (i.e., "Punta Rondinella west area", "Mar Grande–I Lotto", "Mar Grande–II Lotto", "Mar Piccolo"). The plan consisted of a total of 507 cores with variable length and 40 superficial sediment samples.

5. The Case Study of SIN_07 "Taranto"
5.1. Summary of the Characterisation Activities Performed from 2004 to 2015

The activities for the preliminary characterisation of marine and coastal area in the SIN_07 "Taranto" were conducted by ISPRA in the period July 2009–May 2010 [68]. The investigated area includes both the Mar Grande basin and the Mar Piccolo basin. Nevertheless, the southernmost sector of the First Bay in the Mar Piccolo basin (known as "Area 170 ha") was not investigated. The geophysical activities executed in the frame of the characterisation plan included morpho-bathymetric (MultiBeam EchoSounder (MBES) and Side Scan Sonar (SSS)) and seismic surveys (Sub Bottom Profiler (SBP)). In the shallow water areas and in the zone where the navigation was not possible due to the presence of anthropogenic obstacles (i.e., mussel farms), the MBES survey was replaced by a Singlebeam survey. In addition, a magnetometric survey was also carried out to identify war devices on the seafloor. Regarding the sediment quality characterisation, 238 cores in the Mar Grande and 269 in the Mar Piccolo basin with variable length were extracted through a manual core barrel in the mussel farm areas and through a vibrocorer in remain areas. In addition, 40 superficial samples were extracted by a bucket. Approximately 2000 sediment samples from cores and bucket were used to carried out chemical–physical analysis. In particular, particle size, water content, specific weight, pH, redox potential, metals and trace elements, Polychlorobiphenyls (PCB), Organic pesticides, Lead, Copper, Zinc, Vanadium), Organochlorine pesticides, PAHs, Total Hydrocarbons (TPH), Light Hydrocarbons $C \leq 12$, Heavy Hydrocarbons $C > 12$, Total Nitrogen, Total Phosphorus, Cyanides, and Organic Carbon (TOC) were analysed for almost all the samples in both the Mar Grande and Mar Piccolo. For a lower number of samples, further analyses were also performed (Chromium VI, Phenols, Aromatic solvents, Organotin compounds, Dioxins and Furans in part of the samples). In addition, microbiological parameters and ecotoxicological analysis were carried out on representative samples. To evaluate the contamination level, the concentrations of the investigated pollutants were compared with the site-specific reference values and with the "CSC" valid for all the Italian industrial sites (Table 3). The results of the integrated characterisation activities were provided as maps showing the spatial distribution of each parameter elaborated by means of geostatistical methods (Block kriging and Block Co-kriging). Data were interpolated up to the sediment thickness of 2 m in the areas not included in the mussel farm zones and up to 0.50 m for the samples in the mussel farm zones. Results showed that sediments in the Mar Grande are silty sands, sandy silts, and sands, while in the Mar Piccolo basin sediments are mostly silt and sandy silt. The chemical characterisation showed that the contamination in the Mar Grande basin was mostly due to metals and trace elements (Hg and Zn) and Cu, Pb, and As, the presence of which affected at least the first meter of sediment. High concentration levels of Hg were identified in the surface samples (with values even above the national limit in the first 0.50 m) and in the 0.50–1 m layer. In the central part of the basin, a high concentration of Hg was found, even at depths over 1 m. Contamination due to organic compounds was much less evident, both in terms of the extent of the affected area and depth, and this was mainly due to polycyclic aromatic hydrocarbons (PAHs), whose concentration exceeds the site-specific action level, and to TPH, whose concentration, in the same specific areas of the basin in proximity to the coastline, exceeded the value of 1000 mg/kg. In both cases, the contamination affected the first meter of sediment thickness.

The Mar Piccolo environmental state resulted to be very complex due to the presence of a high concentration of both inorganic and organic compounds, with special reference to the First Bay of the basin. The results of the chemical characterisation showed that Hg concentrations exceeded both the site-specific and the national limit in all the analysed surface sediment samples (0–0.50 m) of the First Bay. In the Second Bay, even though the site-specific limit was exceeded in a wide portion of the area, the results were lower in its central part and the easternmost sector. With regard to other metals and traces elements (Zn, Cu, Pb), their concentrations exceeded the site-specific action values both in the First Bay and in the Second Bay up to the 1 m sediment thickness. In the case of Zn, higher concentrations were found even in samples from deeper sediment layers (up to 2 m, when available). Nevertheless, national limits were not exceeded. Contamination from As affected sediment quality only in the First Bay. Considering the organic compounds, PCB concentration exceeded the site-specific action level in the northern sector of the First Bay, where in the shipyard area the contamination also affected the deeper sediment layers, and in the western sector of the Second Bay. National limit was exceeded only in the superficial sediment samples from the First Bay. Even if the TPH contamination affected only some areas mainly in the first bay, their concentration resulted to be higher than the threshold value up to deeper sediment layers (1.5 m). Finally, IPA exceeded the site-specific threshold in the First Bay mainly in the upper layer but, in some limited areas, they reached even the deeper layers. Maps showing the spatial and vertical distribution of organic and inorganic compounds in the Mar Piccolo basin can be consulted in Labianca et al. [40].

The analyses performed by ISPRA in the Mar Piccolo basin were further updated and integrated by the Regional Agency for the Prevention and Protection of the Environment (ARPA Puglia) who has defined and implemented (from May 2013 to April 2014) a technical–scientific programme of activities aimed at supporting the outline a conceptual model of contamination [27]. These analyses allowed the southernmost portion of the First Bay to be characterised; it was excluded from the characterisation performed by ISPRA. The results of program have led to a high number of interdisciplinary papers, which represent a scientific reference for the characterisation of the Mar Piccolo basin ([26] and reference there in).

5.2. Summary of the Characterisation Activities Founded by the Special Commissioner for Urgent Measures of Reclamation, Environmental Improvements, and Redevelopment of Taranto from 2015

The awareness of a widespread environmental risk [24–26], the epidemiological data indicating values above the national average for every type of cause of death [69], and the need to protect a territory characterised by a high socio-economic and geo-environmental relevance led, in 2014, to the definition of the Special Commissioner for the area of Taranto, who promoted an interdisciplinary study for the integrated characterisation of the coastal system in the SIN_07 perimeter. In this framework, new surveys were envisaged to obtain both direct and indirect data necessary for the geological, sedimentological, mineralogical, geochemical, and biological characterisation of the area.

Specifically, in the first phase of the study, started in 2016, the Mar Piccolo basin was investigated, while in the second phase (started in 2017), the survey activities were carried out in the Mar Grande basin. In Figure 7, the navigation lines defined for the acquisition of the geophysical data (Side Scan Sonar-SSS, MultiBeam-MBES, Sub Bottom Profiles-SBP, Sparker-SPK, Magnetometric-MG) are indicated (Figure 7). In the Mar Piccolo area, marine, coastal (land–sea interface), and terrestrial geoelectric surveys were also conducted (Figure 7). The profiles' total length was 6675 m, with an interelectrode spacing of, respectively, 20 m and 5 m for the coastal and terrestrial profiles.

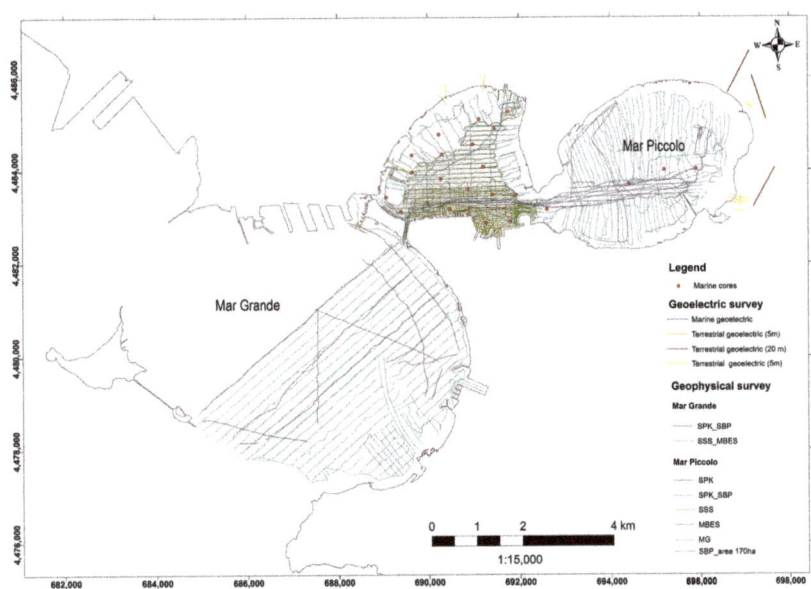

Figure 7. Navigation lines followed for the geophysical and geoelectric surveys carried out in the frame of the activities funded by the Special Commissioner. The positions of the 24 sediment cores carried out in the Mar Piccolo are also indicated.

Regarding the direct analysis, in the period from September 2016 to March 2017, 24 sediment vibrocores were extracted in the Mar Piccolo basin using 1.5 m-long cores (Figure 7) at different sampling depths up to the limit of the argille subappennine informal nit. Each liner was appropriately sectioned so that it could be used for both sedimentological and chemical analyses. Once the cores were transported to the laboratories, a preliminary visual core description was carried out, taking into consideration the following parameters: degree of the drilling process disturbance, colour, lithological and granulometric characteristics, sedimentary structures, accessories (shells type, organic material, presence of glauconite or other minerals, concretions and nodules, archaeological findings) (Figure 8). Extensive physical and chemical analysis in the sediment samples from the First Bay basin were carried out. These activities included the determination of the following parameters: sediment granulometry, redox potential, organic matter, water content, organic and inorganic pollutants (PCBs, PAHs, TPH, TBT, DTB, MBT), metals (Pb, Cd, V, Ni, Cu, Zn, Hg, Cr, Fe, Al, Mn, As and Sn), and Dioxins and Furans. For the definition of the degree of contamination, the concentrations of the pollutants resulting from the chemical analyses were compared with the site-specific action levels established for the SIN_07 "Taranto" and with the national thresholds (CSC) (cf. Table 2).

Detailed description of the technical specifications of the geophysical survey as well as of the analytical procedures for the chemical analyses are reported in Valenzano et al. [33] and Cotecchia et al. [37], respectively. In order to obtain more details on the mineralogical composition of the sediment samples, further analyses were also carried out. These included the acquisition of magnetic susceptibility profiles, the detection of heavy metals in very small concentrations, the X-ray Fluorescence (XRF), X-ray Powder Diffraction (XRPD), and Trasmission Electron Microscope (TEM). Further analysis, such as liquid limit, plasticity index, activity index, soil solid-specific gravity, organic matter, void ratio, water content, and liquidity index, allowed estimate chemo-mechanical proprieties of the sediments.

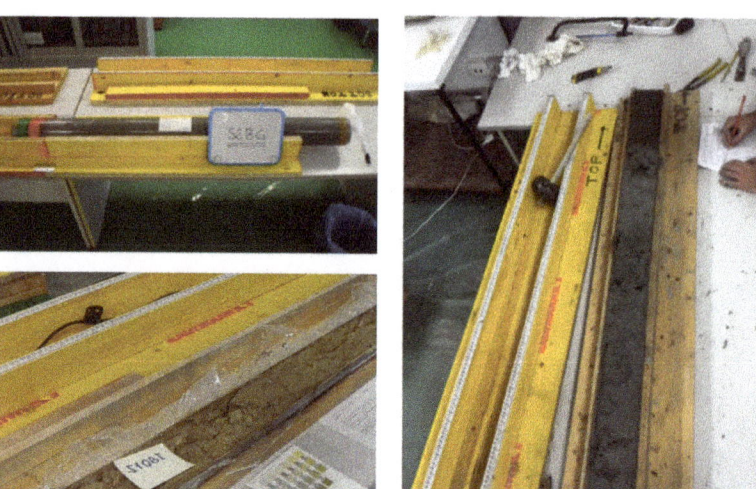

Figure 8. Phase of preliminary description of sediment liners from the Mar Piccolo basin.

The integrated interpretation of the geophysical data acquired in the Mar Piccolo with chronostratigraphic information derived from direct cores and ^{14}C dating [35] allowed the geometrical relationships between sedimentary bodies to be defined, as well as their lateral continuity, thickness, and depth providing scientific support to the definition of the Holocene morpho-sedimentary evolution of basin [33]. In addition, through the qualitative description of the sediment samples, the main *facies* were identified and correlated with the seismic units obtained from the analysis and interpretation of the high-resolution single-channel seismic data (SBP and SPK). Specifically, through the interpretation of the SPK profiles, the upper limit of the carbonate substrate (Calcare di Altamura Fm.) was identified (Figure 9), on which, in discordance, it was possible to recognise a thick clayey succession referable to the informal stratigraphic unit of the argille subappennine (Pliocene–Middle Pleistocene) in heteropia with the Gravina Calcarenite Fm. (Pliocene). The digital model of the carbonate-top surface shown in Figure 9 has been obtained by interpolating available data from SPK interpretation and already available core data [33]. On the other hand, the interpretation of the SBP profiles had highlighted the thicknesses and geometries of the post-Last Glacial Maximum (LGM) units, which develop in the incised-valley morpho-stratigraphic system. As shown in Figure 10, this structure can be followed with good continuity from the Mar Piccolo basin to the Mar Grande basin [33,70].

The interpretation of the acoustic data (MBES and SSS) allowed the high resolution morpho-bathymetric setting of the coastal area to be defined, highlighting the morphological features that can be ascribed to local natural assets (i.e., Citri, [57]). In addition, these data represented a useful support for the identification and mapping of elements and traces from anthropogenic activities, providing, therefore, an indirect assessment of the human footprint in the seafloor. The analysis of the most recent acoustic data will certainly update the indirect and direct surveys already carried out in the past for the Mar Piccolo area [28,34], integrating the same analysis for the Mar Grande basin (Figure 11).

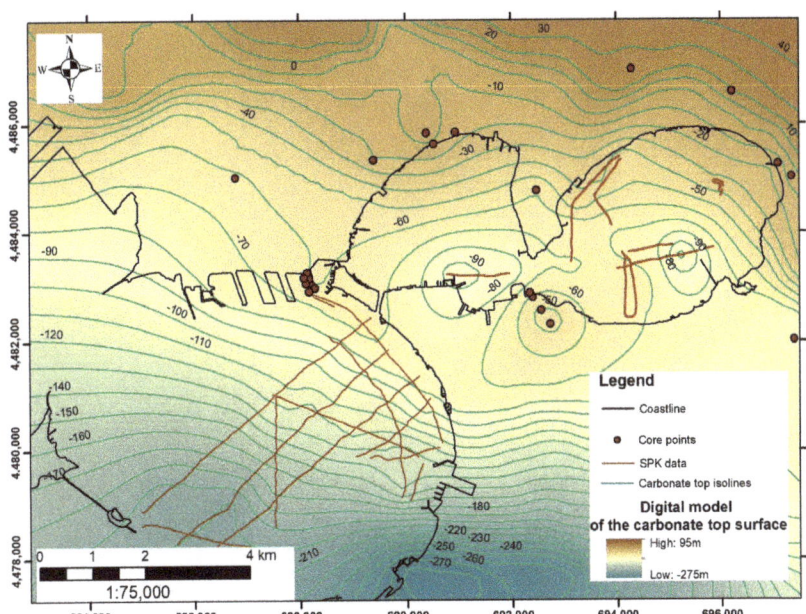

Figure 9. Digital model of the upper limit of the carbonate substrate (Calcare di Altamura Fm.) identified through the analysis and interpolation of SPK data (for the submerged area) and core data. Isolines are referred to local mean sea level.

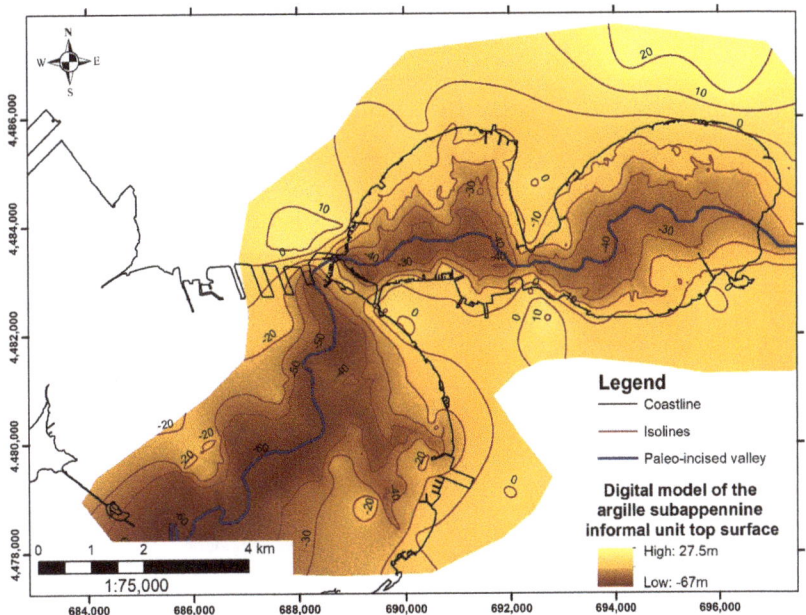

Figure 10. Digital model of the upper limit of the argille subappennine informal unit identified by the analysis of the SBP, SPK data and cores.

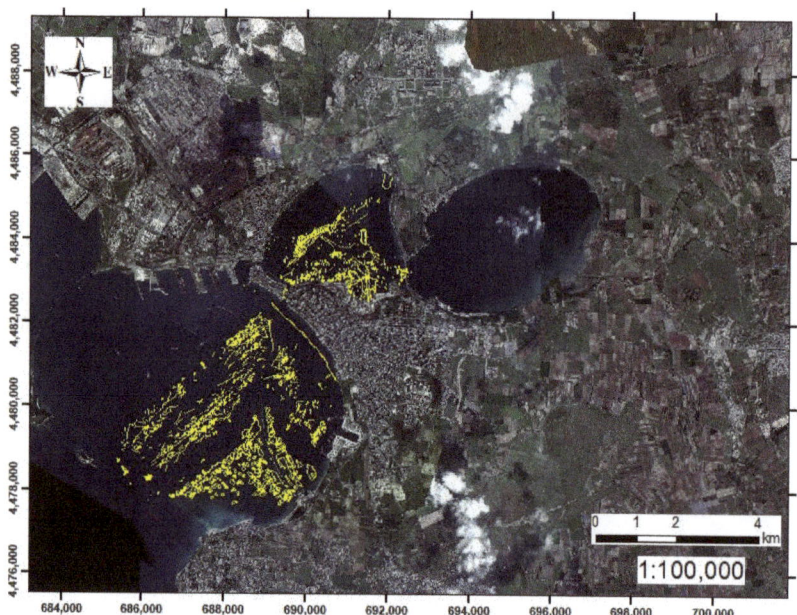

Figure 11. Distribution of anthropogenic traces detected on the seafloor of the Mar Piccolo (first Bay) and Mar Grande basins through the interpretation of SSS and MBES data.

As far as the results obtained from the analysis of chemical parameters are concerned, they showed a substantial environmental criticality in the southern sector of the First Bay (i.e., the area defined as "Area 170 ha"). As regards the distribution of pollutants along the vertical profile of the sediments, it emerged that the highest concentrations characterise the first sediment layer (0–0.50 m). The concentrations of the analysed metals in the first sediment layer are indicated in Table 4.

As can be seen from the analysis of these data, most of the sediment samples present concentration of the inorganic compound higher than the site-specific action values. In surface samples relative to cores S02, S03, S04, S05, S06, S16, there are at least six concentrations of inorganic pollutant higher than the limit values. In sediments from core S03 and S06 the CSC limit value for the Hg is exceed. In Figure 12, the results of the chemical analysis performed on superficial sediments are summarised. Sediment cores are indicated with circles, whose sizes depend on the number of inorganic and organic pollutants exceeding site-specific thresholds.

However, the presence of some pollutants at higher depths, with concentrations even exceeding the threshold limits, infer the occurrence of local mixing phenomena in the more superficial and unconsolidated sediments. In particular, as highlighted by the authors of [71], who analysed and mapped the Organotin compounds concentrations in sediment samples up to 3 m, the greatest thickenings of reworked sediments were detected mostly in the southern and northwestern areas of the First Bay, where the bathymetric data showed a remarkable perturbation of the seafloor mainly ascribed to anthropogenic activities (e.g., dredging and wrecks). Cotecchia et al. [37] provided the analytical results for chemical, geotechnical, and mechanical proprieties evaluated for sediment sampled from six selected cores (S01, S02, S03, S04, S06, S07) from the sea-floor interface up to a depth of approximately 30 m. The results obtained from the analysis of chemical parameters updated the knowledge on contamination level in the Mar Piccolo basin. Nevertheless, no analysis has been carried out for the chemical characterisation of the sediment from the Mar Grande basin.

Table 4. Concentration values calculated for inorganic compounds. (*) indicates the pollutants for which site-specific and national limits are available (cf. Table 2). Values exceeding site-specific action limits are indicated in bold, while values also exceeding the national limits are underlined. Concentrations are expressed in mg/kg dw.

Core	Hg *	Cd *	Pb *	As *	Cr *	Cu *	Ni *	Zn *	Fe	Mn	Sn	Al	V
S01	0.46	0.18	36.38	7.40	44.96	21.97	38.31	49.89	11,857	297	2.58	26,353	48.94
S02	4.70	0.57	**150.38**	**23.42**	60.45	**87.92**	60.83	**293.45**	28,402	484	19.60	36,235	89
S03	<u>**8.33**</u>	1.16	**261.63**	**21.91**	71.06	**76.61**	56.13	**311.43**	24,079	437	15.25	28,950	83.81
S04	4.10	0.63	**129.03**	18.51	**90.63**	**75.17**	**74.27**	**276.63**	34,969	561	15.63	65,743	100.15
S05	3.94	0.69	**213.58**	**21.64**	115	54.82	**82.50**	**218.14**	48,868	732	13.08	102,065	113
S06	<u>**15.36**</u>	0.90	**229.29**	**44.75**	82.85	**83.75**	61.70	**402.90**	25,414	446	18.55	55,449	88.52
S07	0.31	0.28	35.96	10.13	**83.17**	14.83	**68.75**	79.68	32,397	540	2.54	51,116	99.32
S08	0.35	0.26	30.10	15.74	123	27.05	80	102	38,635	671	3.02	85,917	125
S09	<LOQ	0.13	21.67	9.54	103	17.16	64	59	31,776	671	2.31	53,183	84
S10	0.05	0.23	21.60	5.54	74	16.82	**50.56**	52.58	19,473	372	2.19	34,269	79
S11	0.95	0.24	46.60	15.91	150	28.75	**83.41**	121.82	44,018	655	4.24	88,392	121
S12	0.09	0.17	26.78	11.73	100	17.27	**64.47**	72.77	28,658	730	2.82	50,796	93
S13	0.11	0.20	23.37	6.57	95	20.76	**67.13**	71.92	23,178	415	2.66	40,148	93
S14	0.23	0.20	46.75	14.39	99	23.07	**77.17**	90.92	21,952	852	3.23	48,145	131
S15	2.77	0.05	78.78	17.44	81	35.91	**73.22**	125.10	18,456	547	6.05	45,200	129
S16	3.42	0.62	**147.91**	**23.18**	100	**117.00**	**68.78**	**337.85**	33,974	545	88.05	69,798	105
S17	2.35	0.25	**100.34**	15.70	124	37.37	**79.47**	**133.27**	31,413	543	6.38	68,534	138
S18	2.54	0.23	79.10	15.54	**95.62**	37.08	**67.08**	**134.62**	34,745	595	9.67	50,086	100
S19	1.07	0.28	37.92	11.36	151	36.22	**92.24**	107.70	39,844	593	4.57	60,226	118

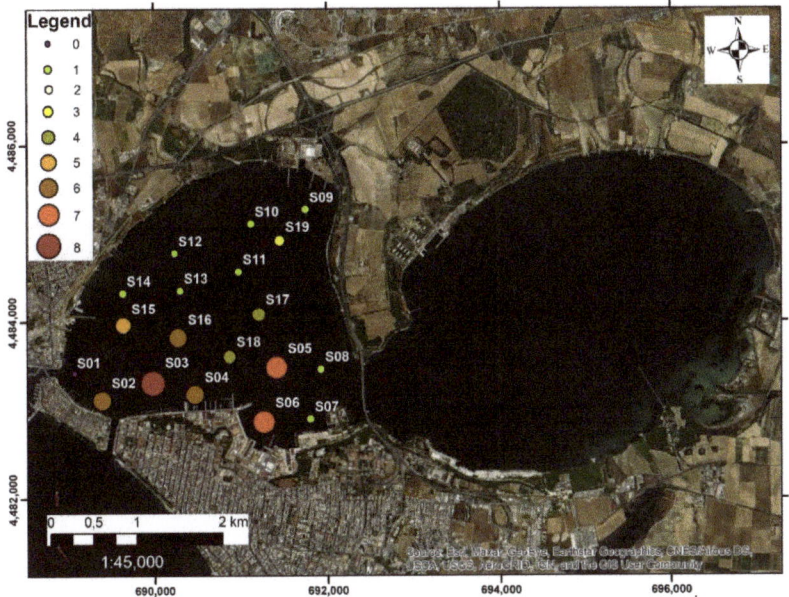

Figure 12. Sediment cores in the First Bay of the Mar Piccolo basin. The size and the colour of the circles are proportional to the number of heavy metals whose concentration in the first layer of sediment (0–0.50 m) has resulted to be above the site-specific action values. S01 does not show any concertation above the site-specific limit, while in the samples from cores S03 and S06, the Hg concentration exceeds the national limit (CSC value).

Comparing the analytical results obtained during the characterisation activities funded by the Special Commissioner with the international sediment quality guidelines ERM and ERL (effects range medium and effects range low), it emerged that the concentrations of As, Cr, Hg, Ni, Pb, Cu, and Zn exceed the ERL values at least in one sample; furthermore,

Hg, Ni and Pb concentrations also exceed the ERM values (Table 5). Specifically, Ni concentrations exceed the ERL values in all the 19 samples, As and Hg concentrations exceed the ERL values in 16 samples (S02, S03, S04, S05, S06, S07, S08, S09, S11, S12, S14, S15, S16, S17, S18, S19 and S01, S02, S03, S04, S05, S06, S07, S08, S09, S11, S14, S15, S16, S17, S18, S19, respectively), Cr concentrations exceed the ERL values in 15 samples (S04, S05, S06, S07, S08, S09, S11, S12, S13, S14, S15, S16, S17, S18, S19), Pb and Cu concentrations exceed the ERL values in 10 samples (S02, S03, S04, S05, S06, S15, S16, S17, S18 and S02, S03, S04, S05, S06, S15, S16, S17, S18, S19, respectively). Finally, Zn concentrations exceed the ERL values in six samples (S02, S03, S04, S05, S06, S16). Considering the ERM values, Hg concentrations are higher in 11 samples (S02, S03, S04, S05, S06, S11, S15, S16, S17, S18, S19), Ni concentrations are higher in all the samples excluding S10 and S10, and Pb concentrations are higher in samples S3 and S6. Nevertheless, it is worth noting that the use of indicators such as ERL and ERM can be considered as a first attempt to link the bulk chemistry with toxicity [72–75]. Chemical concentrations below the ERL value represent a range below which adverse biological effect would rarely be observed; similarity, the ERM values represent a potential range above which adverse effects on biological systems would frequently occur [72].

Table 5. International sediment quality guidelines (ERL and ERM) for some trace elements. All the concentrations are expressed in mg/kg dw. * Reference values provided in [72].

SQGs	As	Cd	Cr	Hg	Ni	Pb	Cu	Zn
ERL *	8.2	1.2	81	0.15	20.09	46.7	34	150
ERM *	70	9.6	370	0.71	51.06	218	270	410
Samples from the First Bay of the Mar Piccolo (min and max values)	5.54 44.75	0.05 1.16	44.96 150	0.05 15.36	38.31 92.24	21.60 261.63	14.83 117.00	49.89 402.90

6. Discussion on Methodological Procedures Available for the Integrated Characterisation of the Coastal Contaminated Sites

As shown in the review of scientific papers available at the national and regional scale [25,28,76–85], the different types of analysis generally envisaged to perform the geo-environmental characterisation of contaminated coastal sites represent a starting point for the assessment of the level of anthropogenic impact on marine environmental matrices. Nevertheless, many of these analyses may support the definition of a geo-morphodynamic model of the investigated area. Based on the review of the achievements of previous analysis carried out in the Taranto area and by the support of expert-based judgments provided by who were involved in the multidisciplinary activities funded by the Special Commissioner, a specific suitability level in the definition of geological model and/or anthropogenic impact has been assigned to each investigation performed for the characterisation of the area in the SIN_07 "Taranto". The suitability has been ranked in three classes, as follows:

- "High suitability" level has been assigned to the investigations that allow reliable data to be obtained, the interpretation of which can be considered independent from other analyses.
- "Medium suitability" has been assigned to those investigations that require the integration of further analysis to be to be properly interpreted. Medium suitability analyses should be coupled with high suitability analyses.
- "Low suitability" has been assigned to investigations that give a minor contribution to the achievement of the main goal of the analysis. Low suitability analyses should be coupled with medium and high suitability analyses.

In Table 6, the suitability levels are indicated considering the definition of the geo-morphodynamic model and anthropogenic impact as main outcomes. The investigation activities have been grouped as it follows: geophysical and geoelectrical surveys, chemical analysis on sediment samples at different depths from the sea–floor interface, and physical, geo-chemical, and bio-chemical analyses on sediments and biota eventually found in them.

Table 6. Different typologies of analyses generally envisaged for the geo-environmental characterisation of contaminated coastal sites. For each of them, the suitability level in supporting the definition of the geological model and the assessment of the anthropogenic impact has been evaluated as high, medium, and low (😀: high suitability; 🙂: medium suitability; 😐: low suitability).

Investigations	Geological Model	Anthropogenic Impact
Geophysical surveys		
Multi/Single Beam	😀	😀
Side Scan Sonar	😀	😀
Sub Bottom Profile	😀	🙂
Sparker	😀	😐
Magnetometric	🙂	🙂
Geoelectric surveys		
Marine geoelectric	😀	😐
Coastal geoelectric	😀	🙂
Terrestrial geoelectric	😀	😐
Chemical analyses on sediments (layer 0–0.5 m)		
Inorganic compounds	😀	😀
Organic compounds	🙂	😀
Additional compounds	🙂	😀
Chemical analyses on sediments (layer 0.5–3 m)		
Inorganic compounds	😀	😀
Organic compounds	🙂	😀
Additional compounds	🙂	😀
Chemical analyses on sediments (deeper layers)		
Inorganic compounds	😀	😀
Organic compounds	🙂	😀
Additional compounds	🙂	😀
Physical analyses on sediments		
Organic matter	🙂	🙂
Grain size	🙂	🙂
Water content	🙂	🙂
Geo-chemical analyses on sediments		
Magnetic susceptibility	🙂	😀
X Ray Fluorescence–XRF	🙂	😀
X Ray Powder Diffraction–XRPD	🙂	😀
SEM	🙂	😀
TEM	🙂	😀
Bio-chemical analyses on biological elements in sediments		
Palynological analysis	😀	😐
Qualitative assessment of macrobenthos	😀	🙂
Radiocarbon dating-^{14}C	😀	😐

From this evaluation, it is evident that several surveys (e.g., acoustic surveys, SBP surveys, concertation of inorganic compounds) allow the acquisition of informative layers that may support both the assessment of site-specific geological features and the environmental characterisation of the investigated site. Furthermore, the indirect geophysical surveys play a significant role in terms of featuring the sediment variability and therefore for orienting the definition of the most suitable sampling grids, which should be tailored according to the site-specific lithological and sedimentological peculiarities. For this reason, they may be considered as priority in the definition of a cost-efficient characterisation program. Several investigations need to be coupled with other surveys to obtain high informative data. By way of example, SBP data, coupled with punctual information on the contaminants' concentrations, allow the spatial correlation among sediments samples and sedimentary units to be defined. Some investigations (e.g., SPK surveys, radiocarbon dating, palynological analysis) are considered to be very useful for the identification of geological, stratigraphical, and paleo-environmental features, but their usefulness in supporting the evaluation of the anthropogenic impacts on marine matrices is considered low. On the other hand, the analysis of the organic compounds is of paramount importance in the definition of the environmental status, but it does not provide any relevant details useful for the analysis of the geo-morph dynamics of the investigated areas.

Furthermore, taking into consideration the Driver-Pressure-State-Impact-Response (DPSIR) model, which represents a *"causal framework for describing the interactions between society and the environment"* (EEA, 1999), the analyses envisaged for the geo-environmental characterisation of coastal contaminated sites may be considered as key factors for the definition of the parameter "State", which identifies the physical, chemical, and biological conditions of the environment and its related matrices.

This aspect can be highlighted in the review paper published by Labianca et al. [40], who have proposed a specific DPSIR model for the Mar Piccolo basin (Figure 13). In their site-specific DPSIR scheme, the authors identified key elements for each parameter of the model, as follows:

- Driving forces: demography, agriculture, industry, landfills and treatment plants;
- Pressures: discharge of nutrients and contaminants, pollution in groundwater, air pollution;
- States: marine sediment characterisation, sea water characterisation, biodiversity monitoring, anthropisation of sea bottom assessment, marine litter identification;
- Impacts: contaminants mobility, effects of contaminants on marine organisms, effect of pollution on human health, eutrophication;
- Responses: political approaches, remediation strategies.

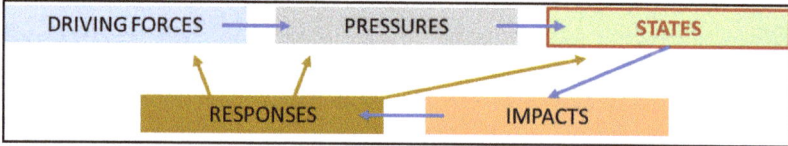

Figure 13. DPSIR framework. The model allows the relationships between the main anthropogenic drivers to be defined, as well as the environmental pressures that they cause and the states of the environmental matrices.

According to the proposed model, operational (short-term) and strategic (medium- and long-term) management responses can interact with forces, pressures, and states. Specifically, while structural and political measures may influence the driving factors, remediation strategies help to reduce environmental degradation and to restore higher environmental quality. Accounting for the above-mentioned parameters, it appears to be clear that the results achieved through the characterisation activities carried out by national and regional scientific entities (ISPRA, ARPA, Special Commissioner, Universities,

CNR) are preparatory for the definition of the current environment state as well as for the identification of the main geomorphological and sedimentological factors that may contribute to the pollutants' accumulation and redistribution. From this perspective, it is considered worthwhile to define a scheme of priority actions to be undertaken in a cost- and operative-effective way.

From the review of the manuals and guidelines drawn up at the national level (cf. Section 2), it emerged that, at present, there are no nationally recognised operative guidelines to refer to for the characterisation of geo-morphodynamic aspects and processes that can induce changes in the coastal environments in the SINs. Therefore, the need to define a scientific protocol of investigations has led to funding a three-year research project supported under a special fund of the Apulia Region financed by the European Union. Activities envisaged under the ongoing research project include: (i) integrated data analysis for the identification of the potential correlation factors between the site-specific geological setting and contamination features; (ii) analysis of geo-morphodynamic factors that may induce changes in the coastal environment resulting in redistribution of the pollutants; (iii) definition of potential environmental risk scenarios. A synthesis of the activities scheduled in the project is reported in Figure 14. The main outcomes of the project will be represented by a set of guidelines specifically defined to address the characterisation activities towards the definition of the geo-morphodynamic model of the investigated area and to support the assessment of environmental and coastal risk scenarios. The project is based on the set of data and information already available for the Taranto area that, at the regional scale, results to be the site for which a higher number of scientific references on its integrated characterisation area available. Nevertheless, to support the definition of the geo-morphological, stratigraphical, and sedimentological features of the Taranto area, new geophysical surveys aimed at the realisation of morpho-topographic and morpho-bathymetric models with high spatial resolution having started and are still in progress (funded by ISPRA in the frame of the national project "CARG" for the realisation of the Geological Map of Italy at the scale 1:50,000 https://www.isprambiente.gov.it/Media/carg/puglia.html, accessed on 3 May 2022).

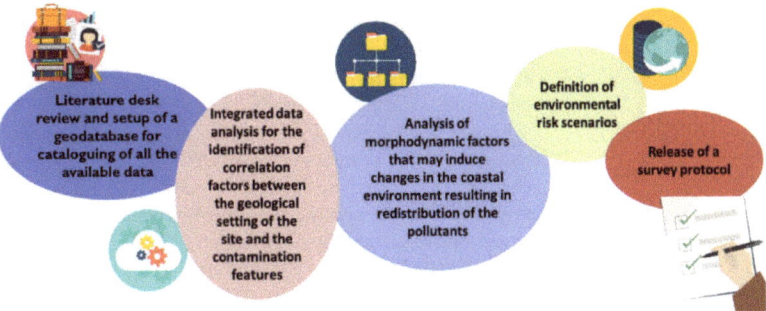

Figure 14. Activities envisaged under the ongoing research project funded by the Apulia Region. Icons are freely downloaded from Freepik platform (freepik.com).

Although the site-specific characteristics (type of coastal system, type of industrial activity) should be considered, it is worth noting that the definition of guidelines should comply with a set of general criteria valid for all types of contaminated coastal sites [10]. Among these, physical parameters such as sediments size and organic content must be accounted for the definition of pollutants accumulation processes. In addition, the geochemical sediment characterisation should include both the surface sediments and the deeper layers to define the vertical distribution of the contaminants. To this aim, it is considered necessary to carry out both in situ cores to support the direct analyses on sediments at different depths and the acquisition of geophysical data to support the spatial correlation among data derived from cores (point data). The storage, processing, and representation of a huge amount of data required for the characterisation of highly contaminated sites cannot disregard the creation of specific geographical multidisciplinary geodatabase which enables the reconstruct of the evolution of the contamination status over time for the different environmental matrices [86].

Furthermore, the definition of a site-specific geo-morphodynamic model cannot disregard the evaluation of physical processes that may induce variations in the coastal setting and enhance the distribution of pollutants. These processes should include both superficial surface processes (e.g., water and solid discharge from river systems) and marine and coastal processes [87–89] (e.g., coastal erosion, coastal flooding, sea level rise). Final considerations about this last aspect are related to the potential effect of climate change on spatial and temporal occurrence of weather-related extreme events (e.g., heavy rainfall, storm surge) that may enhance the magnitude of already occurring processes. As highlighted by [90,91], coastal changes induced by climate and marine processes may undermine remediation and risk management strategies implemented in the contaminated sites and reduce the effectiveness of the original site remediation design.

7. Conclusions

The activities carried out in recent years for the characterisation of marine and coastal areas in the Apulian SINs provided in-depth knowledge of their environmental status. With regard to the SIN_07 "Taranto", the results obtained from the analysis of chemical parameters show a remarkable environmental criticality in the southern sector of the First Bay (in the Mar Piccolo basin), in the area defined as "Area 170 ha". Accounting for the distribution of pollutants along the vertical profile of the sediments, the highest pollutant concentrations characterise the first 50 cm of the samples. Referring to the national limits, Ni and Cr concentrations exceed the site-specific values in 95% and 90% of the sample, respectively, while Zn concentration exceeds the site-specific values in 58%. The concentration of Hg exceeds both the site-specific value (in 47% of the samples) and the national CSC value (11% of the samples). Taking into consideration the international guidelines, 84% of the samples exceed the ERL values for As and Hg, 79% for Cr, 53% for Pb and Cu, and 32% for Zn. Finally, 100% of the samples exceed the ERL value for Ni. Furthermore, the interpretation of the acoustic data acquired during the geophysical surveys has allowed traces of direct and indirect impacts of anthropogenic activities on the seafloor to be identified, which are widely distributed both in the Mar Grande and in the Mar Piccolo.

In this review, based on a critical analysis of the surveys performed in the Apulian SINs, with particular reference to SIN_07 "Taranto", a suitability level has been assigned to each of the performed survey, taking into consideration their usefulness in supporting the definition of the site-specific geo-morphodynamic model and the anthropogenic impact on the environmental matrices. This kind of expert-based evaluation represents the starting point for the definition of a protocol of investigations to be proposed as an operational tool for the selection of the analysis, direct and indirect, to be carried out for a comprehensive characterisation of the geo-morfodynamic setting and environmental state of the coastal contaminated sites. The protocol will support stakeholders in the definition of the conceptual model of the contaminated sites and it will be a useful tool for their environmental restoration. At the regional scale, the guidelines can support the characterisation the five orphan sites approved by MiTE. Furthermore, the National Recovery and Resilience Plan within the Next Generation EU could represent a great opportunity for the definition and implementation of safety measures of the sites to be reclaimed. This last aspect could promote the use of the investigation protocol on a national and international scale.

Author Contributions: Conceptualization, A.R. and G.M.; methodology, A.R. and G.M.; validation, A.R., G.M., G.S. (Giovanni Scicchitano) and M.M.; formal analysis, A.R.; data curation, A.R., F.D.G., G.S. (Giovanni Scardino), A.D.L. and S.L.; writing—original draft preparation, A.R.; writing—review and editing, A.R.; G.M., G.S. (Giovanni Scicchitano), M.M., F.D.G., G.S. (Giovanni Scardino), A.D.L. and S.L.; supervision, G.M.; project administration, A.R. All authors have read and agreed to the published version of the manuscript.

Funding: This research was funded by the Apulia Region (Italy) under the European Regional Development Fund and the European Social Fund (POR Puglia FESR-FSE 2014-2020)-Action 10.4 "Research for Innovation" (REFIN). Research project reference number: FC44BB89. Scientific responsible: Dr. Angela Rizzo.

Data Availability Statement: Data used in this study are derived from the surveys and analysis performed during activities envisaged in "Phase F" and "Phase G" of the Collaboration Agreement "Activities of common interest preparatory to the implementation of the interventions for the reclamation, and requalification of the Mar Piccolo of Taranto" signed in 2015 between the Special Commissioner for urgent measurements of reclamation, environmental improvements, and redevelopment of Taranto, the University of Bari Aldo Moro, and the National Research Council (Scientific responsible of "Phase F" and "Phase G": Prof. Giuseppe Mastronuzzi).

Acknowledgments: The authors are thankful to the Special Commissioner for urgent measurements of reclamation, environmental improvements, and redevelopment of Taranto, Vera Corbelli, who coordinated the characterisation activities in the Taranto area. Scientific activities carried out in the framework of the Collaboration Agreement "Activities of common interest preparatory to the implementation of interventions for the reclamation and requalification of the Mar Piccolo of Taranto" signed with the Special Commissioner were carried out in collaboration with the University of Bari Aldo Moro and the National Research Council. The authors would like to thank Angelo Tursi and Vito Felice Uricchio, as Scientific Managers of the above-mentioned Collaboration Agreement, and Nicola Cardellicchio and Giuseppe Mascolo, as Scientific Managers of the investigation activities related to the "Chemical-physical, microbiological, and ecotoxicological investigations and analyses of sediments in the Mar Piccolo of Taranto" ("Phase H" of the Collaboration Agreement). The authors are also grateful to Biagio Ciuffreda and to Avv. Rosario Arcadio, Port Labour Manager at "Dipartimento di esercizio dei porti di Brindisi e Monopoli" of Autorità di Sistema Portuale del Mare Adriatico Meridionale, who have provided remarkable documents consulted in this review. Finally, the Authors thanks Sabrina Terracciano for her support in the preparation of Figure 11.

Conflicts of Interest: The authors declare no conflict of interest.

References

1. Gómez, J.A. *SUPERFUND: EPA Should Take Additional Actions to Manage Risks from Climate Change Effects*; GAO-20-73; U.S. Government Accountability Office: Washington, DC, USA, 2019; p. 66.
2. Manzoor, R.; Zhang, T.; Zhang, X.; Wang, M.; Pan, J.-F.; Wang, Z.; Zhang, B. Single and Combined Metal Contamination in Coastal Environments in China: Current Status and Potential Ecological Risk Evaluation. *Environ. Sci. Pollut. Res. Int.* **2018**, *25*, 1044–1054. [CrossRef] [PubMed]
3. Minkina, T.M.; Nevidomskaya, D.G.; Pol'shina, T.N.; Fedorov, Y.A.; Mandzhieva, S.S.; Chaplygin, V.A.; Bauer, T.V.; Burachevskaya, M.V. Heavy Metals in the Soil–Plant System of the Don River Estuarine Region and the Taganrog Bay Coast. *J. Soils Sediments* **2017**, *17*, 1474–1491. [CrossRef]
4. Micheli, F.; Halpern, B.S.; Walbridge, S.; Ciriaco, S.; Ferretti, F.; Fraschetti, S.; Lewison, R.; Nykjaer, L.; Rosenberg, A.A. Cumulative Human Impacts on Mediterranean and Black Sea Marine Ecosystems: Assessing Current PressuRes. and Opportunities. *PLoS ONE* **2013**, *8*, e79889. [CrossRef] [PubMed]
5. Civili, F.S. *The Land-Based Pollution of the Mediterranean Sea: Present State and Prospects*; IEMed: New Delhi, India, 2010; p. 5.
6. Lehoux, A.P.; Petersen, K.; Leppänen, M.T.; Snowball, I.; Olsen, M. Status of Contaminated Marine Sediments in Four Nordic Countries: Assessments, Regulations, and Remediation Approaches. *J. Soils Sediments* **2020**, *20*, 2619–2629. [CrossRef]
7. Brand, J.H.; Spencer, K.L.; O'shea, F.T.; Lindsay, J.E. Potential Pollution Risks of Historic Landfills on Low-Lying Coasts and Estuaries. *WIREs Water* **2018**, *5*, e1364. [CrossRef]
8. MITE. Available online: https://bonifichesiticontaminati.mite.gov.it/Sin/istituzione-Perimetrazione/ (accessed on 3 May 2022).
9. ISPRA Siti di Interesse Nazionale (SIN). Available online: https://www.isprambiente.gov.it/it/attivita/suolo-e-territorio/siti-contaminati/siti-di-interesse-nazionale-sin (accessed on 3 May 2022).
10. Ausili, A.; Bergamin, L.; Romano, E. Environmental Status of Italian Coastal Marine Areas Affected by Long History of Contamination. *Front. Environ. Sci.* **2020**, *8*, 34. [CrossRef]

11. Ausili, A.; Romano, E.; Mumelter, E.; Tornato, A. Stato dell'arte Sulle Bonifiche delle Aree Marine e di Transizione Interne ai SIN. In Proceedings of the Workshop "Siti Contaminati. Esperienze negli Interventi di Risanamento" (Universitá di Catania: TEAM PA), Sicily, Italy, 9–11 February 2012; pp. 27–45.
12. Zoumis, T.; Schmidt, A.; Grigorova, L.; Calmano, W. Contaminants in Sediments: Remobilisation and Demobilisation. *Sci. Total Environ.* **2001**, *266*, 195–202. [CrossRef]
13. Fichet, D.; Boucher, G.; Radenac, G.; Miramand, P. Concentration and Mobilization of Cd, Cu, Pb and Zn by Meiofauna Populations Living in Harbour Sediment: Their Role in the Heavy Metal Flux from Sediment to Food Web. *Sci. Total Environ.* **1999**, *243–244*, 263–272. [CrossRef]
14. Linnik, P.M.; Zubenko, I.B. Role of Bottom Sediments in the Secondary Pollution of Aquatic Environments by Heavy-Metal Compounds. *Lakes Reserv. Sci. Policy Manag. Sustain. Use* **2000**, *5*, 11–21. [CrossRef]
15. Spada, L.; Annicchiarico, C.; Cardellicchio, N.; Giandomenico, S.; di Leo, A. Mercury and Methylmercury Concentrations in Mediterranean Seafood and Surface Sediments, Intake Evaluation and Risk for Consumers. *Int. J. Hyg. Environ. Health* **2012**, *215*, 418–426. [CrossRef]
16. Baldrighi, E.; Semprucci, F.; Franzo, A.; Cvitkovic, I.; Bogner, D.; Despalatovic, M.; Berto, D.; Formalewicz, M.M.; Scarpato, A.; Frapiccini, E.; et al. Meiofaunal Communities in Four Adriatic Ports: Baseline Data for Risk Assessment in Ballast Water Management. *Mar. Pollut. Bull.* **2019**, *147*, 171–184. [CrossRef]
17. Bellas, J.; Nieto, Ó.; Beiras, R. Integrative Assessment of Coastal Pollution: Development and Evaluation of Sediment Quality Criteria from Chemical Contamination and Ecotoxicological Data. *Cont. Shelf Res.* **2011**, *31*, 448–456. [CrossRef]
18. Damiani, V.; Bianchi, C.N.; Ferretti, O.; Bedulli, D.; Morri, C.; Viel, M.; Zurlini, G. Risultati di Una Ricerca Ecologica Sul Sistema Marino Costiero Pugliese. *Thalass. Salentina* **1988**, *18*, 153–169.
19. Mastronuzzi, G.; Valletta, S.; Damiani, A.; Fiore, A.; Francescangeli, R.; Giandonato, P.B.; Iurilli, V.; Sabato, L. *Geositi della Puglia*; AA.VV.; Sagraf/Società Italiana di Geologia Ambientale (SIGEA): Capurso, Italy, 2015; ISBN 978-88-906716-8-5.
20. Mastronuzzi, G.; Milella, M.; Parise, M.; Piscitelli, A.; Scardino, G. Patrimonio Culturale e Geositi Dell'area Murgiana. *Geol. Ambiente* **2020**, 43–50.
21. Buongiorno, J.; Intini, M. Sustainable Tourism and Mobility Development in Natural Protected Areas: Evidence from Apulia. *Land Use Policy* **2021**, *101*, 105220. [CrossRef]
22. ARPA. *Mar Piccolo of Taranto—Scientific-Technical Report on the Interaction between the Environmental System and Contaminants Flows from Primary and Secondary Sources*; Technical Report, 1; ARPA: Bari, Italy, 2014; p. 175.
23. ISPRA. *Evaluation of Characterization Results for the Identification of Appropriate Actions for Remediation of Site of National Interest of Taranto*; Technical Report, 1; ISPRA: Bari, Italy, 2010; p. 90.
24. Cardellicchio, N.; Covelli, S.; Cibic, T. Integrated Environmental Characterization of the Contaminated Marine Coastal Area of Taranto, Ionian Sea (Southern Italy). *Environ. Sci. Pollut. Res. Int.* **2016**, *23*, 12491–12494. [CrossRef] [PubMed]
25. Cardellicchio, N.; Annicchiarico, C.; di Leo, A.; Giandomenico, S.; Spada, L. The Mar Piccolo of Taranto: An Interesting Marine Ecosystem for the Environmental Problems Studies. *Environ. Sci. Pollut. Res. Int.* **2016**, *23*, 12495–12501. [CrossRef]
26. Giandomenico, S.; Cardellicchio, N.; Spada, L.; Annicchiarico, C.; Di Leo, A. Metals and PCB Levels in Some Edible Marine Organisms from the Ionian Sea: Dietary Intake Evaluation and Risk for Consumers. *Environ. Sci. Pollut. Res.* **2016**, *23*, 12596–12612. [CrossRef]
27. Trinchera, G.; Ungaro, N.; Blonda, M.; Gramegna, D.; Lacarbonara, M.; Cunsolo, S.; Renna, R. Approfondimento Tecnico-Scientifico Sulle Interazioni tra il Sistema Ambientale ed i Flussi di Contaminanti da Fonti Primarie e Secondarie Nel Mar Piccolo di Taranto. In Proceedings of the ECOMONDO 2015, Rome, Italy, 3–6 November 2015; p. 7.
28. Bracchi, V.; Marchese, F.; Savini, A.; Chimienti, G.; Mastrototaro, F.; Tessarolo, C.; Cardone, F.; Tursi, A.; Corselli, C. Seafloor Integrity of the Mar Piccolo Basin (Southern Italy): Quantifying Anthropogenic Impact. *J. Maps* **2016**, *12*, 1–11. [CrossRef]
29. Bellucci, L.G.; Cassin, D.; Giuliani, S.; Botter, M.; Zonta, R. Sediment Pollution and Dynamic in the Mar Piccolo of Taranto (Southern Italy): Insights from Bottom Sediment Traps and Surficial Sediments. *Environ. Sci. Pollut. Res. Int.* **2016**, *23*, 12554–12565. [CrossRef]
30. di Leo, A.; Annicchiarico, C.; Cardellicchio, N.; Cibic, T.; Comici, C.; Giandomenico, S.; Spada, L. Mobilization of Trace Metals and PCBs from Contaminated Marine Sediments of the Mar Piccolo in Taranto during Simulated Resuspension Experiment. *Environ. Sci. Pollut. Res. Int.* **2016**, *23*, 12777–12790. [CrossRef] [PubMed]
31. Emili, A.; Acquavita, A.; Covelli, S.; Spada, L.; di Leo, A.; Giandomenico, S.; Cardellicchio, N. Mobility of Heavy Metals from Polluted Sediments of a Semi-Enclosed Basin: In Situ Benthic Chamber Experiments in Taranto's Mar Piccolo (Ionian Sea, Southern Italy). *Environ. Sci. Pollut. Res. Int.* **2016**, *23*, 12582–12595. [CrossRef] [PubMed]
32. Kralj, M.; De Vittor, C.; Comici, C.; Relitti, F.; Auriemma, R.; Alabiso, G.; Del Negro, P. Recent Evolution of the Physical-Chemical Characteristics of a Site of National Interest—the Mar Piccolo of Taranto (Ionian Sea)-and Changes over the Last 20 Years. *Environ. Sci. Pollut. Res. Int.* **2016**, *23*, 12675–12690. [CrossRef] [PubMed]
33. Valenzano, E.; Scardino, G.; Cipriano, G.; Fago, P.; Capolongo, D.; De Giosa, F.; Lisco, S.; Mele, D.; Moretti, M.; Mastronuzzi, G. Holocene Morpho-Sedimentary Evolution of the Mar Piccolo Basin (Taranto, Southern Italy). *Geogr. Fis. Din. Quat.* **2018**, *41*, 119–135. [CrossRef]
34. Tursi, A.; Corbelli, V.; Cipriano, G.; Capasso, G.; Velardo, R.; Chimienti, G. Mega-Litter and Remediation: The Case of Mar Piccolo of Taranto (Ionian Sea). *Rend. Fis. Acc. Lincei* **2018**, *29*, 817–824. [CrossRef]

35. Quarta, G.; Fago, P.; Calcagnile, L.; Cipriano, G.; D'Elia, M.; Moretti, M.; Scardino, G.; Valenzano, E.; Mastronuzzi, G. 14C Age Offset in the Mar Piccolo Sea Basin in Taranto (Southern Italy) Estimated on Cerastoderma Glaucum (Poiret, 1789). *Radiocarbon* **2019**, *61*, 1387–1401. [CrossRef]
36. Todaro, F.; Gisi, S.D.; Labianca, C.; Notarnicola, M. Combined Assessment of Chemical and Ecotoxicological Data for the Management of Contaminated Marine Sediments. *Environ. Eng. Manag. J.* **2019**, *18*, 2287–2296.
37. Cotecchia, F.; Vitone, C.; Sollecito, F.; Mali, M.; Miccoli, D.; Petti, R.; Milella, D.; Ruggieri, G.; Bottiglieri, O.; Santaloia, F.; et al. A Geo-Chemo-Mechanical Study of a Highly Polluted Marine System (Taranto, Italy) for the Enhancement of the Conceptual Site Model. *Sci. Rep.* **2021**, *11*, 4017. [CrossRef]
38. Scardino, G.; De Giosa, F.; D'Onghia, M.; Demonte, P.; Fago, P.; Saccotelli, G.; Valenzano, E.; Moretti, M.; Velardo, R.; Capasso, G.; et al. The Footprints of the Wreckage of the Italian Royal Navy Battleship Leonardo da Vinci on the Mar Piccolo Sea-Bottom (Taranto, Southern Italy). *Oceans* **2020**, *1*, 77–93. [CrossRef]
39. Rizzo, A.; Capasso, G.; Corbelli, V.; De Giosa, F.; Lisco, S.N.; Mastronuzzi, G.; Moretti, M.; Scardino, G.; Scicchitano, G.; Valenzano, E.; et al. The Development of a Survey Protocol for the Geomorphodynamic Characterization of Contaminated Sites. In *Le Bonifiche Ambientali nell'ambito della Transizione Ecologica*; Baldi, D., Uricchio, V.F., Eds.; SIGEA: Bari, Italy, 2022; p. 404.
40. Labianca, C.; De Gisi, S.; Todaro, F.; Notarnicola, M. DPSIR Model Applied to the Remediation of Contaminated Sites. A Case Study: Mar Piccolo of Taranto. *Appl. Sci.* **2020**, *10*, 5080. [CrossRef]
41. ISPRA. *Criteri per La Elaborazione di Piani di Gestione Dell'inquinamento diffuso*; ISPRA-Manuali e Linee Guida 146/2017; ISPRA: Rome, Italy, 2017; p. 23.
42. APAT. *Manuale per Le Indagini Ambientali Nei Siti Contaminati*; APAT-Manuali e linee guida 43/2006; ISPRA: Rome, Italy, 2006; p. 202.
43. ISPRA. *Linee Guida per la Determinazione dei Valori di Fondo per i Suoli e per le Acque Sotterranee*; SNPA Linee guida 08/2018; ISPRA: Rome, Italy, 2018; p. 318.
44. APAT/ICRAM. *Manuale per La Movimentazione dei Sedimenti Marini*; ISPRA: Rome, Italy, 2007.
45. Lisi, I.; Feola, A.; Bruschi, A.; di Risio, M.; Pedroncini, A.; Pasquali, D.; Romano, E. *La Modellistica Matematica nella Valutazione Degli Aspetti Fisici Legati alla Movimentazione dei Sedimenti in Aree Marino-Costiere*; ISPRA: Rome, Italy, 2017; p. 144.
46. Araneo, F.; Bartolucci, E.; Vecchio, A. *Synthesis of the Report "Status of Contaminated Sites Management in Italy: Regional Data"*; ISPRA: Rome, Italy, 2021; p. 22.
47. Tropeano, M.; Sabato, L.; Pieri, P. Filling and Cannibalization of a Foredeep: The Bradanic Trough, Southern Italy. *Geol. Soc. Lond. Spec. Publ.* **2002**, *191*, 55–79. [CrossRef]
48. Tropeano, M.; Cilumbriello, A.; Sabato, L.; Gallicchio, S.; Grippa, A.; Longhitano, S.G.; Bianca, M.; Gallipoli, M.R.; Mucciarelli, M.; Spilotro, G. Surface and Subsurface of the Metaponto Coastal Plain (Gulf of Taranto—Southern Italy): Present-Day- vs LGM-Landscape. *Geomorphology* **2013**, *203*, 115–131. [CrossRef]
49. Mastronuzzi, G.; Boccardi, L.; Candela, A.; Colella, C.; Curci, G.; Giletti, F.; Milella, M.; Pignatelli, C.; Piscitelli, A.; Ricci, F.; et al. *Il Castello Aragonese di Taranto in 3D nell'Evoluzione del Paesaggio Naturale*; DIGILABS: Bari, Italy, 2013.
50. Amorosi, A.; Antonioli, F.; Bertini, A.; Marabini, S.; Mastronuzzi, G.; Montagna, P.; Negri, A.; Rossi, V.; Scarponi, D.; Taviani, M.; et al. The Middle–Upper Pleistocene Fronte Section (Taranto, Italy): An Exceptionally Preserved Marine Record of the Last Interglacial. *Glob. Planet. Change* **2014**, *119*, 23–38. [CrossRef]
51. Negri, A.; Amorosi, A.; Antonioli, F.; Bertini, A.; Florindo, F.; Lurcock, P.C.; Marabini, S.; Mastronuzzi, G.; Regattieri, E.; Rossi, V.; et al. A Potential Global Boundary Stratotype Section and PoInt. (GSSP) for the Tarentian Stage, Upper Pleistocene, from the Taranto Area (Italy): Results and Future Perspectives. *Quat. Int.* **2015**, *383*, 145–157. [CrossRef]
52. De Santis, V.; Caldara, M.; Torres, T.; Ortiz, J.E.; Sánchez-Palencia, Y. A Review of MIS 7 and MIS 5 Terrace Deposits along the Gulf of Taranto Based on New Stratigraphic and Chronological Data. *Ital. J. Geosci.* **2018**, *137*, 349–368. [CrossRef]
53. Belluomini, G.; Caldara, M.; Casini, C.; Cerasoli, M.; Manfra, L.; Mastronuzzi, G.; Palmentola, G.; Sanso, P.; Tuccimei, P.; Vesica, P.L. The Age of Late Pleistocene Shorelines and Tectonic Activity of Taranto Area, Southern Italy. *Quat. Sci. Rev.* **2002**, *21*, 525–547. [CrossRef]
54. Lisco, S.; Corselli, C.; De Giosa, F.; Mastronuzzi, G.; Moretti, M.; Siniscalchi, A.; Marchese, F.; Bracchi, V.; Tessarolo, C.; Tursi, A. Geology of Mar Piccolo, Taranto (Southern Italy): The Physical Basis for Remediation of a Polluted Marine Area. *J. Maps* **2016**, *12*, 173–180. [CrossRef]
55. Cotecchia, F.; Lollino, G.; Pagliarulo, R.; Stefanon, A.; Tadolini, T.; Trizzino, R. Hydrogeological Conditions and Field Monitoring of the Galeso Submarine Spring in the Mar Piccolo of Taranto (Southern Italy). In Proceedings of the XI Salt Water Intrusion Meeting, Gdansk, Poland, 14–17 May 1990; pp. 171–208.
56. Zuffianò, L.E.; Basso, A.; Casarano, D.; Dragone, V.; Limoni, P.P.; Romanazzi, A.; Santaloia, F.; Polemio, M. Coastal Hydrogeological System of Mar Piccolo (Taranto, Italy). *Environ. Sci. Pollut. Res. Int.* **2015**, *23*, 12502–12514. [CrossRef]
57. Valenzano, E.; D'Onghia, M.; De Giosa, F.; Demonte, P. Morfologia Delle Sorgenti Sottomarine Dell'area di Taranto (Mar Ionio). *Mem. Descr. Carta Geol. It.* **2020**, *105*, 65–69.
58. di Bucci, D.; Caputo, R.; Mastronuzzi, G.; Fracassi, U.; Selleri, G.; Sansò, P. Quantitative Analysis of Extensional Joints in the Southern Adriatic Foreland (Italy), and the Active Tectonics of the Apulia Region. *J. Geodyn.* **2011**, *51*, 141–155. [CrossRef]

59. Mastronuzzi, G.; Caputo, R.; di Bucci, D.; Fracassi, U.; Iurilli, V.; Milella, M.; Pignatelli, C.; Sansò, P.; Selleri, G. Middle-Late Pleistocene Evolution of the Adriatic Coastline of Southern Apulia (Italy) in Response to Relative Sea-Level Changes. *Geogr. Fis. Din. Quat.* **2011**, *34*, 207–221. [CrossRef]
60. Rossi, D. Note Illustrative della Carta Geologica d'Italia, Alla Scala 1:100,000, Foglio 204 Lecce. *Serv. Geol. It.* **1969**, *23*, 1–24. Available online: http://sgi.isprambiente.it/geologia100k/mostra_foglio.aspx?numero_foglio=204 (accessed on 3 May 2022).
61. Merla, G.; Ercoli, A.; Torre, D. Note Illustrative della Carta Geologica d'Italia, Alla Scala 1:100,000, Foglio 164 Foggia. *Serv. Geol. It.* **1969**, *14*, 1–22. Available online: http://sgi.isprambiente.it/geologia100k/mostra_foglio.aspx?numero_foglio=164 (accessed on 3 May 2022).
62. Caroppo, C.; Giordano, L.; Palmieri, N.; Bellio, G.; Portacci, G.; Hopkins, T.S.; Sclafani, P.; Bisci, A.P. Progress Toward Sustainable Mussel Aquaculture in Mar Piccolo, Italy. *Ecol. Soc.* **2012**, *17*, 10. [CrossRef]
63. Caroppo, C.; Portacci, G. The First World War in the Mar Piccolo of Taranto: First Case of Warfare Ecology? *Ocean. Coast. Manag.* **2017**, *149*, 135–147. [CrossRef]
64. Murgante, B.; Rotondo, F. A Geostatistical Approach to Measure Shrinking Cities: The Case of Taranto, Co. In *Statistical Methods for Spatial Planning and Monitoring*; Montrone, S., Perchinunno, P., Eds.; Springer: Berlin/Heidelberg, Germany, 2012; pp. 119–142. ISBN 978-88-470-2750-3.
65. ISPRA. *Elaborazione e Valutazione Dei Risultati della Caratterizzazione Ai Fini della Individuazione Degli Opportuni Interventi di Messa in Sicurezza e Bonifica Del Sito di Interesse Nazionale di Brindisi CII-El-PU-BR-Area Portuale e Area Costiera*; Relazione-01.11 2011; ISPRA: Rome, Italy, 2011; p. 128.
66. ISPRA. *Valutazione e Rappresentazione dei Risultati della Caratterizzazione ai Fini della Individuazione delle Corrette Modalità di Gestione CII-El-PU-BR_S.Apollinare*; Relazione-01.09 2011; ISPRA: Rome, Italy, 2011; p. 24.
67. ICRAM. *Piano di Caratterizzazione Ambientale Dell'area Marino-Costiera Prospiciente Il Sito di Interesse Nazionale di Manfredonia*; CII-Pr-PU-M-02.09 2004; ICRAM: Rome, Italy, 2004; p. 40.
68. ISPRA. *Elaborazione e Valutazione Dei Risultati della Caratterizzazione Ai Fini della Individuazione Degli Opportuni Interventi di Messa in Sicurezza e Bonifica Del Sito di Interesse Nazionale di Taranto—Mar Grande II Lotto e Mar Piccolo*; CII-El-PU-TA-Mar Grande II Lotto e Mar Piccolo-01.06 2010; ISPRA: Rome, Italy, 2010; p. 90.
69. Pirastu, R.; Comba, P.; Iavarone, I.; Zona, A.; Conti, S.; Minelli, G.; Manno, V.; Mincuzzi, A.; Minerba, S.; Forastiere, F.; et al. Environment and Health in Contaminated Sites: The Case of Taranto, Italy. *J. Environ. Public Health* **2013**, *2013*, 753719. [CrossRef]
70. De Giosa, F.; Lisco, S.N.; Mastronuzzi, G.; Moretti, M.; Rizzo, A.; Scardino, G.; Scicchitano, G.; Valenzano, E.; Capasso, G.; Velardo, R.; et al. La Geologia Marina di Taranto: La Base Fisica per Lo Studio Dell'inquinamento Antropico Nel Settore Settentrionale Del Mar Ionio. In *Abstract Book della Società Geologica Italiana, "La Geologia Marina in Italia, Quarto Convegno dei Geologi Marini Italiani"*; Chiocci, F.L., Budillon, F.B., Ceramicola, S., Gamberi, F., Loreto, M.F., Senatore, M.R., Spagnoli, F., Sulli, A., Eds.; Società Geologica Italiana: Rome, Italy, 2021; p. 74. [CrossRef]
71. Massari, F.; Cotugno, P.; Tursi, A.; Milella, P.; Lisco, S.; Scardino, G.; Scicchitano, G.; Rizzo, A.; Valenzano, E.; Moretti, M.; et al. *Mapping of Organotin Compounds in Sediments of Mar Piccolo (Taranto, Italy) Using Gas Chromatography-Mass Spectrometry Analysis and Geochemical Data*; IEEE: Piscataway, NJ, USA, 2021; pp. 21–26.
72. Long, E.R.; Macdonald, D.D.; Smith, S.L.; Calder, F.D. Incidence of Adverse Biological Effects within Ranges of Chemical Concentrations in Marine and Estuarine Sediments. *Environ. Manag.* **1995**, *19*, 81–97. [CrossRef]
73. Long, E.R. Calculation and Uses of Mean Sediment Quality Guideline Quotients: A Critical Review. *Environ. Sci. Technol.* **2006**, *40*, 1726–1736. [CrossRef]
74. O'Connor, T.P. The Sediment Quality Guideline, ERL, Is Not a Chemical Concentration at the Threshold of Sediment Toxicity. *Mar. Pollut. Bull.* **2004**, *49*, 383–385. [CrossRef]
75. Birch, G.F. A Review of Chemical-Based Sediment Quality Assessment Methodologies for the Marine Environment. *Mar. Pollut. Bull.* **2018**, *133*, 218–232. [CrossRef]
76. Arienzo, M.; Donadio, C.; Mangoni, O.; Bolinesi, F.; Stanislao, C.; Trifuoggi, M.; Toscanesi, M.; di Natale, G.; Ferrara, L. Characterization and Source Apportionment of Polycyclic Aromatic Hydrocarbons (Pahs) in the Sediments of Gulf of Pozzuoli (Campania, Italy). *Mar. Pollut. Bull.* **2017**, *124*, 480–487. [CrossRef] [PubMed]
77. Arienzo, M.; Toscanesi, M.; Trifuoggi, M.; Ferrara, L.; Stanislao, C.; Donadio, C.; Grazia, V.; Gionata, D.V.; Carella, F. Contaminants Bioaccumulation and Pathological Assessment in Mytilus Galloprovincialis in Coastal Waters Facing the Brownfield Site of Bagnoli, Italy. *Mar. Pollut. Bull.* **2019**, *140*, 341–352. [CrossRef] [PubMed]
78. Bonsignore, M.; Salvagio Manta, D.; Oliveri, E.; Sprovieri, M.; Basilone, G.; Bonanno, A.; Falco, F.; Traina, A.; Mazzola, S. Mercury in Fishes from Augusta Bay (Southern Italy): Risk Assessment and Health Implication. *Food Chem. Toxicol.* **2013**, *56*, 184–194. [CrossRef] [PubMed]
79. Bonsignore, M.; Tamburrino, S.; Oliveri, E.; Marchetti, A.; Durante, C.; Berni, A.; Quinci, E.; Sprovieri, M. Tracing Mercury Pathways in Augusta Bay (Southern Italy) by Total Concentration and Isotope Determination. *Environ. Pollut.* **2015**, *205*, 178–185. [CrossRef]
80. Cannata, C.; Cianflone, G.; Vespasiano, G.; Rosa, R. Preliminary Analysis of Sediments Pollution of the Coastal Sector between Crotone and Strongoli (Calabria-Southern Italy). *Rend. Online Soc. Geol. Ital.* **2016**, *38*, 17–20. [CrossRef]
81. Madricardo, F.; Foglini, F.; Campiani, E.; Grande, V.; Catenacci, E.; Petrizzo, A.; Kruss, A.; Toso, C.; Trincardi, F. Assessing the Human FootprInt. on the Sea-Floor of Coastal Systems: The Case of the Venice Lagoon, Italy. *Sci. Rep.* **2019**, *9*, 6615. [CrossRef]

82. Romano, A.; di Risio, M.; Bellotti, G.; Molfetta, M.G.; Damiani, L.; De Girolamo, P. Tsunamis Generated by Landslides at the Coast of Conical Islands: Experimental Benchmark Dataset for Mathematical Model Validation. *Landslides* **2016**, *13*, 1379–1393. [CrossRef]
83. Romano, E.; Bergamin, L.; Ausili, A.; Pierfranceschi, G.; Maggi, C.; Sesta, G.; Gabellini, M. The Impact of the Bagnoli Industrial Site (Naples, Italy) on Sea-Bottom Environment. Chemical and Textural FeatuRes. of Sediments and the Related Response of Benthic Foraminifera. *Mar. Pollut. Bull.* **2009**, *59*, 245–256. [CrossRef]
84. Romano, E.; Bergamin, L.; Finoia, M.G.; Carboni, M.G.; Ausili, A.; Gabellini, M. Industrial Pollution at Bagnoli (Naples, Italy): Benthic Foraminifera as a Tool in Integrated Programs of Environmental Characterisation. *Mar. Pollut. Bull.* **2008**, *56*, 439–457. [CrossRef]
85. Trifuoggi, M.; Donadio, C.; Mangoni, O.; Ferrara, L.; Bolinesi, F.; Nastro, R.A.; Stanislao, C.; Toscanesi, M.; di Natale, G.; Arienzo, M. distribution and Enrichment of Trace Metals in Surface Marine Sediments in the Gulf of Pozzuoli and off the Coast of the Brownfield Metallurgical Site of Ilva of Bagnoli (Campania, Italy). *Mar. Pollut. Bull.* **2017**, *124*, 502–511. [CrossRef]
86. Ciampi, P.; Esposito, C.; Petrangeli Papini, M. Hydrogeochemical Model Supporting the Remediation Strategy of a Highly Contaminated Industrial Site. *Water* **2019**, *11*, 1371. [CrossRef]
87. LeMonte, J.J.; Stuckey, J.W.; Sanchez, J.Z.; Tappero, R.; Rinklebe, J.; Sparks, D.L. Sea Level Rise Induced Arsenic Release from Historically Contaminated Coastal Soils. *Environ. Sci. Technol.* **2017**, *51*, 5913–5922. [CrossRef] [PubMed]
88. Bardos, P.; Spencer, K.L.; Ward, R.D.; Maco, B.H.; Cundy, A.B. Integrated and Sustainable Management of Post-Industrial Coasts. *Front. Environ. Sci.* **2020**, *8*, 86. [CrossRef]
89. Nicholls, R.J.; Beaven, R.P.; Stringfellow, A.; Monfort, D.; Le Cozannet, G.; Wahl, T.; Gebert, J.; Wadey, M.; Arns, A.; Spencer, K.L.; et al. Coastal Landfills and Rising Sea Levels: A Challenge for the 21st Century. *Front. Mar. Sci.* **2021**, *8*, 710342. [CrossRef]
90. Beaven, R.P.; Stringfellow, A.M.; Nicholls, R.J.; Haigh, I.D.; Kebede, A.S.; Watts, J. Future Challenges of Coastal Landfills Exacerbated by Sea Level Rise. *Waste Manag.* **2020**, *105*, 92–101. [CrossRef]
91. Maco, B.; Bardos, P.; Coulon, F.; Erickson-Mulanax, E.; Hansen, L.J.; Harclerode, M.; Hou, D.; Mielbrecht, E.; Wainwright, H.M.; Yasutaka, T.; et al. Resilient Remediation: Addressing Extreme Weather and Climate Change, Creating Community Value. *Remediat. J.* **2018**, *29*, 7–18. [CrossRef]

Article

Quantifying Drivers of Coastal Forest Carbon Decline Highlights Opportunities for Targeted Human Interventions

Lindsey S. Smart [1,*], Jelena Vukomanovic [1,2], Paul J. Taillie [3], Kunwar K. Singh [4,5] and Jordan W. Smith [6]

1. Center for Geospatial Analytics, College of Natural Resources, North Carolina State University, 2800 Faucette Drive, Raleigh, NC 27695, USA; jvukoma@ncsu.edu
2. Department of Parks, Recreation, and Tourism Management, College of Natural Resources, North Carolina State University, 2800 Faucette Drive, Raleigh, NC 27695, USA
3. Department of Wildlife Ecology and Conservation, University of Florida, Gainesville, FL 32611, USA; paultaillie@ufl.edu
4. AidData Global Research Institute, The College of William and Mary, 424 Scotland Street, Williamsburg, VA 23185, USA; kksingh@wm.edu
5. Center for Geospatial Analysis, The College of William and Mary, 400 Landrum Dr., Williamsburg, VA 23185, USA
6. Department of Environment and Society, Utah State University, Logan, UT 84322, USA; jordan.smith@usu.edu
* Correspondence: lssmart@ncsu.edu

Citation: Smart, L.S.; Vukomanovic, J.; Taillie, P.J.; Singh, K.K.; Smith, J.W. Quantifying Drivers of Coastal Forest Carbon Decline Highlights Opportunities for Targeted Human Interventions. *Land* **2021**, *10*, 752. https://doi.org/10.3390/land10070752

Academic Editors: Pietro Aucelli, Angela Rizzo, Rodolfo Silva Casarín and Giorgio Anfuso

Received: 18 June 2021
Accepted: 14 July 2021
Published: 18 July 2021

Publisher's Note: MDPI stays neutral with regard to jurisdictional claims in published maps and institutional affiliations.

Copyright: © 2021 by the authors. Licensee MDPI, Basel, Switzerland. This article is an open access article distributed under the terms and conditions of the Creative Commons Attribution (CC BY) license (https://creativecommons.org/licenses/by/4.0/).

Abstract: As coastal land use intensifies and sea levels rise, the fate of coastal forests becomes increasingly uncertain. Synergistic anthropogenic and natural pressures affect the extent and function of coastal forests, threatening valuable ecosystem services such as carbon sequestration and storage. Quantifying the drivers of coastal forest degradation is requisite to effective and targeted adaptation and management. However, disentangling the drivers and their relative contributions at a landscape scale is difficult, due to spatial dependencies and nonstationarity in the socio-spatial processes causing degradation. We used nonspatial and spatial regression approaches to quantify the relative contributions of sea level rise, natural disturbances, and land use activities on coastal forest degradation, as measured by decadal aboveground carbon declines. We measured aboveground carbon declines using time-series analysis of satellite and light detection and ranging (LiDAR) imagery between 2001 and 2014 in a low-lying coastal region experiencing synergistic natural and anthropogenic pressures. We used nonspatial (ordinary least squares regression–OLS) and spatial (geographically weighted regression–GWR) models to quantify relationships between drivers and aboveground carbon declines. Using locally specific parameter estimates from GWR, we predicted potential future carbon declines under sea level rise inundation scenarios. From both the spatial and nonspatial regression models, we found that land use activities and natural disturbances had the highest measures of relative importance (together representing 94% of the model's explanatory power), explaining more variation in carbon declines than sea level rise metrics such as salinity and distance to the estuarine shoreline. However, through the spatial regression approach, we found spatial heterogeneity in the relative contributions to carbon declines, with sea level rise metrics contributing more to carbon declines closer to the shore. Overlaying our aboveground carbon maps with sea level rise inundation models we found associated losses in total aboveground carbon, measured in teragrams of carbon (TgC), ranged from 2.9 ± 0.1 TgC (for a 0.3 m rise in sea level) to 8.6 ± 0.3 TgC (1.8 m rise). Our predictions indicated that on the remaining non-inundated landscape, potential carbon declines increased from 29% to 32% between a 0.3 and 1.8 m rise in sea level. By accounting for spatial nonstationarity in our drivers, we provide information on site-specific relationships at a regional scale, allowing for more targeted management planning and intervention. Accordingly, our regional-scale assessment can inform policy, planning, and adaptation solutions for more effective and targeted management of valuable coastal forests.

Keywords: aboveground carbon storage; coastal forests; light detection and ranging (LiDAR); remote sensing; satellite imagery; sea level rise

1. Introduction

Coastal forests cover less than 3% of the Earth's surface yet sequester and store as much carbon (C) as their terrestrial counterparts [1,2]. By storing carbon in vegetation and sediments, coastal forests play important roles in biogeochemical carbon (C) cycling and mitigating climate change [3,4]. Coastal forests provide a suite of additional ecosystem services, including storm surge protection, wildlife habitat, water filtration, and recreational opportunities [3,5]. At the terrestrial–ocean interface, coastal forests are disproportionately impacted by pressures from human land use modifications and sea level rise, altering the structure and function of these valuable ecosystems [6]. Sea level rise rates are accelerating, outpacing the average 20th century rates, and leading to increased flooding and erosion, saltwater intrusion, and coastal forest loss [7]. At the same time, rapid population growth, demand for food and fiber, and unconstrained coastal development have caused rapid declines in the health and extent of coastal forests [4]. The interactive effects of sea level rise, land use, and other disturbances (e.g., wildfires) impact the quantity and quality of ecosystem services supplied by coastal forests [8], but regional assessments that account for these effects together are limited [9], and few account for spatial dependencies and nonstationarity in the socio-spatial processes driving ecosystem change. Declines in coastal forest health highlight the need for targeted human interventions. More comprehensive regional assessments that explicitly account for socio-spatial processes are critical to understand how and where different adaptation and management strategies will be most effective.

The exposure of freshwater-dependent coastal forests to saltwater from accelerating rates of sea level rise alters freshwater and terrestrial biota and biogeochemical cycling, subsequently altering the carbon storage potential of coastal forests [10–12]. Coastal forests generally share characteristics of both wetlands and uplands, and thus sequester and store carbon in both aboveground biomass (e.g., vegetation) and in hydric soils that provide significant long-term stores of carbon [13]. Though C burial rates are highly variable, global rates for tropical mangrove forests, for example, are approximately 31–34 TgC y^{-1} compared to 78 TgC y^{-1} in tropical upland forests, despite covering less than 1% of the area covered by tropical forests [13]. These soils are often peat, water-saturated soils that make conditions favorable for anaerobic decomposition and subsequent carbon storage [11]. Wetland drainage and disturbances such as storm surge inundation can cause the rapid collapse of freshwater peats, particularly after vegetation dies off [14]. However, these dynamics and the associated impacts on carbon storage remain poorly understood. The large C stocks stored in coastal forests' vegetation and soils highlight the important, yet largely unexamined, role in the global sequestration and storage of carbon that would otherwise remain as atmospheric CO^2 and exacerbate climate change [13,15–17].

As sea levels rise and storm surge events increase in frequency and severity, saltwater intrusion, the inland movement of seawater via natural and manmade conduits (e.g., streams and irrigation canals), will become increasingly commonplace. Though freshwater-dependent forests can withstand brief periods of inundation and associated saltwater exposure, prolonged periods result in tree mortality and limited regeneration [18–20]. More salt-tolerant herbaceous species move into these areas, altering species composition and ecological function [21,22]. The process is often referred to as marsh migration, and the stands of dead or dying trees that remain are known as 'ghost forests', marking the landward reach of sea level rise impacts [23]. Along the North American Atlantic and Gulf Coasts, this phenomenon is particularly prevalent because sea level rise rates here are three times the global average of ~0.6 mm yr^{-1} [24], due in part to rapid land subsidence [25,26]. Coastal forests will continue to retreat landward, provided barriers such as roads, other infrastructure, or incompatible land uses do not exist [6]. Targeted adaptation and management activities can enhance forest resilience or facilitate marsh migration, but this requires knowledge of the environmental and land use conditions affecting site suitability and disturbance regimes.

While coastal forests are particularly vulnerable to sea level rise long term, land use activities, resource management (e.g., water use) and other disturbances (e.g., wildfires) are near-term threats. Land use change, and the interactive effects between land use change and sea level rise, will largely determine where and how much carbon is stored in coastal forests [27]. Land ownership in coastal landscapes involves private landowners and operators as well as local, regional, state, and federal agencies [23,28]. The land use decisions of individuals and organizations have regional impacts and necessitate approaches that account for socio-spatial processes in coastal planning, adaptation, and management. These processes often give rise to spatial heterogeneity in the relationships between environmental, land use, and disturbance drivers and coastal forest change. For example, if the relationship between coastal forest health and canal density changes across the study area, we might infer that the relationship exhibits spatial nonstationarity and that there are unmeasured social processes driving these variations. To understand where and when specific planning and management activities will be most successful, it is critical to not only quantify the relationships between coastal forest change and associated drivers of change [29] but also evaluate how these relationships might vary across space.

To date, research on carbon flux (e.g., the driving processes and controls of C dynamics) in temperate coastal forests has focused on findings from field studies [2,4,30]. These studies are requisite to an understanding of the fine scale processes driving coastal forest carbon change, but their applicability at regional scales is limited. They can be used, however, in conjunction with other data to scale up C cycling estimates for regional and landscape scale assessments. The use of remote sensing data in aboveground carbon estimation, with its broad coverage and moderate-to-high spectral, spatial, and temporal resolutions has become an important field of research in recent years [31]. Carbon has been successfully estimated using derived spectral metrics from multispectral satellite imagery [32–34], structural metrics from light detection and ranging (LiDAR) data [35–38], or a combination of the two [39–41] in correlative models linking these metrics with field data and has become a common source for carbon estimation at regional scales. Access to multitemporal remote sensing data provides additional opportunities to measure change in carbon stocks over time; however regional analyses assessing carbon dynamics have focused on upland forests (e.g., [42–45]), herbaceous marsh (e.g., [46]), and mangrove forests (e.g., [47,48]) and not temperate coastal forests (however, see [49]). These studies also do not quantify the combined effects of the natural and anthropogenic factors driving carbon variability and change at regional scales, nor do they account for the spatial heterogeneity in the strength of the relationships between drivers and aboveground carbon stocks by implementing appropriate spatial modeling approaches. Traditional nonspatial regression models cannot capture local variation (in geographic space) in relationships between drivers and responses (referred to as spatial non-stationarity) [50]. Local spatial regression methods, however, can account for variation resulting from the location (non-stationarity) or the spatial scale considered (grain size, sampling interval, or spatial extent) and are well suited to quantify the relationships between drivers of carbon change and the change itself [51]. All of these are requisite to cost effective adaptation, conservation, and management [13,52].

To address these knowledge gaps, we quantified spatiotemporal patterns of aboveground carbon change over a decade (2001–2014) across one of the largest, most vulnerable regions of coastal forest in North America using a combination of field measurements, multispectral satellite imagery, and LiDAR data. We developed a suite of sea level rise, land use, and disturbance drivers and applied global and local spatial regression approaches to model relationships between these drivers and aboveground carbon change. We address the following questions: (1) How do the relationships between aboveground carbon decline and sea level rise, land use, and natural disturbance drivers vary across the landscape? (2) What are the relative contributions of land use, natural disturbance, and sea level rise drivers to carbon decline? (3) Using the spatial regression models developed, what are the future potential aboveground carbon declines as sea levels rise? Our results strengthen the case for the value of coastal forests' role in the global carbon cycle and highlight the need for similar research on other carbon pools in coastal forests (e.g., soils). Additionally, understanding the drivers of variability in coastal forests' carbon stores can help managers allocate resources effectively and implement appropriate conservation, adaptation, and management strategies with the potential to address regional carbon management and climate mitigation needs.

2. Materials and Methods

2.1. Study System

The study area is a 4000 km² coastal region located west of the Albemarle-Pamlico Estuary in Eastern North Carolina and buffered from the Atlantic Ocean by a chain of barrier islands (Figure 1). Forested and herbaceous wetlands, together, cover more than 50 percent of the Albemarle-Pamlico Peninsula. Almost half (47%) of the study area lies below 1 m in elevation, making it particularly vulnerable to sea level rise impacts. Forty percent of the peninsula is publicly owned, most of which can be considered 'public-non extractive' land or areas without extractive activities but that may still be managed for biodiversity (e.g., mimicking disturbance events through prescribed fires). 'Public extractive' lands are those lands subject to extractive uses (e.g., logging or mining). Private property (60% of the study area) is a mix of natural forest, agricultural lands, and forestry uses. Agriculture and forestry are two main economic drivers, contributing to the state approximately $367 million in agriculture cash receipts and $342.8 million from forest output per year [53]. Industrial forests and agricultural lands occur in low-elevation areas and are thus highly managed with canals, water control structures, and impoundments.

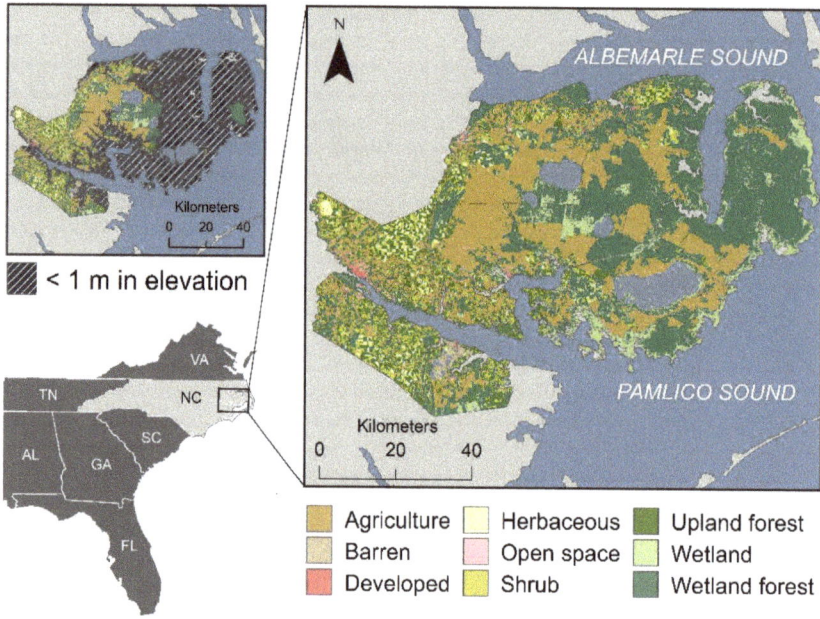

Figure 1. The Albemarle-Pamlico Peninsula is an approximately 4000 km² region in Eastern North Carolina, USA. It is bounded by the Albemarle and Pamlico Sounds, which are separated from the Atlantic Ocean by a chain of barrier islands. Almost half (47%) of the peninsula lies below 1 m in elevation exposing agricultural lands, forestry operations, and natural coastal forests to impacts from sea level rise.

2.2. Mapping Coastal Forest Carbon Declines

We mapped aboveground biomass change between 2001 and 2014 using a combination of field measurements, repeat light detection and ranging (LiDAR) surveys (see Table S1 for specifications) [54,55], and Landsat imagery (see Table S2 for specifications) [49,56]. We quantified vegetation structure and composition with metrics derived from LiDAR and Landsat data for 2001 and 2014 (see Tables S3 and S4 for complete lists of satellite-derived and LiDAR-derived variables). To relate the remote sensing metrics to aboveground biomass, we inventoried the size and density of woody vegetation in 98 plots of 12 m radius [57] across a vegetation gradient from forest to marsh [49]. We calculated total (e.g., woody and herbaceous species) aboveground biomass in megagrams per hectare (Mg ha^{-1}), from these field measurements using established allometric equations [58–60] relating diameter at breast height (dbh) or percent cover to total aboveground biomass. We applied the random forest machine learning algorithm (RF; [61]) to quantify relationships between aboveground biomass field measurements and remotely sensed metrics and predict aboveground biomass across the entire study area for 2001 and 2014. RF is an ensemble learning method for classification and regression that functions by constructing many decision trees for model training and then generating mean predictions or the mode of classes across the individual trees. The algorithm corrects for decision trees' tendency to overfit their training datasets through the bootstrap approach [61]. Using 1000 permutations of the final fitted RF model, we tested model significance, performed validation withholding 30% of the training data, and generated metrics of predictor variable importance. We used the model improvement ratio (MIR) function to select the best predictor variables among the suite of candidate variables, following the methods found in [42]. We then quantified aboveground biomass change by comparing the output maps of aboveground biomass for 2001 and 2014 (Figures 2a and 3a). The RF models performed well (2014 biomass: $R^2_{adj.}$ = 0.78, RMSE = 15.0 Mg ha^{-1}, % RMSE = 9.5; 2001 biomass: $R^2_{adj.}$ = 0.75, RMSE = 18.3 Mg ha^{-1}, % RMSE = 12.6) [49] (Figure S1). Metrics that contributed most to overall predictive power included mean and median vegetation height, highlighting the importance of vegetation structure for estimating aboveground biomass (Figure S2). We converted aboveground biomass change to carbon loads assuming a carbon concentration of 0.5 [62]. We developed a binary 30-m resolution map of carbon declines, where a value of 1.0 represented any cell with a decrease in carbon over the 13-year period, and a value of 0.0 was assigned to any cell with either no change or an increase in carbon storage. We excluded from the binary map non-vegetated cells and cells classified as developed or agriculture in 2001 or 2014.

Using variogram analysis, we identified 450 m as an appropriate level of aggregation for the response variable (Figure S3). We overlaid the binary carbon decline maps with a 450 × 450-m regular grid and calculated the fraction of the percentage of vegetated area within each modeling unit experiencing carbon declines (Figure 3b).

2.3. Selecting and Mapping Driver Variables

We developed a suite of variables that we expected to be drivers of carbon declines and classified them into different representative classes—land use, natural disturbance, and sea level rise drivers (for complete list, see Table 1).

2.3.1. Land Use Drivers

Land use activities and modifications, such as farming, forestry, and even land abandonment, influence carbon storage and flux on the landscape. Agricultural and forestry pressures (as quantified by agricultural pressure and harvest intensity described below) can alter soil properties, increase nutrient runoff, and change vegetation composition, all of which influence carbon pools both in the areas actively managed and on spatially adjacent or nearby lands [4,9]. Land abandonment (e.g., fallow lands) can also change carbon pools by allowing for natural regeneration, providing buffers from storms for nearby lands, and increasing productivity on site or nearby [51]. Because of their potential to influence carbon fluxes, we included several land use-related variables, derived from spatial datasets, as described in more detail below.

Table 1. Suite of spatially explicit predictor variables explored within the land use, natural disturbance, and sea level rise categories. All variables were developed as 30-m resolution raster datasets and aggregated using a 450 × 450-m grid.

Description		Base Data	Year	Data Source
	Land Use Drivers			
Agricultural pressure	Gravity model, number of neighboring agriculture cells within a search distance and weighted by distance	CDL, Cropscape	2001, 2014	NASS
Fallow pressure	Gravity model, number of neighboring fallow cells within a search distance and weighted by distance	CDL, Cropscape	2001, 2014	NASS
Harvest intensity	Threshold applied to between-year mean NDVI change to create binary outputs of harvest/acute event, summed across years	NDVI	2001–2014	Google Earth Engine
	Natural Disturbance Drivers			
Time since fire	Using fire perimeters and year of fire, assigned number of years since fire across the landscape	Vector file	Current	MTBS
	Sea Level Rise Drivers			
Connected canal density	Line density of only those canals connected to open water in study area, with influence at 1000 m distance	NHD	Current	USGS
Distance to estuarine shoreline	Euclidean distance (km) to estuarine shoreline	NHD	Current	USGS
Fast storm surge	Fast storm surge categories	Vector file	Current	NC Onemap
Flow accumulation	Flow accumulation	DEM	2014	Derived
Flow direction	Flow direction	DEM	2014	Derived
Salinity	Inverse distance weighting interpolation of average salinity (ppt) from 2001 to 2014	STORET	2001–2014	EPA
Mean precipitation deviation	Yearly (2001–2014) average precipitation difference (mm) from historical norm (1990–2000)		2001–2014	Google Earth Engine
MHHW adjusted elevation *	Elevation (m) adjusted by current mean higher high water (MHHW)	MHHW, DEM	2014	Derived
Minimum temperature deviation	Yearly (2001–2014) minimum temperature difference (degrees Celsius) from historical norm (1990–2000)		2001–2014	Google Earth Engine
Slow storm surge	Slow storm surge categories	Vector file	Current	NC Onemap

Note: Monitoring Trends in Burn Severity Program (MTBS), National Agricultural Statistics Service (NASS), US Environmental Protection Agency (EPA), US Geological Survey (USGS) digital elevation model (DEM), National Hydrography Dataset (NHD), normalized difference vegetation index (NDVI). * Mean higher high water (MHHW) is the average of the higher high water height of each tidal day over the National Tidal Datum Epoch. The station on Duck Pier in North Carolina, USA, was used as our reference and for the current National Tidal Datum Epoch (1983–2001); their MHHW is 1.24 m [63].

Agricultural and Fallow Pressure.

To account for pressures from agriculture or land abandonment that might influence carbon, we extracted all agricultural lands from the 30-m resolution 2001 National Land Cover Dataset [64] raster and applied a gravity modeling technique, computing for each cell the number of neighboring agricultural cells within a search distance, weighted by a distance decay function [65,66]. This variable accounts for the effect of agricultural lands on nearby forest carbon changes, with more proximate agricultural lands having a stronger influence, as controlled by a coefficient for the distance decay function. For land abandonment, we applied the same methods, extracting instead the fallow land use class from the USDA National Agricultural Statistics Service's Cropscape dataset [67].

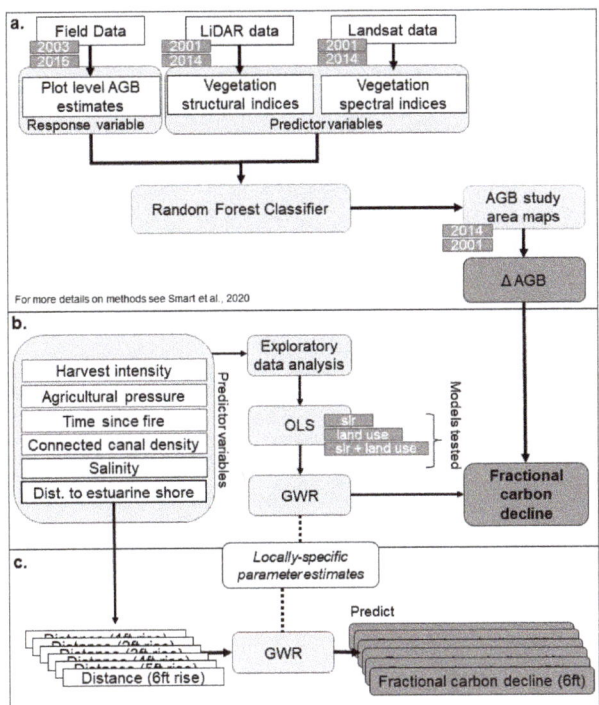

Figure 2. Schematic describing the three main analyses that include (**a**) mapping aboveground biomass (AGB) change and calculating fractional carbon decline (Section 2.2; for more details on methods [49]); (**b**) exploratory data analysis to select uncorrelated predictor variables and modeling fractional carbon decline as a function of sea level rise (slr), land-use, or a combination of drivers using nonspatial (ordinary least squares; OLS) and spatial (geographically weighted regression; GWR) regression models (Sections 2.3 and 2.4); and (**c**) using the locally specific parameter estimates established from the GWR model to predict future fractional carbon declines by holding all other variables constant and updating the distance to estuarine shoreline variable (Section 2.5).

Harvest Intensity.

To identify possible forestry-related land use modifications, we developed a 'harvest intensity' metric to quantify changes related to industrial timber harvesting and clearing. We selected the pre-processed annual average Landsat-based Enhanced Vegetation Index (EVI) for every year between 2001 and 2014 from Google Earth Engine [68]. We calculated year-to-year EVI differences and applied a threshold to the average change values to differentiate acute or harvest events from normal variability in EVI based on approximately one standard deviation from the average change values across all year-to-year differences (+/−0.12). We created the 'harvest intensity index' by summing all binary rasters created after threshold application; each value in the raster represented the number of times that a cell experienced an acute event during the study period. We removed acute events due to wildfires from the harvest intensity metric because wildfires were accounted for in a separate predictor variable.

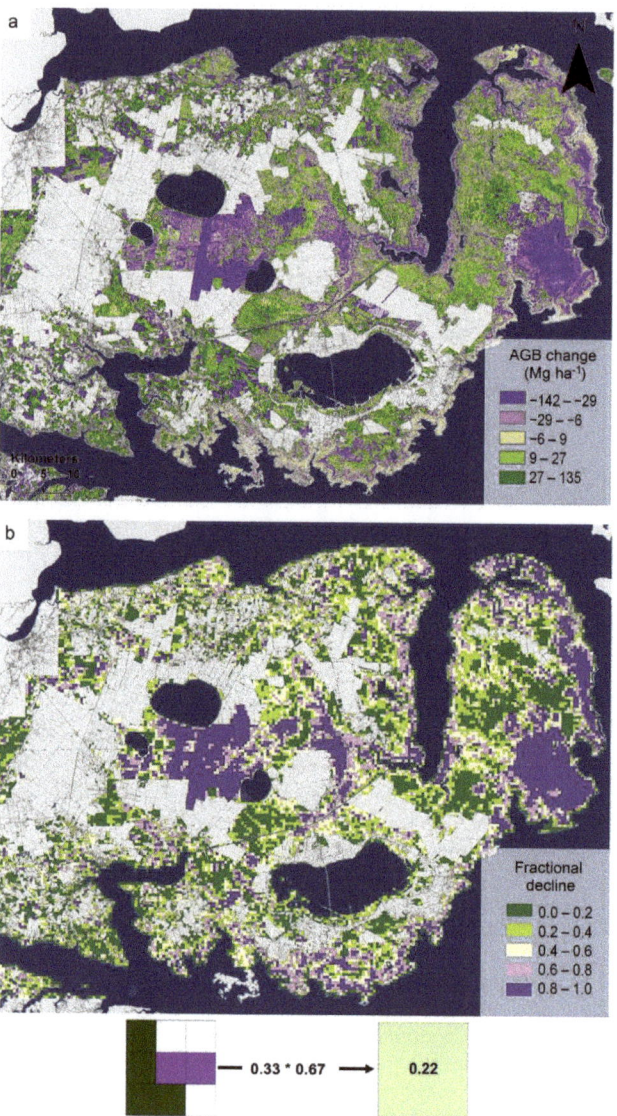

Figure 3. (**a**) Aboveground biomass change, measured as Mg ha $^{-1}$, between 2001 and 2014, for the Albemarle-Pamlico study area in Eastern North Carolina. Areas excluded from the model are in light grey. Water bodies and the Albemarle and Pamlico Sounds are dark blue. (**b**) Fractional carbon decline, calculated from (**a**) by converting 30-m grid cells to carbon loads, developing a binary 30-m resolution map by assigning carbon declines values of 1, and aggregating data to 450 × 450-m grid. Fractional carbon decline is the fraction (see inset schematic; purple cells divided by green cells; 0.33) of the percentage of vegetation area (colored cells divided by total number of cells; 0.67) within each 450 × 450-m modeling unit experiencing decline.

2.3.2. Sea Level Rise Drivers

Sea level rise and associated climatic factors can drive changes in carbon storage and flux through a variety of mechanisms [13,14]. To capture general trends or changes in climate, we included climatic variables such as mean precipitation and temperature deviations over the last decade. As climate changes, extreme weather events, such as flooding and hurricanes, are expected to become more frequent and severe [7,25]. Though acute, these events can have lasting impacts on coastal carbon storage—both aboveground and in soils—by altering vegetation and soil composition. These impacts can be compounded by accelerating rates of sea level rise [19,20,29]. To address the potential for these factors to influence carbon flux, we included drivers related to where hurricane storm surges occur (e.g., slow and fast storm surge variables), where water flows and accumulates on the landscape (e.g., elevation, flow direction and accumulation), and proximity to bodies of water (e.g., distance to estuarine shoreline).

Saltwater intrusion, resulting from a combination of both acute events and gradual sea level rise, facilitates changes in vegetation composition and subsequent carbon storage. Long term saltwater exposure in freshwater-dependent forests leads to osmotic stress and eventual mortality [14,20,21]. More salt-tolerant herbaceous species will replace tree species, changing carbon storage capacity, particularly aboveground. Hurricanes and even droughts (by shifting the freshwater–saltwater interface landward) can bring saltwater inland into areas not adapted to saltwater exposure [10,12]. Drainage networks, comprised of canals and ditches, though built to keep water off coastal lands, can serve as conduits during storms, to move water inland [51]. Because salt influences aboveground and soil carbon storage, we included spatial data representing the likelihood of salt impacts by accounting for potential pathways (e.g., the density of canals connected to the estuarine shoreline) and overall salinity values in the adjacent estuary (interpolated from water quality monitoring stations across the landscape). Other factors such as elevation, hurricane storm surge extents, and distance to the estuarine shoreline also indirectly influence where saltwater exposure occurs on the landscape.

Connected Canal Density.

We incorporated hydrological connectivity between coastal forests and the adjacent estuary by extracting artificial drainage features from the US Geological Survey's National Hydrography Dataset [69]. We removed any isolated canals not connected by other canals or natural water features to the estuarine shoreline. We then calculated the density of linear canal features surrounding each 30-m raster cell using multiple search radii values, ultimately selecting a 1000-m search radius for the final predictor variable.

Salinity.

We used the US Environmental Protection Agency's STOrage and RETrieval (STORET) database to extract average salinity values (in parts per thousand; ppt) for 3500+ observations from 40 water quality monitoring sites in the Albemarle and Pamlico Sounds between 2001 and 2014 [70]. To interpolate a surface from these point data, we performed an inverse distance weighting interpolation at 30 m resolution [71].

2.3.3. Natural Disturbance Drivers

Frequent, low intensity fires have historically maintained these landscapes [72]. However, catastrophic fires can alter carbon pools by burning the rich organic soils in addition to the vegetation. Though regeneration occurs naturally and rapidly post-fire, interactions between fire, extreme weather events, and climate change can alter carbon storage and flux in unanticipated ways. We included a variable that measures the length of time since a wildfire for the study period to capture these changes in carbon storage and flux.

Time Since Fire.

We used the spatially explicit fire perimeters developed by the Monitoring Trends in Burn Severity Program (MTBS; [73]) and extracted the year of fire attribute to assign the number of years since fire for every fire polygon across the study area.

2.4. Statistical Modeling Approach

We mapped all predictor variables at a 30-m spatial resolution and then aggregated cell values to the spatial modeling unit (450 × 450-m grid cells) via majority (for nominal data) or mean (for continuous data) aggregation methods and derived values for each grid cell. Prior to model fitting, we evaluated predictor variables for multicollinearity and selected a set of uncorrelated (r < 0.5) variables from our initial list (Figure S4). We scaled all variables to values between 0.0 and 1.0 using a minimum–maximum normalization, which retained the original distribution spread but allowed us to evaluate relative contributions of different drivers.

We modeled the relationships between fractional carbon declines and predictor variables using ordinary least squares regression (OLS), a conventional global regression technique, and geographically weighted regression (GWR), a local regression method (Figure 2b). GWR detects heterogeneity in data relationships across geographic space via the fitting of individual localized ordinary least squares regressions [74]. This allows for local variations in rates of change. Coefficients in the model are not global estimates but are specific to a particular location and based on observations taken at sample points near that location [51,75]. The regression equation is then:

$$y_i = \alpha_{i0} + \sum_{k=1,m} \alpha_{ik} x_{ik} + \varepsilon_i \qquad (1)$$

where y_i is the ith observation of the dependent variable, x_{ik} is the ith observation of the kth independent variable, ε_i is the error term, and α_{ik} is the value of the kth parameter at location 1. Estimates of are then based on observations taken at sample points close to i. A weighting function is then employed so that for each point i there is a bump of influence around i in such a way that those observations near i have more influence in the estimation of parameters than those farther away. In ordinary least squares, the sum of the squared differences of predicted and observed y_i is minimized in the coefficient estimates. In weighted least squares, a weighting factor is applied to each squared difference before minimizing so that the inaccuracy of some predictions carries more of a penalty than others. In weighted regression models the values of the weighting factor are constant so that only one calibration is carried out. In geographically weighted regression, the weighting factor varies with location i so that a different calibration exists for every point in the study area. The geographical weighting is implemented through a spatial kernel function. Common choices for weighting function include Gaussian, bisquare, or tricubic [52]. In addition to the weighting function, a bandwidth is needed that determines the rate of distance decay for each of the data weightings. A large bandwidth smooths parameter estimates, approaching the estimates provided by the global regression, and a small bandwidth tends to sharpen them. The bandwidth can be user-defined or determined using cross-validation (CV) or Akaike's information criterion (AIC) and can be either a fixed distance or a set number of nearest neighbors (adaptive bandwidth). To minimize potential edge effects, we used an adaptive Gaussian kernel function with an optimal bandwidth selected based on a least squares CV that minimized the squared error. All statistical analyses were carried out using R statistical software and the GWR analyses were performed using the 'spgwr' package for R [76,77].

We examined the variance explained by the global and local regression models, Moran's I tests of model residuals, and the statistical significance and relative importance of predictor variables. We also quantified spatial dependencies and nonstationarity in the relationships between the response and predictor variables by mapping residuals and testing the interquartile ranges of the local coefficient estimates provided by GWR. Additionally, we used the studentized Breusch–Pagan test for heteroscedasticity to identify non-constant variance in the errors. We derived a pseudo-significance test for predictor variables by calculating t-values for each local regression and mapping t-values spatially to identify locations that are statistically significant at alpha = 0.05 [51].

2.5. Predicting Future Coastal Forest Carbon Declines from Sea Level Rise

We used spatial models of future sea level rise inundation from the National Oceanic and Atmospheric Administration (NOAA) to forecast future forest carbon decline [78].

The data depict the potential inundation of coastal areas resulting from a projected 0.3 to 1.8 m sea level rise above current mean higher high water (MHHW) conditions. The values represent the projections for the intermediate to high global mean sea level (GMSL) scenario, which predicts sea levels under Representative Concentration Pathway (RCP) 8.5 from the Intergovernmental Panel on Climate Change (IPCC). This is considered the high-emissions scenario and is referred to as the business-as-usual scenario, which suggests that it is a likely outcome if society does not make concerted efforts to cut greenhouse gas emissions. The model is a function of static elevation and hydrologic connectivity but does not account for other factors such as vertical accretion or subsidence rates, nor does it account for storm surge potential. We calculated total coastal forest carbon loss, measured in teragrams of carbon (TgC), through 2100 by combining our 2014 aboveground carbon maps and the GMSL inundation models. To predict future coastal forest declines in areas not inundated, we updated the distance to the estuarine shoreline predictor variable used in our GWR model using the new estuarine shoreline extent under inundation levels between 0.3 and 1.8 m. Holding all other predictor variables constant, we re-ran the GWR model and examined the resulting predicted fractional carbon declines on the remaining study area (Figure 2c).

3. Results

3.1. Statistical Modeling of Coastal Forest Declines

The global regression model explained 39% of the variation in fractional aboveground carbon declines. Statistically significant predictor variables included in the final model were agricultural pressure, connected canal density, time since fire, harvest intensity, distance to shoreline, and salinity ($p < 0.05$). We found negative relationships between fractional aboveground carbon declines and agricultural pressure, time since fire, and distance to estuarine shoreline. We found positive relationships between fractional aboveground carbon declines and canal density, harvest intensity, and salinity (Table 2).

Table 2. (a) Coefficient and standard errors for predictor variables used in the global ordinary least squares (OLS) regression model. (b) Coefficient mean (and range) and standard error mean (and range) for the local geographically weighted regression (GWR) model. + denotes nonstationarity in the relationship between predictor variable and response in the GWR model as calculated by the Breusch–Pagan statistic.

	(a) OLS		(b) GWR	
	Coefficient	Standard Error	Coefficient Mean (Range)	Standard Error Mean (Range)
Intercept ****+	0.891	0.009	0.825 (0.120, 1.765)	0.054 (0.026, 0.273)
Agricultural pressure ****+	−0.431	0.006	−0.387 (−1.140, −0.075)	0.029 (0.015, 0.232)
Connected canal density ****+	0.084	0.013	0.168 (−1.308, 1.629)	0.075 (0.026, 0.198)
Time since fire ****+	−0.471	0.007	−0.398 (−1.250, 0.507)	0.048 (0.014, 0.311)
Harvest intensity ****+	0.160	0.012	0.186 (−0.225, 2.573)	0.058 (0.028, 0.238)
Distance to shoreline *+	−0.015	0.008	−0.119 (−1.872, 0.895)	0.101 (0.030, 0.375)
Salinity ****+	−0.049	0.008	0.003 (−5.128, 5.210)	0.117 (0.017, 1.000)

OLS: R^2 = 0.39, AIC = −2413.6 | GWR R^2 = 0.56, AIC = −9981.7, RSS = 952.1. * Statistically significant at $p < 0.05$; *** Statistically significant at $p < 0.001$.

Land use and disturbance metrics were the most important predictors of variation in fractional carbon decline within the global model, with agricultural pressure explaining the most variation, followed by the time-since-fire variable (Figure 4). Moran's I tests indicated autocorrelation in the response variable, even at the aggregated 450-m resolution (Moran's I = 0.75, z-score = 167.77, p-value < 0.0001). The relationship between the predictor variables and the response variable exhibited nonstationarity (Breusch–Pagan statistic = 4092.4, p-value < 0001).

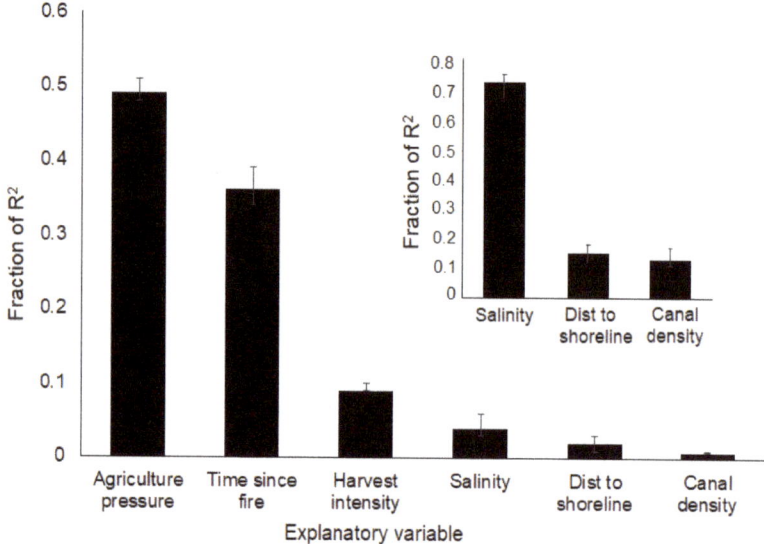

Figure 4. Relative importance for each predictor variable used in the final model (R^2 = 0.39) with 95% bootstrap confidence intervals using the method from [79]. Metrics are normalized to sum to 100%. Inset shows the relative importance of predictor variables for a model that includes only the sea level rise drivers (R^2 = 0.05).

Predictive performance improved with the local regression model, with a lower AIC and higher quasi-global R^2. The GWR model explained 52% of the variance, and quasi-global R^2 values ranged from 0.08 to 0.79, with more than 50% of the landscape having a local R^2 value greater than the global R^2 of 0.39 (Figure 5a). Standard errors ranged from a low of 0.006 to a high of 0.06 (Figure 5b). Coefficients for all predictor variables exhibited patterns of spatial non-stationarity according to tests for spatial dependencies (Figure 6).

We found a statistically significant negative relationship between fractional carbon decline and agricultural pressure, canal density, and distance to the estuarine shoreline in the GWR. The range of local coefficient values and statistical significance varied across space, indicating spatial dependencies in the data. On average, we found statistically significant positive relationships between fractional carbon declines and harvest intensity, time since fire, and salinity with spatial heterogeneity in local regression coefficients and statistical significance.

For 52% of the study area, land use and disturbance drivers had the strongest regression coefficients (13% of the total landscape for agricultural pressure, 13% for harvest intensity, and 26% for time since fire), and for 48% of the study area the sea level rise variables had the largest regression coefficients (2% of the total landscape for canal density, 20% for distance to estuarine shoreline, and 27% for salinity). In the southeast portion of the study area, the sea level rise drivers best explained variation in carbon declines (e.g., Figure 6c,d). In the northeast portion of the study area, both land use/disturbance drivers and sea level rise drivers contribute to declines (Figure 6). To the interior, relationships were mixed, with land use drivers most important in areas dominated by agricultural and forestry operations and sea level rise drivers generally most important in interior areas connected to the estuary via canals (e.g., Figure 6b–f).

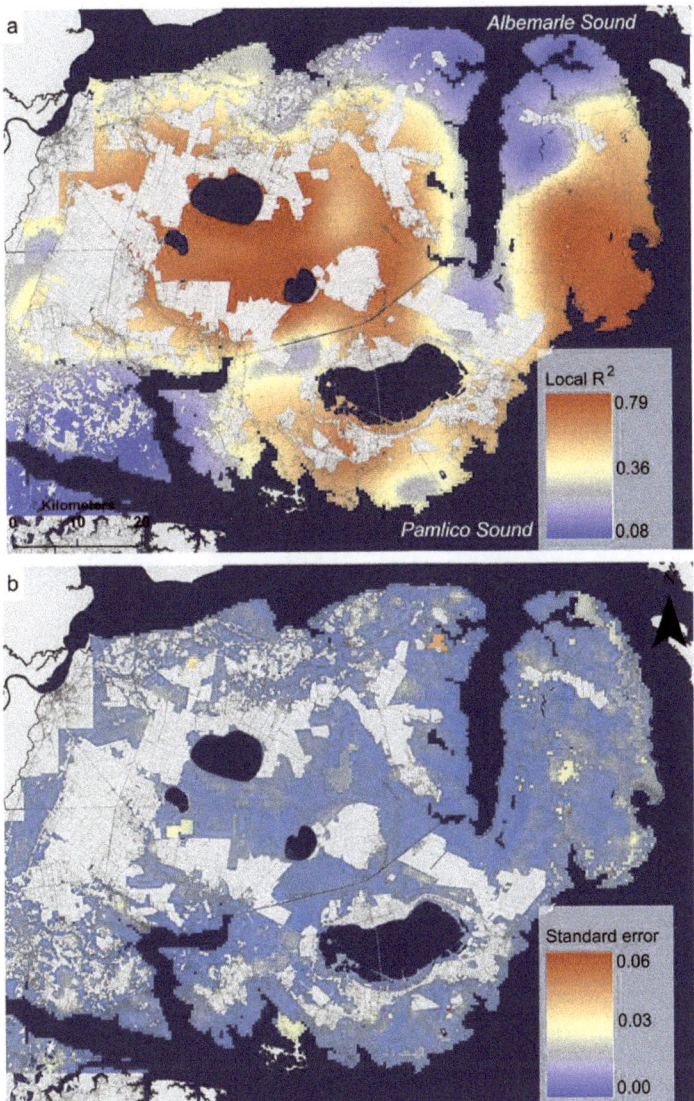

Figure 5. (a) Quasi-global R^2 from geographically weighted regression (GWR) model for the Albemarle-Pamlico study area in Eastern North Carolina, USA. Areas excluded from the model are in light grey. Water bodies and the Albemarle and Pamlico Sounds are dark blue. (b) Quasi-global R^2 standard error values from the GWR model.

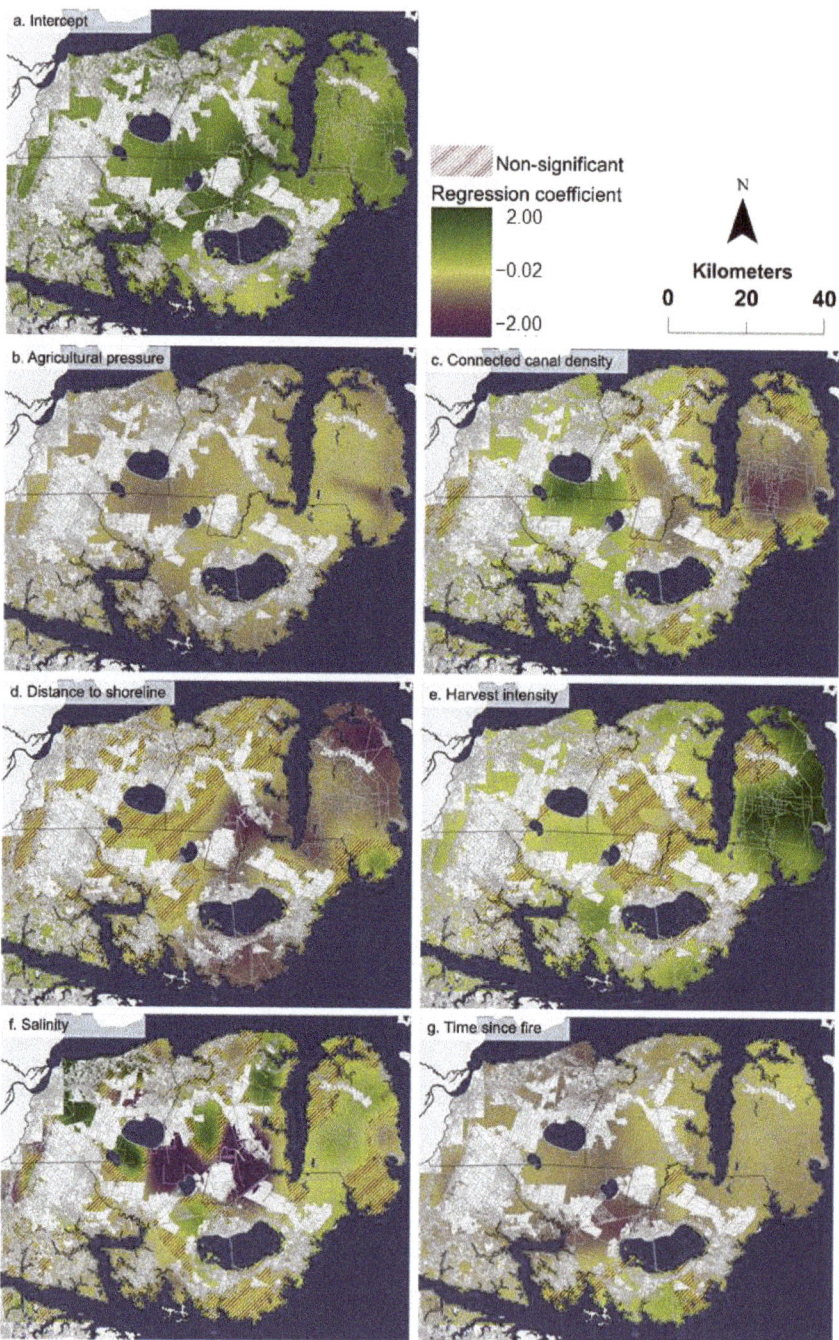

Figure 6. (**a**–**g**) Coefficients for intercept and predictor variables used in the final geographically weighted regression model for the Albemarle-Pamlico study area in Eastern North Carolina, USA. Areas excluded from the model are in light grey. Water bodies and the Albemarle and Pamlico Sounds are dark blue. Red diagonal lines indicate locations where the predictor variables were not statistically significant in the local regression models according to the pseudo-significance tests.

3.2. Future Coastal Forest Declines

We estimated a minimum inundation extent of 1094 km² for a 0.3 m rise in sea level and a maximum extent of 2914 km² for a 1.8 m rise in sea level (Figure 7). Associated losses in total aboveground carbon ranged from 2.9 ± 0.1 TgC (0.3 m rise) to 8.6 ± 0.3 TgC (1.8 m rise). On the remaining landscape, we calculated potential declines increasing from 29% to 32% between a 0.3 and 1.8 m rise in sea level. The greatest changes in carbon declines are visible in each scenario adjacent to the shoreline and surrounding the lakes in the study area's interior (Figure 8).

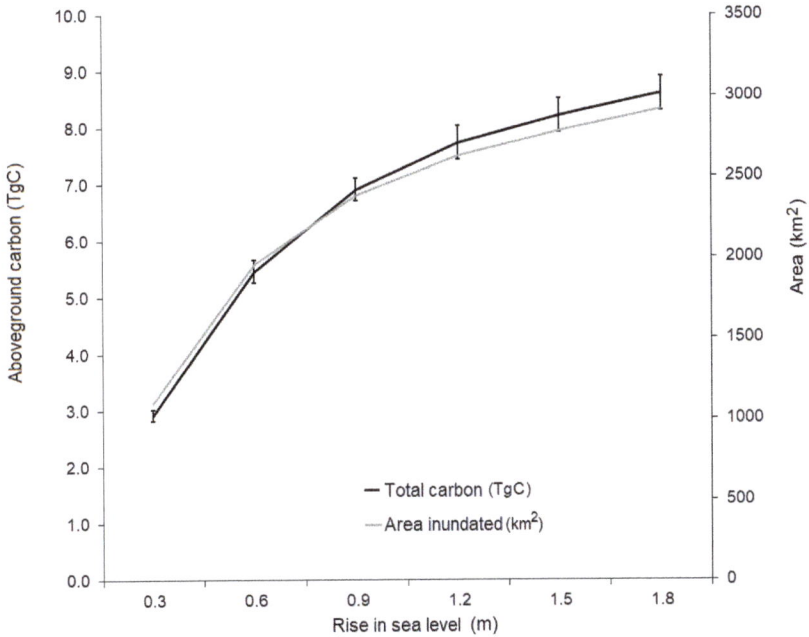

Figure 7. Total area (km²) inundated (grey line; right axis) and total aboveground carbon, measured in teragrams of carbon (TgC), lost (black line; left axis) and associated standard errors from sea level rise inundation. The sea level rise inundation model, from the National Oceanic and Atmospheric Administration, represents inundation under the intermediate–high sea level rise (slr) scenario (ranging from 0.3 to 1.8 m) or Representative Concentration Pathway 8.5.

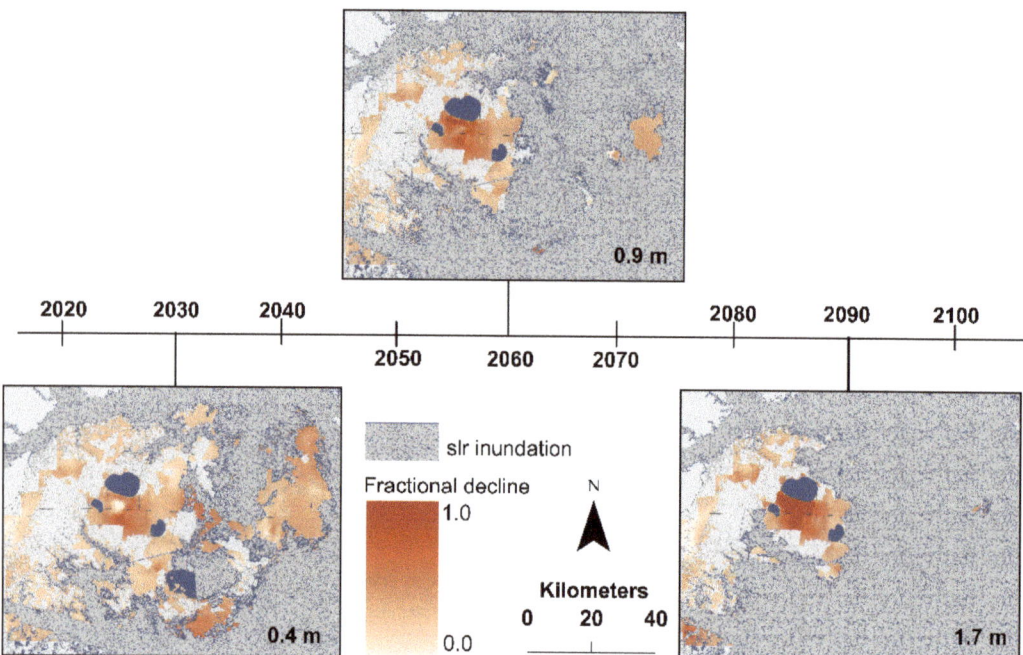

Figure 8. Fractional carbon declines associated with sea level rise inundation using the National Oceanic and Atmospheric Administration's Sea level rise (slr) inundation models for the intermediate–high sea level rise inundation scenario (values range from 0.3 to 1.8 m). Fractional carbon declines estimated using the geographically weighted regression model with distance to estuarine shoreline predictor variable updated with incremental inundation between 0.3 and 1.8 m.

4. Discussion

Sea level rise, land use activities, and disturbances act synergistically in low-lying coastal forests to alter their structure and function, affecting the ecosystem services they provide. A regional scale understanding of where and how these forests have changed and where to expect future change is critical for prioritizing funding for management and establishing resilient coastal communities. Spatial heterogeneity in the drivers of coastal forest degradation, however, make quantification at regional scales difficult. To address this challenge, we examined the drivers of coastal forest degradation using a local regression approach that accounts for these spatial processes. Using geographically weighted regression (GWR), we examined the relationship between coastal forest fractional carbon declines and land use, disturbance, and sea level rise drivers. We found the patterns of influence to be spatially heterogeneous, providing local insights at regional scales, highlighting the benefit of using a local regression approach to inform management and resource allocation activities more effectively [50,80].

In a low-lying coastal region highly exposed to impacts from sea level rise, we observed that land use and disturbance metrics were the strongest drivers of fractional carbon declines. Agricultural pressure, harvest intensity, and time since fire explained most of the variation in the global model. Interestingly we found a negative relationship between agricultural pressure and fractional carbon declines. Because agricultural lands are generally located at higher elevations in the interior of the study area, the adjacent coastal forests are likely less susceptible to the effects of saltwater intrusion and sea level rise than lower elevation coastal forests near the shore. Alternatively, water control structures, tide gates, and other measures intended to protect the productivity of large-scale agricultural operations may also benefit the coastal forests in these higher elevation areas. In addition to agriculture, approximately 13% of the study area's inland coastal forests are managed as loblolly pine (*Pinus taeda*) plantations for timber harvest revenues, which removes a significant amount of aboveground carbon [81]. Though stand-replacing disturbances (e.g., harvests) lead to a net forest C loss, replanting as part of the rotational cycle will eventually replace the lost carbon. However, if trees do not reach maturity and peak primary productivity before subsequent harvests [4,9], increased harvest intensity could result in net carbon losses as was demonstrated in the positive relationship between harvest intensity and aboveground carbon declines in our results.

Natural disturbance events play an important role in coastal forest carbon dynamics. Both the local and global regression models indicated that the time-since-fire metric contributed significantly to the explanatory power of the model, explaining 36% of the model variation. The more recent the fire event, the higher the fractional carbon declines. Despite a low prevalence of prescribed fire and wildfire in our study area compared to historical estimates, fire remains one of the most important drivers of carbon dynamics on the APP [80]. During the study period, several catastrophic wildfires occurred on the Albemarle-Pamlico Peninsula. Though severe wildfires result in aboveground carbon declines, many of the region's tree species are adapted to fire and regenerate rapidly after fire disturbances [72]. However, if saltwater exposure occurs post-fire, regeneration may be limited by salinity intolerance of the saplings. Accordingly, fire disturbances and saltwater exposure can act synergistically to accelerate forest retreat and facilitate marsh migration of salt-tolerant herbaceous species into the area [19,20].

Although our global regression model indicated sea level rise metrics contributed minimally to the overall explanatory power of the model (explaining only 6.8% of model variation), our local regression model identified places on the landscape where these metrics played a strong and significant role in fractional carbon declines. Distance to the estuarine shoreline, salinity, and connected canal density were important predictors of fractional carbon declines in the northeast and eastern portions of our study area. That there are high levels of spatial heterogeneity in these relationships reinforces the need to consider local spatial regression approaches in dynamic coastal systems. Sea level rise factors are a critical and significant driver of fractional carbon declines, but only near the shoreline and in areas predicted to have high salinities. Unlike the global regression model, our local spatial regression approach can be used to identify the extent and strength of influence for each of these sea level rise drivers in a spatially explicit way. Because the changes associated with saltwater intrusion and sea level rise are likely to be permanent, our approach can be used to identify areas unlikely to recover or return to coastal forest, particularly as sea level rise accelerates. This information can then be used to make more informed decisions about policy, mitigation, and the prioritization of limited resources.

Our results indicate that by 2090, the study area could lose approximately 8.6 TgC of aboveground carbon because of inundation from sea level rise, with the remaining landscape experiencing, on average, 32% fractional declines in carbon. These estimates, however, do not account for future vegetation growth or succession, nor do they consider a potential shift in carbon storage, from aboveground to belowground [2]. This highlights the need to consider impacts on other carbon pools in coastal forests (e.g., soils) and landscape context because shifts in carbon storage in one location alter the amount of carbon stored in another location. Considering these aspects and incorporating potential non-linear relationships, feedbacks, and interactions between natural and anthropogenic drivers via mechanistic or process-based models would provide additional support for targeted adaptation and management activities.

Our study underscores the importance of human interventions in determining future coastal forest resilience. The spatial variation in driving forces of coastal forest degradation has significant implications for prioritizing management activities. Our results indicate that along the shoreline and in the eastern portion of the study area the best management decision might be to facilitate marsh migration, whereas inland there may be more opportunities to work with landowners in conserving or restoring coastal forests.

Management of drainage networks, originally built to drain wetlands and lower water tables for agricultural and forestry operations, is another important consideration [81,82]. In the near term these features help mitigate flooding; however, our results from lower elevation coastal forests adjacent to the estuaries suggest that without effective canal and ditch management, these networks act as conduits for saltwater intrusion from extreme weather events and gradual sea level rise [81]. In addition to water management, proactive coastal forest management activities such as lengthened harvest cycles, restricted harvests, and selective harvesting have the potential to enhance the carbon storage and coastal resilience [9]. Selective harvesting and tree species diversification on plantations will also play an important role as sea levels rise and saltwater intrusion increasingly affects forestry operations. Timber harvest revenues can be sustained by harvesting the most at-risk timber (e.g., timber most affected by inundation and saltwater exposure). The common practice of planting a single timber species on plantations increases the likelihood of widespread mortality during extreme storm and flooding events, particularly if that single species is salt-intolerant. Planting a more diverse set of species with a range of salinity tolerances will decrease the severity of impact from extreme storm events and produce added co-benefits associated with increased diversity [83].

5. Conclusions

Our analysis illustrates several key points. First, in a low-lying coastal region highly exposed to impacts from sea level rise, we observed that land use (explaining more than 50% of model variation) and disturbance metrics (explaining 36% of model variation) were the strongest drivers of fractional carbon declines. Second, although our global regression model indicated sea level rise metrics contributed minimally to the overall explanatory power of the model (explaining only 6.8% of model variation), our local regression model identified places on the landscape where these metrics played a strong and significant role in fractional carbon declines. Finally, our results indicate that by 2090, the study area could lose approximately 8.6 TgC of aboveground carbon because of inundation from sea level rise, with the remaining landscape experiencing, on average, 32% fractional declines in carbon. By measuring coastal forest degradation via aboveground carbon storage, we provide an important link to ecosystem services. Because communities may have diverse values for coastal forests, understanding when, where, and how ecosystem services in coastal forests are changing helps managers prioritize resources for management, adaptation, and restoration of coastal ecosystems [84]. Management decisions must consider these trade-offs among ecosystem services as well as current and future susceptibility to land use modifications and sea level rise.

Our approach quantified drivers of coastal forest decline in a spatially explicit way, highlighting the ways that human land use activities, climate, and natural disturbances can act synergistically to alter coastal forest structure and function. By mapping the spatially varying relationships between drivers and carbon declines, we identified geographic locations best suited for targeted management activities. However, to fully understand how these management activities will be implemented, and through what mechanisms, we need a greater understanding of landowner preferences, their evaluation of trade-offs among ecosystem services, their behaviors, and risk perceptions. These are requisite for reducing the uncertainty in coastal forest persistence as sea levels rise.

Supplementary Materials: The following are available online at https://www.mdpi.com/article/10.3390/land10070752/s1, Figure S1: Observed vs. predicted biomass, Figure S2: Predictor variable permutation importance in RF classification, Figure S3: Variogram, Figure S4: Correlation matrix, Table S1: LiDAR flight specifications, Table S2: Landsat imagery specifications, Table S3: Random Forest classification predictor variables; Table S4: Random Forest classification LiDAR predictor variables.

Author Contributions: Conceptualization, L.S.S. and J.W.S.; methodology, L.S.S., J.W.S., K.K.S., P.J.T.; formal analysis, L.S.S.; resources, J.W.S., J.V.; data curation, P.J.T., L.S.S.; writing—original draft preparation, L.S.S.; writing—review and editing, J.W.S., J.V., P.J.T., K.K.S.; funding acquisition, J.W.S. All authors have read and agreed to the published version of the manuscript.

Funding: This research was funded by a joint fellowship through NOAA's North Carolina Sea Grant Program and NASA's North Carolina Space Grant Program (NOAA award # NA14OAR4170073, NASA award #NNX15AH81H). Funding from a College of Natural Resources Innovation Grant at NC State University was also used to support this research.

Institutional Review Board Statement: Not applicated.

Informed Consent Statement: Not applicated.

Data Availability Statement: Data generated from this study are available upon request.

Acknowledgments: We are grateful to Joshua Randall from the Center for Geospatial Analytics for edits and comments that greatly improved the quality of this manuscript. We thank Georgina Sanchez of NC State University who provided technical insights during the spatial regression modeling efforts. Additionally, Douglas Newcomb of the US Fish and Wildlife Service provided valuable comments on methods and analyses.

Conflicts of Interest: The authors declare no conflict of interest.

References

1. Costanza, R.; de Groot, R.; Sutton, P.; Van der Ploeg, S.; Anderson, S.J.; Kubiszewski, I.; Farber, S.; Turner, R.K. Changes in the global value of ecosystem services. *Glob. Environ. Chang.* **2014**, *26*, 152–158. [CrossRef]
2. Krauss, K.W.; Noe, G.B.; Duberstein, J.A.; Conner, W.H.; Stagg, C.L.; Cormier, N.; Jones, M.C.; Bernhardt, C.E.; Graeme Lockaby, B.; From, A.S.; et al. The role of the upper tidal estuary in wetland blue carbon storage and flux. *Glob. Biogeochem. Cycles* **2018**, *32*, 817–839. [CrossRef]
3. Barbier, E.B.; Hacker, S.D.; Kennedy, C.; Koch, E.W.; Stier, A.C.; Silliman, B.R. The value of estuarine and coastal ecosystem services. *Ecol. Monogr.* **2011**, *81*, 169–193. [CrossRef]
4. Aguilos, M.; Mitra, B.; Noormets, A.; Minick, K.; Prajapati, P.; Gavazzi, M.; Sun, G.; McNulty, S.; Li, X.; Domec, J.C.; et al. Long-term carbon flux and balance in managed and natural coastal forested wetlands of the Southeastern USA. *Agric. Forest Meteorol.* **2020**, *288*, 108022. [CrossRef]
5. Arkema, K.K.; Verutes, G.M.; Wood, S.A.; Clarke-Samuels, C.; Rosado, S.; Canto, M.; Rosenthal, A.; Ruckelshaus, M.; Guannel, G.; Toft, J.; et al. Embedding ecosystem services in coastal planning leads to better outcomes for people and nature. *Proc. Natl. Acad. Sci. USA* **2015**, *112*, 7390–7395. [CrossRef]
6. Enwright, N.M.; Griffith, K.T.; Osland, M.J. Barriers to and opportunities for landward migration of coastal wetlands with sea-level rise. *Front. Ecol. Environ.* **2016**, *14*, 307–316. [CrossRef]
7. Kopp, R.E.; Kemp, A.C.; Bittermann, K.; Horton, B.P.; Donnelly, J.P.; Gehrels, W.R.; Hay, C.C.; Mitrovica, J.X.; Morrow, E.D.; Rahmstorf, S. Temperature-driven global sea-level variability in the Common Era. *Proc. Natl. Acad. Sci. USA* **2016**, *113*, E1434–E1441. [CrossRef] [PubMed]
8. Jones, M.C.; Bernhardt, C.E.; Krauss, K.W.; Noe, G.B. The impact of late Holocene land use change, climate variability, and sea level rise on carbon storage in tidal freshwater wetlands on the southeastern United States coastal plain. *J. Geophys. Res. Biogeosci.* **2017**, *122*, 3126–3141. [CrossRef]
9. Law, B.E.; Hudiburg, T.W.; Berner, L.T.; Kent, J.J.; Buotte, P.C.; Harmon, M.E. Land use strategies to mitigate climate change in carbon dense temperate forests. *Proc. Natl. Acad. Sci. USA* **2018**, *115*, 3663–3668. [CrossRef]
10. Ardón, M.; Morse, J.L.; Colman, B.P.; Bernhardt, E.S. Drought-induced saltwater incursion leads to increased wetland nitrogen export. *Glob. Chang. Biol.* **2013**, *19*, 2976–2985. [CrossRef]
11. Herbert, E.R.; Boon, P.; Burgin, A.J.; Neubauer, S.C.; Franklin, R.B.; Ardón, M.; Hopfensperger, K.N.; Lamers, L.P.; Gell, P. A global perspective on wetland salinization: Ecological consequences of a growing threat to freshwater wetlands. *Ecosphere* **2015**, *6*, 1–43. [CrossRef]
12. Ardón, M.; Helton, A.M.; Bernhardt, E.S. Drought and saltwater incursion synergistically reduce dissolved organic carbon export from coastal freshwater wetlands. *Biogeochemistry* **2016**, *127*, 411–426. [CrossRef]
13. Mcleod, E.; Chmura, G.L.; Bouillon, S.; Salm, R.; Björk, M.; Duarte, C.M.; Lovelock, C.E.; Schlesinger, W.H.; Silliman, B.R. A blueprint for blue carbon: Toward an improved understanding of the role of vegetated coastal habitats in sequestering CO_2. *Front. Ecol. Environ.* **2011**, *9*, 552–560. [CrossRef]
14. Henman, J.; Poulter, B. Inundation of freshwater peatlands by sea level rise: Uncertainty and potential carbon cycle feedbacks. *J. Geophys. Res. Biogeosci.* **2008**, *113*. [CrossRef]
15. Chmura, G.L.; Anisfeld, S.C.; Cahoon, D.R.; Lynch, J.C. Global carbon sequestration in tidal, saline wetland soils. *Glob. Biogeochem. Cycles* **2003**, *17*. [CrossRef]
16. Donato, D.C.; Kauffman, J.B.; Murdiyarso, D.; Kurnianto, S.; Stidham, M.; Kanninen, M. Mangroves among the most carbon-rich forests in the tropics. *Nat. Geosci.* **2011**, *4*, 293. [CrossRef]

17. Loder, A.L.; Finkelstein, S.A. Carbon accumulation in freshwater marsh soils: A synthesis for temperate North America. *Wetlands* **2020**, *40*, 1173–1187. [CrossRef]
18. Williams, K.; Ewel, K.C.; Stumpf, R.P.; Putz, F.E.; Workman, T.W. Sea-level rise and coastal forest retreat on the west coast of Florida, USA. *Ecology* **1999**, *80*, 2045–2063. [CrossRef]
19. Poulter, B.; Qian, S.S.; Christensen, N.L. Determinants of coastal treeline and the role of abiotic and biotic interactions. *Plant Ecol.* **2009**, *202*, 55–66. [CrossRef]
20. Desantis, L.R.; Bhotika, S.; Williams, K.; Putz, F.E. Sea-level rise and drought interactions accelerate forest decline on the Gulf Coast of Florida, USA. *Glob. Chang. Biol.* **2007**, *13*, 2349–2360. [CrossRef]
21. Brinson, M.M.; Christian, R.R.; Blum, L.K. Multiple states in the sea-level induced transition from terrestrial forest to estuary. *Estuaries* **1995**, *18*, 648–659. [CrossRef]
22. Moorhead, K.K.; Brinson, M.M. Response of wetlands to rising sea level in the lower coastal plain of North Carolina. *Ecol. Appl.* **1995**, *5*, 261–271. [CrossRef]
23. Kirwan, M.L.; Gedan, K.B. Sea-level driven land conversion and the formation of ghost forests. *Nat. Clim. Chang.* **2019**, *9*, 450–457. [CrossRef]
24. Sallenger, A.H.; Doran, K.S.; Howd, P.A. Hotspot of accelerated sea-level rise on the Atlantic coast of North America. *Nat. Clim. Chang.* **2012**, *2*, 884–888. [CrossRef]
25. Sweet, W.V.; Marra, J.J. *2015 State of U.S. Nuisance Tidal Flooding. Supplement to State of the Climate: National Overview for May 2016*; National Oceanic and Atmospheric Administration, National Centers for Environmental Information: Asheville, NC, USA, 2016; p. 5. Available online: https://www.ncdc.noaa.gov/monitoring-content/sotc/national/2016/may/sweet-marra-nuisance-flooding-2015.pdf (accessed on 1 December 2018).
26. Karegar, M.A.; Dixon, T.H.; Engelhart, S.E. Subsidence along the Atlantic Coast of North America: Insights from GPS and late Holocene relative sea level data. *Geophys. Res. Lett.* **2016**, *43*, 3126–3133. [CrossRef]
27. Poulter, B.; Halpin, P.N. Raster modelling of coastal flooding from sea-level rise. *Int. J. Geogr. Inf. Sci.* **2008**, *22*, 167–182. [CrossRef]
28. Field, C.R.; Dayer, A.A.; Elphick, C.S. Landowner behavior can determine the success of conservation strategies for ecosystem migration under sea-level rise. *Proc. Natl. Acad. Sci. USA* **2017**, *114*, 9134–9139. [CrossRef]
29. White, E.; Kaplan, D. Restore or retreat? Saltwater intrusion and water management in coastal wetlands. *Ecosyst. Health Sustain.* **2017**, *3*, e01258. [CrossRef]
30. Stagg, C.L.; Baustian, M.M.; Perry, C.L.; Carruthers, T.J.; Hall, C.T. Direct and indirect controls on organic matter decomposition in four coastal wetland communities along a landscape salinity gradient. *J. Ecol.* **2018**, *106*, 655–670. [CrossRef]
31. Fatoyinbo, T.; Feliciano, E.A.; Lagomasino, D.; Lee, S.K.; Trettin, C. Estimating mangrove aboveground biomass from airborne LiDAR data: A case study from the Zambezi River delta. *Environ. Res. Lett.* **2018**, *13*, 025012. [CrossRef]
32. Klemas, V.V. Remote sensing of landscape-level coastal environmental indicators. *Environ. Manag.* **2001**, *27*, 47–57. [CrossRef] [PubMed]
33. DeFries, R.S.; Houghton, R.A.; Hansen, M.C.; Field, C.B.; Skole, D.; Townshend, J. Carbon emissions from tropical deforestation and regrowth based on satellite observations for the 1980s and 1990s. *Proc. Natl. Acad. Sci. USA* **2002**, *99*, 14256–14261. [CrossRef] [PubMed]
34. Byrd, K.B.; O'Connell, J.L.; Di Tommaso, S.; Kelly, M. Evaluation of sensor types and environmental controls on mapping biomass of coastal marsh emergent vegetation. *Remote Sens. Environ.* **2014**, *149*, 166–180. [CrossRef]
35. Lefsky, M.; Warren, B.; Hardings, D.J.; Parker, G.G.; Ackery, S.A.; Gower, S.T. Lidar remote sensing of above-ground biomass in three biomes. *Environ. Res.* **2002**, *11*, 393–399. [CrossRef]
36. Lim, K.; Treitz, P.; Baldwin, K.; Morrison, I.; Green, J. Lidar remote sensing of biophysical properties of tolerant northern hardwood forests. *Can. J. Remote Sens.* **2003**, *29*, 658–678. [CrossRef]
37. Anderson, J.; Martin, M.E.; Smith, M.-L.; Dubayah, R.O.; Hofton, M.A.; Hyde, P.; Peterson, B.E.; Blair, J.B.; Knox, R.G. The use of waveform lidar to measure northern temperate mixed conifer and deciduous forest structure in New Hampshire. *Remote Sens. Environ.* **2006**, *105*, 248–261. [CrossRef]
38. Fatoyinbo, T.E.; Simard, M. Height and biomass of mangroves in Africa from ICESat/GLAS and SRTM. *Int. J. Remote Sens.* **2013**, *34*, 668–681. [CrossRef]
39. Sexton, J.O.; Bax, T.; Siqueira, P.; Swenson, J.J.; Hensley, S. A comparison of lidar, radar, and field measurements of canopy height in pine and hardwood forests of southeastern North America. *For. Ecol. Manag.* **2009**, *257*, 1136–1147. [CrossRef]
40. Riegel, J.B.; Bernhardt, E.; Swenson, J. Estimating above-ground Carbon Biomass in a Newly Restored Coastal Plain Wetland Using Remote Sensing. *PLoS ONE* **2013**, *8*, e68251. [CrossRef]
41. Kulawardhana, R.W.; Popescu, S.C.; Feagin, R.A. Fusion of lidar and multispectral data to quantify salt marsh carbon stocks. *Remote Sens. Environ.* **2014**, *154*, 345–357. [CrossRef]
42. Hudak, A.T.; Strand, E.K.; Vierling, L.A.; Byrne, J.C.; Eitel, J.U.H.; Martinuzzi, S.; Falkowski, M.J. Quantifying aboveground forest carbon pools and fluxes from repeat LiDAR surveys. *Remote Sens. Environ.* **2012**, *123*, 25–40. [CrossRef]
43. Økseter, R.; Bollandsås, O.M.; Gobakken, T.; Næsset, E. Modeling and predicting aboveground biomass change in young forest using multi-temporal airborne laser scanner data. *Scand. J. For. Res.* **2015**, *30*, 458–469. [CrossRef]
44. Cao, L.; Coops, N.C.; Innes, J.L.; Sheppard, S.R.J.; Fu, L.; Ruan, H.; She, G. Estimation of forest biomass dynamics in subtropical forests using multi-temporal airborne LiDAR data. *Remote Sens. Environ.* **2016**, *178*, 158–171. [CrossRef]

45. Hudak, A.T.; Fekety, P.A.; Kane, V.R.; Kennedy, R.E.; Filippelli, S.K.; Falkowski, M.J.; Tinkham, W.T.; Smith, A.M.; Crookston, N.L.; Domke, G.M.; et al. A carbon monitoring system for mapping regional, annual aboveground biomass across the northwestern USA. *Environ. Res. Lett.* **2020**, *15*, 095003. [CrossRef]
46. Byrd, K.B.; Windham-Myers, L.; Leeuw, T.; Downing, B.; Morris, J.T.; Ferner, M.C. Forecasting tidal marsh elevation and habitat change through fusion of Earth observations and a process model. *Ecosphere* **2016**, *7*, e01582. [CrossRef]
47. Hamilton, S.E.; Friess, D.A. Global carbon stocks and potential emissions due to mangrove deforestation from 2000 to 2012. *Nat. Clim. Chang.* **2018**, *8*, 240–244. [CrossRef]
48. Lagomasino, D.; Fatoyinbo, T.; Lee, S.; Feliciano, E.; Trettin, C.; Shapiro, A.; Mangora, M.M. Measuring mangrove carbon loss and gain in deltas. *Environ. Res. Lett.* **2019**, *14*, 025002. [CrossRef]
49. Smart, L.S.; Taillie, P.J.; Poulter, B.; Vukomanovic, J.; Singh, K.K.; Swenson, J.J.; Mitasova, H.; Smith, J.W.; Meentemeyer, R.K. Aboveground carbon loss associated with the spread of ghost forests as sea levels rise. *Environ. Res. Lett.* **2020**, *15*, 104028. [CrossRef]
50. Foody, G.M. Spatial nonstationarity and scale-dependency in the relationship between species richness and environmental determinants for the sub-Saharan endemic avifauna. *Glob. Ecol. Biogeogr.* **2004**, *13*, 315–320. [CrossRef]
51. Fotheringham, S.; Brundson, C.; Charlton, M. *Geographically Weighted Regression & Associated Techniques*; John Wiley & Sons: West Sussex, UK, 2003.
52. Mason, S.A.; Olander, L.P.; Grala, R.K.; Galik, C.S.; Gordon, J.S. A practice-oriented approach to foster private landowner participation in ecosystem service conservation and restoration at a landscape scale. *Ecosyst. Serv.* **2020**, *46*, 101203. [CrossRef]
53. Bhattachan, A.; Jurjonas, M.D.; Moody, A.C.; Morris, P.R.; Sanchez, G.M.; Smart, L.S.; Taillie, P.J.; Emanuel, R.E.; Seekamp, E.L. Sea level rise impacts on rural coastal social-ecological systems and the implications for decision making. *Environ. Sci. Policy* **2018**, *90*, 122–134. [CrossRef]
54. National Oceanic and Atmospheric Administration (NOAA). 2001 NCFMP Lidar: Phase. 2012. Available online: https://chs.coast.noaa.gov/htdata/lidar1_z/geoid18/data/1397/ (accessed on 1 December 2016).
55. National Oceanic and Atmospheric Administration (NOAA). 2014 NCFMP Lidar: Phase. 2014. Available online: https://chs.coast.noaa.gov/htdata/lidar1_z/geoid18/data/4954/ (accessed on 1 December 2016).
56. US Geographical Survey (USGS). 2018 GloVis Data Warehouse. Available online: http://glovis.usgs.gov (accessed on 15 December 2016).
57. Taillie, P.J.; Moorman, C.E.; Poulter, B.; Ardón, M.; Emanuel, R.E. Decadal-scale vegetation change driven by salinity at leading edge of rising sea level. *Ecosystems* **2019**, *22*, 1918–1930. [CrossRef]
58. Smith, W.B.; Brand, G.J. Allometric biomass equations for 98 species of herbs, shrubs, and small trees. In *Research Note NC-299*; US Dept. of Agriculture, Forest Service, North Central Forest Experiment Station, 299: St. Paul, MN, USA, 1983.
59. Jenkins, J.C.; Chojnacky, D.C.; Heath, L.S.; Birdsey, R.A. National-scale biomass estimators for United States tree species. *For. Sci.* **2003**, *49*, 12–35.
60. Castillo, J.M.; Leira-Doce, P.; Rubio-Casal, A.E.; Figueroa, E. Spatial and temporal variations in aboveground and belowground biomass of Spartina maritima (small cordgrass) in created and natural marshes. *Estuar. Coast. Shelf Sci.* **2008**, *78*, 819–826. [CrossRef]
61. Breiman, L. Random Forests. *Mach. Learn.* **2001**, *45*, 5–32. [CrossRef]
62. Martin, A.R.; Doraisami, M.; Thomas, S.C. Global patterns in wood carbon concentration across the world's trees and forests. *Nat. Geosci.* **2018**, *11*, 915–920. [CrossRef]
63. National Ocean Service. Tides and Currents: Bench Mark Sheet for 8651370 2017, Duck NC. Available online: https://tidesandcurrents.noaa.gov/benchmarks.html?id=8651370 (accessed on 1 January 2018).
64. MRLC. Multi-Resolution Land Characteristics Consortium: National Land Cover Database (NLCD) 2001. 2017. Available online: https://www.mrlc.gov/data?f%5B0%5D=category%3Aland%20cover&f%5B1%5D=year%3A2016 (accessed on 1 January 2017).
65. Meentemeyer, R.K.; Tang, W.; Dorning, M.A.; Vogler, J.B.; Cunniffe, N.J.; Shoemaker, D.A. Futures: Multilevel simulations of emerging urban–rural landscape structure using a stochastic patch-growing algorithm. *Ann. Assoc. Am. Geogr.* **2013**, *103*, 785–807. [CrossRef]
66. GRASS Development Team. *Geographic Resources Analysis Support System (GRASS) Software, Version 7.2*; Open Source Geospatial Foundation: Chicago, IL, USA, 2017. Available online: https://grass.osgeo.org/ (accessed on 1 January 2017).
67. US Department of Agriculture (USDA). National Agricultural Statistics Service Cropland Data Layer: Published Crop-Specific Data Layer. 2018. Available online: https://nassgeodata.gmu.edu/CropScape/ (accessed on 1 December 2017).
68. Google Earth Engine. A Planetary-Scale Platform for Earth Science Data & Analysis. 2018. Available online: https://earthengine.google.com/ (accessed on 1 December 2017).
69. US Geological Survey (USGS). National Hydrography Geodatabase: The National Map Viewer. 2013. Available online: http://nhd.usgs.gov/data.html (accessed on 15 May 2016).
70. US Environmental Protection Agency (US EPA). 2017 STORage and RETrieval Warehouse/Water Quality Exchange. Available online: www3.epa.gov/storet/bck/dbtop.html (accessed on 1 January 2017).
71. Emadi, M.; Baghernejad, M. Comparison of spatial interpolation techniques for mapping soil pH and salinity in agricultural coastal areas, northern Iran Arch. *Agron. Soil Sci.* **2014**, *60*, 1315–1327. [CrossRef]

72. Frost, C.C. Presettlement fire regimes in southeastern marshes, peatlands, and swamps. In Proceedings of the 19th Tall Timbers Fire Ecology Conference: Fire in Wetlands: A Management Perspective, Tallahassee, FL, USA, 1995; Tall Timbers Research Inc.: Tallahassee, FL, USA, 1995. Available online: http://talltimbers.org/wp-content/uploads/2014/03/Frost1995_op.pdf (accessed on 23 August 2016).
73. US Forest Service (USFS); US Geological Survey (USGS). Monitoring Trends in Burn Severity (MTBS) Data Access: Fire Level Geospatial Data. 2017. Available online: http://mtbs.gov/direct-download (accessed on 1 May 2017).
74. Harris, P.; Fotheringham, A.S.; Crespo, R.; Charlton, M. The use of geographically weighted regression for spatial prediction: An evaluation of models using simulated data sets. *Math. Geosci.* **2010**, *42*, 657–680. [CrossRef]
75. Brunsdon, C.; Fotheringham, S.; Charlton, M. Geographically weighted regression. *J. R. Stat. Soc. Ser. D* **1998**, *47*, 431–443. [CrossRef]
76. R Core Development Team. *R: A Language and Environment for Statistical Computing*; R Foundation for Statistical Computing: Vienna, Austria, 2018. Available online: https://www.R-project.org/ (accessed on 10 January 2018).
77. Bivand, R.; Yu, D.; Nakaya, T.; Garcia-Lopez, M.A.; Bivand, M.R. Packag 'Spgwr'. R Software Package. 2017. Available online: https://cran.r-project.org/web/packages/spgwr/index.html (accessed on 10 January 2018).
78. NOAA Office for Coastal Management (NOAA OCM). Sea Level Rise Data Download. 2017. Available online: https://coast.noaa.gov/slrdata/ (accessed on 1 December 2017).
79. Lindeman, R.H.; Merenda, P.F.; Gold, R.Z. *Introduction to Bivariate and Multivariate Analysis*; Scott, Foresman: Glenview, IL, USA, 1980.
80. Su, S.; Li, D.; Xiao, R.; Zhang, Y. Spatially non-stationary response of ecosystem service value changes to urbanization in Shanghai, China. *Ecol. Indic.* **2014**, *45*, 332–339. [CrossRef]
81. Bhattachan, A.; Emanuel, R.E.; Ardon, M.; Bernhardt, E.S.; Anderson, S.M.; Stillwagon, M.G.; Ury, E.A.; Bendor, T.K.; Wright, J.P. Evaluating the effects of land-use change and future climate change on vulnerability of coastal landscapes to saltwater intrusion. *Elem. Sci. Anthr.* **2018**, *6*, 62. [CrossRef]
82. Poulter, B.; Goodall, J.L.; Halpin, P.N. Applications of network analysis for adaptive management of artificial drainage systems in landscapes vulnerable to sea level rise. *J. Hydrol.* **2008**, *357*, 207–217. [CrossRef]
83. Tully, K.; Gedan, K.; Epanchin-Niell, R.; Strong, A.; Bernhardt, E.S.; BenDor, T.; Mitchell, M.; Kominoski, J.; Jordan, T.E.; Neubauer, S.C.; et al. The invisible flood: The chemistry, ecology, and social implications of coastal saltwater intrusion. *BioScience* **2019**, *69*, 368–378. [CrossRef]
84. Rodrigues, J.G.; Conides, A.J.; Rivero Rodriguez, S.; Raicevich, S.; Pita, P.; Kleisner, K.M.; Pita, C.; Lopes, P.F.; Alonso Roldáni, V.; Ramos, S.S.; et al. Marine and coastal cultural ecosystem services: Knowledge gaps and research priorities. *One Ecosyst.* **2017**, *2*, e12290. [CrossRef]

MDPI
St. Alban-Anlage 66
4052 Basel
Switzerland
Tel. +41 61 683 77 34
Fax +41 61 302 89 18
www.mdpi.com

Land Editorial Office
E-mail: land@mdpi.com
www.mdpi.com/journal/land

www.ingramcontent.com/pod-product-compliance
Lightning Source LLC
LaVergne TN
LVHW070225100526
838202LV00015B/2092